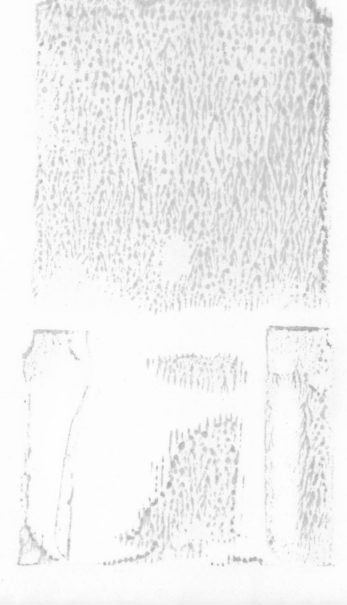

Reagents for Organic Synthesis

Reagents for Organic Synthesis

VOLUME 3

Mary Fieser

Research Fellow in Chemistry
Harvard University

Louis F. Fieser

Sheldon Emery Professor of Organic Chemistry, Emeritus
Harvard University

WILEY-INTERSCIENCE

A DIVISION OF JOHN WILEY & SONS

NEW YORK · LONDON · SYDNEY · TORONTO

Copyright© 1972, by John Wiley & Sons, Inc.

All rights reserved. Published simultaneously in Canada.

Library of Congress Catalog Card Number: 66-27894

ISBN 0-471-25879-2

Printed in the United States of America.

10 9 8 7 6 5 4 3 2

PREFACE

This volume, covering literature received in Cambridge in the period 1969–70, contains 711 references to 273 reagents discussed in the first or second volume and 289 references to 142 reagents reviewed now by us for the first time. There is little sign of any slackening in the need for reagents for synthesis or in the ingenuity of chemists in devising new ones.

We are indebted to Research Corporation for a grant in support of the project and to colleagues for providing information pertaining to their own work.

Miss Theodora S. Lytle typed the manuscript and drew the formulas, where appropriate, with use of a ring-formula stencil.[1] A new photograph by David A. Lang shows Shio Pooh as he is today; his pedigree lists his date of birth as August 10, 1952.

Cambridge, Massachusetts
February 18, 1971

MARY FIESER
LOUIS F. FIESER

[1]Chemist's triangle, available from Reinhold Publishing Corporation.

CONTENTS

Reagents for Organic Synthesis

Introduction

Arrangement. Suppliers mentioned in the text are listed in a section placed before the indexes and easily located by an indenture.

For enhanced usefulness the book is provided not only with a subject and an author index but also with an index of types, that is, types of reactions or types of compounds, for example: acetylation, bromination, cycloaddition, decarboxylation, or: acetonides, benzyne precursors, carbene precursors. Listed alphabetically under each such entry are all the reagents which figure in the operation or group cited, whether as prime reactant, catalyst, solvent, scavenger, etc. A given reagent may fit appropriately in two or more categories. When a reagent does not fit easily into a reasonable category, we leave it unclassified rather than make a forced assignment. With so many reagents available as oxidants and for use as reducing reagents, it seems out of the question to attempt to indicate in the index of types further details about these general reactions.

Names and spelling. One guideline we have followed is the rule recently adopted by *Organic Syntheses* that when an ester, ether, or peroxide contains two or more alkyl, aryl, or acyl groups the name must indicate the number of such groups:

Formula	*Correct*	*Incorrect*
$(CH_3)_2O$	Dimethyl ether	Methyl ether
$(C_2H_5O)_2SO_2$	Diethyl sulfate	Ethyl sulfate
$(C_6H_5)_2O$	Diphenyl ether	Phenyl ether
$(CO_2CH_3)_2$	Dimethyl oxalate	Methyl oxalate
$CH_2(CO_2C_2H_5)_2$	Diethyl malonate	Ethyl malonate
$(C_6H_5COO)_2$	Dibenzoyl peroxide	Benzoyl peroxide
$HC(OC_2H_5)_3$	Triethyl orthoformate	Ethyl orthoformate
$(C_2H_5O)_4C$	Tetraethyl orthocarbonate	Ethyl orthocarbonate

That the situation previously was highly confused is evident from the following entries in the index of *Org. Syn., Coll. Vol.*, **4**: "Diethyl oxalate" and "Diethyl malonate" (both correct), but "Ethyl orthoformate" and "Ethyl orthocarbonate" (both incorrect). The following entry is describable as a double error: "Triethyl orthoformate, *see* Ethyl orthoformate." To locate all references to a given ester, it is thus necessary to search under two names. We urge suppliers to revise their catalogs in accordance with the rule cited. In this book we do not even list, with cross references, names which we consider to be incorrect.

Similar reform in the nomenclature of polyhalogen compounds may come some day, but for the present we consider it imprudent to do more than make a start.

Thus the correct names for BF_3 and for $ClCH_2CH_2Cl$ surely are boron trifluoride and ethylene dichloride, and we feel no restraint from using them. However, although the names methylene chloride for CH_2Cl_2 and aluminum chloride for $AlCl_3$ seem incorrect, we cannot bring ourselves to break with tradition and employ other names.

Abbreviations. Short forms of abbreviations of journal titles are as follows:

Accounts of Chemical Research
Journal of the American Chemical Society
Analytical Letters
Angewandte Chemie
Angewandte Chemie, international Edition in English
Annalen der Chemie
Annales de chimie (Paris)
Australian Journal of Chemistry
Chemische Berichte (formerly Berichte der deutschen chemischen Gesellschaft)
Bulletin de la société chimique de France
Canadian Journal of Chemistry
Carbohydrate Research
Chemical Communications
Chemical and Pharmaceutical Bulletin Japan
Acta Chemica Scandinavica
Chemistry and Industry
Chemical Reviews
Collection of Czechoslovak Chemical Communications
Comptes rendus hebdomadaires des séances de l'académie des sciences
Gazzetta Chimica Italiana
Helvetica Chimica Acta
Inorganic Synthesis
Journal of Chemical Education
Journal of the Chemical Society (London)
Journal of Heterocyclic Chemistry
Journal of Medicinal Chemistry
Journal of Organic Chemistry
Journal of Organometallic Chemistry
Journal fur praktische Chemie
Monatschefte für Chemie
Organic Syntheses
Organic Syntheses, Collective Volume
Proceedings of the Chemical Society
Records of Chemical Progress
Recueil des travaux chimique des Pays-Bas (The Netherlands)

The book by one of us, **Organic Experiments**, 2nd Ed., D. C. Heath and Co., Boston (1968), is referred to as **Org. Expts.**

Abbreviations

Ac	Acetyl
AcOH	Acetic acid
BuOH	Butanol
Bz	Benzoyl
CAN	Ceric ammonium nitrate
Cathyl	Carboethoxy
Cb	Carbobenzoxy
DABCO	1,4-Diazabicyclo[2,2,2]octane
DCC	Dicyclohexylcarbodiimide
DDQ	2,3-Dichloro-5,6-dicyano-1,4-benzoquinone
Diglyme	Diethylene glycol dimethyl ether
Dimsyl sodium	Sodium methylsulfinylmethide
DMA	Dimethylacetamide
DMF	Dimethylformamide
DMSO	Dimethyl sulfoxide
DNF	2,4-Dinitrofluorobenzene
DNP	2,4-Dinitrophenylhydrazine
EtOH	Ethanol
Glyme	1,2-Dimethoxyethane
HMPT	Hexamethylphosphoric triamide
MeOH	Methanol
MMC	Magnesium methyl carbonate
Ms	Mesyl, CH_3SO_2
NBA	N-Bromoacetamide
NBS	N-Bromosuccinimide
Ph	Phenyl
Phth	Phthaloyl
PPA	Polyphosphoric acid
PPE	Polyphosphate ester
Py	Pyridine
THF	Tetrahydrofurane
TMEDA	N,N,N,N-Tetramethylenediamine
Triglyme	Triethylene glycol dimethyl ether
Trityl	$(C_6H_5)_3C-$
Ts	Tosyl, $p\text{-}CH_3C_6H_4SO_2-$
TsCl	Tosyl chloride
TsOH	Tosic acid, $p\text{-}CH_3C_6H_4SO_3H$
TTFA	Thallium(III) trifluoroacetate

A

Acetic-formic anhydride, 1, 4; 2, 10–12.

   Preparation. Krimen's[1] preparation of the reagent has now been published.

[1]L. I. Krimen, *Org. Syn.*, **50**, 1 (1970)

Acetone, 1, 1110; 2, 13.

   Tetramethyl bismethylenedioxy steroids. The bismethylenedioxy (BMD) group has outstanding advantages for the protection of the dihydroxyacetone side chain of adrenocortical steroids (**1**, 401). The original papers (ref. 21) noted that a ketonic group at C_{11} and even groups in ring A retarded hydrolysis of the protective group. However, tetramethyl bismethylenedioxy steroids (2), prepared in 60–65% yield by reaction of the steroid with acetone catalyzed by perchloric acid, are free from this disadvantage; derivatives of steroids whose BMD derivatives were stable to formic acid or glacial acetic acid at 100° for 6 hours were readily hydrolyzed by 50% acetic acid at steam bath temperature for 3–4 hours.[1]

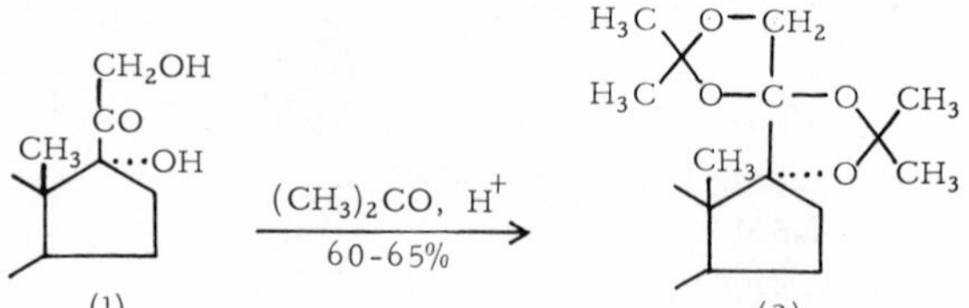

[1]A. Roy, W. D. Slaunwhite, and S. Roy, *J. Org.*, **34**, 1455 (1969)

3-Acetyl-1,5,5-trimethylhydantoin, Mol. wt. 160.18, m.p. 126–127°.

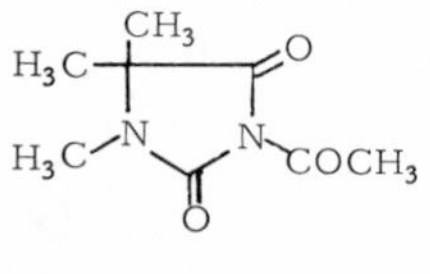

   Preparation. The reagent is prepared in 89% yield by the reaction of 1,5,5-trimethylhydantoin with acetic anhydride.

   Acetylation of phenols. A typical acetylation is conducted as follows: 2-naphthol is heated for 12 hours at 80° in anhydrous acetonitrile with the reagent; 2-naphthyl acetate is obtained in quantitative yield. Only phenolic groups are acetylated. Thus 17β-estradiol is converted by the reagent into estradiol 3-acetate (60% yield).[1]

[1]O. O. Orazi and R. A. Corral, *Am. Soc.*, **91**, 2162 (1969)

Acrolein, 2, 15.

Reaction with trialkylboranes. The reaction of trialkylboranes with acrolein (**2,** 15) and with methyl vinyl ketone (**2,** 284–185) yields, respectively, aldehydes and methyl ketones. However, the reaction is not directly applicable to all α,β-unsaturated carbonyl compounds. Thus *trans*-crotonaldehyde (1) and mesityl

oxide (2) fail to react with triethylborane at temperatures of 25–125° for up to 24 hours. Brown *et al.*[1] investigated the possibility that a free-radical mechanism is involved and indeed found that the reaction is inhibited by addition of 4 mole % of galvinoxyl, an efficient scavenger of radicals (**1,** 409). Apparently the α,β-unsaturated ketones which react readily either do not require a catalyst for initiation or are catalyzed by mere traces of catalyst. Brown and Kabalka[2] then found that inert α,β-unsaturated carbonyl groups do react with trialkylboranes in the presence of catalytic amounts of diacyl peroxides (diacetyl peroxide[3] was found best) or under photochemical activation. In addition the reaction is induced readily and conveniently by introduction of small quantities of air (oxygen).[4] It is essential to add the air slowly and in controlled amounts to minimize oxidation of the organoborane. Yields are highest in alcohol solvents, particularly isopropyl alcohol. With this technique, the reaction is widely general for synthesis of β-substituted aldehydes and ketones:

[1] G. W. Kabalka, H. C. Brown, A. Suzuki, S. Honma, A. Arase, and M. Itoh, *Am. Soc.,* **92,** 710 (1970)

[2] H. C. Brown and G. W. Kabalka, *ibid.,* **92,** 712 (1970)

[3] Available as a 25% solution in dimethyl phthalate from the Lucidol Division, Wallace and Tiernan, Inc.

[4] H. C. Brown and G. W. Kabalka, *Am. Soc.,* **92,** 714 (1970)

Adamantane (3). Mol. wt. 136.33, m.p. 205–210° (sublimes), 268–270° (sealed capillary completely immersed in the melting-point bath fluid). Supplier: Aldrich, 100 g, $17.00.

Preparation A. In the method of Schleyer *et al.*,[1] 200 g. of *endo*-dicyclopentadiene (1) purified by distillation is hydrogenated in ether in the presence

(1)   (2)   (3) Adamantane

of platinum oxide, and the resulting *endo*-tetrahydrodicyclopentadiene (2) is isomerized to adamantane (3) by heating it with aluminum chloride at 150–180° for 8–12 hours (yield 15%).

Preparation B. Ault and Kopet[2] describe as "an introductory organic experiment" a procedure for the preparation of adamantane by isomerization of *endo*-tetrahydrodicyclopentadiene by the method of Schleyer but the product is isolated from a hexane extract of the reaction mixture as the beautifully crystalline inclusion complex with thiourea.[3] As noted in **1**, 1164, the ratio in this complex is 3.4 molecules of host per molecule of guest hydrocarbon. Although the reaction time is only 1 hour, the yield (15%) is about the same as that reported for a reaction time of 8–12 hours.

[1]P. von R. Schleyer, M. M. Donaldson, R. D. Nicholas, and C. Cupas, *Org. Syn.*, **42**, 8 (1962).
[2]A. Ault and R. Kopet, *J. Chem. Ed.*, **46**, 612 (1969)
[3]S. Landa and S. Hola, *Chem. Listy*, **51**, 2325 (1957); *Czech. Commun.*, **24**, 93 (1959)

Alumina, 1, 19–20; 2, 17.

Addition of diazomethane to ketosteroids. Alumina-catalyzed (Woelm, neutral, activity 1) reaction of diazomethane with 5α-androstane-17β-ol-3-one (1) gives as the major product the epoxide (2).[1] A minor product is probably a ring-enlarged

(1)   (2)

ketone. The corresponding 5β-isomer gives a small amount of A-homo-5β-androstane-17β-ol-4-one and another unidentified ketone as well as an epoxide. Catalysis with methanol rather than alumina gives complex mixtures. Hence use of alumina favors epoxide formation, but the A/B ring fusion is also a factor.

[1]P. A. Hart and R. A. Sandmann, *Tetrahedron Letters*, 305 (1969)

Aluminum amalgam, 1, 20–21.

Pinacol reduction. Pinacol reduction is generally carried out with amalgamated magnesium or aluminum in benzene. Schreibmann[1] reports that higher yields are obtained with amalgamated aluminum in either methylene chloride or THF, in which the aluminum pinacolates are very soluble. They are thus easily separated from excess metal before hydrolysis to the pinacol.

[1]A. A. P. Schreibmann, *Tetrahedron Letters*, 4271 (1970)

Aluminum bromide, 1, 22–23; **2**, 19–21.

Preparation of diamantane (3). Schleyer *et al.*[1] have developed a convenient, high yield preparation of diamantane[2] from the readily available dimer of norbornadiene, Bisnor-S (1) (**2**, 123–124).[3] The hydrocarbon is hydrogenated to a

$$\text{(1, } C_{14}H_{16}) \xrightarrow[\substack{\text{HOAc, HCl} \\ 96\%}]{2\ H_2,\ PtO_2} C_{14}H_{20} \xrightarrow[65\%]{\text{AlBr}_3,\ CS_2} \text{(3, } C_{14}H_{20})$$

(1, $C_{14}H_{16}$)　　　　　　(2)　　　　　　(3, $C_{14}H_{20}$)

tetrahydro derivative (2) of unknown structure. This is isomerized in high yield to diamantane by the $AlBr_3$ "sludge" catalyst used previously for synthesis of adamantane and triamantane (**2**, 20).

[1]T. M. Gund, V. Z. Williams, Jr., E. Osawa, and P. von R. Schleyer, *Tetrahedron Letters*, 3877 (1970).
[2]Originally named congressane: C. Cupas, P. von R. Schleyer, and D. J. Trecker, *Am. Soc.*, **87**, 917 (1965); V. Z. Williams, Jr., P. von R. Schleyer, G. J. Gleicher, and L. B. Rodewald, *ibid.*, **88**, 3862 (1966).
[3]The correct IUPAC name is heptacyclo[8.4.0.0^{2,12}.0^{3,7}.0^{4,9}.0^{6,8}.0^{11,13}]tetradecane.

Aluminum chloride, 1, 24–34; **2**, 21–23.

Selective ether cleavage (**1**, 30–31). Seshadri[1] noted that the methoxyl group adjacent to the formyl group of 2,3,4-trimethoxy-6-methylbenzaldehyde (1) was selectively cleaved in quantitative yield by demethylation with anhydrous aluminum chloride in ether at room temperature in 48 hours. This reaction was

$$\text{(1)} \xrightarrow{} \text{Al-complex} \xrightarrow[\text{Quant.}]{H_2O,\ HCl} \text{(2)}$$

(1)　　　　　　　　　　(2)

used[2] in one step of a new simplified synthesis of 2,3-dimethoxy-5,6-dimethylbenzoquinone (4, aurantiocladin). The substrate (1) was partially demethylated to (2), but in this case the yield was only 43%. The remaining steps involved reduction to (3) and oxidation with Fremy's salt to (4).

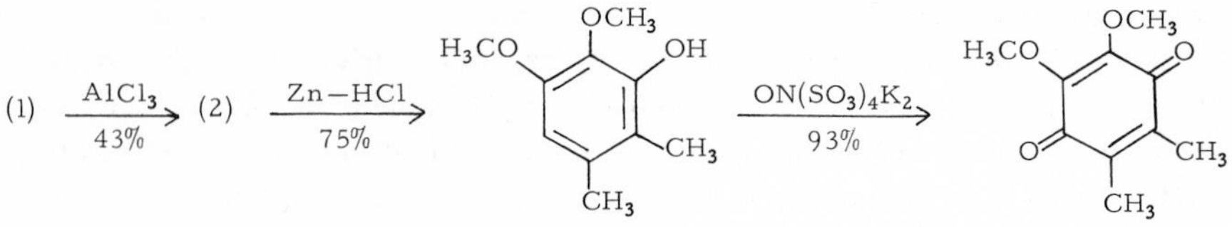

(1) $\xrightarrow[43\%]{AlCl_3}$ (2) $\xrightarrow[75\%]{Zn-HCl}$ (3) $\xrightarrow[93\%]{ON(SO_3)_4K_2}$ (4)

Modified Darzens reaction. The original Darzens synthesis of α,β-unsaturated ketones[3] involved the addition of acid chlorides to cyclohexenes in the presence of aluminum chloride or stannic chloride followed by dehydrohalogenation. For example, the addition of acetyl chloride to cyclohexene affords 1-acetyl-2-chlorocyclohexane, which on dehydrohalogenation (dimethylaniline) gives methyl cyclohexenyl ketone.

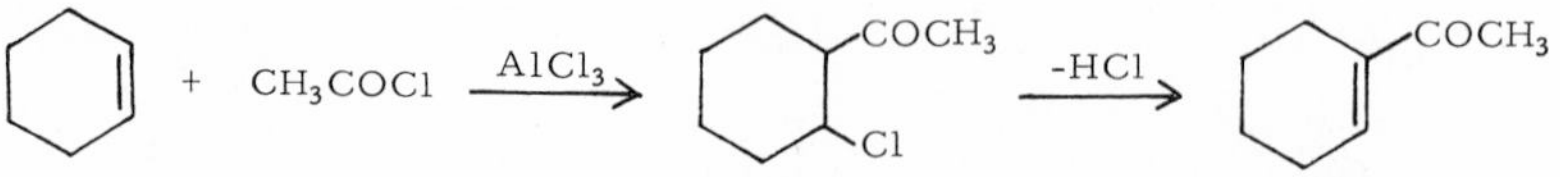

A modification of this reaction provides a useful route to β-tetralones. Burck-halter and Campbell[4] found that phenylacetyl chlorides condense with ethylene under the influence of aluminum chloride (CS_2, ice cooling) to give β-tetralones in good yield. The intermediate β-chloroethyl ketones are cyclized under the reaction conditions. In the original procedure 6-methoxy-2-tetralone was obtained

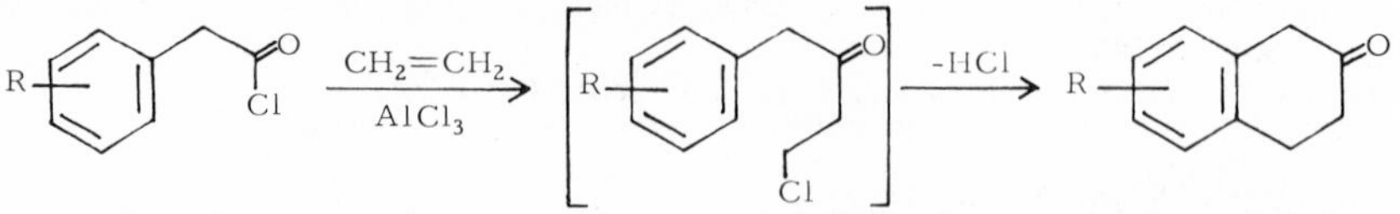

in this way in 56% yield. If methylene chloride, in which the aluminum chloride complex of the acid chloride is soluble, is used rather than carbon disulfide, the yield is improved to 60–68%.[5]

Modest *et al.*[6] have used this procedure for preparation of a number of chloro-substituted 2-tetralones; because of the deactivating effect of the halogen substituents the reaction was allowed to proceed overnight.

Diels-Alder catalyst, **1**, 31–32; **2**, 21–22. The Diels-Alder reaction of 1-carbo-methoxypyrrole (1) with dimethyl acetylenedicarboxylate catalyzed by aluminum chloride gives 2,3,7-tricarbomethoxy-7-azanorbornadiene (2) in 90% yield. The uncatalyzed reaction gives yields of 35–40%. The improved yield is attributed to the milder reaction conditions and the fact that formation of a complex of (2) with $AlCl_3$ renders the reaction irreversible. The reaction is the first step in an im-

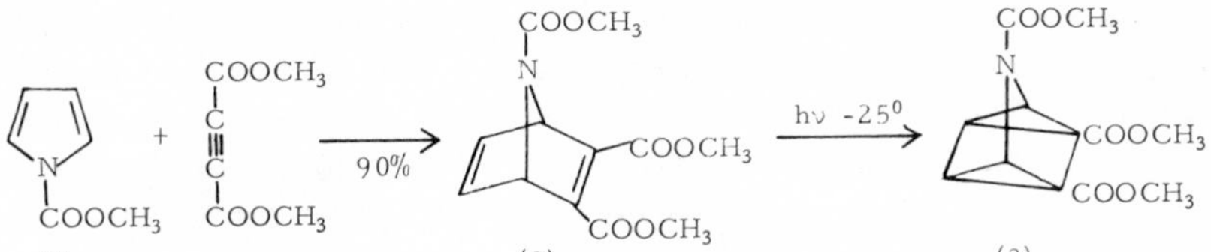

(1)   +   (2)   →   (3)

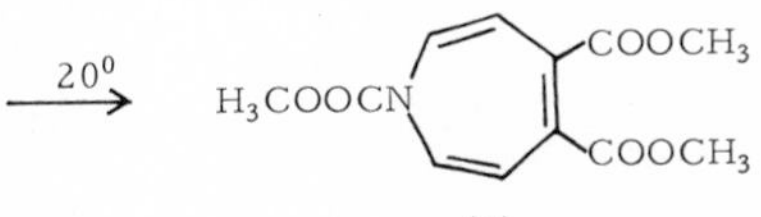

$\xrightarrow{90\%}$ $\xrightarrow{h\nu \ -25^0}$

$\xrightarrow{20^0}$ (4)

proved synthesis of 1,4,5-tricarbomethoxyazepine (4) by way of the 3-azaquadri-cyclane (3).[7]

[1]T. R. Seshadri and G. B. Venkatasubramanian, *J. Chem. Soc.*, 1660 (1959)
[2]H. Aquila, *Ann.*, **721**, 220 (1969)
[3]G. Darzens, *Compt. rend.*, **150**, 707 (1910)
[4]J. H. Burckhalter and J. R. Campbell, *J. Org.*, **26**, 4232 (1961)
[5]J. J. Sims, L. H. Selman, and M. Cadogan, procedure submitted to *Org. Syn.* 1969
[6]A. Rosowsky, J. Battaglia, K. K. N. Chen, and E. J. Modest, *J. Org.*, **33**, 4288 (1968)
[7]R. C. Bansal, A. W. McCulloch, and A. G. McInness, *Canad. J. Chem.*, **47**, 2391 (1969)

Aluminum hydride, 1, 34–35; **2**, 23–24.

Hydrogenolysis of enamines.[1] Hydrogenation of an enamine double bond is readily achieved both catalytically and with formic acid and sodium borohydride. However, enamines react with aluminum hydride mainly by hydrogenolysis to give olefins. Good yields of olefins are obtained from pyrrolidine enamines of both acyclic and cyclic ketones. Thus 6-methyl-1-N-pyrrolidinocyclohexene (1) is converted into 3-methylcyclohexene (2) in 85% yield. Only 11% of the product of hydrogenation, 2-methyl-1-N-pyrrolidinocyclohexane (3), is formed. Mixed hydrides also give some products of hydrogenolysis, but hydrogenation products predominate. Thus the reaction of (1) with AlH_2Cl gives 42% of (2) and 57% of (3). Morpholine enamines give somewhat lower yields of olefins than pyrrolidine enamines.

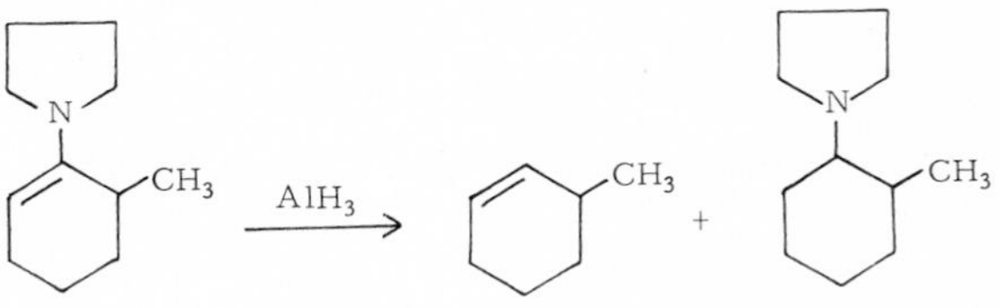

(1) $\xrightarrow{AlH_3}$ (2, 85%)   +   (3, 11%)

Selective reduction of the carbonyl group of Δ^2-*cyclopentenones.*[2] The reduction of α,β-unsaturated ketones to the corresponding carbinols by complex hydrides is accompanied with concomitant saturation of the double bond. This undesired reaction is particularly noted in the case of Δ^2-cyclopentenones. However, use of aluminum hydride (inverse addition) produces the unsaturated carbinols in satisfactory yields.

Examples:

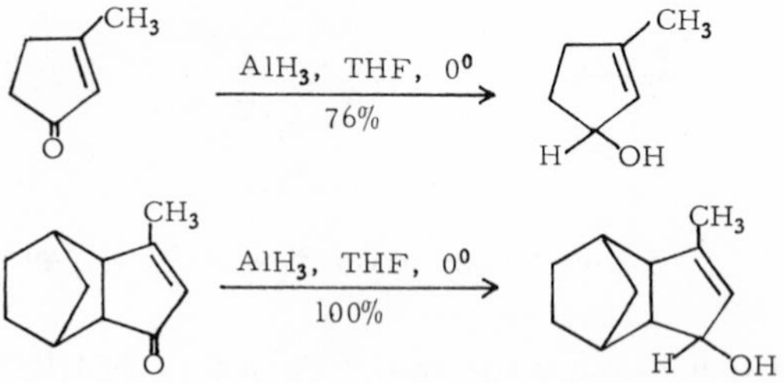

Lithium trimethoxyaluminum hydride (**1**, 625; **2**, 252–253) is almost as selective.

[1]J. M. Coulter, J. W. Lewis, and P. P. Lynch, *Tetrahedron*, **24**, 4489 (1968)
[2]H. C. Brown and H. M. Hess, *J. Org.*, **34**, 2206 (1969)

Aluminum isopropoxide, 1, 35–37.

Evidence from molecular weight determinations indicating that aluminum isopropoxide aged in benzene solution consists largely of the tetramer (I), whereas freshly distilled molten material is trimeric (II)[1] is fully confirmed by NMR spectroscopy.[2]

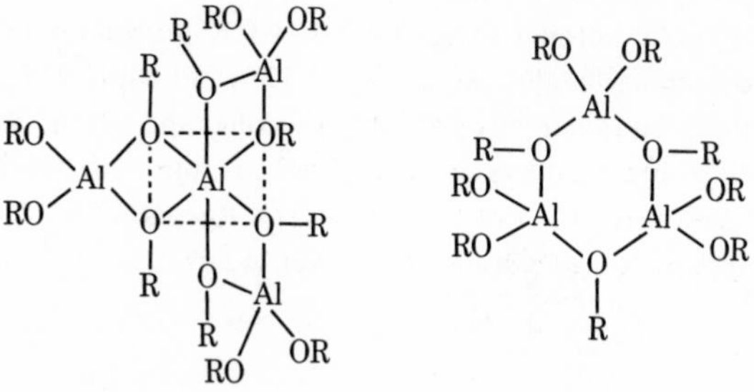

I   II

[1]V. J. Shiner, D. Whittaker, and V. P. Fernandez, *Am. Soc.*, **85**, 2318 (1963)
[2]I. J. Worrall, *J. Chem. Ed.*, **46**, 510 (1969)

1-Aminobenzotriazole, 1, 37.

Preparation. The method cited on **1**, 482, amination of benzotriazole by hydroxylamine-O-sulfonic acid, has the disadvantage that both 1- and 2-aminobenzotriazole are formed and separation requires chromatography on silica gel. Campbell and Rees[1] report an alternative synthesis of the 1-amino isomer from o-nitroaniline (1) in 59% overall yield. The starting material is diazotized and

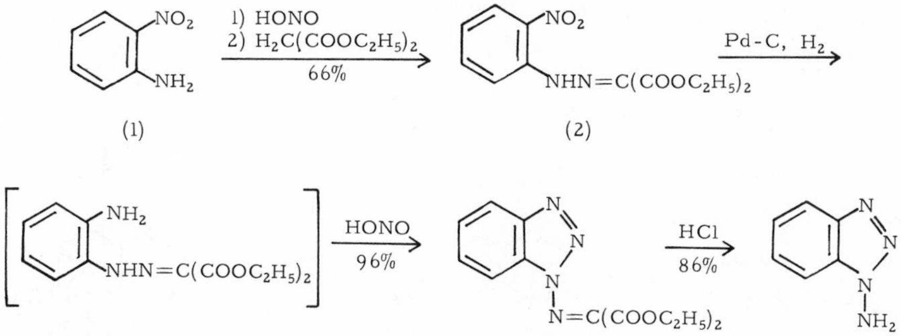

coupled with diethyl malonate to give (2); hydrogenation gives the amine (3), usually not isolated, but diazotized directly to the triazole (4). Hydrolysis with cold concd. hydrochloric acid gives 1-aminobenzotriazole.

Generation of benzyne (see also **1**, 560). Campbell and Rees[1] point out that the generation of benzyne from this precursor by oxidation with lead tetraacetate differs from other methods in that, in the absence of a trapping agent, the dimer, diphenylene, is formed in yields as high as 83%. The trimer, triphenylene, is formed in less than 0.5% yield. Addition reactions of benzyne generated in this way with 1,3-dienes have been discussed.[2]

Benzyne is also formed almost quantitatively on oxidation of 1-aminobenzo-triazole with NBS (2 equivalents); in the presence of tetracyclone, 1,2,3,4-tetraphenylnaphthalene is obtained in 88% yield. In the absence of a trap, o-dibromobenzene is obtained in 52% yield. No dimer is formed. Bromine is known to be an efficient scavenger of benzyne generated from this precursor.[3]

[1]C. D. Campbell and C. W. Rees, *J. Chem. Soc.*, (C), 742 (1969)
[2]*Idem, ibid.*, (C), 748 (1969)
[3]*Idem, ibid.*, (C), 752 (1969)

1-Amino-(S)-2-[(R)-1-hydroxyethyl]indoline, 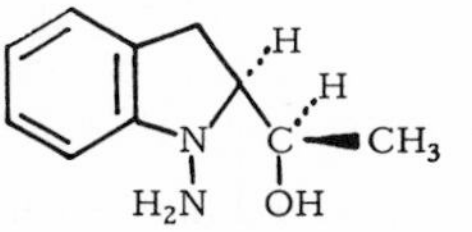   Mol. wt. 178.23

(14)

Corey *et al.*[1] prepared this levorotatory chiral reagent by the stereospecific synthesis and resolution formulated thus:

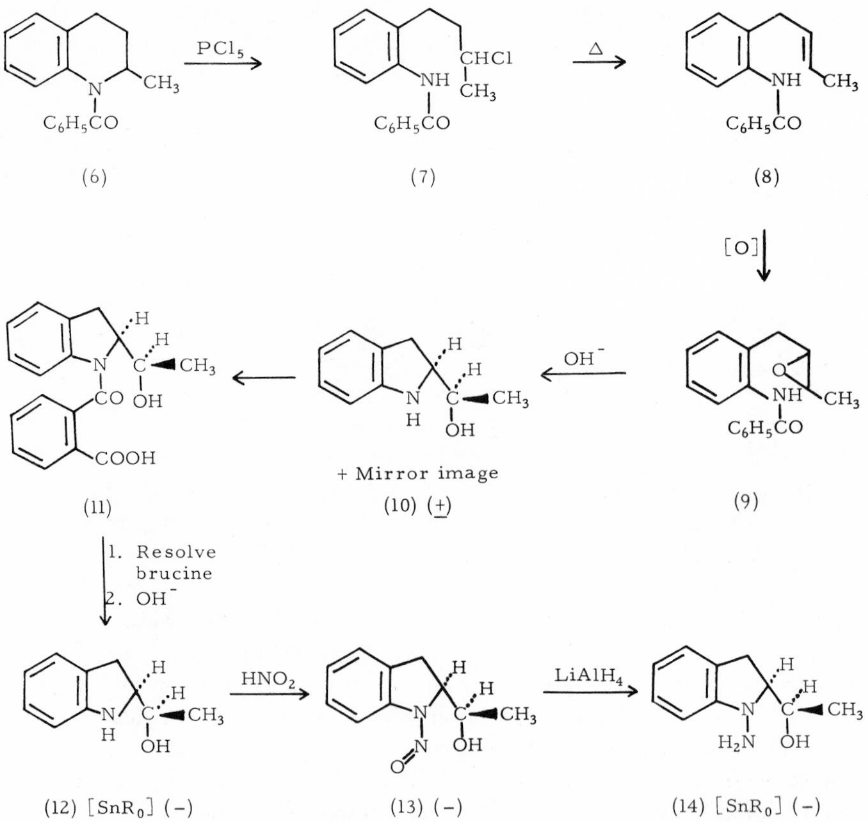

(6) → PCl₅ → (7) → Δ → (8)

[O] ↓

(11) ← (10) (±) + Mirror image ← OH⁻ ← (9)

(11)

1. Resolve brucine
2. OH⁻
↓

(12) [SnR₀] (−) → HNO₂ → (13) (−) → LiAlH₄ → (14) [SnR₀] (−)

The reagent was applied to the asymmetric synthesis of a number of α-amino acids.

[1]E. J. Corey, H. S. Sachdev, J. Z. Gougoutas, and W. Saenger, *Am. Soc.*, **92**, 2488 (1970)

(S)-1-Amino-2-hydroxymethylindoline, Mol. wt. 164.20.

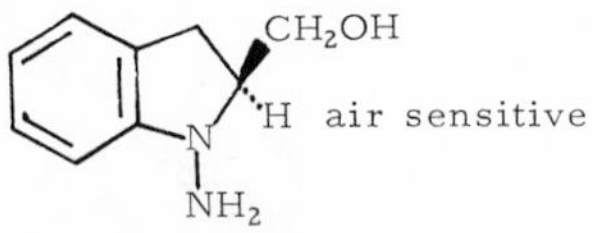

CH₂OH
'H air sensitive
NH₂

(S)-9

The Greek word chiral means hand; a chiral reagent is one capable of turning plane-polarized light to either hand, right or left. Corey *et al.*[1] devised chemical processes for the asymmetric synthesis of α-amino acids which allow regeneration of the chiral reagent. The reagent formulated and designated (S)-9 was prepared and resolved as follows:

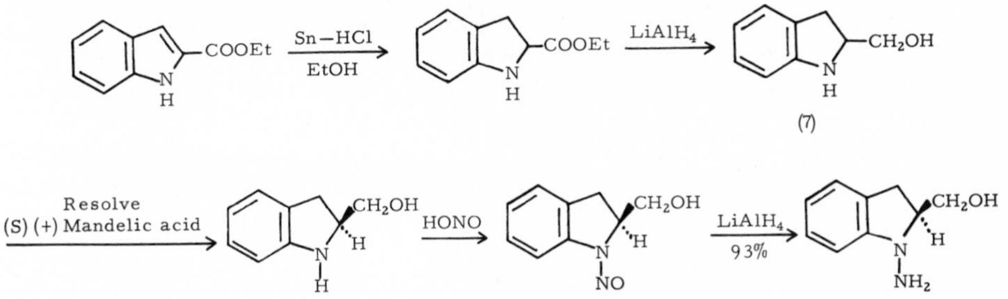

(7)

(S)-9

The indoline (7) obtained on LiAlH$_4$ reduction, after resolution, was dextro-rotatory (+60.5°) and shown to have the (S)-configuration by means of a chemical correlation.

The pathway for the asymmetric synthesis of an amino acid from the hydroxy-hydrazine derivative (S)-9 and an α-keto ester is as follows:

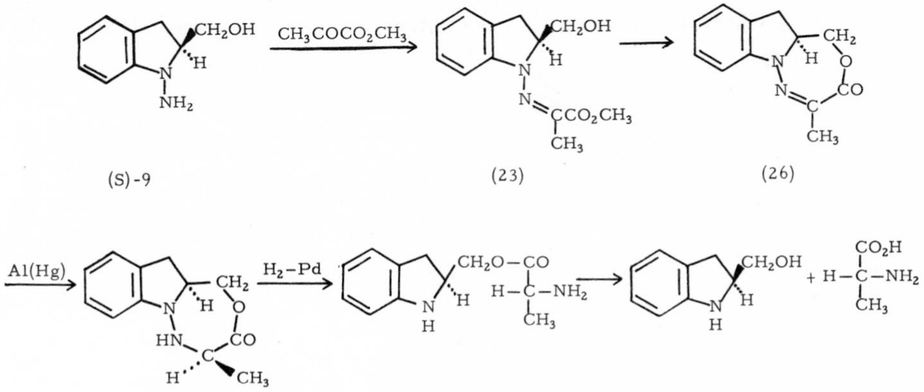

(S)-9   (23)   (26)

(29)   (32)   (S)-18   (35)

Starting from the hydrazine reagent (S)-9 and methyl pyruvate, reaction in methanol solution at room temperature gave the hydrazone (23), and this was cyclized by heating under anhydrous conditions in benzene with sodium meth-oxide to form the hydrazono lactone (26), a crystalline, levorotatory substance. The next step in the desired sequence, reduction of the hydrazono lactone (26), proved to be unexpectedly difficult but was accomplished by chemical reduction with aluminum amalgam under carefully controlled conditions. The hydrazino lactone (29) could be purified by recrystallization and obtained as a single dia-stereoisomer which on hydrogenolysis in dimethoxyethane–water containing hydrochloric acid afforded the amino acid ester (32). Finally, either acid-catalyzed or base-catalyzed hydrolysis of (32) afforded alanine (35) of high purity.

[1]E. J. Corey, R. J. McCaully, and H. S. Sachdev, *Am. Soc.*, **92**, 2476 (1969)

2-Amino-2-methyl-1-propanol, $(CH_3)_2C(NH_2)CH_2OH$, **1**, 37.

Protection of —*COOH*. This 2-aminoethanol derivative reacts readily with carboxylic acids to form 2-oxazolines (1).[1] Meyers *et al.*[2] found that the carboxylic

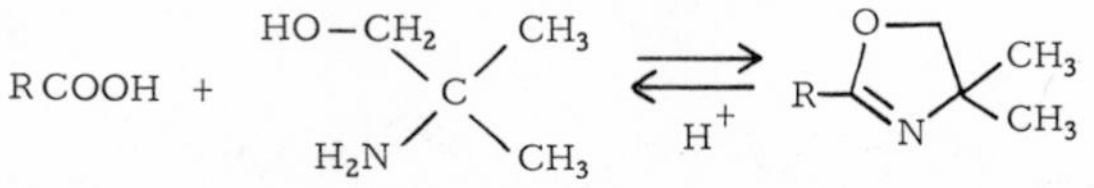

$$R\,COOH \; + \; \underset{H_2N}{\overset{HO-CH_2}{\diagdown}}\underset{CH_3}{\overset{CH_3}{C}} \; \underset{H^+}{\overset{}{\rightleftharpoons}} \; R-\overset{O-}{\underset{N}{\diagup}}\overset{CH_3}{\underset{CH_3}{\diagdown}}$$

(1)

acid can be readily regenerated by acid hydrolysis and that the 2-oxazoline system is inert to the Grignard reagent. He then showed that a functionally substituted carboxylic acid can be protected as the oxazoline and then subjected to the Grignard reaction. In the final step the carboxyl function is regenerated. For example, the tetralonecarboxylic acid (2) was heated with a slight excess of 2-amino-2-methyl-1-propanol to give the oxazoline (3). Reaction with phenylmagnesium bromide (2.0 equivalents) and magnesium bromide (1.0 equivalent) in THF gave the tertiary alcohol (4). Treatment with 3 N HCl regenerated the carboxylic acid (5); as expected, the tertiary hydroxyl group was lost during the hydrolytic step.

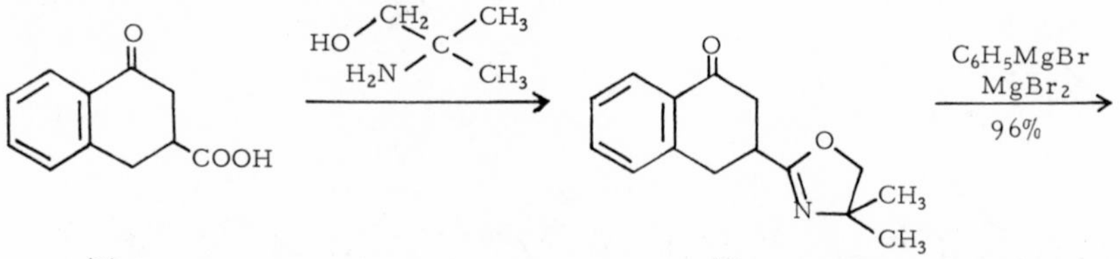

(2)   (3)

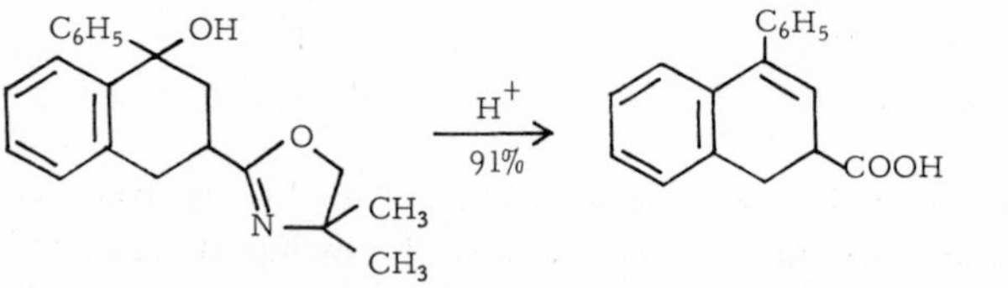

(4)   (5)

The procedure has also been applied to halobenzoic acids.[3] Thus *p*-bromobenzoic acid (6) was converted into the corresponding 2-oxazoline (7) as usual, and then into the Grignard derivative (8). This is, in effect, an aryl Grignard reagent with a protected carboxyl function. Reaction with an electrophile, for example benzonitrile, gives (9). The final step is hydrolysis. If carried out in 5–7% ethanolic sulfuric acid, the ester (10b) is obtained; if carried out in an aqueous medium, the free acid (10a) is obtained.

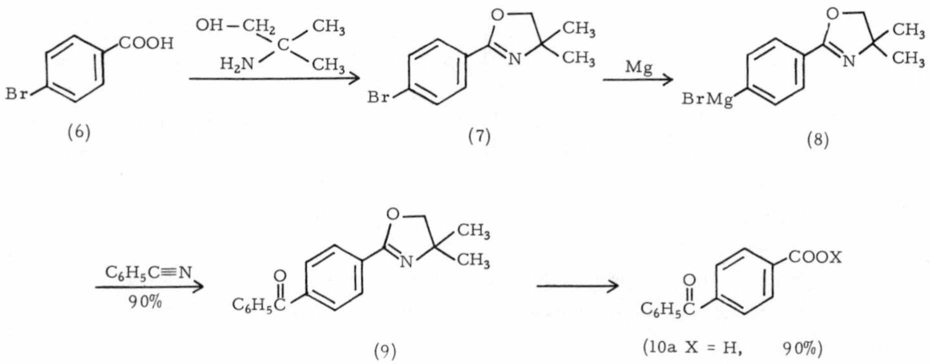

(6)   (7)   (8)

$$\xrightarrow[90\%]{C_6H_5C\equiv N}$$

(9)

(10a X = H,   90%)
(10b X = C₂H₅, 92%)

[1] J. W. Cornforth, *Heterocyclic Compounds*, **5**, 386 (1957)
[2] A. I. Meyers and D. L. Temple, Jr., *Am. Soc.*, **92**, 6644 (1970)
[3] *Idem, ibid.*, **92**, 6646 (1970)

Ammonium persulfate–Silver nitrate (see Potassium persulfate–Silver nitrate, 1, 954; 2, 348–349).

Selective oxidation of ethylbenzenes. Ethylbenzenes can be oxidized to the corresponding acetophenones by this combination of reagents.[1] A 2:1 molar ratio of persulfate to hydrocarbon is optimum, and silver ion is required only in catalytic amounts.

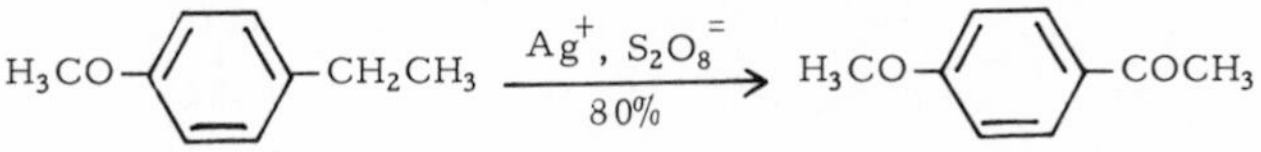

$$H_3CO-\!\!\!\left\langle\!\!\bigcirc\!\!\right\rangle\!\!-CH_2CH_3 \xrightarrow[80\%]{Ag^+,\ S_2O_8^=} H_3CO-\!\!\!\left\langle\!\!\bigcirc\!\!\right\rangle\!\!-COCH_3$$

[1] F. A. Daniher, *Org. Prep. Proc.*, **2**, 207 (1970)

***t*-Amyl chloroformate, 1, 40; 2, 24–25.**

t-Amyloxycarbonylamino acids. In the early papers of Sakakibara (**2**, 24) direct synthesis of Aoc-amino acids with *t*-amyl chloroformate was reported to be unsatisfactory since strictly anhydrous conditions are required. Sakakibara[1] now reports that the derivatives can be directly prepared by the reaction of *t*-amyl chloroformate and the free amino acid under Schotten-Baumann conditions (aqueous sodium hydroxide) if an inert water-soluble organic solvent such as THF, isopropanol, or methanol is added. The diluent increases the rate of acylation in comparison with that of decomposition of the reagent. The same procedure was found satisfactory for the preparation of many *t*-butyloxycarbonylamino acids.

[1] S. Sakakibara, I. Honda, K. Takada, M. Miyoshi, T. Ohnishi, and K. Okumura, *Bull. Chem. Soc. Japan*, **42**, 809 (1969)

Argentic picolinate [Ag(pic)$_2$], $\left(\text{[pyridine ring]} COO \right)_2 Ag$.   Mol. wt. 349.88.

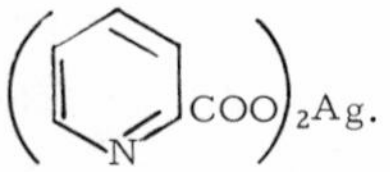

Preparation.[1] The reagent is prepared by the reaction of picolinic acid (pyridine-2-carboxylic acid) with silver nitrate and potassium persulfate. The material is stable on storage in the dark at room temperature for several months.

Oxidation.[1] The reagent oxidizes primary alcohols in good yield to aldehydes; it can also oxidize aldehydes further to carboxylic acids. Ketones are obtained in high yield from secondary alcohols. Oxidation is most rapid in DMSO but proceeds satisfactorily in water. Hydroquinone is oxidized to *p*-benzoquinone in 89% yield.

[1]T. G. Clarke, N. A. Hampson, J. B. Lee, J. R. Morley, and B. Scanlon, *Canad. J. Chem.*, **47**, 1649 (1969)

Arsenic trichloride (tribromide), AsX$_3$, Mol. wt. 181.28 (314.66). Suppliers: Alfa Inorganics, British Drug Houses, Ltd.

Halogenation of nucleosides.[1] The reaction of 2′,3′-O-isopropylideneuridine (1) with arsenic trichloride and arsenic tribromide in DMF gives 5′-desoxy-5′-chloro(bromo)-2′,3′-O-isopropylideneuridine (2) in 50–60% yield. Arsenic triiodide gives poor results. The reaction with uridine itself shows high degree of specificity for the 5′-position. Dods and Roth reasoned that the actual reagents

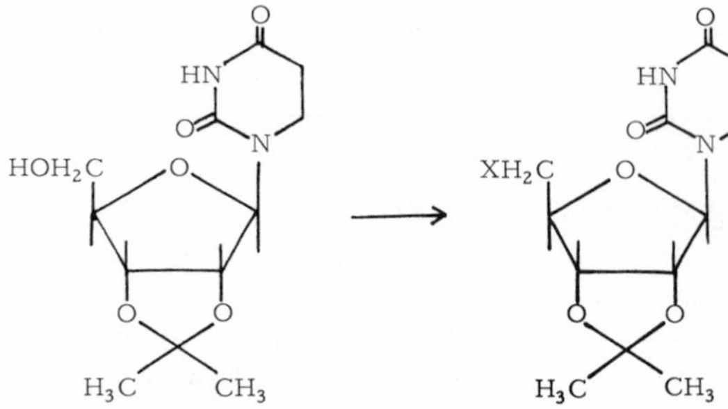

(1)   (2)

are probably chloromethylenedimethylammonium chloride and bromomethylenedimethylammonium bromide, Vilsmeier reagents, [(CH$_3$)$_2$N$^+$=CHX]X$^-$. They prepared these in the standard way from DMF and SOCl$_2$ or SOBr$_2$ and found that these reagents gave the above reaction, but in improved yields (90%).

[1]R. F. Dods and J. S. Roth, *J. Org.*, **34**, 1627 (1969)

Aryldiazonium tetrahaloborates, 1, 43–44; **2**, 25. Correction to **2**, 25: The last sentence should read: If decomposition occurs near room temperature, the salt should *never* be allowed to dry completely and should be used as soon as possible.

B

Benzenesulfonyl azide, $C_6H_5SO_2N_3$. Mol. wt. 183.19, decomposes at about 105°. Crude material detonates violently on heating.

Preparation. The reagent is prepared by the reaction of excess sodium azide in water–ethanol with benzenesulfonyl chloride. In an early procedure the latter reagent was dissolved in ethanol before addition to the sodium azide solution.[1] In a more recent preparation, benzenesulfonyl chloride is added directly to the ethanolic solution of sodium azide.[2]

Reactions. The thermal decomposition of benzenesulfonyl azide gives radicals which react with aromatic hydrocarbons to give products of amidation.[1,3]

Like phenyl azide, benzenesulfonyl azide undergoes addition reactions with strained double bonds. Thus it reacts with norbornene (1, bicyclo[2.2.1]-2-heptene) with loss of nitrogen to give the azetidine (2) in nearly quantitative yield.[2] The reaction is considered to involve the intermediate (a). In contrast the reaction

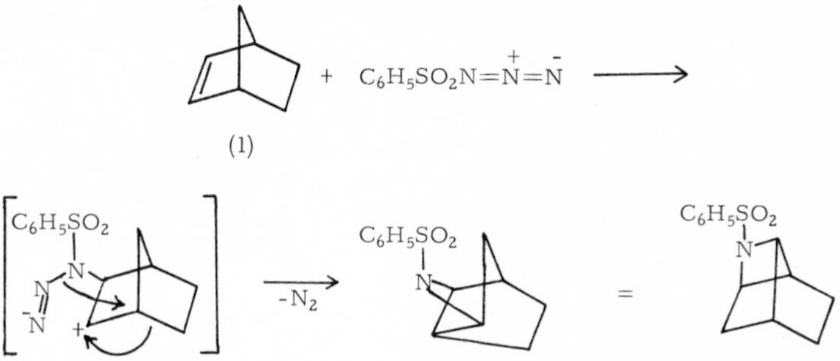

(1)

(a)   (2)

of (1) with phenyl azide gives a triazoline rather than an azetidine. The reaction of benzenesulfonyl azide, however, takes a different course when applied to bicyclo [2.2.1]-6-heptene-*endo-cis*-2,3-dicarboxylic anhydride (3). In this case the main product is the more hindered *endo*-aziridine (4); the corresponding *exo*-aziridine is a minor product.[4] The reaction apparently involves an unstable *exo*-triazole

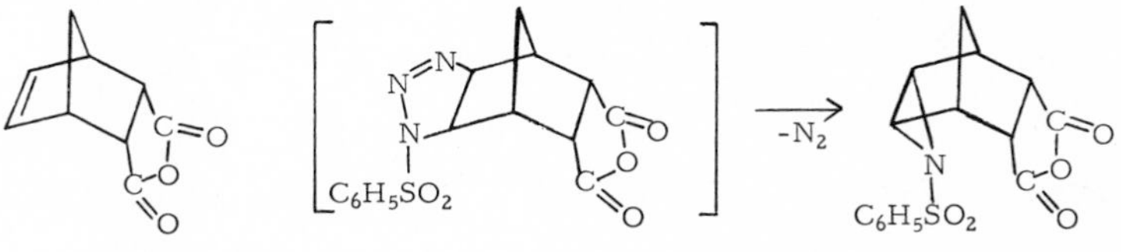

(3)   (b)   (4)

adduct (b), which rearranges with loss of N_2 to the observed *endo*-aziridine (4) and the isomeric *exo*-aziridine.

[1]O. C. Derner and M. T. Edmison, *Am. Soc.*, **77**, 70 (1955)
[2]L. H. Zalkow and A. C. Oehlschlager, *J. Org.*, **28**, 3303 (1963)
[3]J. F. Heacock and M. T. Edmison, *Am. Soc.*, **82**, 3460 (1960)
[4]A. C. Oehlschlager and L. H. Zalkow, *Canad. J. Chem.*, **47**, 461 (1969)

Benzenesulfonyl bromide, $C_6H_5SO_2Br$. Mol. wt. 164.17, b.p. 76–78°/0.17 mm.

Preparation.[1] A suspension of 82.0 g. (0.50 mole) of Eastman's benzenesulfinic acid sodium salt in 0.5 l. of benzene was stirred vigorously at room temperature[2] and bromine was added until the bromine color persisted (25 ml., 0.5 mole).[3]

A small amount of sodium benzenesulfinate was then added to react with the excess bromine. Filtration removed sodium bromide, and benzene was used to wash the filtrate (51 g. dry weight, 0.5 mole). Solvent was first removed from the filtered solution and combined washings on a rotary evaporator and finally by vacuum distillation. Vacuum distillation of the crude product (b.p. 76–78° at 0.17 mm.) gave 106.3 g. (92.2%) of benzenesulfonyl bromide.[4]

[1]M. S. Singer, Ph.D. Dissertation, University of Colorado, 1966; S. J. Cristol, J. K. Harrington, and M. S. Singer, *Am. Soc.*, **88**, 1529 (1966); M. S. Singer and S. J. Cristol, procedure submitted to *Org. Syn.* 1969.
[2]Elaborate apparatus is not necessary. An Erlenmeyer flask shaken manually is adequate.
[3]Because the large amount of solid material causes difficulty in mixing, ample time should be allowed for the bromine color to dissipate.
[4]Many applications of benzenesulfonyl bromide are compatible with benzene as solvent. In such cases, isolation is unnecessary and it may be assumed that benzenesulfonyl bromide was formed quantitatively. For most uses there appears to be no real advantage in the final vacuum distillation. The procedure has the merit of simplicity and high yield and takes only a few minutes.

N-Benzenesulfonylformimidic acid ethyl ester, $C_2H_5OCH{=}NSO_2C_6H_5$. Mol. wt. 213.26, m.p. 61°.

Preparation.[1] The reagent is prepared in 92% yield by sulfuric acid-catalyzed reaction of benzenesulfonamide and triethyl orthoformate.

Primary amines.[1] The reagent reacts with Grignard compounds in THF to give sulfonamides of primary amines in 70–90% yield.

$$2 \ RMgBr \ + \ C_2H_5OCH{=}NSO_2C_6H_5 \ \longrightarrow \ \overset{\displaystyle R}{\underset{\displaystyle R}{H\overset{|}{C}NHSO_2C_6H_5}}$$

[1]H. Stetter and D. Theisen, *Ber.*, **102**, 1641 (1969)

β-Benzoylpropionic acid, **2**, 26, ref. 1. Mol. wt. 178.19; m.p. 115–118°. Supplier: Aldrich.

Protective group for nucleosides.[1] Stepwise synthesis of oligonucleotides utilizing 2-cyanoethyl phosphate (**1**, 172–173) requires protection of the 3'-hydroxyl group which can be cleaved under essentially neutral conditions, since a methoxytrityl ether is sensitive to acid and a β-cyanoethyl phosphoric ester is sensitive to alkali. The β-benzoylpropionyl group meets these requirements, since it is quantitatively cleaved by dilute solutions of hydrazine hydrate in pyridine–acetic acid. The esters are prepared by condensation with DCC (dicyclohexylcarbodiimide).

[1]Definitive paper: R. L. Letsinger and P. S. Miller, *Am. Soc.*, **91**, 3356 (1969)

Benzyltriethylammonium   chloride, $[(C_2H_5)_3\overset{+}{N}CH_2C_6H_5]Cl^-$. Mol. wt. 227.78. Supplier: Eastman.

Carbanion formation. Mąkosza[1] has used the combination of benzyltriethylammonium chloride and sodium hydroxide as the base for alkylation of phenylacetonitrile with alkyl halides and with halonitrobenzenes. The same combination was found to be superior to phenyllithium (used previously[2]) for alkylation of Reissert's compound (N-benzoyl-1,2-dihydroisoquinaldonitrile).[3]

$$\text{(structure)} \xrightarrow[78\%]{C_6H_5CH_2Cl - NaOH} \text{(structure)}$$

[1]M. Mąkosza, *Tetrahedron Letters*, 673 (1969)
[2]J. Weinstock and V. Boekelheide, *Org. Syn., Coll. Vol.*, **4**, 641 (1963)
[3]M. Mąkosza, *Tetrahedron Letters*, 677 (1969)

Birch reduction, **1**, 54–56; **2**, 27–29.

Reduction of α,β-unsaturated ketones. House *et al.*[1] have compared the reduction of α,β-unsaturated ketones under standard Birch conditions (Li + *t*-BuOH in NH₃–THF) and with sodium or lithium and *t*-BuOH in hexamethylphosphoric triamide (**1**, 430–431; **2**, 208–210)–THF. In general, more of the less stable epimer is obtained with Na–HMPT. Factors which favor formation of the less stable epimer include the use of low temperatures (−33 to −78°), the presence of a proton donor (*t*-BuOH), use of a cosolvent such as THF, and the presence of excess metal throughout the reduction. Thus the enone (1) is reduced in either case to a mixture of (2) and (3). Under Birch conditions the less stable epimer (2) is obtained in 6–12% yield; with HMPT, (2) can be obtained in 14–20% yield.

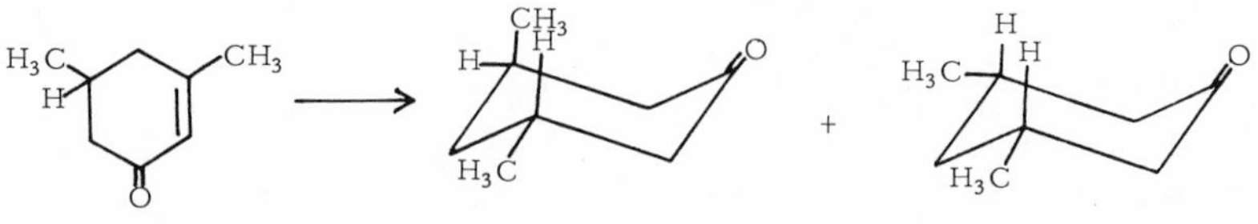

(1)   (2)   (3)

[1]H. O. House, R. W. Giese, K. Kronberger, J. P. Kaplan, and J. F. Simeone, *Am. Soc.*, **92**, 2800 (1970)

Bis(acrylonitrile)nickel(O),

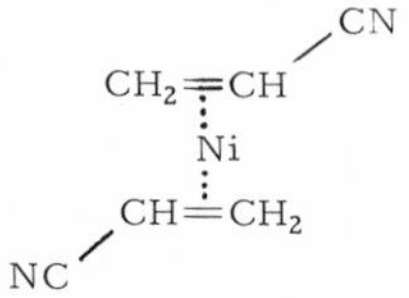

Mol. wt. 164.82; red crystals decomposing into nickel and acrylonitrile at about 100°. Pyrophoric; must be handled with complete exclusion of oxygen.

Preparation. The complex is prepared by the reaction of nickel carbonyl with acrylonitrile.[1]

[2 + 2] Cycloaddition.[2] In the presence of this nickel catalyst, methylenecyclopropanes undergo an unusual cycloaddition across carbon–carbon double bonds. Thus methylenecyclopropane (1) when heated in a sealed tube (60°, 48 hrs.) with excess methyl acrylate in the presence of bis(acrylonitrile)nickel(O) gives the 1 : 1 adduct, methyl 3-methylenecyclopentanecarboxylate (2), in 82% yield. Methyl vinyl ketone or acrylonitrile is also a suitable substrate. The reaction provides a useful synthesis of methylenecyclopentane derivatives.

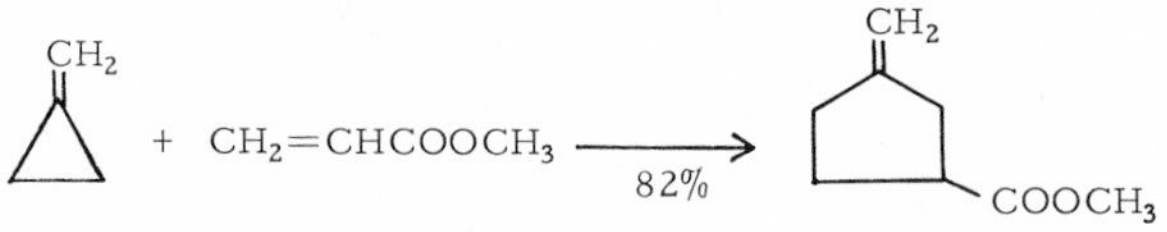

(1)   (2)

[1]G. N. Schrauzer, *Am. Soc.*, **81**, 5310 (1959); *idem, Ber.*, **94**, 642 (1961)
[2]R. Noyori, T. Odagi, and H. Takaya, *Am. Soc.*, **92**, 5780 (1970)

Bisbenzenesulfenimide, $(C_6H_5S)_2NH$. Mol. wt. 233.35, m.p. 129–130°.
    Preparation:[1] $2C_6H_5SCl + NH_3 \rightarrow (C_6H_5S)_2NH + 2HCl$.
    Primary alkylamines.[2] Lithium bisbenzenesulfenimide (prepared by the reaction of bisbenzenesulfenimide and *n*-butyllithium) reacts with alkyl bromides or

alkyl *p*-toluenesulfonates to give N-alkyl bisbenzenesulfenimides; these are cleaved by 3 *N* hydrochloric acid to give a primary amine. Either THF or dimethoxyethane is a suitable solvent. Overall yields are generally in the range 40–80%.

$$(C_6H_5S)_2NLi \xrightarrow{\text{ROTs or RBr}} (C_6H_5S)_2NR \xrightarrow{\text{HCl}} RNH_2 \cdot HCl + 2\ C_6H_5SCl$$

The sequence provides a useful alternative to the Gabriel synthesis particularly for preparation of amines containing nitrile, ester, or amide groups, which are hydrolyzed to a carboxylic acid group when the phthaloyl group is removed.

[1]H. Lecher, *Ber.*, **58**, 409 (1925)
[2]T. Mukaiyama and T. Taguchi, *Tetrahedron Letters*, 3411 (1970)

Bis-3,4-dichlorobenzoyl peroxide, (1). Mol. wt. 380.02, m.p. 139° dec.
   Preparation by dropwise addition of a toluene solution of 3,4-dichlorobenzoyl chloride with stirring and ice cooling to an aqueous solution of sodium peroxide.[1] The precipitated peroxide is collected, washed with water, air dried, and purified by dissolving it in chloroform and precipitating by the addition of methanol.
   Arylation, for example of benzene.[1] The peroxide (0.1 mole) is added to a boiling solution of 3 g. of *m*-dinitrobenzene in benzene and the solution is refluxed

for 40 hours (the small amount of nitro compound increases the yield by about 30%). The solvent is removed from the red solution until the volume is about 200 ml., the 3,4-dichlorobenzoic acid is crystallized in two crops and removed, and 3,4-dichlorodiphenyl is obtained by distillation at 146–150°/2 mm.

[1]D. H. Hey and M. J. Perkins, *Org. Syn.*, **49**, 44 (1969)

Bis(dimethylamino)methane   (**N,N,N′,N′-Tetramethylaminomethane**), $CH_2[N-(CH_3)_2]_2$. Mol. wt. 102.18, b.p. 85°. Supplier: Aldrich.
   Modified Mannich reaction. In a synthesis of some diaza steroids, Taylor and Shvo[1] found that the allylic protons at C_{19} of the diaza steroid (1) have considerable active methylene character. Thus treatment of (1) with bis(dimethylamino)

methane in acetic anhydride led to a vigorous exothermic reaction to give the *exo*-methylene derivative (2) in 75% yield. The same product was obtained in lower yield under normal Mannich conditions using dimethylamine hydrochloride and formaldehyde. Taylor and Shvo suggest that this modified Mannich reaction may be useful in synthesis of olefins which result from pyrolysis of preformed Mannich bases.

[1] E. C. Taylor and Y. Shvo, *J. Org.*, **33**, 1719 (1968)

1,8-Bis(dimethylamino)naphthalene, Mol. wt. 214.30, m.p. 47–48°. Supplier: Aldrich.

This substance is prepared by alkylation of 1,8-diaminonaphthalene with dimethyl sulfate. Alder *et al.*[1] report that it is a very strong base with pK_a 12.34. It is thus stronger than normal aliphatic amines. This strong basicity is due to the fact that 1,8-bis(dimethylamino)naphthalene is very strained and the strain is relieved by protonation. On the other hand, it is only weakly nucleophilic.

[1] R. W. Alder, P. S. Bowman, W. R. P. Steele, and D. R. Winterman, *Chem. Commun.*, 723 (1968)

Bis-3-methyl-2-butylborane (Disiamylborane, Sia_2BH), **1**, 57–59; **2**, 29. Supplier: Alfa Inorganics.

Selective reductions. Brown *et al.*[1] have reported an extended study on use of disiamylborane as a reducing agent. The reagent reacts rapidly with unhindered primary and secondary alcohols, but tertiary alcohols react only sluggishly, if at all. Consequently, primary and unhindered secondary hydroxyl groups of a polyol containing hindered secondary and/or tertiary hydroxyl groups can be protected as the corresponding disiamylborin. Oxidation with alkaline hydrogen peroxide liberates the protected hydroxyl group.

Carbonyl groups are reduced selectively in the presence of many other substituents: carboxylic acid, ester, nitrile, and nitro groups. In this respect the reagent exhibits selectivity comparable to sodium borohydride but should be useful when it is desirable to avoid alkaline conditions.

Another important application is for reduction of γ-lactones to hydroxyaldehydes. Thus γ-butyrolactone is reduced to γ-hydroxybutyraldehyde in 74% yield:

Reduction of γ-lactones with diborane proceeds to the diol.

Another interesting application is reduction of tertiary amides to aldehydes:

$$RC\overset{O}{\underset{N(CH_3)_2}{\diagup\diagdown}} \xrightarrow{Sia_2BH} RC\overset{OBSia_2}{\underset{H}{|}}N(CH_3)_2 \xrightarrow{H_2O} RCHO$$

This reduction has been achieved previously with lithium triethoxyaluminum hydride (**1**, 625).

Hydroboration of propargyl chlorides. The reaction of 1-chloro-2-heptyne (1) with disiamylborane in THF at 0° proceeds readily to the monohydroboration stage (2); protonolysis with glacial acetic acid gives isomerically pure *cis*-1-chloro-2-heptene (3). Zweifel *et al.*[2] reasoned that (2) might undergo β-elimination on treatment with base to give an allene, and this reaction was indeed realized in high yield (4). The reaction constitutes a general method for synthesis of terminal

$$\underline{n}\text{-}C_4H_9C\equiv CCH_2Cl \xrightarrow{Sia_2BH} \underset{H}{\overset{\underline{n}\text{-}C_4H_9}{\diagup}}C=C\underset{BSia_2}{\overset{CH_2Cl}{\diagdown}} \xrightarrow[83\%]{CH_3COOH} \underset{H}{\overset{\underline{n}\text{-}C_4H_9}{\diagup}}C=C\underset{H}{\overset{CH_2Cl}{\diagdown}}$$

$$(1) \qquad\qquad\qquad (2) \qquad\qquad\qquad\qquad (3)$$

$$\downarrow NaOH$$

$$\left[\underset{H}{\overset{\underline{n}\text{-}C_4H_9}{\diagup}}C=C\underset{\underset{Sia\quad Sia}{\diagup\diagdown}{B-OH}}{\overset{CH_2-Cl}{\diagup}} \right] \xrightarrow{83\%} \underline{n}\text{-}C_4H_9CH=C=CH_2$$

$$(4)$$

allenes (65–75% yield). The substituted propargyl chlorides are available by the following sequence:

$$RC\equiv CH \xrightarrow{CH_3Li} RC\equiv CLi \xrightarrow{(CH_2O)_n} RC\equiv CCH_2OH \xrightarrow[(C_4H_9)_3N]{SOCl_2} RC\equiv CCH_2Cl$$

[1]H. C. Brown, D. B. Bigley, S. K. Arora, and N. M. Yoon, *Am. Soc.*, **92**, 7161 (1970)
[2]G. Zweifel, A. Horng, and J. T. Snow, *ibid.*, **92**, 1427 (1970)

Bis(trimethylsilyl)acetamide, $\quad\underset{CH_3\overset{|}{C}=NSi(CH_3)_3}{OSi(CH_3)_3}$ (BTSA) $\qquad$ (**1**, 61; **2**, 30).

Protection of a hydroxyl group. Trimethylsilyl ethers have found wide use for gas-liquid chromatography and also to some extent for mass spectroscopy (**2**, 208). The first use of a trimethylsilyl ether as a protecting group for a sterically hindered hydroxyl group was reported only in 1968.[1] In a synthesis of an isomer of ecdysone (3), 20R-hydroxy-22-desoxyecdysone, from (1), a degradation product of crustecdysone, the C_2 and C_3 hydroxyl groups were protected as the

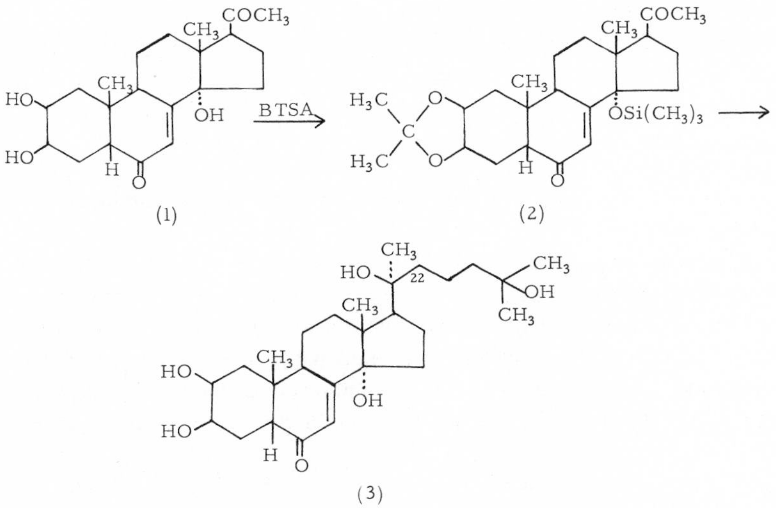

(1)   (2)

(3)

acetonide and the C_{14} hydroxyl group was protected as the trimethylsilyl ether, prepared from bis(trimethylsilyl)acetamide in DMF at 78°. After completion of the Grignard reaction to add the side chain, the protective groups were both removed by treatment with 0.1 N hydrochloric acid in 10% aqueous THF at room temperature. The silyl ether group was hydrolyzed more slowly than the acetonide group.

[1]M. N. Galbraith, D. H. S. Horn, E. Middleton, and R. J. Hackney, *Chem. Commun.*, 466 (1968)

9-Borabicyclo[3.3.1]nonane (9-BBN), **2**, 31.

Synthesis of aldehydes by carbonylation. Brown's original synthesis of aldehydes by reaction of trialkylboranes with carbon monoxide in the presence of lithium trimethoxyaluminum hydride (*see* Carbon monoxide, **2**, 60) suffered from the disadvantage that only one of the three alkyl groups was utilized; thus the maxi-

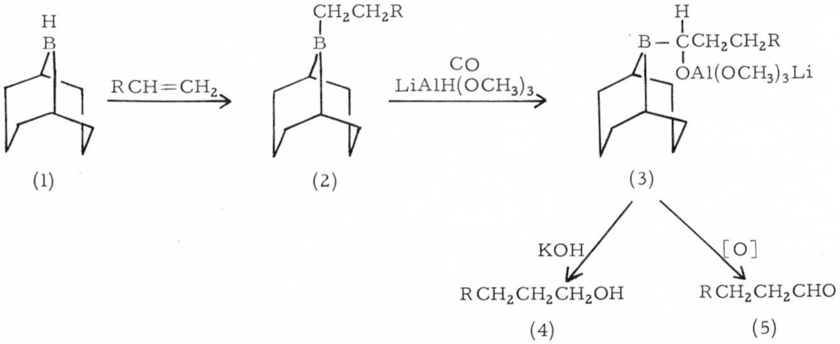

mum yield was 33%. This difficulty has now been circumvented by use of 9-BBN (1), which reacts with olefins readily to form B-alkyl derivatives (2). These have now been found to react rapidly with carbon monoxide in the presence of lithium trimethoxyaluminum hydride to give an intermediate (3) which can be hydrolyzed to the methylol derivative (4) or oxidized to the aldehyde (5).[1]

Examples:

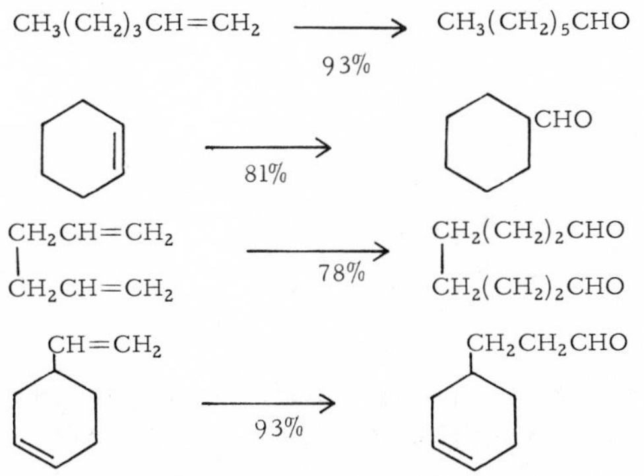

$$CH_3(CH_2)_3CH=CH_2 \longrightarrow CH_3(CH_2)_5CHO$$
$$93\%$$

81%

$$\begin{matrix} CH_2CH=CH_2 \\ | \\ CH_2CH=CH_2 \end{matrix} \xrightarrow[78\%]{} \begin{matrix} CH_2(CH_2)_2CHO \\ | \\ CH_2(CH_2)_2CHO \end{matrix}$$

93%

The last example shows that selective hydroboration of an acyclic double bond in the presence of a cyclic double bond can be achieved.

The reaction is also applicable to olefins of the type >C=C<.[2]

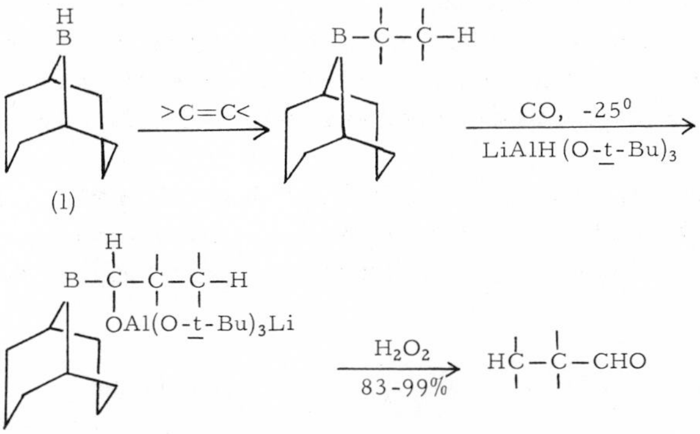

(1)

Lithium tri-t-butoxyaluminum hydride, (**1**, 620–625; **2**, 251–252), a mild reducing agent, is preferred because it does not reduce many functional groups. In the case of olefins which do not contain functional groups, lithium trimethoxyaluminum hydride (**1**, 625; **2**, 252–253) can be used.

Examples:

10-Undecenenitrile → 10-Cyanododecanal (99%)
Allyl benzoate → 4-Benzoxybutanal (89%)
Ethyl vinyl acetate → 4-Carboethoxybutanal (83%)

Carboethoxymethylation. As in the carbonylation reaction cited above, the original two-carbon-atom homologation reaction of trialkylboranes with ethyl bromoacetate (**2**, 192–193) had the defect that only one alkyl group of the tri-alkylborane is utilized. This disadvantage has now been eliminated by use of 9-BBN as the hydroboration reagent. Thus the B-alkyl-9-borabicyclo[3.3.1] nonane (B-R-9-BBN, **2**), prepared as above, reacts with ethyl bromoacetate under the influence of potassium *t*-butoxide to give the homologated ester (3) in

$$
\begin{array}{ccc}
\underset{(1)}{\text{H–B}} & \xrightarrow{\text{RCH=CH}_2} & \underset{(2)}{\text{CH}_2\text{CH}_2\text{R–B}} \xrightarrow[\underset{=}{t\text{-BuOK}}]{\text{BrCH}_2\text{COOC}_2\text{H}_5} \text{RCH}_2\text{CH}_2\text{CH}_2\text{COOC}_2\text{H}_5 \\
& & \quad\quad\quad\quad\quad\quad\quad (3)
\end{array}
$$

50–80% yield.[3] The sequence is applicable to ethyl dibromoacetate and ethyl dichloroacetate to give an α-halocarboxylic acid; yields in this case are somewhat higher (70–90%). This variation has another advantage in that it is not so subject to steric hindrance as the reaction of relatively hindered trialkylboranes with ethyl dihaloacetates.

Stereochemistry. Brown[4] has shown that the carbonylation reaction and the carboethoxymethylation reaction are highly stereospecific in that the original boron–carbon bond is converted into a carbon–carbon bond with retention of configuration. These reactions thus are similar to oxidation of organoboranes with alkaline hydrogen peroxide and amination with hydroxylamine-O-sulfonic acid.

α-Alkylation of ketones. Brown, Rogić, and Rathke[5] reported in 1968 that the reaction of α-bromo ketones with trialkylboranes under the influence of potassium *t*-butoxide in THF gives the corresponding α-alkyl derivatives. The method has

the advantage over the classical method of alkylation of ketones with an alkyl halide and a strong base in that it results in monoalkylation. However, the method has two drawbacks: only one alkyl group of the organoborane is utilized, and the reaction fails with branched-chain boranes. Both of these flaws are eliminated by use of B-alkyl-9-borabicyclo[3.3.1]nonanes (B-R-9-BBN) prepared as above.[6] Maximum yields are obtained by simultaneous addition of the α-bromo ketone and potassium *t*-butoxide to the B-R-9-BBN at 0°.

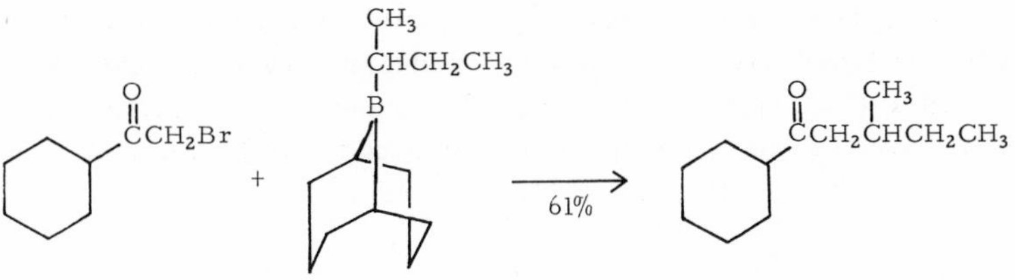

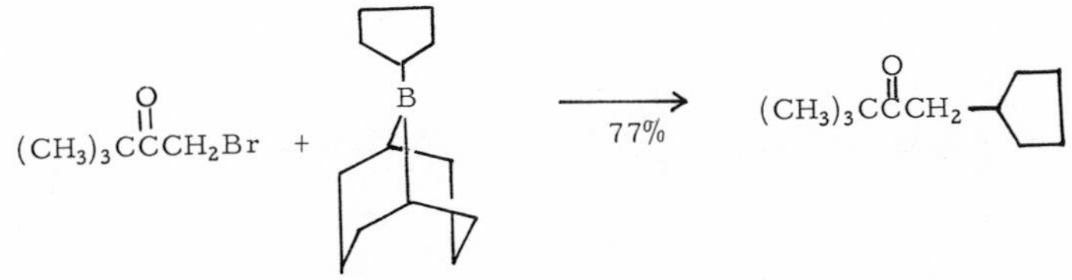

Synthesis of cyclopropane derivatives. The cyclization of hydroborated allylic chlorides was discovered by Hawthorne and Dupont.[7] The major difficulty is that two isomeric boron derivatives are formed in roughly equal amounts, and only one is converted into a cyclopropane. Use of the more selective hydro-borating reagent disiamylborane (**1**, 57–59; **2**, 29; this volume) circumvented this

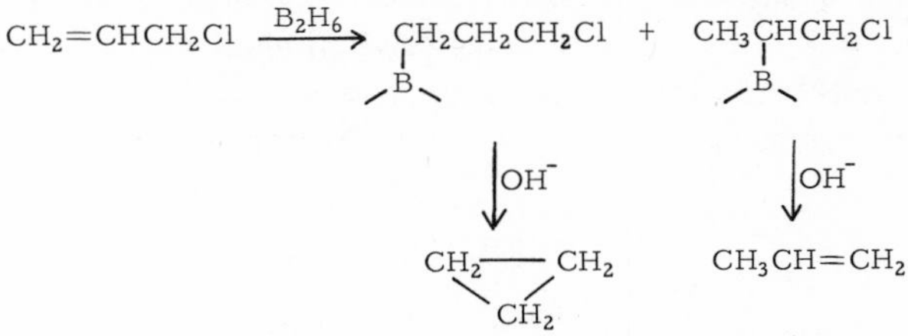

difficulty, but loss of the chlorine substituent was observed as a competing reaction.

Both difficulties are resolved by use of B-(γ-chloropropyl)-9-borabicyclo[3.3.1] nonanes, available by the reaction of 9-BBN with an allylic chloride.[8]

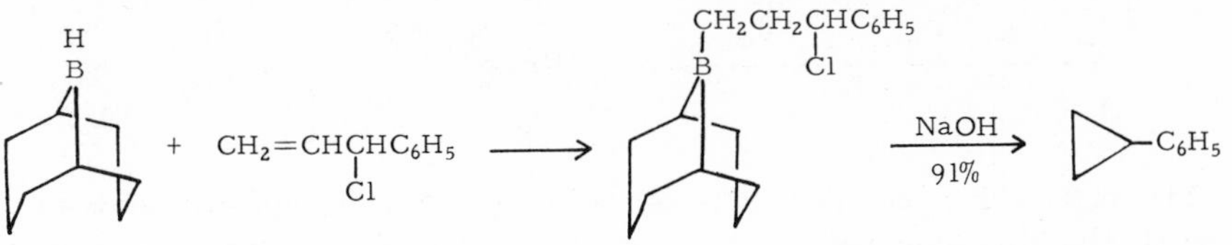

Other examples:

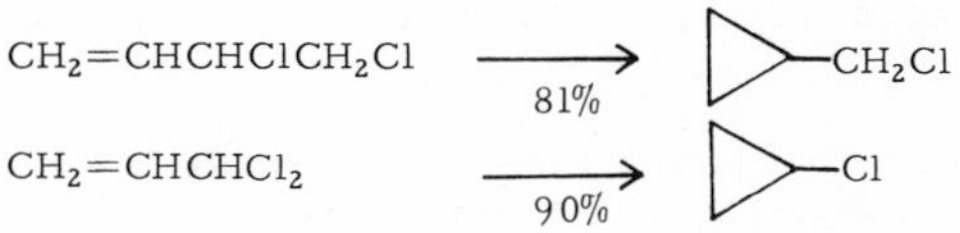

A convenient synthesis of cyclopropanol involves the reaction of propargyl bromide with 2 moles of 9-BBN in THF; sodium hydroxide or methyllithium is then added. B-Cyclopropyl-9-borabicyclo[3.3.1]nonane is obtained in 70% yield. Oxidation with hydrogen peroxide (sodium acetate) in the usual way affords cyclopropanol (65% yield) and cis-cyclooctane-1,5-diol.[9]

$$HC{\equiv}CCH_2Br \xrightarrow{\text{9-BBN}} HCCH_2CH_2Br \xrightarrow{\text{NaOH or MeLi}}$$

$$\xrightarrow{H_2O_2, \text{ NaOAc}} \quad \triangleright\text{—OH} \quad + \quad \underline{cis\text{-1,5-Cyclooctanediol}}$$

The method has been extended to the synthesis of cyclobutanol. In this case the tosylate of 3-butyne-1-ol, $HC{\equiv}CCH_2CH_2OTs$, was the starting material and ring closure was effected with methyllithium. Oxidation to cyclobutanol was carried out *in situ*; yield 65%.[9]

α-Arylation of ketones and esters. B-Alkyl- and B-aryl-9-BBN can be synthesized readily from the corresponding organolithium derivatives and 9-BBN. This procedure is useful in the case of alkyl groups not available by hydroboration, and especially to aryl groups. Thus 9-BBN is treated with phenyllithium in

(1) $+ C_6H_5Li \longrightarrow$ $[\text{BHC}_6\text{H}_5]$ Li $\xrightarrow[{-H_2, \ -CH_3SO_3Li}]{CH_3SO_3H, \ 0°}$ BC_6H_5 (2)

$$\xrightarrow[50\%]{\substack{CH_2BrCOOC_2H_5 \\ THF, \ t\text{-BuOK}, \ t\text{-BuOH}}} C_6H_5CH_2COOC_2H_5 \ + \quad BO\text{-}t\text{-Bu} \quad + \ KBr$$

(3)

THF at 0°. This is immediately followed by dropwise addition of methanesulfonic acid. The B-phenyl-9-BBN (2) is obtained in 85% yield. This product (2) is then allowed to react with ethyl bromoacetate (**2**, 192–193), ethyl dibromoacetate (**2**, 195), and various bromoketones as described previously to give α-aryl derivatives such as (3).[10]

[1] H. C. Brown, E. F. Knights, and R. A. Coleman, *Am. Soc.*, **91**, 2144 (1969)
[2] H. C. Brown and R. A. Coleman, *ibid.*, **91**, 4606 (1969)

[3]H. C. Brown and M. M. Rogić. *ibid.*, **91**, 2146 (1969)
[4]H. C. Brown, M. M. Rogić, M. W. Rathke, and G. W. Kabalka, *ibid.*, **91**, 2150 (1969)
[5]H. C. Brown, M. M. Rogić. and M. W. Rathke, *ibid.*, **90**, 6218 (1968)
[6]H. C. Brown, M. M. Rogić, H. Nambu, and M. W. Rathke, *ibid.*, **91**, 2147 (1969)
[7]M. F. Hawthorne and J. A. Dupont, *ibid.*, **80**, 5830 (1958); M. F. Hawthorne, *ibid.*, **82**, 1886 (1960)
[8]H. C. Brown and S. P. Rhodes, *ibid.*, **91**, 2149 (1969)
[9]H. C. Brown and S. P. Rhodes, *ibid.*, **91**, 4306 (1969)
[10]H. C. Brown and M. M. Rogić, *ibid.*, **91**, 4304 (1969)

Boric acid, 1, 63–66; 2, 32.

Protection of hydroxyl groups. Although boric acid esters have been known for some time, these have not been used to any extent for protection of hydroxyl groups. Fanta and Erman[1] have found this means of protection useful, especially in a synthesis of dihydro-β-santalol (8). The esters are prepared in quantitative yield by refluxing the alcohol with one third molar equivalent of boric acid in benzene with constant removal of the water formed. They are readily hydrolyzed in an aqueous medium, neutral, acidic, or basic, during workup. Thus the alcohol (1) was esterified (2) and then hydrobrominated in the presence of benzoyl peroxide (anti-Markownikoff). The bromohydrin (3) was then esterified with boric acid (4), and this ester was used to alkylate the sodium enolate of 3-methylnorcamphor (5). The completing step was a Wittig reaction, carried out on the borate ester (7) of (6) to give dihydro-β-santalol (8).

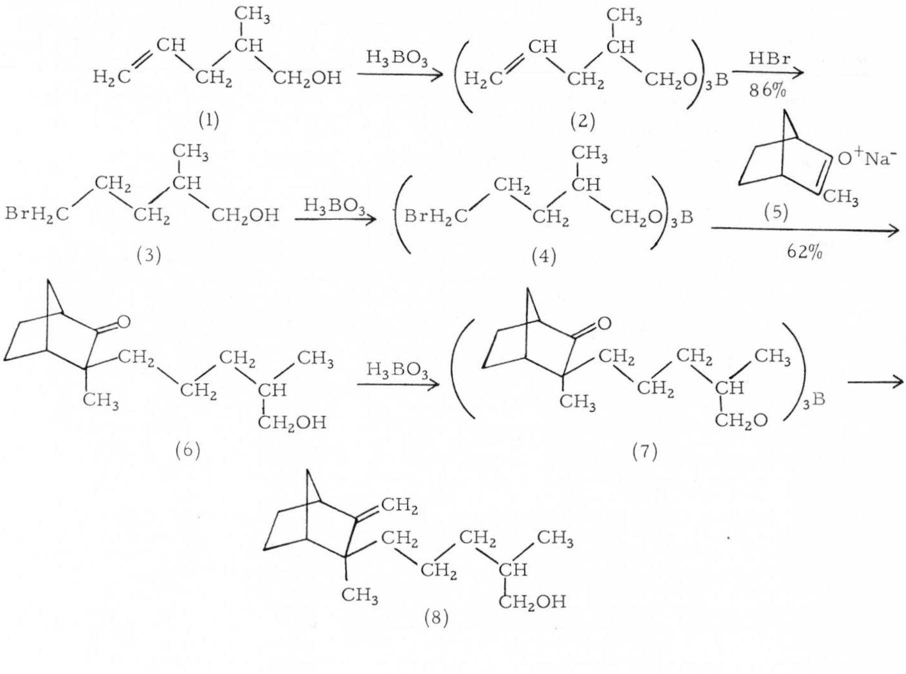

When the Wittig reaction was run directly on (6), a substantial amount of the isomer (10) was also obtained. This product is considered to arise through (9), formed by an intramolecular Meerwein-Ponndorf reaction involving the hindered carbonyl group and the free hydroxyl group.

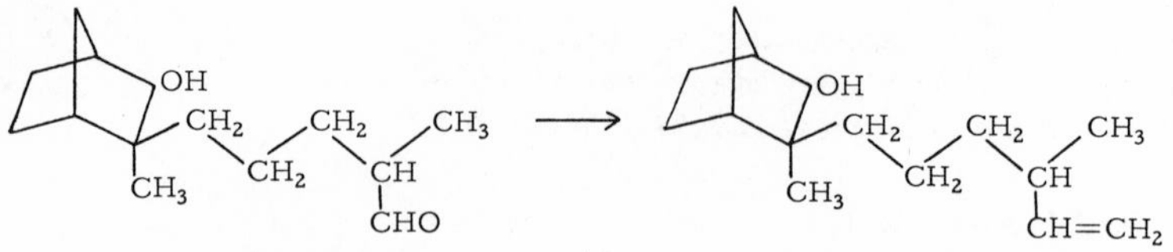

(9)   (10)

[1]W. I. Fanta and W. F. Erman, *Tetrahedron Letters*, 4155 (1969)

Boron tribromide, 1, 66–67; **2,** 33–34.

Demethylation. McOmie and West[1] report use of reagent of 99.9% purity from Koch-Light Laboratories Ltd., Colnbrook, Bucks, England, for demethylation of 3,3'-dimethoxydiphenyl. A solution of the reagent in methylene chloride is added

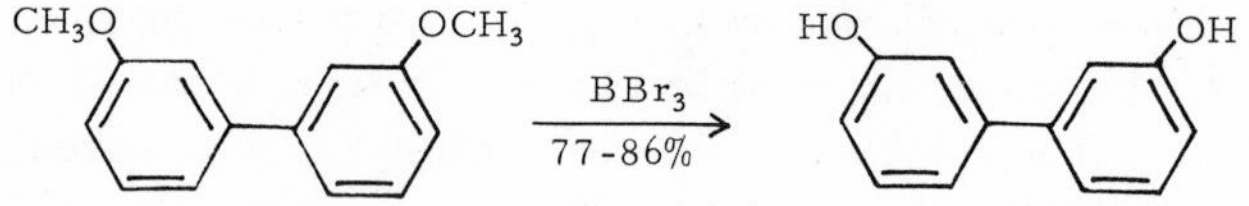

with stirring to a solution of the diether in methylene chloride at −80°. Boron tribromide was the reagent of choice for the final step (demethylation) in the synthesis of the naturally occurring macrolide zearalenone (**2,** 34).

Boron tribromide in methylene chloride or benzene is recommended for demethylation of polymethoxybenzophenones, where use of hydrogen bromide in acetic acid is often accompanied by cyclodehydration of the resulting polyhydroxybenzophenones to give xanthones. However, unexpected difficulties were encountered in the case of 2,2',3',6-tetramethoxybenzophenone (1), where the major product (76% yield) was the monomethyl ether (2), the desired tetra-

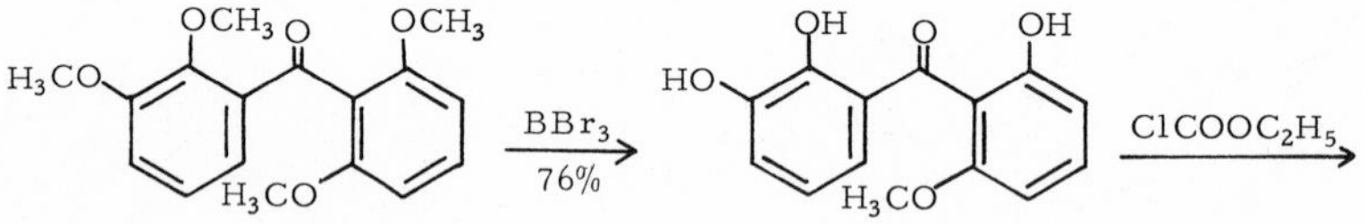

(1)   (2)

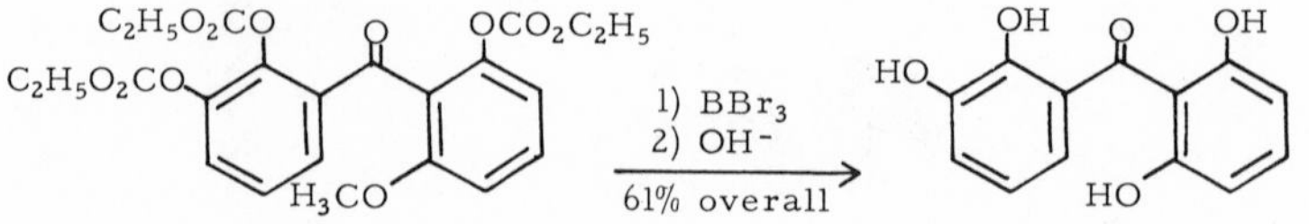

(3)   (4)

hydroxybenzophenone (4) being obtained in only 16% yield. It was reasoned that the inertness of the fourth methoxyl group was due to formation of a planar com-

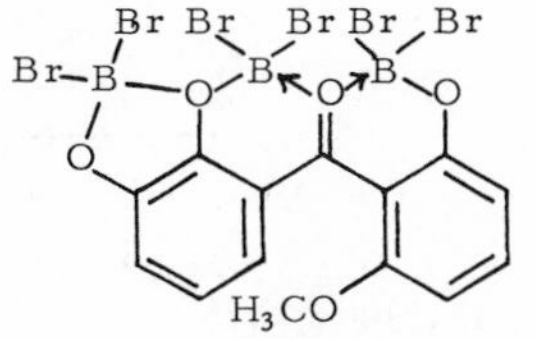

(a)

plex (a), in which the methoxyl group is sterically inaccessible to attack by the reagent. The difficulty was circumvented by conversion of (2) into the tricathylate (3), which cannot form a complex of type (a), treatment of (3) with the reagent, and alkaline hydrolysis of the protective groups. The desired 2,2',3',6-tetrahydroxy-benzophenone (4) was obtained in 61% overall yield.[2]

Canadian chemists[3] report that attempted demethylation of the dimethyl ether of pinosylvin (1) with boron tribromide gave the monodemethylated product in

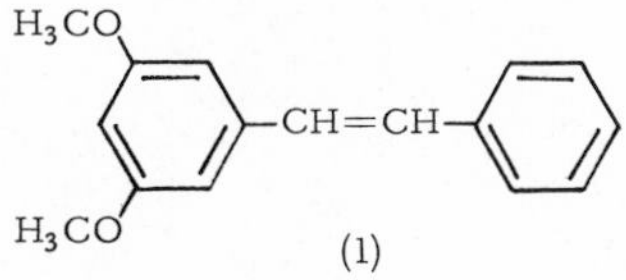

(1)

90% yield. Further attempts to cleave the second methyl ether gave only good yields of recovered ether. Apparently polymethoxy aromatics are completely demethylated only when the methoxyl groups are adjacent or in different rings. They achieved complete demethylation by fusion with pyridine hydrochloride (50% yield).

[1]J. F. W. McOmie and D. E. West, *Org. Syn.*, **49**, 50 (1969).
[2]H. D. Locksley and I. G. Murray, *J. Chem. Soc.*, (C), 392 (1970)
[3]F. W. Bachelor, A. A. Loman, and L. R. Snowdon, *Canad. J. Chem.*, **48**, 1554 (1970)

Boron trichloride, 1, 67–68; **2**, 34–35.

Diketo-1,3-diazetidines. Aryl isocyanates (1) are converted into allophanic acid chlorides (2) in about 50% yield when treated with boron trichloride (N_2 stream). These acid chlorides cyclize to diketo-1,3-diazetidines (3) when treated with pyridine (toluene or chloroform the solvent).[1]

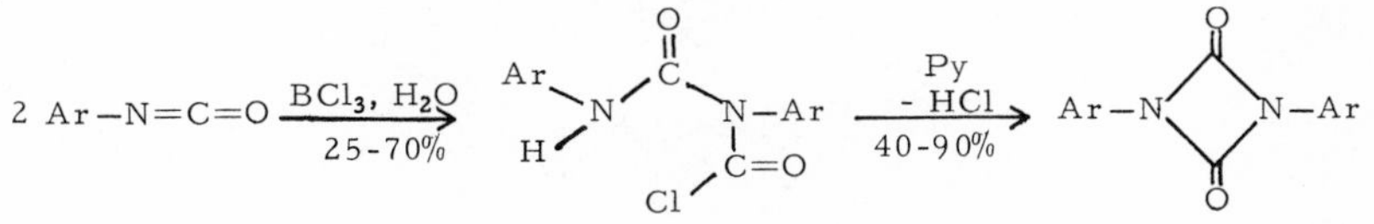

(1)   (2)   (3)

Condensation of carbonyl compounds with ketene.[2] Aldehydes or ketones condense with ketene in the presence of an equivalent of boron trichloride to give α,β-unsaturated acid chlorides:

[1]H. Helfert and E. Fahr, *Angew. Chem., internat. Ed.,* **9**, 372 (1970)
[2]P. I. Paetzold and S. Kosma, *Ber.,* **103**, 2003 (1970)

Boron trifluoride, 1, 68–69.

Cyclization of 4-arylolefins. Boron trifluoride in benzene solution effects cycloalkylation of 4-arylolefins in over 80% yield.[1] This step was used in an improved synthesis of 1,5-dimethylnaphthalene:

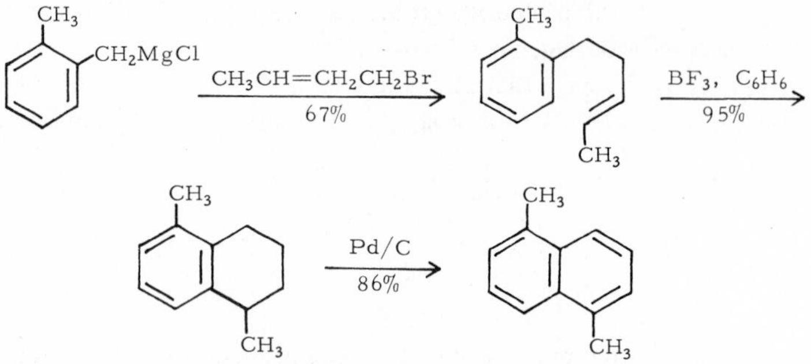

This method was also used for the first synthesis of 1,3,6,8-tetramethyltriphenylene as shown.

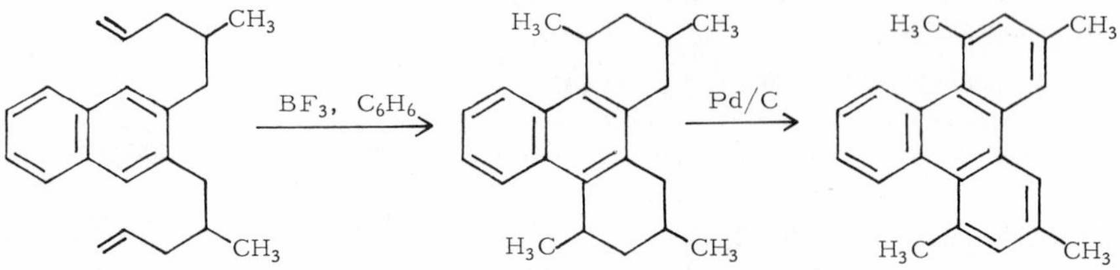

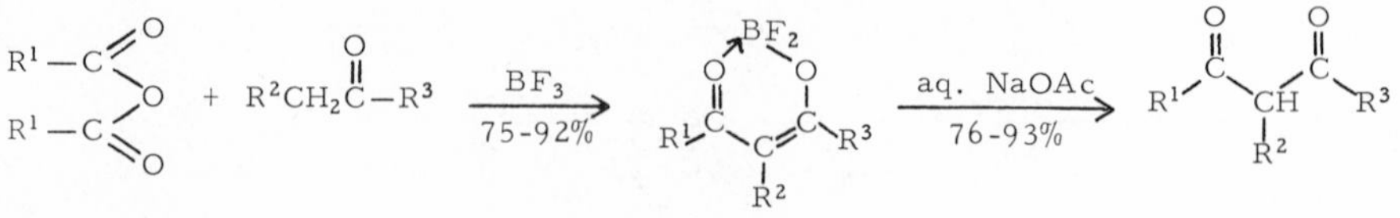

β-Diketones. The Meerwein[2] synthesis of β-diketones (4) involves the reaction of carboxylic acid anhydrides (1) with ketones (2) with boron trifluoride as condensing agent. The yields can be improved if the intermediate boron difluoride complexes (3) are isolated by pouring the reaction mixture into cold water. The complexes are then hydrolyzed by refluxing in aqueous sodium acetate solution.[3]

[1]P. Canonne and A. Regnault, *Canad. J. Chem.*, **47**, 2837 (1969)
[2]H. Meerwein and D. Vossen, *J. prakt. Chem.*, [2], **141**, 149 (1934)
[3]A. N. Sagredos, *Ann.*, **700**, 29 (1966)

Boron trifluoride etherate, 1, 70–72; **2**, 35–36.

Esterification. An esterifying reagent can be prepared easily by mixing boron trifluoride etherate in an alcohol. This reagent is both thorough and mild. Thus *p,p'*-diphenyldicarboxylic acid, which is inert to diazomethane, is converted into the corresponding dimethyl ester in 40% yield when refluxed in 10% boron trifluoride etherate in methanol for three days. The very sensitive 1,4-dihydrobenzoic acid was successfully esterified by this method. The ethyl ester was obtained in 81% yield when the acid was refluxed for 20 hours in anhydrous ethanol containing boron trifluoride etherate.[1]

Intramolecular acylation (**2**, 35). The bridged decalindione (1) has been converted into 8-acetoxy-4-twistanone (2) in one step by treatment with boron trifluoride etherate, acetic acid, and acetic anhydride. The product is readily

(1)   (2)

convertible into twistane itself and is also useful for synthesis of twistane derivatives containing a function at the bridgehead.[2]

[1]J. L. Marshall, K. C. Erickson, and T. K. Folsom, *Tetrahedron Letters*, 4011 (1970)
[2]A. Bélanger, Y. Lambert, and P. Deslongchamps, *Canad. J. Chem.*, **47**, 795 (1969)

Boron trioxide, B_2O_3. Mol. wt. 69.64. Supplier: Research Organic/Inorganic Chem. Corp.

Boron trioxide (excess) has been recommended as the dehydrating reagent in the condensation of a 1,3-diketone and an aldehyde.[1] Piperidine is used as catalyst.

[1]M. Sekiya and K. Suzuki, *Chem. Pharm. Bull. Japan*, **18**, 1530 (1970)

Bromine, Br_2. Mol. wt. 159.83.

Oxidative cleavage of ethers. Bromine in an aqueous acetate buffer at pH 5 is an effective oxidant for ethers containing an α-hydrogen. Under strongly acidic conditions, bromination products are formed. Primary alkyl groups are converted into carboxylic acids and secondary alkyl groups into ketones. Yields are generally over 80%.[1]

Examples:

$$(CH_3CH_2CH_2)_2O \xrightarrow[ca. \ 100\%]{Br_2, \ H_2O} 2 \ CH_3CH_2COOH$$

$$(C_6H_5CH_2)_2O \xrightarrow[98\%]{Br_2, \ H_2O} 2 \ C_6H_5CHO$$

[1]N. C. Deno and N. H. Potter, *Am. Soc.*, **89**, 3550 (1967)

Bromine azide, **2**, 37–38.

Hassner[1] has reviewed the addition of bromine azide to olefins.

[1]A. Hassner, *Accts. Chem. Res.*, **4**, 9 (1971)

p-**Bromophenacyl bromide**, **1**, 77. Additional suppliers: Aldrich; Pfalz and Bauer.

Protection of acids and phenols. Carboxylic acids can be protected as the *p*-bromophenacyl esters, prepared by the reaction of the carboxylate anion with *p*-bromophenacyl bromide in water or DMF. The protecting group is removed at room temperature by zinc in glacial or aqueous acetic acid. The new method is comparable to Woodward's protection of carboxylic acids as β-trichloroethyl esters, which are also cleaved by zinc (*see* Trichloroethanol, this volume). γ-Lactones are converted into γ-hydroxy esters.

Phenols are converted into the ethers by reaction with *p*-bromophenacyl bromide in refluxing acetone and dry potassium carbonate for 1–2 hours. The phenacyl group is removed by zinc in acetic acid.[1]

[1]J. B. Hendrickson and C. Kandall, *Tetrahedron Letters*, 343 (1970)

N-Bromosuccinimide, **1**, 78–80; **2**, 40–42.

Bromohydrins. Dalton *et al.*[1] have prepared bromohydrins by reaction of olefins with NBS in aqueous DMSO. The reaction involves stereospecific Markownikoff addition; thus *trans*-1-phenyl-1-propene is converted into *erythro*-2-bromo-1-phenylpropanol in 92% yield. The exact nature of the reagent is not clear, but labeling experiments show incorporation of oxygen from DMS^{18}O. The reaction fails with highly hindered olefins and with olefins bearing electron-withdrawing groups on the double bond.

The method has been used by Sisti[2] to prepare the halohydrin (2) from 9-ethylidenefluorene (1).

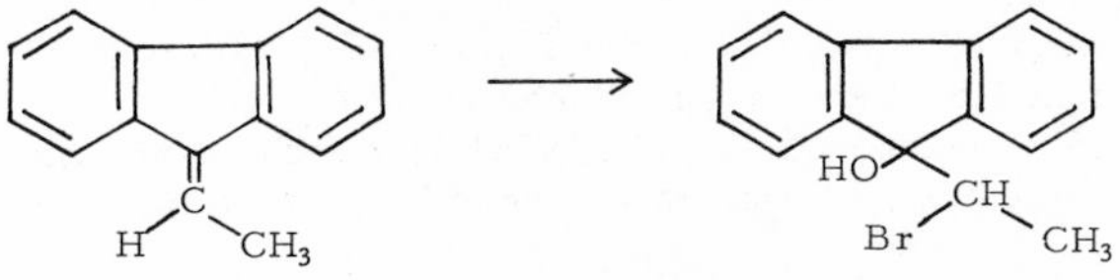

(1)　　　　　　　　　　　　　　(2)

α-Bromo acids. α-Bromo acids have generally been prepared by the Hell-Volhard-Zelinsky reaction (reaction of acids with halogen and phosphorus). Although an acyl halide is generally considered an intermediate, bromination of acid chlorides proceeds with difficulty. Gleason and Harpp[3] now report that NBS in the presence of catalytic amounts of hydrogen bromide effects bromination of acyl halides in high yield. Carbon tetrachloride is used as solvent. The

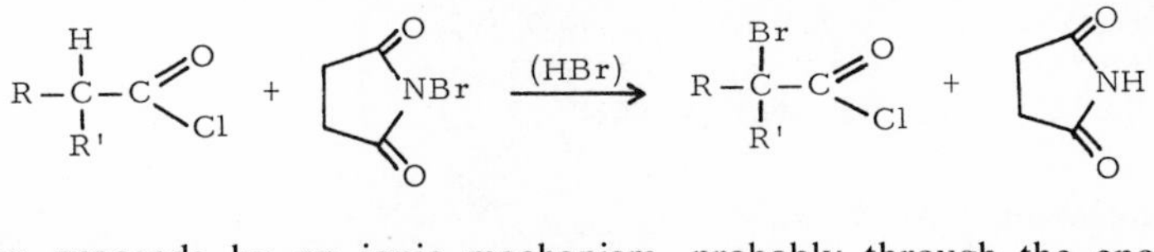

reaction proceeds by an ionic mechanism, probably through the enol. This method has the advantage that it leads to an α-bromo acid chloride, from which a variety of acyl derivatives can be prepared.

Allylic oxidation. Allylic methylene groups are oxidized to carbonyl groups in high yield by N-bromosuccinimide on irradiation with visible light. The olefin (1 mole) is dissolved in THF, glyme, or diglyme containing 1–10% of water, and is stirred for 1 hour, at room temperature with the reagent (2.5 moles) and calcium carbonate (ca. 2 moles) while being irradiated. α-Amyrin acetate (1) is converted into α-amyrenonyl acetate (2) in 98% yield. Cholesteryl acetate gives 7-ketocholesteryl acetate in 81% yield. Yields are variable and considerably lower in the absence of irradiation.[4]

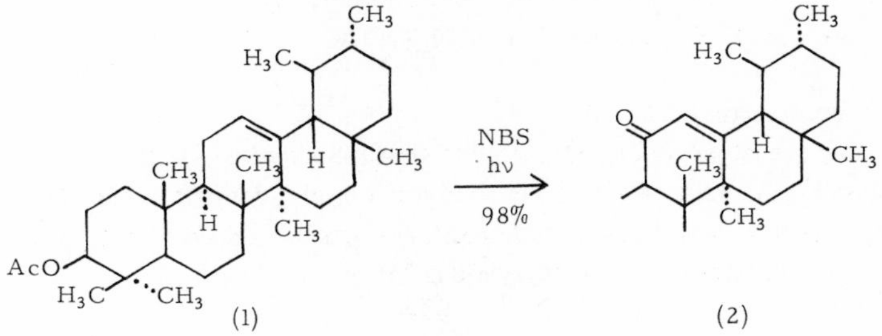

(1)　　　　　　　　　　　　　　(2)

Sulfide oxidation. Sulfides can be oxidized to sulfoxides in anhydrous methanol as solvent with 1 equiv. of either N-chlorosuccinimide or N-bromosuccinimide in 65–93% yield. In the case of diaryl sulfides the oxidation can be controlled to give either the sulfoxide or the sulfone.[5]

[1]D. R. Dalton, V. P. Dutta, and D. C. Jones, *Am. Soc.*, **90**, 5498 (1968)
[2]A. J. Sisti, *J. Org.*, **35**, 2670 (1970)
[3]J. G. Gleason and D. N. Harpp, *Tetrahedron Letters*, 3431 (1970)
[4]B. W. Finucane and J. B. Thomson, *Chem. Commun.* 1220 (1969)
[5]R. Harville and S. F. Reed, Jr., *J. Org.*, **33**, 3976 (1968); the method is an extension of that of W. Tagaki, K. Kikukawa, K. Ando, and S. Oae, *Chem. Ind.*, 1624 (1964)

t-Butyl azidoformate, 1, 84–85; **2**, 44–45.

Preparation. The procedure of Insalaco and Tarbell (**2**, 44–45, ref. 1a) has now been published.[1]

N-Protection of amino acids. For the synthesis of *t*-butyloxycarbonyl-*β*-benzyl-L-aspartic acid by the reaction of the amino acid derivative with *t*-butyl azidoformate, Polzhofer[2] found that use of triethylamine as base gives a higher yield (75%) than use of magnesium oxide (40%). The yield using *t*-butyl 4-nitrophenylcarbonate as reagent is 27%.[3]

[1]M. A. Insalaco and D. S. Tarbell, *Org. Syn.*, **50**, 9 (1970)
[2]K. P. Polzhofer, *Tetrahedron Letters*, 2305 (1969)
[3]L. Benoiton, *Canad. J. Chem.*, **40**, 570 (1962)

t-Butyl carbazate, 1, 85; **2**, 46.

A new procedure for the preparation of *t*-butyl carbazate (2) involves hydrazinolysis of *t*-butyl ethylcarbonate (1). The yield is rather low, but the method is convenient.[1]

$$(CH_3)_3COH + ClCOOC_2H_5 \xrightarrow[65\%]{t-BuOK} (CH_3)_3COCOOC_2H_5 \xrightarrow[43\%]{\substack{excess \\ H_2NNH_2}} (CH_3)_3COCONHNH_2$$

$$(1) \qquad\qquad (2)$$

[1]M. Muraki and T. Mizoguchi, *Chem. Pharm. Bull. Japan*, **18**, 217 (1970)

t-Butylcarbonic diethylphosphoric anhydride, 2, 46.

Preparation. The preparation has now been published.[1]

[1]M. A. Insalaco and D. S. Tarbell, *Org. Syn.*, **50**, 9 (1970)

t-Butyl chloroacetate, 1, 86.

Glycidic esters. Wynberg *et al.*[1] state that the simplest synthesis of adamantane-2-carboxylic acid (4) involves conversion of adamantanone (1) into the glycidic ester (2), pyrolysis to adamantane-2-carboxaldehyde (3), and finally chromic acid oxidation. The overall yield is 71%.

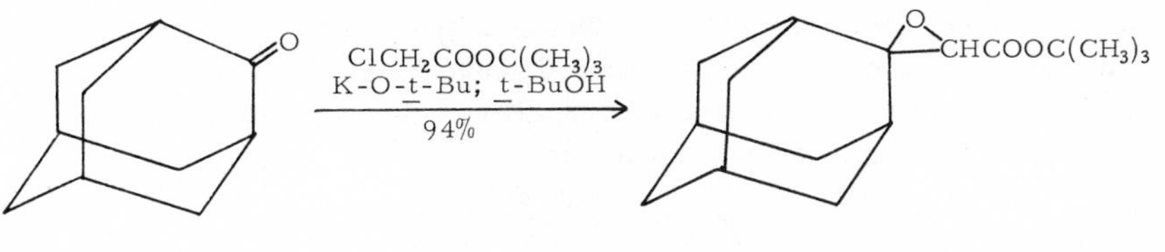

(1)   (2)

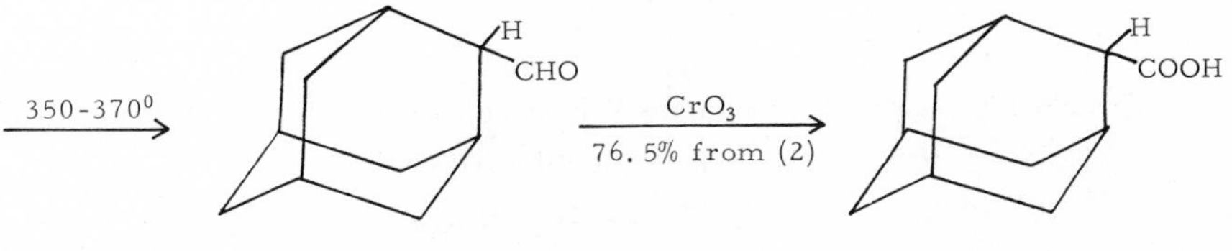

(3)   (4)

[1]J. Scharp, H. Wynberg, and J. Strating, *Rec. trav.*, **89**, 18 (1970)

t-Butyl chromate, **1**, 86–87; **2**, 48.

Oxidation of spiroethers to spirolactones. The steroidal spiroether (1) is converted into the spirolactone (2) in 48% yield by oxidation with *t*-butyl chromate

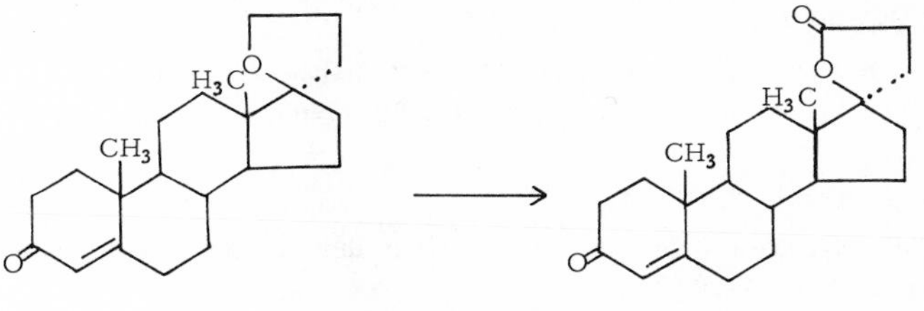

(1)   (2)

(CCl_4, HOAc, Ac_2O). The lactone (2) was also obtained to some extent by oxidation with ruthenium tetroxide, but extensive side reactions were also noted.[1]

[1]G. F. Reynolds, G. H. Rasmusson, L. Birladeneanu, and G. E. Arth, *Tetrahedron Letters*, 5057 (1970)

t-Butyl hydroperoxide, **1**, 88–89, **2**, 49–50.

Alkene synthesis.[1] Alkenes can be prepared from *vic*-dicarboxylic acids by conversion into the diacyl chloride followed by treatment with *t*-butyl hydroperoxide. The resulting di-*t*-butyl peresters (1) are decomposed when heated in solution in *t*-butylbenzene (N_2 or Ar) to give olefins. Use of *p*-cymene (Wiberg

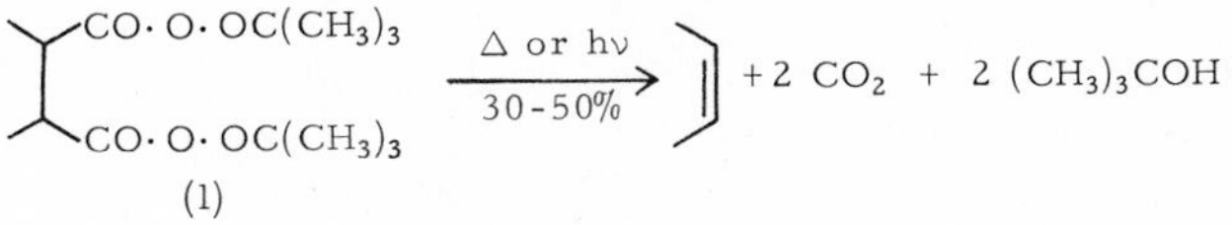

(1)

decarboxylation) led to formation of saturated hydrocarbons as well as alkenes. The decomposition can also be induced photochemically in the case of unstable olefins. For example, hydrocarbon (2) was obtained photolytically in 40% yield; thermolysis gave the isomer (3).

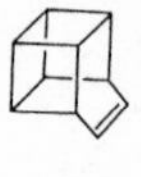

(2) 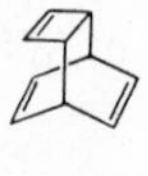   (3)

Aromatic acylation. Italian chemists[2] have generated acyl radicals by reaction of an aldehyde with *t*-butyl hydroperoxide and ferrous sulfate:

$$(CH_3)_3COOH + Fe^{2+} \longrightarrow (CH_3)_3CO\cdot + FeOH^{2+}$$

$$(CH_3)_3CO\cdot + RCHO \longrightarrow (CH_3)_3COH + R\dot{C}O$$

The acyl radicals thus produced acylate heteroaromatic bases. Thus benzothiazole is converted into 2-acetylbenzothiazole when acetaldehyde is used as the radical precursor. Aromatics such as N,N-dimethylaniline do not react, and hence this new type of free-radical acylation is opposite to the usual electrophilic aromatic acylation.

[1]E. N. Cain, R. Vukov, and S. Masamune, *Chem. Commun.*, 98 (1969)
[2]T. Caronna, G. P. Gardini, and F. Minisci, *ibid.*, 201 (1969)

t-Butyl hypochlorite, 1, 90–94; 2, 50.

The new preparative procedure (2, 50) is described by M. J. Mintz and C. Walling, *Org. Syn.*, **49**, 9 (1969)

3-Butyne-2-one (Acetylacetylene), $HC{\equiv}CCOCH_3$. Mol. wt. 68.07, b.p. 84°/752 mm.

Preparation. The reagent has been prepared by oxidation of the corresponding carbinol with chromium trioxide (sulfuric acid).[1]

Reaction with trialkylboranes. 3-Butyne-2-one, like acrolein (this volume), does not react spontaneously with trialkylboranes, but in the presence of catalytic amounts of oxygen it reacts readily to give α,β-unsaturated methyl ketones (2)

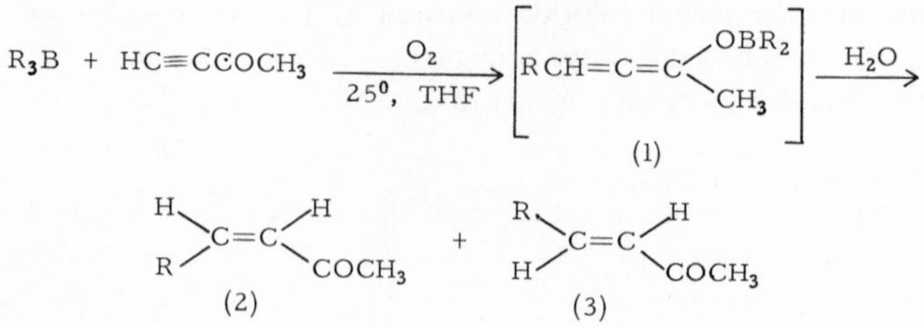

and (3) by way of an allenic intermediate (1), which is readily hydrolyzed by water, which should be present in the reaction mixture to achieve good yields of the final product.[2]

Examples:

$$CH_3CH_2CH=CH_2 \xrightarrow[72\%]{} CH_3(CH_2)_3CH=CHCOCH_3$$

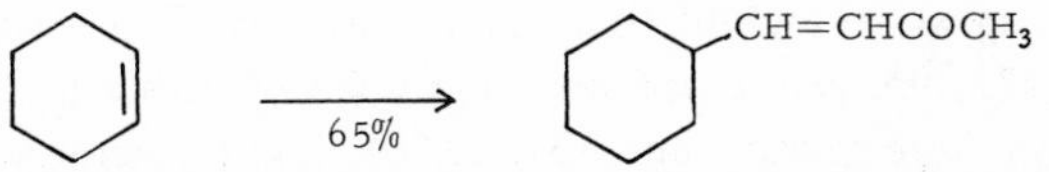

[1]K. Bowden, I. M. Heilbron, E. R. H. Jones, and B. C. L. Weedon, *J. Chem. Soc.*, 39 (1946)
[2]A. Suzuki, S. Nozawa, M. Itoh, H. C. Brown, G. W. Kabalka, and G. W. Holland, *Am. Soc.*, **92**, 3503 (1970)

C

Calcium carbide, CaC_2. Mol. wt. 64.10. Suppliers: Fisher, Research Organic/ Inorganic Chem. Corp.

Condensing agent.[1] A 500-ml. flask fitted with a Soxhlet extractor large enough to hold a 43 × 123 mm. extraction thimble is charged with 1 mole each of cyclohexanone and cyclopentanone, and 32 g. (0.5 mole) of calcium carbide is placed in the extraction thimble. The mixture is heated to boiling and cycled through the

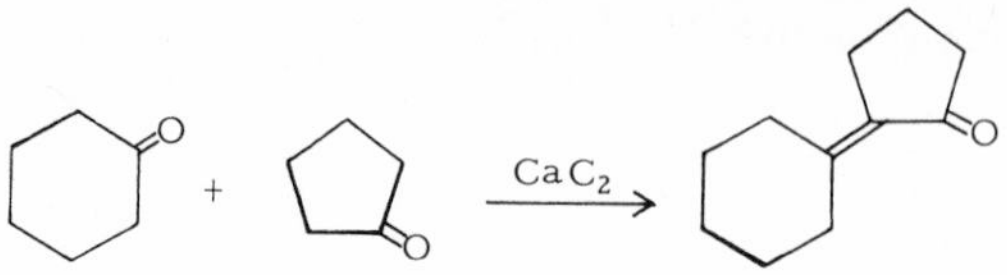

extraction thimble for 4 hours. The product may be purified by gas chromatography with a column of 20% Dow-710 on Chromosorb W at 225°. The distillate is typically 33% 2-cyclopentylidenecyclopentanone, 64% 2-cyclohexylidene-cyclopentanone, and 3% 2-cyclohexylidenecyclohexanone.

[1] D. R. Sands, *Org. Syn.*, submitted 1970

Carbomethoxymethylenetriphenylphosphorane, **1**, 112–114; **2**, 60. Replace ref. 3a (**2**, 60) by G. R. Pettit, B. Green, A. K. Das Gupta, P. A. Whitehouse, and J. P. Yardley, *J. Org.*, **35**, 1381 (1970).

Carbon dioxide, CO_2. Supplier, *see* **2**, 204.

Carboxylation of ketones. In the Robinson-Cornforth[1] formal total synthesis of epiandrosterone two relays were utilized. The second was the Köster-Logemann[2] ketone (1), available as a by-product of the oxidation of cholesteryl acetate dibromide. Based on model experiments with alicyclic ketones, (1) was converted

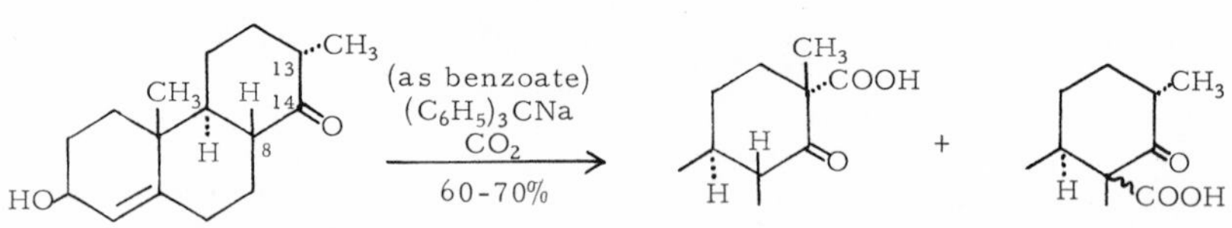

into the benzoate and then treated in benzene solution with tritylsodium in ether. The resulting enolates were then run onto chopped solid carbon dioxide. Two keto acids (isolated as the methyl esters) were obtained in 60–70% yields. Unfortunately the desired keto acid (2) was the minor product. Even so, the synthesis

(4)   (5)   (6)

of epiandrosterone was successfully completed. This method was also used by Wenkert[3] in a study of conversion of the tricyclic ketone (4) into resin acids. Unfortunately the desired product (6) is the minor product of carboxylation.

The procedure was used successfully in the synthesis of gibberellin C from the dienone group (7).[4] The ketone group at C_8 was protected by ketalization, carboxylation was conducted by the Robinson-Cornforth method, the resulting acid esterified with diazomethane, and finally the diketo diester (8) was obtained after

1) CH_2OH
   CH_2OH, H^+
2) $(C_6H_5)_3CNa$, CO_2
3) CH_2N_2
4) H_2O, H^+

11. 4%

(7)   (8)

deketalization in 11.4% yield. The low yield may be due to steric hindrance by the C_{10}-carbomethoxy group; the carbomethoxy group introduced at C_1 has the desired α-orientation. In this case carbomethoxylation was unsuccessful using methyl carbonate and sodium hydride or sodium amide, carbon dioxide and lithium amide, methyl chloroformate and tritylsodium.

[1]H. M. E. Cardwell, J. W. Cornforth, S. R. Duff, H. Holtermann, and R. Robinson, *J. Chem. Soc.*, 361 (1953)
[2]H. Köster and W. Logemann, *Ber.*, **73**, 298 (1940)
[3]E. Wenkert and B. G. Jackson, *Am. Soc.*, **81**, 5601 (1959)
[4]K. Mori, M. Shiozaki, N. Itaya, M. Matsui, and Y. Sumiki, *Tetrahedron*, **25**, 1293 (1969)

Carbon monoxide, 2, 60, 204.

Carbonylation of organoboranes. Review.[1] Brown[2,4] has developed an interesting synthesis of angularly substituted polycyclic alcohols. Hydroboration and isomerization of the linear triene 1,4,8-nonatriene (1) gives the bicyclic borane (2) in 41% yield. Carbonylation at 1000 psi and 150° in the presence of ethylene glycol followed by oxidation with alkaline hydrogen peroxide gives a mixture of *cis*- and *trans*-9-decalol (3), in which the former isomer predominates. Actually it is not necessary to isolate the intermediate (2); indeed the yield of (3) can be improved to 73% overall in this way. There are two points of interest in this synthesis. In

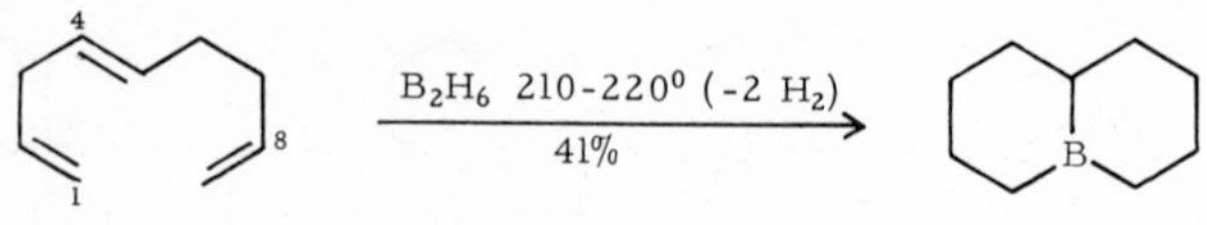

(1)   (2)

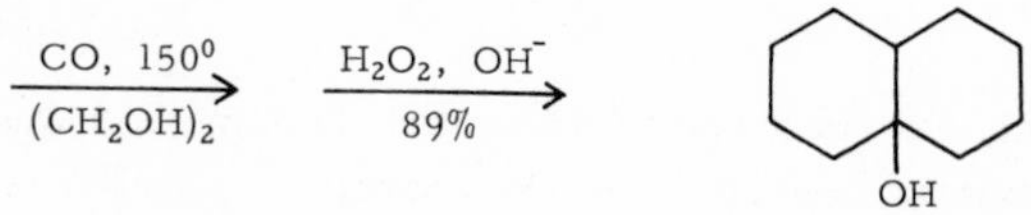

(3)

cis:trans = 4:1

the synthesis of the borane, the boron atom is placed at the tertiary position of the carbon chain, presumably because of the tendency to form the fused-ring system. In the carbonylation reaction, a tertiary carbon atom migrates from boron to carbon.

By the same procedure, *cis*-8-hydrindanol (6) has been obtained in 33% yield from 1,3,7-octatriene (4), readily available by dimerization of 1,3-butadiene.

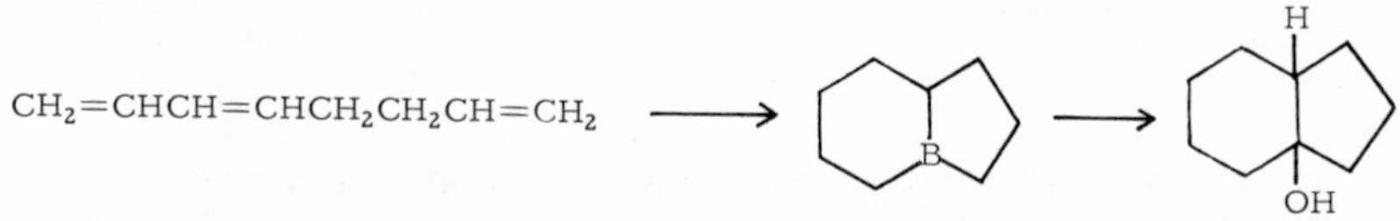

$CH_2=CHCH=CHCH_2CH_2CH=CH_2$ ⟶

(4)   (5)   (6)

Carbonylation of *cis,cis,trans*-perhydro-9b-boraphenalene (7) provides the *cis,cis,trans*-alcohol (8) in 70% yield.[3] Presumably each transfer of a bond from

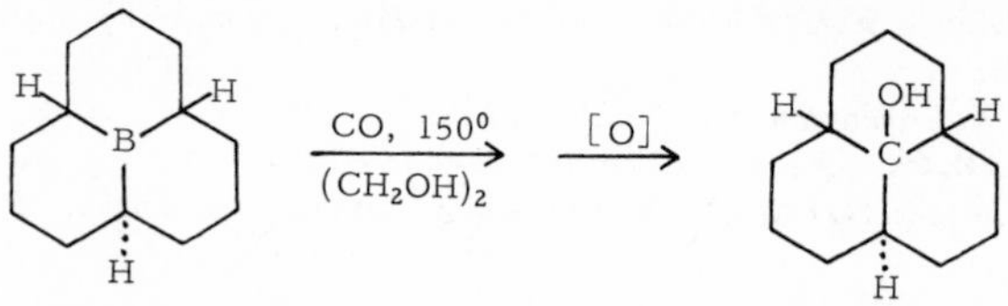

(7)   (8)

boron to carbon proceeds with retention. Unexpectedly, carbonylation of the *cis,cis,cis*-perhydro-9b-boraphenalene (9) gives the highly strained *cis,cis,cis*-alcohol (10). This result requires that carbonylation take place predominantly from the more hindered side of the borane intermediate to give the more strained of the two possible alcohols.[4]

Isocyanates.[5] The pyrolysis of an aromatic azide at 160–180° in an autoclave under a carbon monoxide pressure of 200–300 atm. affords the corresponding

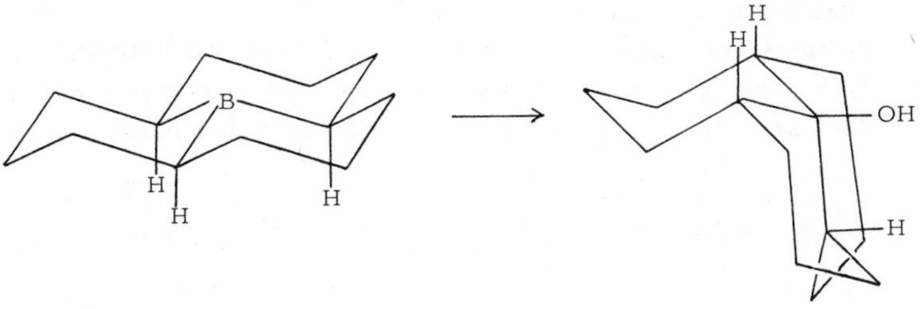

(9)   (10)

isocyanate in about 50% yield. Benzene or 1,1,2-trichloro-1,2,2-trifluoroethane is used as solvent. The reaction may involve decomposition to a nitrene, ArN, followed by reaction with carbon monoxide.

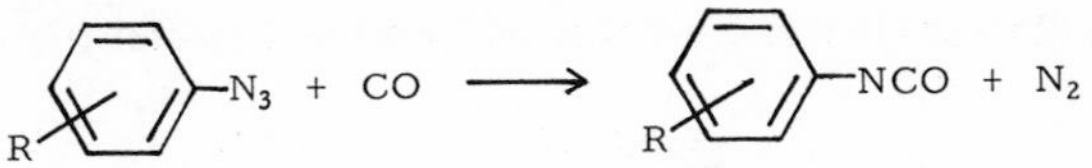

$$R = 2\text{-}CH_3,\ 4\text{-}CH_3,\ 4\text{-}NO_2,\ 4\text{-}OCH_3,\ 4\text{-}Cl,\ 2\text{-}C_6H_5$$

[1] H. C. Brown, *Accts. Chem. Res.*, **2**, 65 (1969)
[2] H. C. Brown and E. Negishi, *Am. Soc.*, **91**, 1224 (1969)
[3] *Idem., ibid.*, **89**, 5478 (1967)
[4] H. C. Brown and W. C. Dickason, *ibid.*, **91**, 1226 (1969)
[5] R. P. Bennett, and W. B. Hardy, *ibid.*, **90**, 3295 (1968)

Carbon tetrachloride

Aromatic chloro compounds. Photolysis of aromatic iodides has been used for synthesis of triphenyls, phenanthrenes, and organophosphorus and organoboron compounds.[1] If the photolysis is carried out in carbon tetrachloride with 3000-Å light[2] for about five hours, the corresponding chloro compound is obtained in 50–95% yield.[3]

$$Ar\!-\!I \xrightarrow[50\text{-}85\%]{CCl_4,\ h\nu} ArCl$$

[1] R. K. Sharma and N. Kharasch, *Angew. Chem., internat. Ed.*, **7**, 36 (1968)
[2] Rayonet photochemical reactor (The Southern New England Ultraviolet Co.)
[3] F. Kienzle and E. C. Taylor, *J. Org.*, **35**, 528 (1970)

Caro's acid, 1, 118–119.

Oxidation of aldehydes. Aldehydes are oxidized by Caro's acid in the presence of alcohols to give esters of the corresponding carboxylic acids. Yields are high.[1]

$$RCHO + H_2SO_5 \xrightarrow{R'OH} RCOOR' + H_2O + H_2SO_4$$

[1] A. Nishihara and I. Kubota, *J. Org.*, **33**, 2525 (1968)

Ceric ammonium nitrate (CAN), **1**, 120–121; **2**, 63–65.

Oxidation of 1-pentanol. 1-Pentanol (1) is oxidized to 2-methyltetrahydro-furane (2) in about 20% yield by CAN.[1] The same result has been obtained also with lead tetraacetate in comparable yield.[2] However, higher yields are obtained

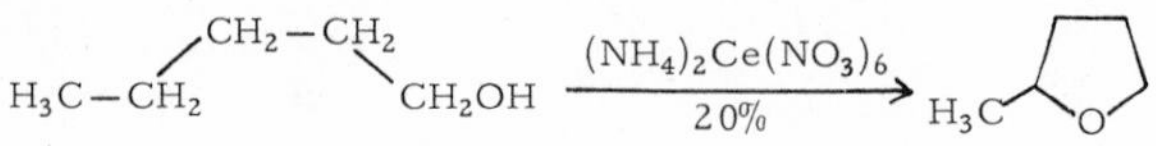

with silver oxide and bromine or mercuric oxide and iodine (*see* Silver oxide, this volume).

Oxidative cleavage of bicyclic alcohols. Oxidation of either *exo*- or *endo*-norbornanol (1) with 2 equiv. of ceric ammonium nitrate in 50% aqueous acetonitrile at 50° gives three products of oxidative cleavage, 3- and 4-cyclopentene-acetaldehydes (2) and (3) and 3-nitratocyclopentaneacetaldehyde (4).[3] Oxidation

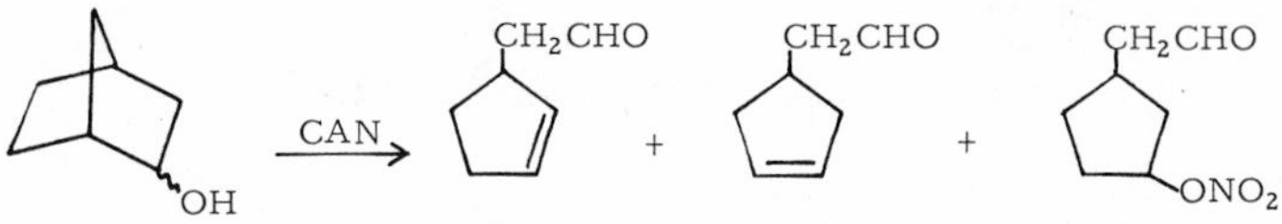

of bicyclo[2.2.2]-2-octanol under the same conditions gave three products, 4-cyclohexeneacetaldehyde (6) and *cis*- and *trans*-4-nitratocyclohexaneacetalde-hyde, (7) and (8). Oxidation of either borneol or isoborneol (9) gave the same

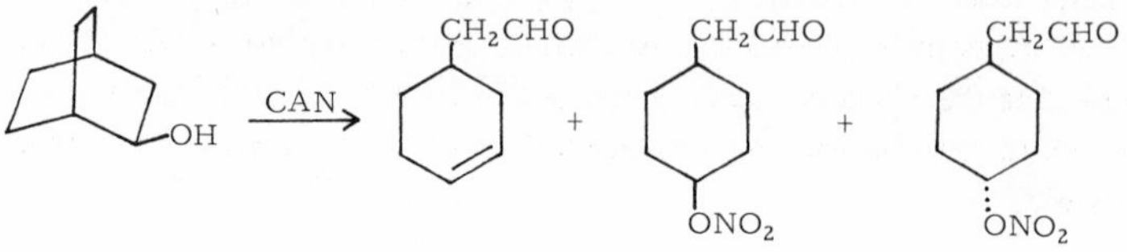

reaction mixture, which contains at least ten products (glpc analysis). However, the major product in both cases is α-campolenic aldehyde (10). This last oxidation is the most convenient way to prepare (10).

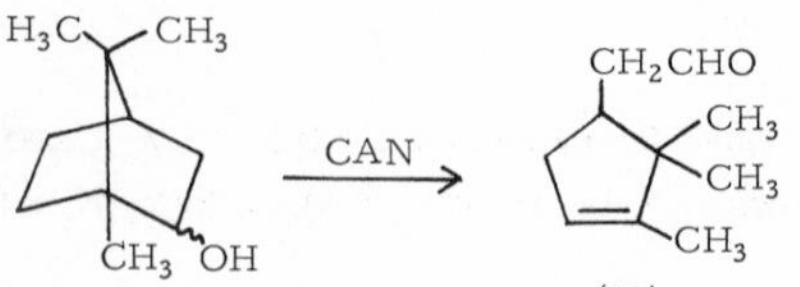

43% from borneol
55% from isoborneol

Oxidation of oximes and semicarbazones. Aldehydes and ketones can be re-generated from their oximes or semicarbazones by oxidation with excess ceric ammonium nitrate at −40 to 0° in methanol, acetonitrile, or acetic acid. Yields are in the range 70–90%.[4]

[1] W. S. Trahanovsky, M. G. Young, and P. M. Nave, *Tetrahedron Letters*, 2501 (1969)
[2] R. E. Partch, *J. Org.*, **30**, 2498 (1965); M. Lj. Mihailović *et al.*, *Tetrahedron*, **22**, 955 (1966)
[3] W. S. Trahanovsky, P. J. Flash, and L. M. Smith, *Am. Soc.*, **91**, 5068 (1969)
[4] J. W. Bird and D. G. M. Diaper, *Canad. J. Chem.*, **47**, 145 (1969)

Chloral, CCl_3CHO, **1**, 122.

Precursor of dichlorocarbene. Treatment of chloral with potassium *t*-butoxide at about 0° leads to dichlorocarbene, as shown by the reaction with an olefin.[1]

$$CCl_3CHO + KO\text{-}\underline{t}\text{-}Bu + >C=C< \longrightarrow >C\underset{\underset{CCl_2}{\diagdown\diagup}}{\text{———}}C< + HCOO\text{-}\underline{t}\text{-}Bu + KCl$$

Yields of 1,1-dichlorocyclopropanes are about 50%. If sodium methoxide is used as base, a temperature of 50° is necessary for reaction. ω,ω,ω-Trichloroaceto-phenone can be used in place of chloral; treatment with potassium *t*-butoxide in the presence of cyclohexene at −10 to 0° gives 7,7-dichloronorcarane (60% yield) and *t*-butyl benzoate (87% yield).

[1] F. Nerdel, H. Dahl, and P. Weyerstahl, *Tetrahedron Letters*, 809 (1969)

Chloramine, **1**, 122–125; **2**, 65–66.

α-Diazoketones (**1**, 122–123). Wheeler and Meinwald[1] have described in detail a procedure for conversion of the α-oximino ketone (2) into the α-diazo ketone (3) by reaction of chloramine, generated *in situ* from sodium hypochlorite (3.0 *M*), sodium hydroxide, and ammonia. The generation is exothermic and cooling is required to maintain a temperature of 20°. The procedure includes the

photochemical Wolff rearrangement to ring-contracted carboxylic acids (4 and 5).

[1] T. N. Wheeler and J. Meinwald, *Org. Syn.*, submitted 1969

Chloranil, 1, 125–127; **2,** 66–67.

Dehydrogenation. The most convenient route[1] to benzobicyclo[2.2.2]octadiene (3) involves dimerization of cyclohexa-1,3-diene (1) to dicyclohexadiene (2). The reaction is best carried out by heating the monomer at 178–186° in a stainless steel autoclave for 24 hours. The dimer is then dehydrogenated to the desired product (3) with chloranil in xylene at 130–140°. Lower yields are obtained with DDQ in

place of chloranil. Hydrocarbon (3) was first obtained by cycloaddition of benzyne to (1), but the yield is low and the isolation tedious.[2]

[1]K. Kirahonoki and Y. Takano, *Tetrahedron,* **25,** 2417 (1969)
[2]H. E. Simmons, *Am. Soc.,* **83,** 1657 (1961)

Chloroacetyl chloride, 1, 130.

Aziridines.[1] Primary amines (1) can be converted into aziridines by a two-step reaction. Reaction with chloroacetyl chloride yields an N-chloroacetyl derivative (2), which upon reduction with aluminum hydride (**2,** 23–24) forms a β-chloro-amine (3) which cyclizes to an aziridine (4).

[1]Y. Langlois, H. P. Husson, and P. Potier, *Tetrahedron Letters,* 2085 (1969)

1-Chlorobenzotriazole, 2, 67. Supplier: Aldrich.

Ref. 1: definitive papers, C. W. Rees and R. C. Storr, *J. Chem. Soc.,* (C), 1474, 1478 (1969).

A detailed description of the preparation has been submitted to *Org. Syn.*[1] The reagent can be readily regenerated from benzotriazole hydrochloride, produced in oxidation with the reagent, by treatment with sodium hypochlorite.

Oxidations.[1] The reagent compares favorably with N-haloamides and chromic acid for oxidation of secondary alcohols. Cyclohexanol is oxidized to cyclo-hexanone in 90–95% yield at room temperature. Less easily oxidized alcohols may require higher temperatures to initiate the free-radical reaction. The oxidation may become too vigorous if cooling is insufficient or if run on a large scale.

Ethyl quinoline-2-hydrazocarboxylate is oxidized to ethyl quinoline-2-azocar-boxylate in 90% yield:

Phenylhydroxylamine is converted into nitrosobenzene in 50% yield:

$$C_6H_5NHOH \longrightarrow C_6H_5NO$$

Sulfides → *sulfoxides*. Like *t*-butyl hypochlorite, this positive-chlorine compound oxidizes sulfides to sulfoxides without formation of sulfones. The reaction is carried out in methanol at −78°. Yields are usually high.[2] Note that oxidation of sulfides of the type $RSCH_2R'$ with positive-halogen compounds results in part or entirely in substitution at the α-methylene group.[3]

[1]M. Keating, C. W. Rees, and R. C. Storr, *Org. Syn.*, submitted 1969
[2]W. D. Kingsbury and C. R. Johnson, *Chem. Commun.*, 365 (1969)
[3]L. Skattebøl, B. Boulette, and S. Solomon, *J. Org.*, **32**, 3111 (1967)

β-Chloroethyl vinyl ketone, CH_2=$CHCOCH_2CH_2Cl$. Mol. wt. 118.56, b.p. 25°/0.08 mm.

Preparation. The reagent is prepared in a two-step sequence involving acylation of ethylene with β-chloropropionyl chloride (aluminum chloride) followed by monodehydrochlorination with sodium carbonate.[1]

$$CH_2=CH_2 \ + \ ClCH_2CH_2COCl \xrightarrow{\text{AlCl}_3} ClCH_2CH_2COCH_2CH_2Cl \xrightarrow[\text{37\% overall}]{\text{Na}_2\text{CO}_3} CH_2=CHCOCH_2CH_2Cl$$

Annelation.[1] Use of divinyl ketone, CH_2=$CHCOCH$=CH_2, in annelation reactions has the disadvantage that bisalkylation competes with monoalkylation. β-Chloroethyl vinyl ketone, however, can be used as the synthetic equivalent. Thus the sodium salt of the diketone (1) reacts smoothly with β-chloroethyl vinyl ether to give the Michael addition product (2), which on dehydrohalogenation in glyme gives (3) in overall yield of 60% (pure). This product reacts with *t*-butyl

(1)   (2)   (3)   (4)

acetoacetate in *t*-butanol (potassium *t*-butoxide catalyst) to give a gummy product, which on treatment with TsOH–HOAc at 78° for 3 hrs. gives the tricyclic dione (4), in a sequence which involves twofold cyclodehydration, ester cleavage, and β-keto acid decarboxylation.

[1]S. Danishefsky and B. H. Migdalof, *Am. Soc.*, **91**, 2806 (1969)

Chloroiridic acid, 1, 131–132; **2**, 67–68.

Henbest reduction.[1] Under Henbest's conditions (**1**, 131–132), 3-ketosteroids are reduced predominantly to the axial alcohols; that is, 3-keto-5α-steroids are reduced to 3α-hydroxy-5α-steroids, and 3-keto-5β-steroids are reduced to 3β-hydroxy-5β-steroids. The reduction is highly selective for 3-keto groups; thus keto groups at 6, 7, 11, 12, 17, 17α (D-homo), and 20 are not reduced. The only complicating feature is that the pregnane-20-one side chain is partially epimerized at C_{17}. Freshly prepared mixtures of chloroiridic acid, trimethyl phosphite, and 90% aqueous propanol-2 cause little reaction at first; the rate then increases over 12–16 hours of boiling. This difficulty was eliminated by preheating the reagent solution for about 14 hours before use.

[1]P. A. Browne and D. N. Kirk, *J. Chem. Soc.*, (C), 1653 (1969)

2-Chloromethyl-4-nitrophenyl phosphorodichloridate,
B.p. 165–167°/0.2 mm. Mol. wt., 304.47.

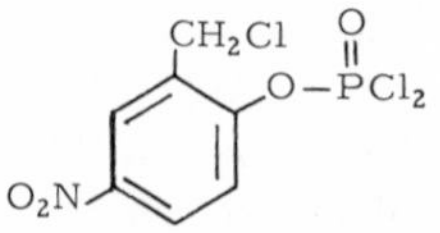

Preparation.[1] The reagent is prepared in 71% yield by refluxing 4-nitro-2-chloromethylphenol[2] with phosphoryl chloride.

Phosphorylation.[1] Alkyl(phenyl) dihydrogen phosphates (3) are prepared conveniently in fair to good yield by the following sequence. The reaction of an alcohol (or phenol) with the reagent (1) in THF in the presence of pyridine at −20° and subsequent hydrolysis with water and a trace of pyridine at 15° yields the diester (2). The O-protective group in (2) is then eliminated by hydrolysis with pyridine–water at 80° to give (3).

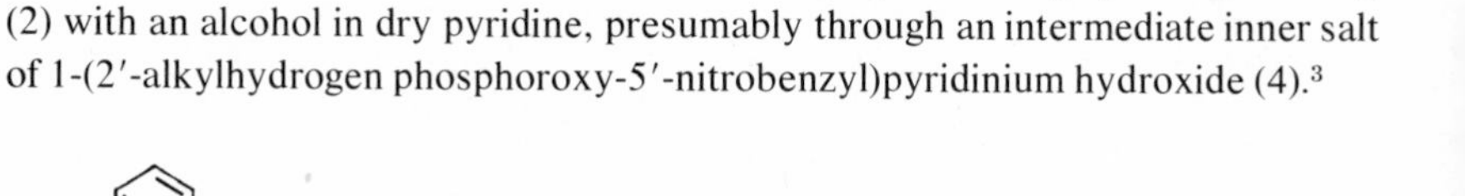

(1)   (2)   (3)

Mixed diesters of phosphoric acid (5) can be prepared in high yield by treating (2) with an alcohol in dry pyridine, presumably through an intermediate inner salt of 1-(2′-alkylhydrogen phosphoroxy-5′-nitrobenzyl)pyridinium hydroxide (4).[3]

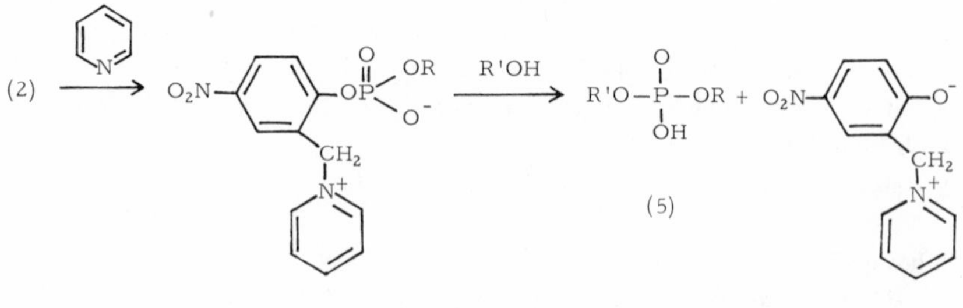

(4)

[1]T. Hata, Y. Mushika, and T. Mukaiyama, *Am. Soc.*, **91**, 4532 (1969)
[2]C. A. Buehler, F. K. Kirchner, and G. F. Deebel, *Org. Syn.*, **20**, 59 (1940)
[3]T. Hata, Y. Mushika, and T. Mukaiyama, *Tetrahedron Letters*, 3505 (1970)

1-Chloro-3-pentanone, $CH_3CH_2COCH_2CH_2Cl$. Mol. wt. 120.58, b.p. 62–65°/15 mm.

Preparation. The chloroketone is prepared by condensation of propionyl chloride with ethylene with aluminum chloride as condensing reagent in methylene chloride[1] or chloroform[2] as solvent.

Annelation. The Robinson annelation of (−)-dihydrocarvone (1) to give the ketol (2) was originally carried out using the Mannich base 1-diethylamino-3-pentanone methiodide and sodamide in ether–pyridine.[3] Halsall[1] improved the

(1)   (2)

yield to 45% by using 1-chloro-3-pentanone with sodium hydride as base and THF as solvent. It has been reported recently[4] that the yield can be improved further (51%) by use of an excess of the chloroketone and by lowering the temperature from 25° to 0–5°. In either case starting ketone is recovered. A note states that use of ethyl vinyl ketone gives a moderate yield of (2) mixed with the $5\alpha,10\alpha$-diastereoisomer in a 7:3 ratio.

[1]T. G. Halsall, D. W. Theobald, and R. B. Walshaw, *J. Chem. Soc.*, 1029 (1964)
[2]R. B. Woodward *et al.*, *Am. Soc.*, **74**, 4224 (1952)
[3]F. J. McQuillin, *J. Chem. Soc.*, 528 (1955)
[4]A. G. Hortmann, J. E. Martinelli, and Y. Wang, *J. Org.*, **34**, 732 (1969)

m-**Chloroperbenzoic acid, 1,** 135–139; **2,** 68–69.

Preparation. The procedure cited (**2,** 68, ref. 1a) has now been published.[1]

Epoxidation (**1,** 136–137, ref. 5). The procedure for epoxidation of 1,2-dimethyl-1,4-cyclohexadiene has now been published.[2]

Allene epoxides. Synthesis of allene epoxides is difficult because of the tendency to undergo further epoxidation and to isomerize to the corresponding cyclopropanone (**2,** 308–309). The first known one was obtained by Camp and Green (**2,** 309, ref. 3g) by oxidation of 1,3-di-*t*-butylallene with *m*-chloroperbenzoic acid. Crandall and Machleder[3] have now reported successful monoepoxidation of 1,1,3-tri-*t*-butylallene (1), also using *m*-chloroperbenzoic acid. The epoxide (2) is the main product but is accompanied by an oxetanone (3), undoubtedly formed via an intermediate diepoxide. The epoxide (2) is extraordinarily stable and shows no tendency to isomerize to a cyclopropanone.

$$(1) \longrightarrow (2) \quad + \quad (3)$$
$$10:1$$

Unstable tosylates. Tosylates have usually been prepared by reaction of the alcohol with tosyl chloride in pyridine. However, the method is not generally suitable for hindered alcohols or very reactive alcohols. Coates and Chen[4] have reported a new method that is suitable for hindered or unstable tosylates. The alcohol is converted into the *p*-toluenesulfinate ester[5] by reaction with *p*-toluene-sulfinyl chloride[6] in ether containing 1 equivalent of pyridine; yield 70–85%. The sulfinates are then oxidized to the sulfonates by *m*-chloroperbenzoic acid in methylene chloride at 0°, yield 75–88%. The tosylates of *t*-butyl alcohol, 2-bicyclo-

[3.1.1]heptyl alcohol, 1-adamantanol, and other alcohols, were prepared in this way in good yield.

Tertiary amine N-oxides.[7] N-Oxides of tertiary amines can be prepared in high yield by reaction of the amine with *m*-chloroperbenzoic acid in chloroform at 0–25°; the amine N-oxide *m*-chlorobenzoate is formed, from which the pure N-oxide is obtained by passage through a column of alumina.

[1] R. N. McDonald, R. N. Steppel, and J. E. Dorsey, *Org. Syn.*, **50**, 15 (1970)
[2] L. A. Paquette and J. H. Barrett, *ibid.*, **49**, 62 (1969)
[3] J. K. Crandall and W. H. Machleder, *J. Heterocyclic Chem.*, **6**, 777 (1969)
[4] R. M. Coates and J. P. Chen, *Tetrahedron Letters*, 2705 (1969)
[5] J. W. Wilt, R. G. Stein, and W. J. Wagner, *J. Org.*, **32**, 2097 (1967)
[6] F. Kurzer, *Org. Syn., Coll. Vol.*, **4**, 937 (1963)
[7] J. C. Craig and K. K. Purushothaman, *J. Org.*, **35**, 1721 (1970)

_p_-Chlorophenyl chlorothionoformate,   Mol. wt. 207.08, b.p. 82°/0.8 mm.

Preparation. The reagent is prepared in 81% yield by the reaction of *p*-chloro-phenol in 1 *N* NaOH solution with thiophosgene in chloroform.[1]

Nitriles.[2] Aldoximes are dehydrated to nitriles rapidly and at room tempera-ture in 60–70% yield by the reagent in the presence of pyridine. The process

(1)

probably involves formation and decomposition of an O-acyl derivative (1). Methyl chlorothionoformate[3] can also be used in the same way.

[1]D. L. Garmaise, A. Uchiyama, and A. F. McKay, *J. Org.*, **27**, 4509 (1962)
[2]D. L. J. Clive, *Chem. Commun.*, 1014 (1970)
[3]D. Martin and W. Mucke, *Ber.*, **98**, 2059 (1965)

Chloroplatinic acid–Triethylsilane, H_2PtCl_4–$(C_2H_5)_3SiH$.

Hydrogenation catalyst. When ethanolic solutions of chloroplatinic acid (Alfa Inorganics) and triethylsilane (Pierce Chemical Co.) in the ratio 1:7–9 are warmed, a brownish suspension is formed. Centrifugation gives a solid which is essentially a finely divided platinum. This material is a highly active hydrogenation catalyst. It is more active than Adams catalyst for hydrogenation of alkenes, of alkynes (particularly if reduction only to the alkene is desired), of $N \equiv N$ bonds (azobenzenes), of $C \equiv N$ bonds (benzylideneazine), and of nitrobenzenes to anilines. However, Adams catalyst is more effective for hydrogenation of aromatic rings. The new catalyst is also more active than the Brown and Brown platinum catalyst (**1**, 890–891) in the cases in which the two catalysts were compared except for hydrogenolysis of fluoro- and chlorobenzene.[1]

[1]C. Eaborn, B. C. Pant, E. R. A. Peeling, and S. C. Taylor, *J. Chem. Soc.*, (C), 2823 (1969)

Chlorosulfonyl isocyanate (N-Carbonylsulfamoyl chloride), 1, 117–118; 2, 70. Review.[1]

Reaction with olefins.[2] The reagent reacts with an olefin to give a β-lactam-N-sulfonyl chloride and a β,γ-unsaturated N-chlorosulfonylamide; the former product usually predominates except in the case of certain aromatic olefins. Thus isobutylene reacts to give the lactam (1) in 70% yield and the amide (2) in 30% yield. Under the conditions of workup the minor product (2) is readily hydrolyzed

$$(CH_3)_2C{=}CH_2 \; + \; ClSO_2N{=}C{=}O \longrightarrow$$

(1) 70%

(2) 30%

51–53% overall | 3 NaOH

(3)

and the β-lactam-N-sulfonyl chloride (1) is easily isolated. On pH-controlled hydrolysis with sodium hydroxide this is converted into 4,4-dimethyl-2-azetidinone (3).

The N-chlorosulfonyl β-lactams can also be reduced in very high yield to β-lactams by an aqueous solution of sodium sulfite (Na_2SO_3) and ether, the aqueous phase is kept slightly basic by addition of potassium hydroxide during reduction.[3]

Allenes react similarly.[4] Thus 3-methyl-1,2-butadiene reacts to give 1-chlorosulfonyl-3-methylene-4,4-dimethyl-2-azetidinone (4) and 2-carboxamido-3-methyl-1,3-butadiene (5) in the yields indicated.

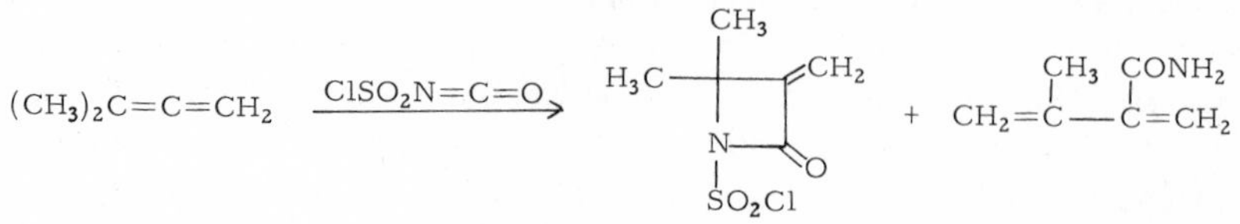
(CH3)2C=C=CH2
ClSO2N=C=O
CH3
H3C
CH2
N
O
SO2Cl
+
CH3 CONH2
CH2=C—C=CH2

(4) 20% (5) 36%

Reaction with acetylenes. Acetylenes react with chlorosulfonyl isocyanate to give 1 : 1 adducts in high yield. For example, 3-hexyne (1) gives the adduct (5) in 96% yield. The formation is presumed to involve cycloaddition (2), electrocyclic ring opening (3), 1,5-sigmatropic halogen shift (4), and finally electrocyclic ring closure to the adduct (5). Aqueous hydrolysis of (5) affords 3-hexanone (6).[5]

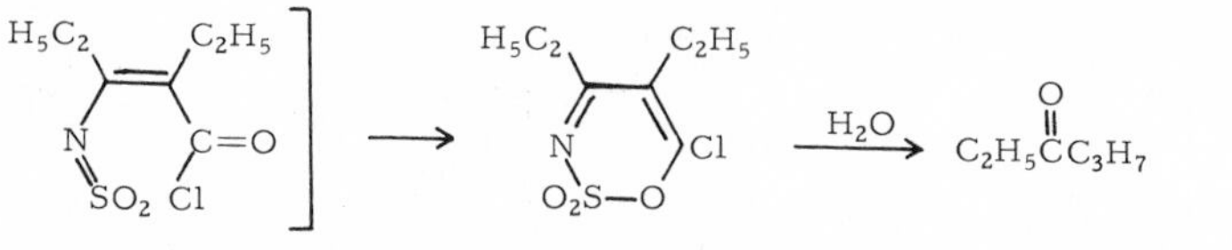
C2H5C≡CC2H5
(1)
ClSO2N=C=O
H5C2
C2H5
ClO2S
N
O
H5C2
C2H5
N
C
SO2Cl
O

(2) (3)

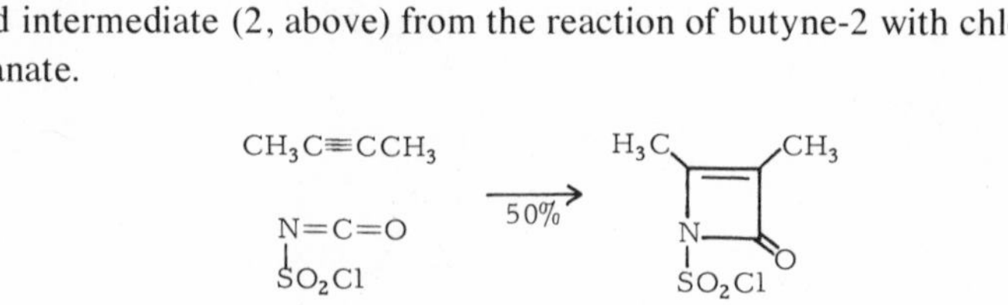
H5C2
C2H5
N
C=O
SO2
Cl
H5C2
C2H5
N
Cl
O2S—O
H2O
O
C2H5CC3H7

(4) (5) (6)

Clauss and Jensen[6] have isolated a cycloadduct (7) corresponding to the postulated intermediate (2, above) from the reaction of butyne-2 with chlorosulfonyl isocyanate.

CH3C≡CCH3
N=C=O
SO2Cl
50%
H3C
CH3
N
O
SO2Cl

(7)

Kampe[7] was unable to isolate a primary addition product from the reaction of phenylacetylene and the reagent.

Synthesis of nitriles (2, 70). Lohaus has published two procedures which illustrate the use of chlorosulfonyl isocyanate for the preparation of nitriles. One, the preparation of 2,4-dimethoxybenzonitrile,[8] illustrates the reaction of the reagent with aromatic compounds that readily undergo electrophilic substitution. Thus reaction of resorcinol dimethyl ether (1) with chlorosulfonyl isocyanate in methylene chloride gives the amide N-sulfonyl chloride (2), which on treatment with an amide[9] gives 2,4-dimethoxybenzonitrile (3) in 95–96% yield with a purity of 98%.

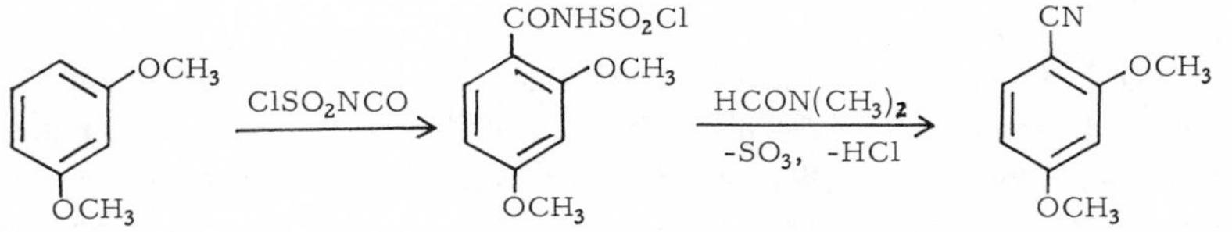

The other procedure[10] illustrates a general method for conversion of carboxylic acids into the corresponding nitriles. Treatment of cinnamic acid (4) with chlorosulfonyl isocyanate gives the carboxylic acid amide N-sulfonyl chloride (5). As above, treatment of (5) with DMF gives cinnamonitrile (6) in 78–87% yield.

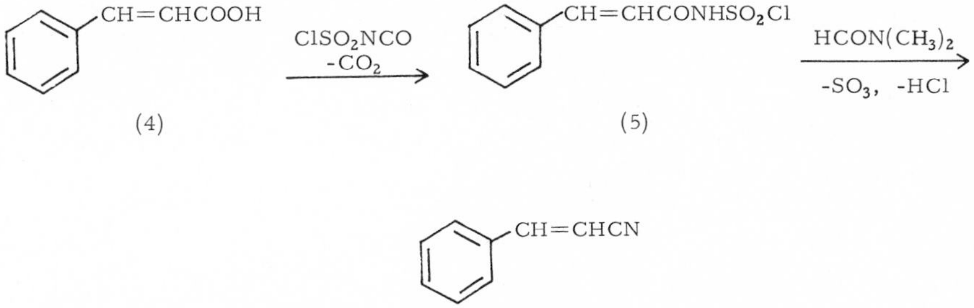

[1]R. Graf, *Angew. Chem., internat. Ed.*, 7, 172 (1968)

[2]R. Graf, *Org. Syn.*, 46, 51 (1966)

[3]T. Durst and M. J. O'Sullivan, *J. Org.*, 35, 2043 (1970)

[4]E. J. Moriconi and J. F. Kelly, *Am. Soc.*, 88, 3657 (1966)

[5]E. J. Moriconi, J. G. White, R. W. Franck, J. Jansing, J. F. Kelly, R. A. Salomone, and Y. Shimakawa, *Tetrahedron Letters*, 27, 1970

[6]K. Clauss and H. Jensen, *ibid.*, 119 (1970)

[7]K.-D. Kampe, *ibid.*, 123 (1970)

[8]G. Lohaus, *Org. Syn.*, 50, 52 (1970)

[9]DMF is preferred because of low molecular weight, high solvent power, and miscibility with water.

[10]G. Lohaus, *Org. Syn.*, 50, 18 (1970)

Chromic acid, 1, 142–144; **2,** 70–71.

Jones reagent (**1,** 142–143; **2,** 70–71). Spring *et al.*[1] were unable to isolate the expected bisnor-keto acid (2) from the complex mixture obtained on ozonolysis

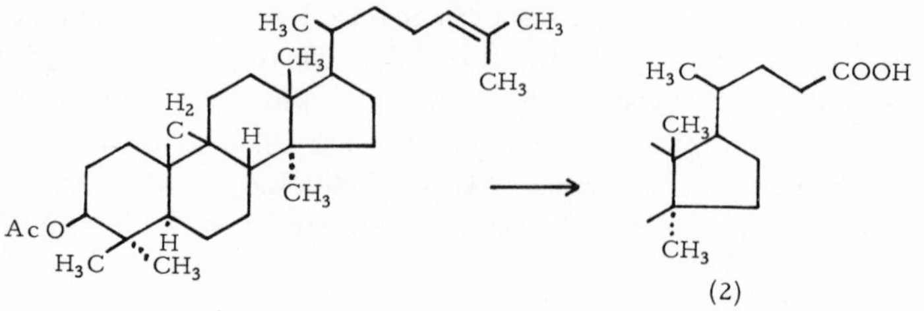

of cycloartenyl acetate (1) under conventional conditions (decomposition of the ozonide with water). However, if the crude ozonide is treated with Jones reagent, the acid is obtained in excellent yield. The procedure is general; thus adipic acid is obtained from cyclohexene in 80–85% yield and sebacic acid is obtained from undecylenic acid in 95% yield.[2]

Oxidation of a cyclopropanol. The cyclopropanol (1) was obtained as an intermediate in a total synthesis of prostaglandins E_2 and $F_{2\alpha}$. It was not oxidized by mercuric, thallic, or lead IV acetates, but was successfully cleaved with 1 equivalent of chromic acid and 0.05 equivalent of ceric ammonium nitrate in water-acetic acid at 20° (1 hour). A mixture of three products was obtained, one of which was the desired hydroxy aldehyde (2).[3]

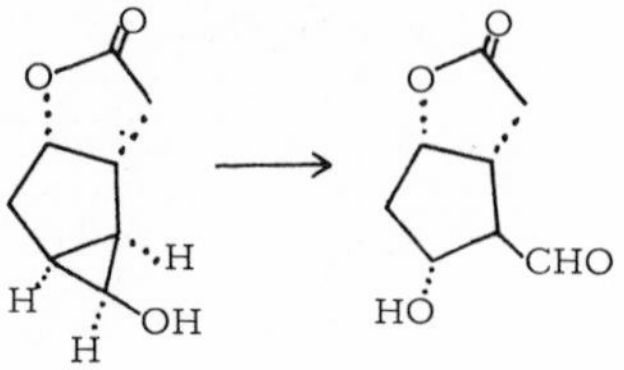

[1]H. R. Bentley, J. A. Henry, D. S. Irvine, and F. S. Spring, *J. Chem. Soc.*, 3673 (1953)
[2]A. S. Narula and S. Dev, *Tetrahedron Letters*, 1733 (1969)
[3]E. J. Corey, Z. Arnold, and J. Hutton, *ibid.*, 307 (1970)

Chromic anhydride, 1, 144–147; **2,** 72–75.

Oxidation of androstenolone acetate dibromide. Ciba chemists in 1952[1] reported that bromination of androstenolone acetate (1) in glacial acetic acid followed by addition of chromic anhydride and then debromination with zinc led to 3β-acetoxy-14α-hydroxyandrostenolone (2) in about 25% yield. The reaction is of considerable interest since a number of natural 14α-hydroxylated steroids are known. In more recent work Sykes and Kelly[2] found that a catalyst

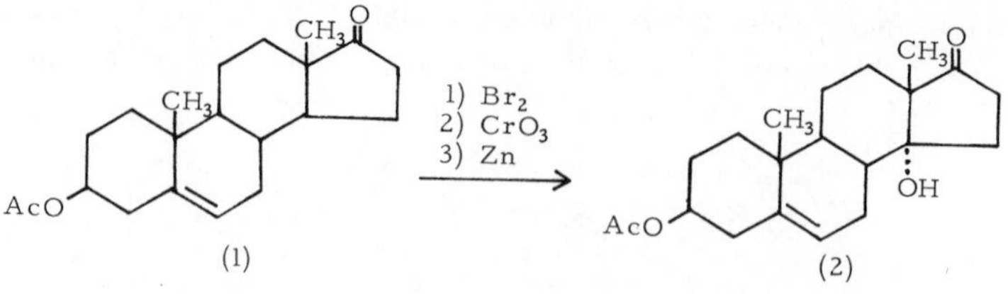

(1)   1) Br₂ 2) CrO₃ 3) Zn →   (2)

such as hydrobromic acid or perchloric acid improves the reaction, but even so reported that chromyl acetate (1, 151; 2, 78–79) is a more useful oxidant. However, more recently, it has been found[3] that addition of 1.5–2% of water markedly improves the yields in the original procedure of André *et al.* Yields of 44% (based on consumed starting material) are reported. As the percentage of water is increased, yields diminish.

CrO₃–pyridine complex (Sarett reagent), **1**, 145–146; **2**, 74–75. Ratcliffe and Rodehorst[4] find that the reagent can be prepared *in situ* directly in methylene chloride. Chromium trioxide (60 mmol.) is added to a stirred solution of pyridine (120 mmol.) in methylene chloride. The deep burgundy solution is stirred for 15 minutes at room temperature and then a solution of the alcohol (10 mmol.) in methylene chloride is added. A black deposit separates, and the product is isolated from the filtrate. One main advantage is that the complex can be prepared safely in this way without fire hazard.

Dauben *et al.*[5] report that Sarett reagent as prepared by Collins, Hess, and Frank (2, 74, ref. 11a) is useful for allylic oxidation of cyclohexenes and unsaturated steroids. Yields of enones are 48–95% (based on recovered starting material) and are generally higher than those obtained with *t*-butyl chromate or chromic anhydride in acetic acid. The following generalizations have been formulated. Allylic methyl groups are not oxidized readily (example 1). If more than one allylic methylene group is present in a conformationally flexible molecule, attack of both groups is observed (example 1). However, if the molecule is rigid, as in a steroid, selectivity is observed (example 2). Attack at an allylic methine position

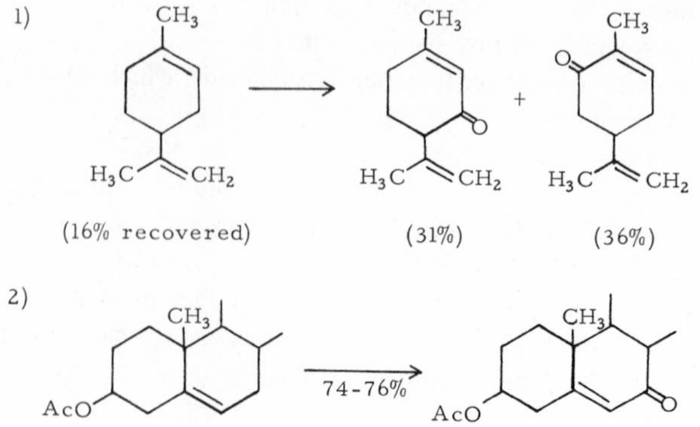

1)   (16% recovered)   →   (31%)   +   (36%)

2)   → 74–76%

yields the isomeric enone when possible (example 3). Evidently in this example there is a marked preference for abstraction of the tertiary allylic hydrogen.

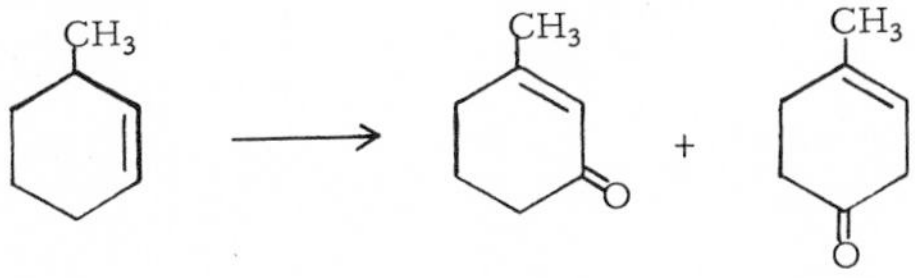

(14% recovered)   (68%)   (10%)

Reagent prepared by the modification of Collins, Hess, and Frank (**2**, 74, ref. 11a) was used by Ireland *et al.*[6] in the synthesis of 5,6-dimethyl-*trans*-5,9-decadienal (4), starting with the readily available *trans*-dibromide (1).

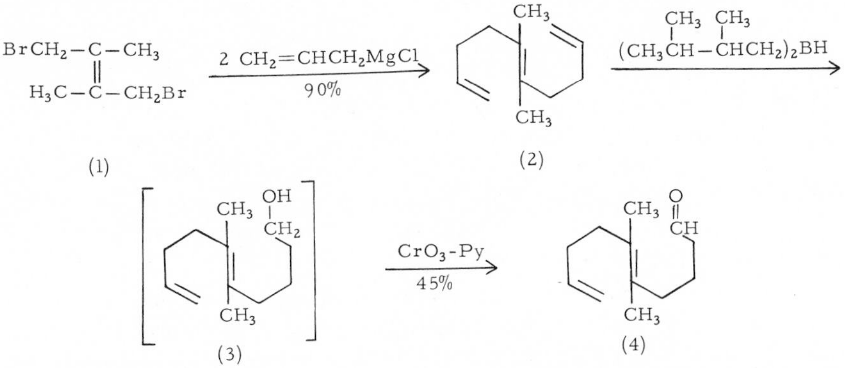

Anhydrous CrO₃–Acetic acid, **2**, 72–74. *Oxidative demethylation* (**2**, 72–73). Angyal and James[7] have pointed out that oxidative demethylation with chromium trioxide in glacial acetic acid is potentially very useful in the carbohydrate field, since the conditions are mild and the hydroxyl groups can be protected by acylation. Esters are not attacked, but glycosides and acetals are unstable to the reagent. Benzyl ethers are oxidized by the reagent to benzoic esters. The reaction is useful for debenzylation of compounds which contain groups that are reducible by catalytic hydrogenation or by dissolving metals.

The reagent is more selective than boron trichloride, which removes most other O-substituents as well.

Chromic anhydride–Dimethylformamide, **1**, 147. A note from the Sterling-Winthrop Research Institute[8] warns that a fire resulted on attempting the Snatzke oxidation of a secondary alcohol to a ketone on a large scale. Chromic oxide was ground in a mortar and added in a nitrogen atmosphere to ice-cooled dimethylformamide. As long as powdered anhydride was added in small portions, it went into solution, but, as soon as some lumps were added, a puff of flame flashed from the surface of the DMF along with white smoke, which quickly subsided. The DMF did not ignite.

[1] A. F. St. André *et al.*, *Am. Soc.*, **74**, 5506 (1952)
[2] P. J. Sykes and R. W. Kelly, *J. Chem. Soc.*, (C), 2346 (1968)
[3] C. M. Hol, M. G. J. Bos, and H. J. C. Jacobs, *Tetrahedron Letters*, 1157 (1969)
[4] R. Ratcliffe and R. Rodehorst, *J. Org.*, **35**, 4000 (1970)
[5] W. G. Dauben, M. Lorber, and D. S. Fullerton, *ibid.*, **34**, 3587 (1969). The warning is stressed that the CrO_3 must be added to the pyridine rather than vice versa; pyridine may ignite if added to CrO_3. Efficient stirring is necessary.
[6] R. E. Ireland, M. I. Dawson, J. Bordner, and R. E. Dickerson, *Am. Soc.*, **92**, 2568 (1970)
[7] S. J. Angyal and K. James, *Carbohydrate Res.*, **12**, 147 (1970)
[8] H. Newmann, *Chem. Eng. News*, p. 4, July 6, 1970

Chromium(II)·amine complexes.

Reduction. Chromous salts (chromous acetate, chromous chloride, chromous sulfate) have been useful for reduction of *vic*-dihalides, α-haloketones, allylic and benzyl halides, and α-ketoepoxides (**1**, 147–151; **2**, 75–78). Kochi and Mocadlo[1] found that complexing chromous ion with various amines (ethylenediamine was used mostly) greatly enhances the activity. The reducing agent is prepared by adding an aqueous solution of chromous perchlorate[2] to DMF containing ethylenediamine. A clear blue-purple homogeneous solution results, to which the substrate is then added. The reagent has been called for short Cr(II)en, but the actual composition is probably Cr(II)en$_2$ for the most part. The complex is also soluble in DMSO. The complexed ion will even reduce alkyl halides, the reactivity decreasing in the following orders: tertiary > secondary > primary and iodides > bromides > chlorides. The reagent also reduces aryl and vinyl halides and epoxides and episulfides in high yield.[3] 1-Naphthyl bromide is reduced to naphthalene in 93–96% yield.[4]

Like zinc, the new reagent effects *trans*-elimination of dibromides; however, *erythro*- and *threo*-α-hydroxy bromides give the same mixture of *cis*- and *trans*-olefins.[5] Likewise reduction of either *cis*- or *trans*-4-*t*-butyl-1-chloro-1-methyl-cyclohexane (1 or 2) affords predominantly $(62 \pm 6\%)$ *cis*-4-*t*-butyl-1-methyl-cyclohexane (3).[6] If butanethiol is added to the reduction (method of Barton and

(1)   or   (2) $\xrightarrow{\text{Cr(II)en}}$

(3)   +   (4)

Basu, **1**, 148, ref. 5), the more stable *trans*-isomer (4) becomes the predominant product (91% ± 1%) from either chloride. The lack of stereospecificity is attributed to a radical mechanism.

Okamura *et al.*[7] have found that the reagent is useful for selective reduction of bridgehead chlorine in the bicyclo[2.2.1]-Δ²-heptenyl system. For example, reduction of (1) with a large excess of chromous perchlorate–ethylenediamine gave (2) in 63% yield.

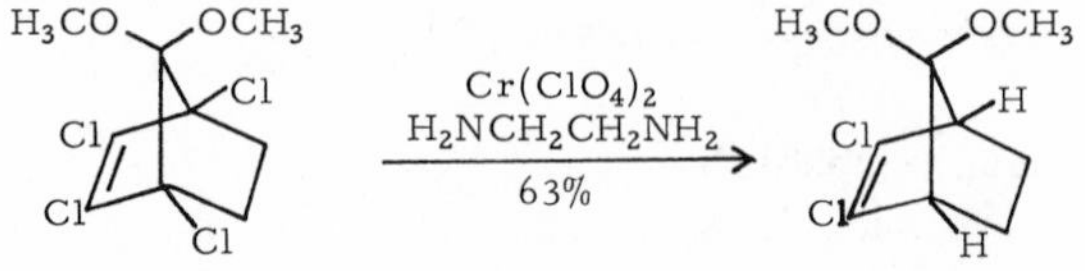

(1)   (2)

Synthesis of previtamin D₃. The reagent, bis(ethylenediamine)chromium(II), was employed by a British group[8] in the first total synthesis of previtamin D₃, the immediate product of irradiation of 7-dehydrocholesterol. Condensation of the lithium derivative of the trimethylsilyl ether of the enyne (2) with the chloroketone (1) gives the chlorohydrin (3). This is converted into the enyene (4) by reaction with bis(ethylenediamine)chromium(II). Semihydrogenation with Lindlar's catalyst completes the synthesis. The overall yield from (1) is over 25%.

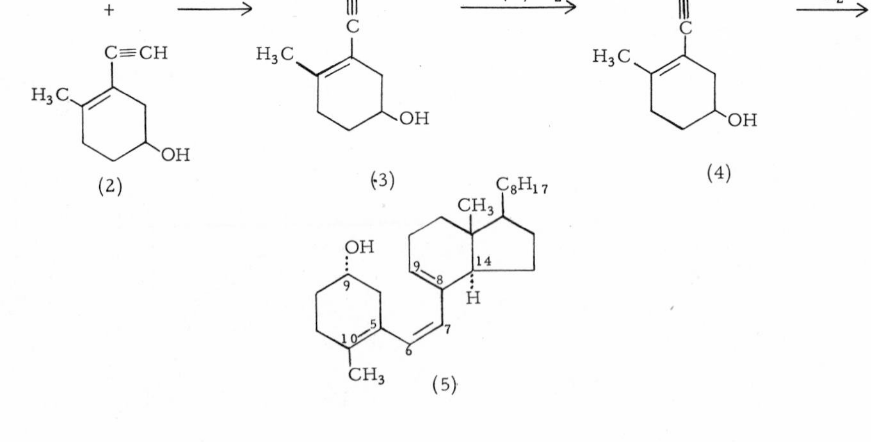

A salient feature of (5) is the central *cis*-double bond. This stereochemistry is best obtained by semihydrogenation. Previous work had indicated that enyenes of type (4) are difficult to prepare by dehydration methods. For example, dehydration of a tertiary alcohol such as (3, Cl replaced by H) gives mixtures in which the $\Delta^{8(14)}$-isomer predominates over the desired $\Delta^{8(9)}$-isomer. The present method assures the desired intermediate (4).

[1]J. K. Kochi and P. E. Mocadlo, *Am. Soc.*, **88**, 4094 (1966)
[2]Prepared from pure chromium metal (United Mineral and Chemical Corp., 129 Hudson St., New York) and dilute perchloric acid.
[3]J. K. Kochi, D. M. Singleton, and L. J. Andrews, *Tetrahedron*, **24**, 3503 (1968)
[4]R. S. Wade and C. E. Castro, *Org. Syn.*, submitted 1970
[5]J. K. Kochi and D. M. Singleton, *Am. Soc.*, **90**, 1582 (1968)
[6]R. E. Erickson and R. K. Holmquist, *Tetrahedron Letters*, 4209 (1969)
[7]W. H. Okamura, J. F. Monthony, and C. M. Beechan, *ibid.*, 1113 (1969)
[8]J. Dixon, P. S. Littlewood, B. Lythgoe, and A. K. Saksena, *Chem. Commun.*, 993 (1970)

Chromous acetate, 1, 147–149; 2, 75–76.

Conversion of ketoximes into ketones. Corey and Richman[1] have reported a new procedure for regeneration of ketones from their oximes. The oxime is first converted into the O-acetate by treatment with acetic anhydride at 20°; this derivative is then reductively deoximated by treatment with >2 molar equivalents of chromous acetate in 9:1 THF–water at 25–65°.[2] The reaction presumably involves reductive fission of the oxime N—O linkage to give an imine which is then rapidly hydrolyzed. Yields are in the range 75–85%. It is noteworthy that the reaction occurs more readily with acetoximes of conjugated ketones than with those of nonconjugated ketones and that the reaction occurs readily with acetoximes of hindered ketones such as camphor.

The reaction has been used in a three-step conversion of olefins into ketones. For example, nitrosyl chloride is added to cyclooctene (1) and the resulting 2-chlorocyclooctanone oxime[3] (2) is then acetylated and treated with chromous acetate at 65° for 16 hours. Cyclooctanone (3) is obtained in 88% yield from (2).

The reaction also has been used for transposition of a carbonyl group. Oxidation of propiophenone (4), followed by reduction with sodium borohydride and acetylation, gives the α-acetoxy acetoxime (5), which on treatment with excess chromous acetate affords phenylacetone (6).

[1]E. J. Corey and J. E. Richman, *Am. Soc.*, **92**, 5276 (1970)
[2]Free oximes can be used, but the reaction is slow and requires high temperatures.
[3]M. Ohno, N. Naruse, and I. Terasawa, *Org. Syn.*, **49**, 27 (1969)

Chromous chloride, 1, 149–150; **2**, 76–77.

Reductions with chromium(II) salts, mainly the chloride and the acetate, have been reviewed by Hanson and Premuzic.[1]

Reduction of halides. Chromous chloride (prepared by reduction of chromic chloride with zinc powder) has been used to reduce the *exo* chlorine atom of (1), prepared by Diels-Alder reaction between cyclopentadiene and thiophosgene followed by oxidation of the sulfide group to the sulfone.[2]

Reduction of enediones. Δ^4-Cholestene-3,6-dione (1) is reduced by chromous chloride in either THF or acetone to 5β-cholestanone-3,6-dione (2).[3] Note that reduction with zinc and acetic acid gives the more stable 5α-cholestane-3,6-

dione.[4] Δ^4-Cholestene-4-one and Δ^5-3β-acetoxycholestene-7-one are not reduced by $CrCl_2$ under the same conditions. However, $\Delta^{1,4}$-cholestadiene-3,6-dione is readily reduced to Δ^1-5β-cholestene-3,6-dione.

Reduction of quinones. Benzoquinones are reduced in good yield to the hydroquinones by chromous chloride or acetate.[5] In the 1,4-naphthoquinone series, however, yields are variable.

β-Chlorocarbamates. Chromous chloride promotes the addition of N-chlorocarbamates (ROCONHCl) to olefins to give β-chlorocarbamates in good yield.[6] The reaction is illustrated for the addition of N-chlorourethane (1) to cyclohexene (2). The main product is a mixture of the *cis*- and *trans*-β-chlorocarbamates (3) and (4), in which the *cis*-isomer predominates. The highest yields of β-chlorocarbamates are obtained by using an excess of N-chlorourethane. (Compare the thermal addition of N,N-dichlorourethane to olefins, **2**, 121–122).

Chromous chloride also promotes the addition of N-chlorocarbamates to enol ethers.[7] Thus N-chlorourethane adds to dihydropyrane (8) in methanol to give the

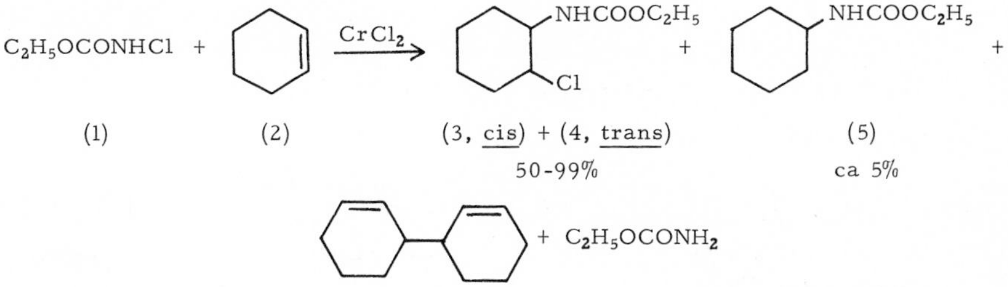

$C_2H_5OCONHCl$ + (cyclohexene) $\xrightarrow{CrCl_2}$ (cyclohexane with NHCOOC$_2$H$_5$ and Cl) + (cyclohexane with NHCOOC$_2$H$_5$) +

(1)　　　　　(2)　　　　　(3, cis) + (4, trans)　　　　　(5)

　　　　　　　　　　　　　　　　50-99%　　　　　ca 5%

(bicyclohexenyl) + $C_2H_5OCONH_2$

(6)　　　　　(7)

ca.1-4%

adducts shown. The addition can be utilized for preparation of α-amino ketones. Thus N-chlorourethane adds to 1-methoxycyclohexene (12) in methanol to give

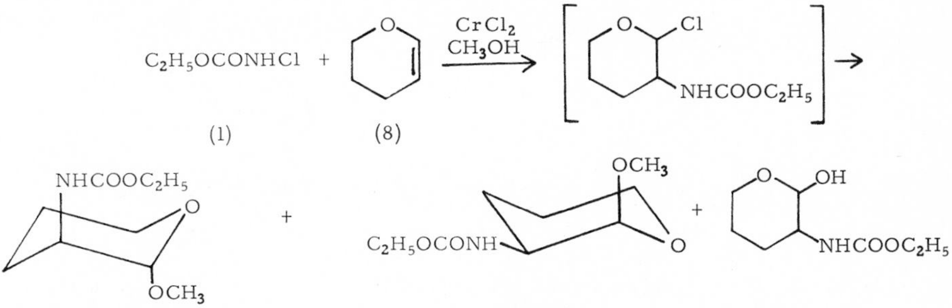

$C_2H_5OCONHCl$ + (8) $\xrightarrow[CH_3OH]{CrCl_2}$ [(intermediate with O, Cl, NHCOOC$_2$H$_5$)] →

(1)　　　　　(8)

(9) 77%　　　　　+　　　　　(10) 3%　　　　　(11) 5%

the α-carbamido ketal (13), which on hydrolysis gives the amino ketone derivative (14).

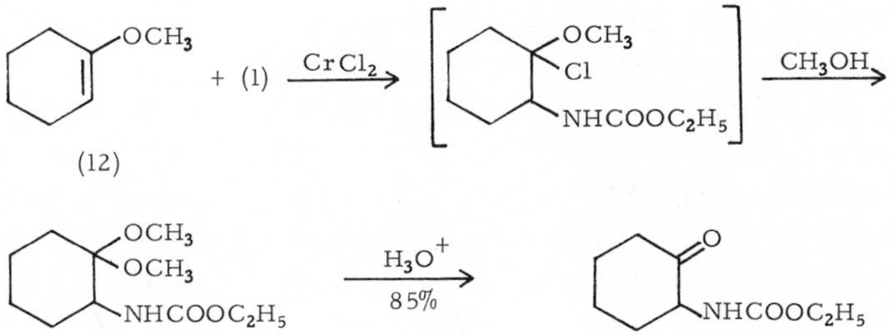

(12) + (1) $\xrightarrow{CrCl_2}$ [(intermediate OCH$_3$, Cl, NHCOOC$_2$H$_5$)] $\xrightarrow{CH_3OH}$

(13) $\xrightarrow[85\%]{H_3O^+}$ (14)

[1] J. R. Hanson and E. Premuzic, *Angew. Chem., internat. Ed.*, **7**, 247 (1968)
[2] C. R. Johnson, J. E. Keiser, and J. C. Sharp, *J. Org.*, **34**, 860 (1969)
[3] J. R. Hanson and E. Premuzic, *J. Chem. Soc.*, (C), 1201 (1969)
[4] A. Windaus, *Ber.*, **39**, 2249 (1906)
[5] J. R. Hanson and S. Mehta, *J. Chem. Soc.*, (C), 2349 (1969)
[6] J. Lessard and J. M. Paton, *Tetrahedron Letters*, 4883 (1970)
[7] J. Lessard, H. Driguez, and J. P. Vermes, *ibid.*, 4887 (1970)

Chromous sulfate, 1, 150–151; **2,** 77–78.

Preparation (**2,** 77). The procedure cited for the preparation and the use for reduction of diethyl fumarate has now been published.[1]

[1]A. Zurqiyah and C. E. Castro, *Org. Syn.*, **49,** 98 (1969)

Chromyl chloride, 1, 151–152; **2,** 79.

Oxidation of styrenes. Freeman *et al.*[1] have examined the chromyl chloride oxidation of the three styrenes (1), (4), and (7). Under conditions employed,

$$C_6H_5CH\overset{t}{=}CHCH_3 \xrightarrow[\text{0-5}^0,\ CH_2Cl_2]{CrO_2Cl_2} C_6H_5CH_2COCH_3\ +\ C_6H_5CHO$$

$$(1) \qquad\qquad\qquad (2,\ 40\%) \qquad (3,\ 16\text{-}24\%)$$

$$\underset{\underset{CH_3}{|}}{C_6H_5C}=CH_2 \xrightarrow{\hspace{2cm}} \underset{\underset{CH_3}{|}}{C_6H_5CHCHO}\ +\ \underset{\underset{CH_3}{|}}{C_6H_5C}=O$$

$$(4) \qquad\qquad\qquad (5,\ 60\%) \qquad (6,\ \text{trace})$$

$$(C_6H_5)_2C=CH_2 \xrightarrow{\hspace{2cm}} (C_6H_5)_2CHCHO\ +\ (C_6H_5)_2CO$$

$$(7) \qquad\qquad\qquad (8,\ 63\%) \qquad (9,\ 3.4\%)$$

reductive decomposition of the Étard complex *in situ* with zinc and water, the main products are carbonyl compounds formed by rearrangement, (2), (5), and (8). Cleavage products, (3), (6), and (9), are formed to a minor extent. The yield of the carbonyl compound is highest when one side of the double bond is disubstituted as in (7). The suggested mechanism involves electrophilic attack of the reagent on the double bond to give a resonance stabilized carbonium ion-like intermediate, which rearranges to the observed product.

[1]F. Freeman, R. H. DuBois, and N. J. Yamachika, *Tetrahedron*, **25,** 3441 (1969)

Copper acetylacetonate, 2, 81.

Intramolecular cyclization of bis-α-diazoketones. The decomposition of 1,7-bisdiazoheptane-2,6-dione (1) with this copper chelate in boiling benzene under high dilution conditions affords cyclohept-2-ene-1,4-dione (2) in 32% yield.[1] A lower yield (20%) is obtained if copper bronze is used as catalyst.

$$(CH_2)_3 \underset{\diagdown\ COCHN_2}{\overset{\diagup\ COCHN_2}{}} \xrightarrow{32\%} (CH_2)_3$$

$$(1) \qquad\qquad\qquad\qquad (2)$$

The method has been used to synthesize 4-hydroxytropone (γ-tropolone). Thus catalytic decomposition of 1,7-bis(diazo)-4-chloroheptane-2,6-dione (3) leads to 6-chlorocyclohept-2-ene-1,4-dione (4); loss of HCl on treatment with weak bases affords 4-hydroxytropone (5).

(3)   (4)   (5)

[1]J. Font, F. Serratosa, and J. Valls, *Chem. Commun.*, 721 (1970)

Copper powder (salts) (1, 157–158; 2, 82–84).

Decarboxylation. Cohen and Schambach[1] report that the usual copper metal–quinoline decarboxylation procedure is much slower than that using the cupric or cuprous salt of the acid (N_2, 180–200°). They also note that the rate is increased markedly by certain chelating agents such as 2,2'-dipyridyl (Eastman, Aldrich) and 1,10-phenanthroline (Eastman, Aldrich). The latter is more effective, but about twice as expensive.

α-Diazoketones (2, 82–84). The intramolecular cyclization of α-ketocarbenoids generated by copper-catalyzed decomposition of α-diazoketones continues to be invaluable for synthesis of unusual ring systems, particularly those related to natural products. Thus the key step in the synthesis of ($\pm$)-aristolone (2) involved the copper(II) sulfate-catalyzed intramolecular cyclization of the diazoketone (1). ($\pm$)-6,7-Epiaristolone (3) was obtained as a minor product.[2]

(1)   (2) 42%   (3) 20%

Two groups have reported syntheses of the bicyclo[3.2.1]octane ring system found in certain tetracyclic diterpenes in which the key step is an intramolecular carbenoid insertion reaction. Thus decomposition of the diazoketone (4) in the presence of copper(II) sulfate gives the cyclopropyl ketone (5) in 48% yield. Acid hydrolysis of the acetal grouping is accompanied by concomitant fragmentation to the substituted bicyclo[3.2.1]octanone derivative (6).[3]

In a second route[4] to this ring system the α-diazoketone (7) was cyclized to the cyclopropane derivative (8). Cleavage of (8) with dry hydrogen chloride in chloroform gave the bridged-ring compound (9). Note that the bond in conjunction with the aromatic ring and the keto group is cleaved preferentially. Catalytic hydrogenation afforded the saturated tetracyclic ketone (10). The *trans* stereochemistry of (10) is assigned by analogy.

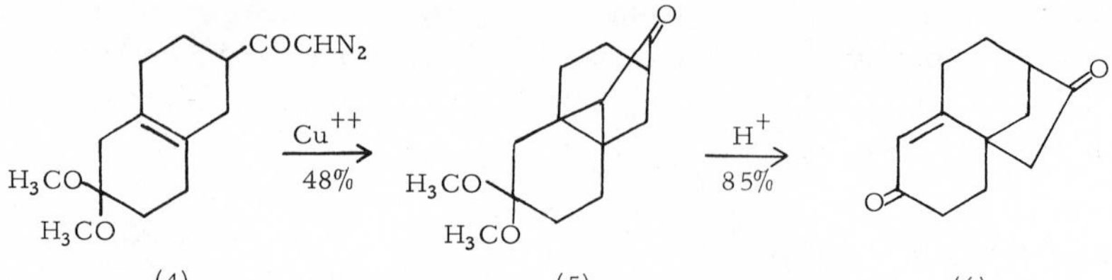

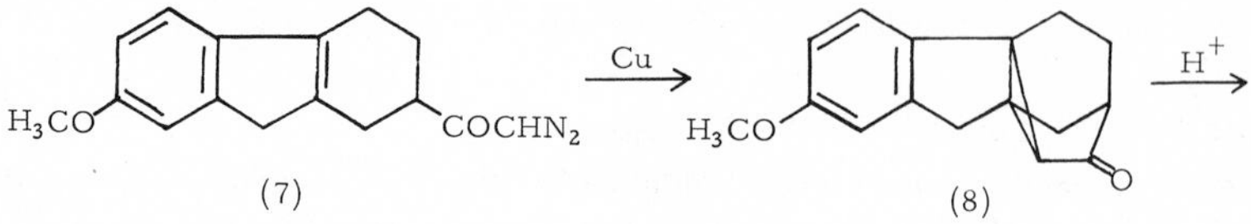

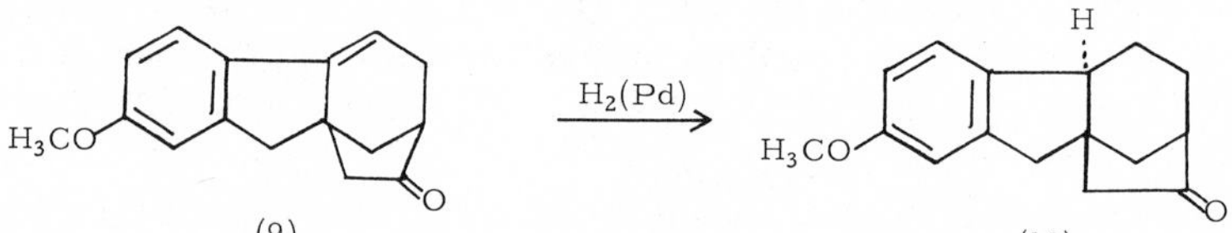

The copper-catalyzed degradation of diazoketones of type (11) in boiling cyclohexane affords bicyclic ketones (12) as a mixture of C_7-epimers; the starting materials are obtained by the Arndt-Eistert synthesis.[5]

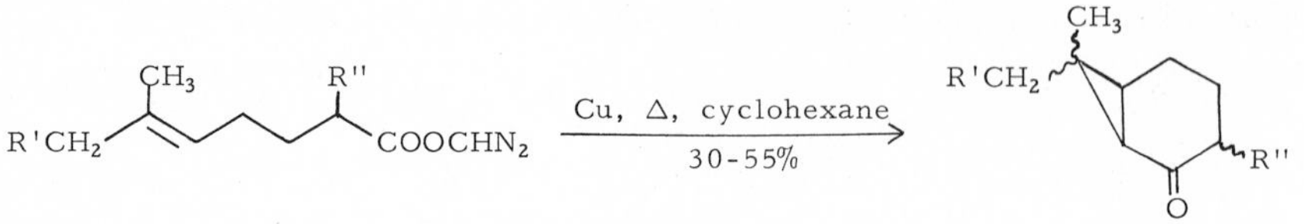

A key step in the synthesis of the *dl*-sesquicarene (15) involved the reaction of the diazoketone (13) with 2 equiv. of anhydrous cupric sulfate to give the bicyclic ketone (14) in 60% yield (overall from the acid).[6]

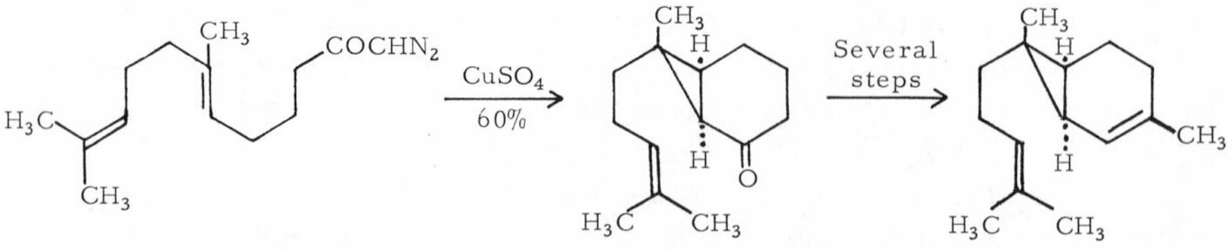

[1]T. Cohen and R. A. Schambach, *Am. Soc.*, **92**, 3189 (1970)
[2]E. Piers, R. W. Britton, and W. de Waal, *Canad. J. Chem.*, **47**, 831, 4299 (1969)
[3]D. J. Beames and L. N. Mander, *Chem. Commun.*, 498 (1969)
[4]S. K. Dasgupta, R. Dasgupta, S. R. Ghosh, and U. R. Ghatak, *ibid.*, 1253 (1969)
[5]K. Mori and M. Matsui, *Tetrahedron*, **25**, 5013 (1969)
[6]E. J. Corey and K. Achiwa, *Tetrahedron Letters*, 1837 (1969)

Cupric acetate — amine complexes.

Oxygenation. Brackman *et al.*[1] found that cupric acetate complexed with an amine (pyridine was used) in the presence of a base such as triethylamine functions as a homogeneous catalyst in methanol for the air oxidation of Δ^5-cholestenone to Δ^4-cholestene-3,6-dione in 75% yield. The reaction is applicable to α,β- and β,γ-aldehydes and ketones; for example:

$$CH_3CH=CHCHO \xrightarrow[60-75\%]{CH_3OH} OHCCH(OCH_3)CH_2CHO$$

Straight-chain aliphatic aldehydes with at least three carbon atoms are oxidized by the same system. The carbon atoms, starting with the carbonyl carbon atom, are successively split off as methyl formate or formic acid until acetaldehyde, which is stable in this system, is formed. Ketones are oxidized only slowly.[2]

$$CH_3(CH_2)_nCHO + nO_2 + nCH_3OH \longrightarrow CH_3CHO + nHCOOCH_3 + nH_2O$$

Van Rheenen[3] has found that a modified version of Brackman's procedure is useful for oxygenation of branched aldehydes to ketones. Cupric acetate complexed with 2,2'-bipyridyl or 1,10-phenanthroline was used as catalyst, DABCO (**2**, 99–101) was used as base, and DMF was used as solvent. Under these conditions 3-ketobisnor-Δ^4-cholene-22-al (1) is oxidized to progesterone (2) in 90% yield. DABCO or 3-quinuclidinol was found to be much more satisfactory than

(1)   (2)

trimethylamine as the base; alcohols in general were less satisfactory as solvent than DMF or HMPT.

Other examples:

Cyclohexanecarboxaldehyde → Cyclohexanone (66%)
Diphenylacetaldehyde → Benzophenone (94%)

The original paper should be consulted for a proposed mechanism.

Briggs *et al.*[4] used Van Rheenen's procedure in the last step in a degradation of the side chain of lanosterol; no yield was reported.

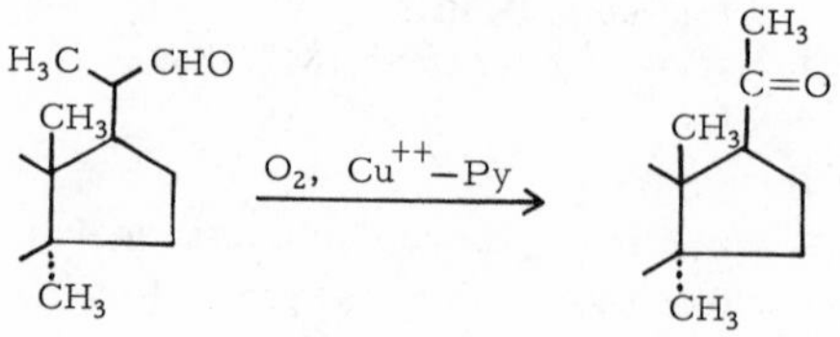

[1] H. C. Volger and W. Brackman, *Rec. trav.*, **84**, 579 (1965); H. C. Volger, W. Brackman, and J. W. F. M. Lemmers, *ibid.*, **84**, 1203 (1965)

[2] W. Brackman, C. J. Gaasbeek, and P. J. Smit, *ibid.*, **85**, 437 (1966)

[3] V. Van Rheenen, *Tetrahedron Letters*, 985 (1969); *idem*, *Chem. Commun.*, 314 (1969)

[4] L. H. Briggs, J. P. Bartley, and P. S. Rutledge, *Tetrahedron Letters*, 1237 (1970)

Cupric chloride, 1, 163; **2**, 84–85.

Oxidation of hydrazo compounds (**2**, 84–85). The procedure cited has now been published.[1]

[1] P. G. Gassman and K. T. Mansfield, *Org. Syn.*, **49**, 1 (1969)

Cupric fluoroborate, $Cu(BF_4)_2$. Mol. wt. 237.18. Suppliers: Alfa Inorganics, Pfalz and Bauer, ROC/RIC.

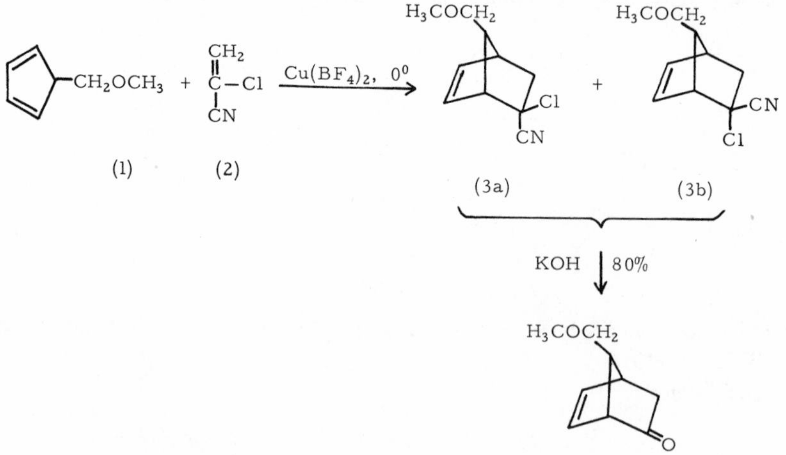

Diels-Alder reaction. The first step in a new stereo-controlled synthesis of prostaglandins $F_{2\alpha}$ and E_2 involved a Diels-Alder reaction between 2-chloro-acrylonitrile (2) and 5-methoxymethyl-1,3-cyclopentadiene (1).[1] This diene readily undergoes a 1,5-hydrogen shift to give 1-methoxymethyl-1,3-cyclopentadiene. In order to accelerate the Diels-Alder reaction without appreciable concurrent isomerization of the diene, cupric fluoroborate was added as catalyst. The use of this salt was suggested by the fact that copper(II) salts markedly catalyze decarboxylation of cyano-substituted α,β-unsaturated malonic acid derivatives in pyridine.[2] The effect is considered to be due to complex formation between Cu(II) ion and the nitrogen of the cyano function.

The reaction gives a mixture of stereoisomers (3a and 3b), which on treatment with hot aqueous potassium hydroxide in DMSO gave the *anti*-bicyclic ketone (4) in high yield.

[1]E. J. Corey, N. M. Weinshenker, T. K. Schaaf, and W. Huber, *Am. Soc.*, **91**, 5675 (1969)
[2]E. J. Corey, *ibid.*, **75**, 1163 (1953)

Cuprous bromide, 1, 165–166, **2**, 90–91.

Gabriel reaction. Review.[1] The Gabriel reaction is ordinarily not applicable to aryl halides; however, under catalysis with cuprous bromide or iodide and in boiling DMA (dimethylacetamide) as solvent, aryl bromides or iodides (but not chlorides) react with potassium phthalimide to give phthalimido compounds in 35–95% yield.[2]

[1]M. S. Gibson and R. W. Bradshaw, *Angew. Chem., internat. Ed.*, **7**, 919 (1968)
[2]R. G. R. Bacon and A. Karim, *Chem. Commun.*, 578 (1969)

Cuprous chloride, 1, 166–169; **2**, 91–92.

Dehalogenation.[1] Cuprous chloride dissolves in DMSO to give a colorless clear solution, which reacts with certain benzylic and allylic halides to give products of Wurtz-type condensation.

Examples:

$$C_6H_5CHCl_2 \xrightarrow[60\%]{} C_6H_5CHClCHClC_6H_5$$
$$\text{(meso)}$$

$$(C_6H_5)_2CCl_2 \xrightarrow[\text{quant.}]{} (C_6H_5)_2C{=}C(C_6H_5)_2$$

vic-Dehalogenation occurs with benzal bromide and dichlorodiphenylmethane to give *trans*-stilbene and tetraphenylethylene, respectively.

The reagent reacts with olefin–dibromocarbene adducts to give allenes, but yields are poor if the reaction occurs at all.

Cyclopropanation of olefins. Three laboratories[2–4] reported independently in 1960–61 that copper-catalyzed decomposition of diazomethane gives a species that reacts with benzene to give tropilidene (1, 16% yield) and with cyclohexene to give norcarane (2, 24% yield). The reaction was then developed into a useful synthetic method by Doering and Roth,[5] who found that cuprous chloride is an excellent catalyst. The method has the advantage that no insertion products are formed such as those obtained when diazomethane is decomposed photo-

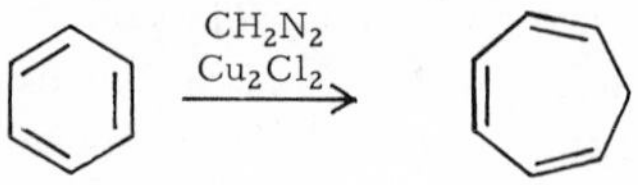

(1)

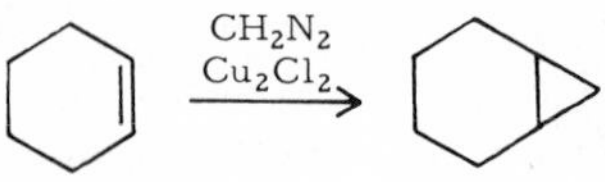

(2)

chemically. Application of the reaction to tropilidene (1) leads to a simple synthesis of (3) and (4); some di- and tricyclopropanated derivatives are also formed.

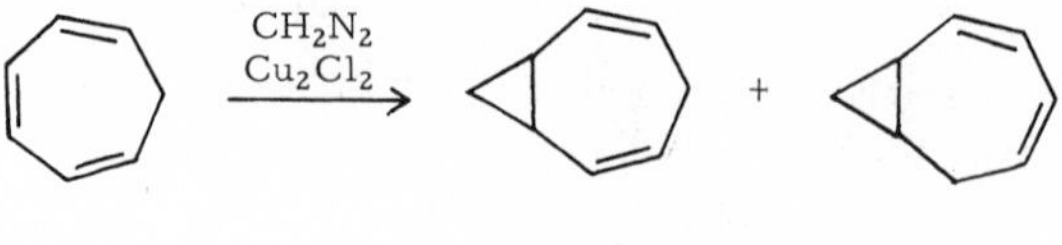

(1)   (3)   (4)

Doering and Roth applied the reaction also to acyclic olefins.
  Other examples:

(5)[6]

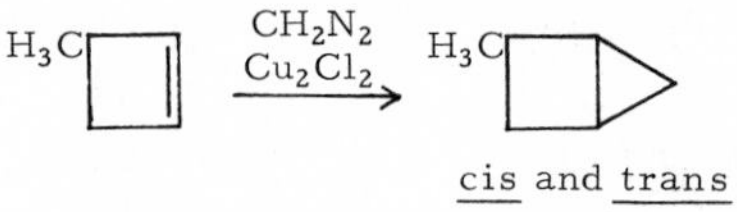

cis and trans

(6)[7]

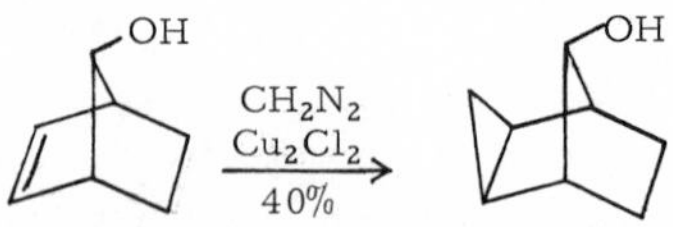

40%

(7)[8]

50-65%

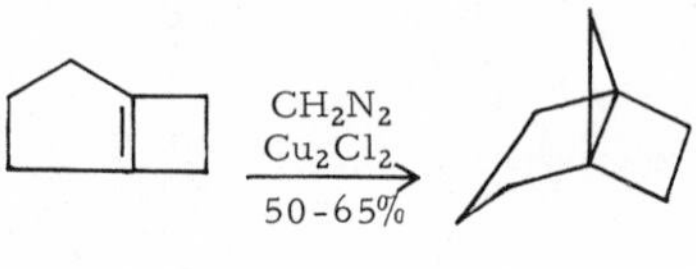

  In example (6), it is noteworthy that the product formed has the *exo*-configuration of the cyclopropyl group. In this case the Simmons-Smith reagent gives a very low yield.

  Ullmann ether synthesis. The original Ullmann ether synthesis[9] involved melting the salt of a phenol with an aryl bromide in the presence of copper metal. Yields are low. Williams *et al.*[10] found that the reaction can be carried out at lower temperatures by using as solvent pyridine, which forms a complex with copper salts (cuprous chloride preferred), which provides catalysis for the reaction; reflux temperature is then sufficient.

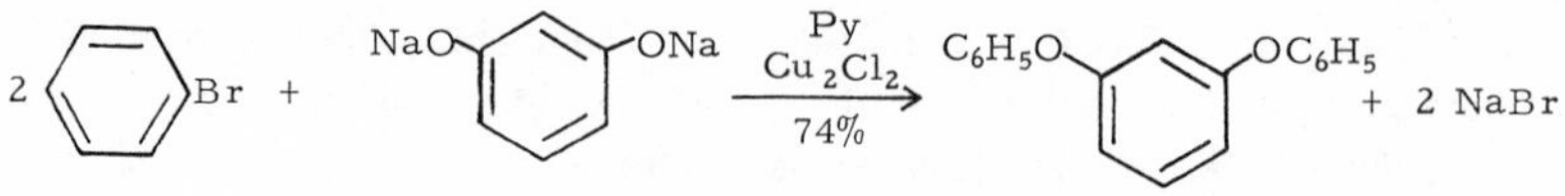

74%

The modified procedure was used successfully for synthesis of ±-dihydro-*o*-methylsterigmatocystine (4), a carcinogenic mold metabolite containing a dihydro-furobenzofurane ring system. Thus (1) was condensed with (2) in refluxing pyridine in the presence of cuprous chloride to give, after hydrolysis, (3). Ring closure was then effected with oxalyl chloride to give the desired natural product (4).[11]

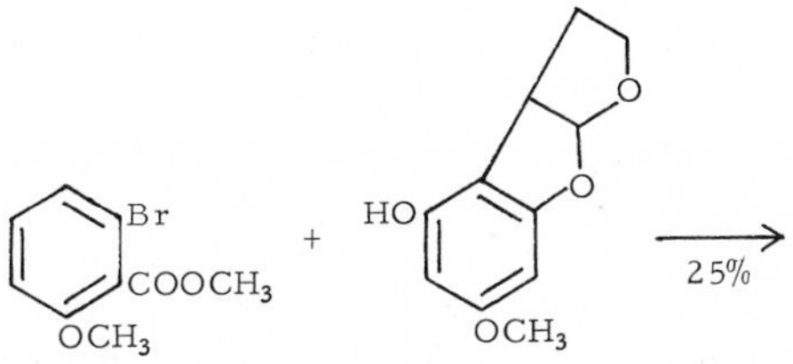

Coupling of terminal acetylenes (**2**, 91, ref. 17a). The cuprous chloride-catalyzed coupling of phenylethynylmagnesium bromide with propargyl bromide to form 1-phenyl-1,4-pentadiyne has now been published.[12]

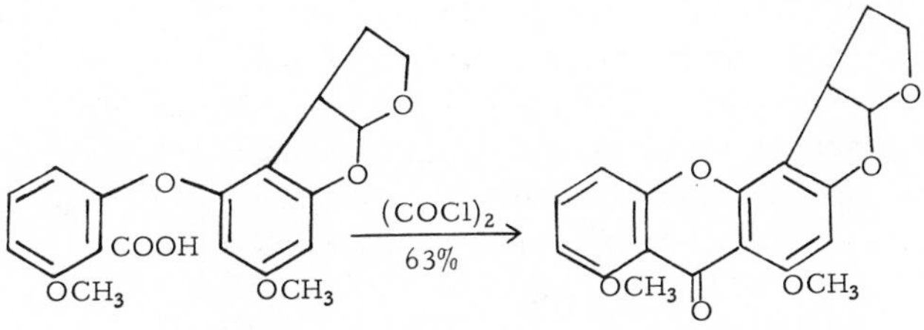

[1] H. Nozaki, T. Shirafuji, and Y. Yamamoto, *Tetrahedron*, **25**, 3461 (1969)
[2] E. Müller, H. Fricke, and W. Rundel, *Z. Naturforsch.*, **15b**, 753 (1960)
[3] G. Wittig and K. Schwarzenbach, *Ann.*, **650**, 1 (1961)
[4] P. P. Gaspar, Doctoral Thesis, Yale, 1961
[5] W. von E. Doering and W. R. Roth, *Tetrahedron*, **19**, 715 (1963)
[6] J. P. Chesick, *Am. Soc.*, **84**, 3250 (1962)
[7] R. E. Pincock and J. I. Wells, *J. Org.*, **29**, 965 (1964)
[8] P. G. Gassman, A. Topp, and J. W. Keller, *Tetrahedron Letters*, 1093 (1969)
[9] F. Ullmann and P. Sponagel, *Ann.*, **350**, 83 (1906)
[10] A. L. Williams, R. E. Kinney, and R. F. Bridger, *J. Org.*, **32**, 2501 (1967)
[11] M. J. Rance and J. C. Roberts, *Tetrahedron Letters*, 277 (1969)
[12] H. Taniguchi, I. M. Mathai, and S. I. Miller, *Org. Syn.*, **50**, 97 (1970)

Cuprous iodide, CuI, **1**, 169; **2**, 92.

Intramolecular diazo olefin cyclization. Corey and Achiwa[1] have described a simple four-step synthesis of sesquicarene (5) from *cis-trans*-farnesol (1),[2] in which the key step is an intramolecular addition reaction of an unsaturated diazo

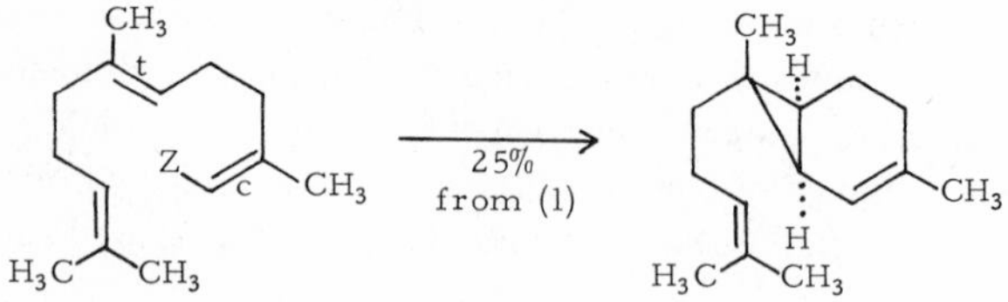

(1), Z = CH$_2$OH
(2), Z = CHO
(3), Z = CH=NNH$_2$
(4), Z = CH=N$_2$

(5)

compound. The alcohol (1) is oxidized to farnesal (2) with activated manganese dioxide; this is converted into the hydrazone (3) by hydrazine and triethylamine. The hydrazone was isolated and oxidized to the diazo compound (4) by activated manganese dioxide. This was dissolved in THF and added to a suspension of cuprous iodide in THF at 35°. dl-Sesquicarene was obtained in 25% yield from (1). Use of cupric fluoroborate in cyclohexane gave a fair yield (16%), but cuprous chloride, copper powder, and lithium bromide were not satisfactory.

Corey and Achiwa[3] have reported recently that mercuric iodide is an excellent catalyst not only for conversion of the cis,trans-diazo compound (4) into (5) but for generation of (5) from the trans,trans-isomer of (4). Thus it is possible to synthesize sesquicarene from the commercially available farnesol in 35% overall yield without separation of the trans,trans- and cis,trans-isomers.

This type of intramolecular addition reaction had previously been observed in low yield on photolysis by Büchi and White.[4] A mixture of the two isomeric diazo compounds (6) was photolyzed; the hydrocarbon fraction contained (±)-thujopsene (7) in low yield; the main component was the cyclopropene (8).

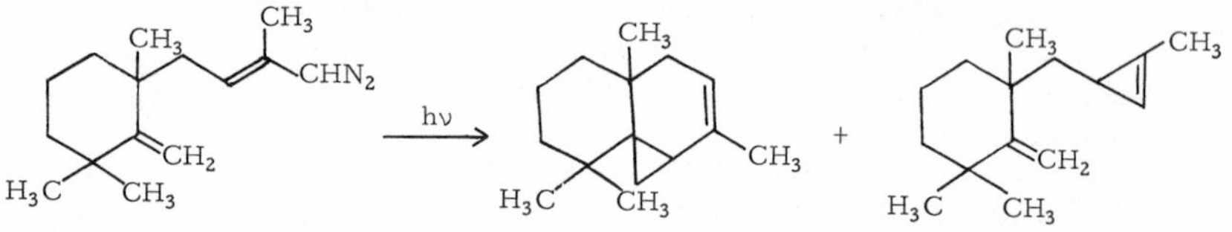

(6, mixture of 2 isomers)   (7, 4%)   (8, 10%)

The cuprous iodide procedure has also been used in a simplified synthesis of dl-sirenin (11) from the diazo compound (9). Thus, when (9) is slowly added to a stirred suspension of an equivalent amount of cuprous iodide in THF at 35°, the bicyclic ester (10) is obtained in about 50% yield.[3] This ester has been converted into dl-sirenin (11) by Rapoport[5] in two steps: allylic oxidation with selenium dioxide followed by reduction of the aldehyde and ester functions with mixed hydride.

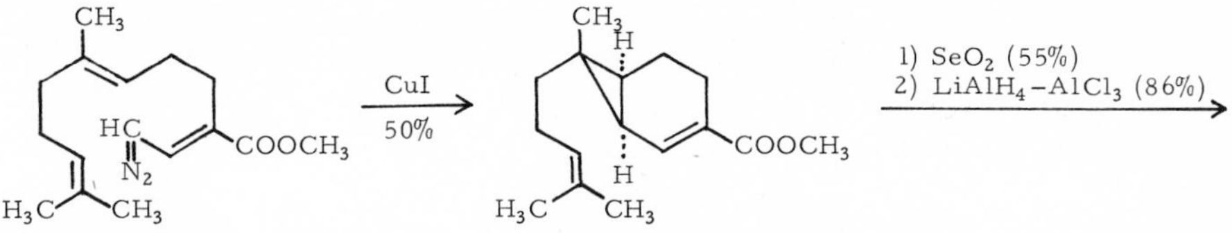

(9)   (10)

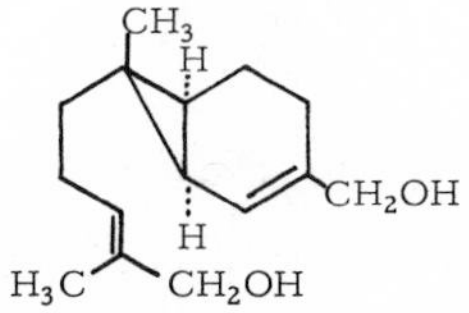

1) SeO_2 (55%)
2) $LiAlH_4-AlCl_3$ (86%)

(11)

[1]E. J. Corey and K. Achiwa, *Tetrahedron Letters*, 3257 (1969)
[2]Commercial farnesol (Givandan Corp., Clifton, N.J.) is a mixture of 70% *trans,trans*-isomer and 28% *cis,trans*-isomer. They can be separated by fractionation in a spinning-band column.
[3]E. J. Corey and K. Achiwa, *Tetrahedron Letters*, 2245 (1970)
[4]G. Büchi and J. D. White, *Am. Soc.*, **86**, 2884 (1964)
[5]J. J. Plattner, U. T. Bhalerao, and H. Rapoport, *ibid.*, **91**, 4933 (1969)

Cyanamide, 1, 170.

vic-Halocyanoamines can be prepared in 20–65% yield by simultaneous addition of cyanamide and positive halogen (NBS, dichlorourethane, *t*-butyl hypochlorite) to a double bond. The adducts are converted into N-cyanoaziridines in high yield by treatment with alkali.[1]

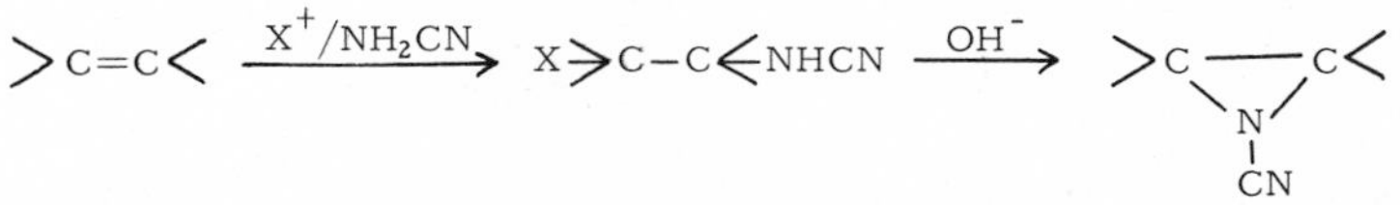

[1]K. Ponsold and W. Ihn, *Tetrahedron Letters*, 1125 (1970)

Cyanogen azide, 1, 173–174; **2**, 92–93.

Preparation.[1] A solution of cyanogen bromide in acetonitrile is stirred with powdered sodium azide for 2 hrs.; the supernatant, containing the reagent, is removed from NaBr via syringe. Although the pure reagent is highly explosive, **solutions** are apparently safe to handle.

Ring expansion. McMurry[1] has extended the work of Marsh and Hermes (ref. 1, **1**, 174) on the reaction of cyanogen azide with olefins and conjugated dienes and found that it can lead to ring enlargement. Thus the diene (1) reacts with reagent in acetonitrile containing a high concentration of lithium perchlorate followed by acid hydrolysis to give two ring-expanded products (3) and (4). Ring expansion also occurs with exocyclic monoolefins. Thus methylenecyclohexane is converted into cycloheptanone in 80% yield.

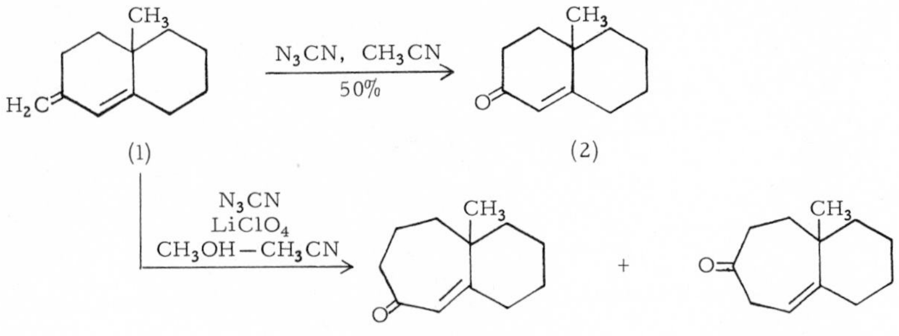

[1]J. E. McMurry, *Am. Soc.*, **91**, 3676 (1969)

Cyanuric chloride, 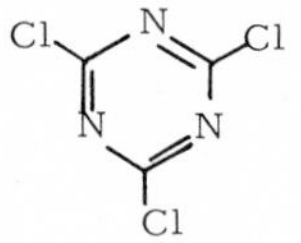 Mol. wt. 184.41, m.p. 145–148°. Suppliers: Eastman, Aldrich, MCB, others.

Hydrochlorination of alcohols.[1] Primary, secondary, and tertiary alcohols are converted into alkyl chlorides when heated somewhat below the boiling point with an excess of cyanuric chloride under anhydrous conditions.

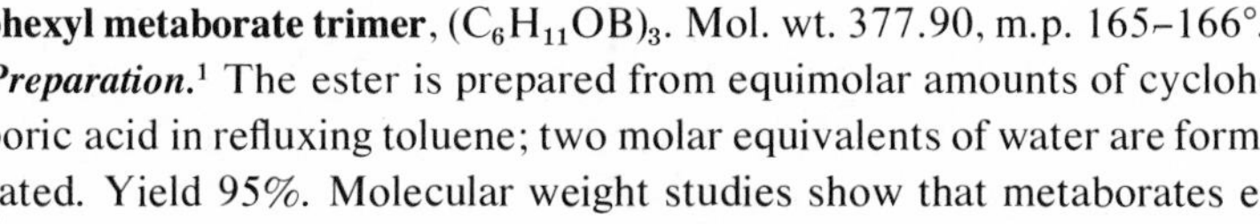

[1]S. R. Sandler, *J. Org.*, **35**, 3967 (1970)

Cyclohexyl metaborate trimer, $(C_6H_{11}OB)_3$. Mol. wt. 377.90, m.p. 165–166°.
   Preparation.[1] The ester is prepared from equimolar amounts of cyclohexanol and boric acid in refluxing toluene; two molar equivalents of water are formed and separated. Yield 95%. Molecular weight studies show that metaborates exist as trimers:

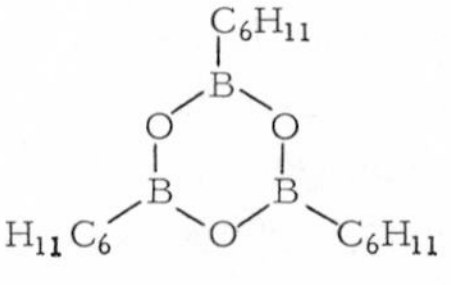

   Use. See **Tetralin hydroperoxide**, this volume.

[1]G. L. O'Connor and H. R. Nace, *Am. Soc.*, **77**, 1578 (1955)

D

Dehydroabietylamine, 1, 183; **2,** 97.

Resolution. (±)-Chlorofluoroacetic acid has been resolved by way of the dehydroabietylamine salts. Use of brucine was unsuccessful.[1]

[1]G. Bellucci, G. Berti, A. Borraccini, and F. Macchia, *Tetrahedron*, **25,** 2979 (1969)

trans,trans-**1,4-Diacetoxybutadiene, 1,** 183–185.

Preparation (**1,** 183–184). The procedure of Carlson and Hill[1] has been published.

[1]R. M. Carlson and R. K. Hill, *Org. Syn.*, **50,** 24 (1970)

Diazoacetaldehyde, 2, 101–102.

Reaction with trialkylboranes. The reagent reacts with trialkylboranes to give aldehydes with a two-carbon extension (ethanalation) of olefins:

$$R_3B \ + \ N_2CHCHO \ \xrightarrow[\text{H}_2\text{O}]{-N_2} \ RCH_2CHO$$

Yields are in the range 77–98% with organoboranes derived from 1-alkenes; yields are lower with sterically hindered trialkylboranes. The hydrolytic step is conveniently accomplished by addition of water to the solution of the organoborane in THF prior to addition of diazoacetaldehyde.[1]

[1]J. Hooz and G. F. Morrison, *Canad. J. Chem.*, **48,** 868 (1970)

Diazoethane, CH_3CHN_2. Mol. wt. 56.07.

Preparation.[1] The reagent is prepared in ether solution by treatment of N-ethyl-N-nitrosourea[2] with potassium hydroxide.

Ring expansion. In early work[3] the ring expansion of cyclohexanone to 2-methylcycloheptanone by reaction with ethereal diazoethane was found to proceed slowly and in poor yield. The results were confirmed by Marshall and Partridge.[4] These chemists noted, however, that Mosettig[5] had observed that methanol accelerates the reaction of cyclohexanone with diazomethane. They then carried out the ring expansion of cyclohexanone (1) with diazoethane in 20% ethanol in diethyl ether as solvent. The reaction was complete within 2 hours and gave a 9:1 mixture of 2-methylcycloheptanone (2) and the oxide (3) in over 90% yield. They also carried out the reaction on a number of 4-substituted cyclohexanones.

The reaction was used in the first step of a synthesis of the hydroazulenic alcohol bulnesol (4).[4]

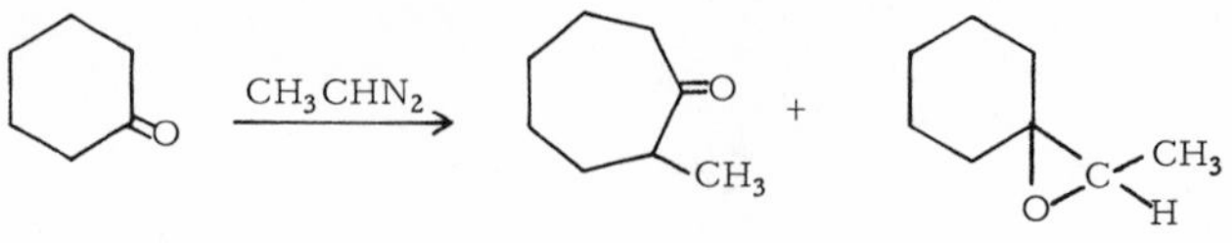

(1)   (2)   (3)

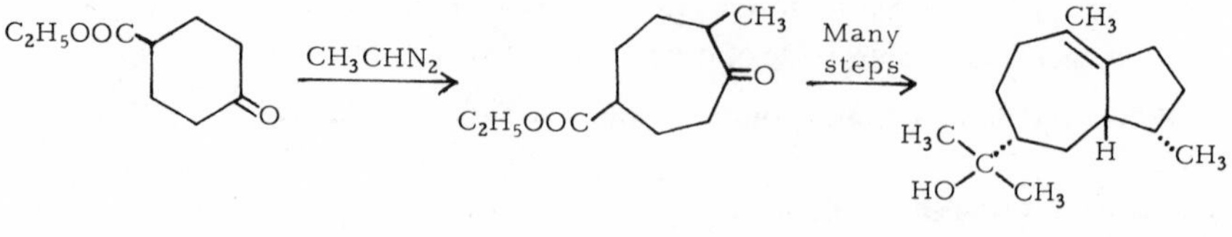

(4)

[1] J. A. Marshall and J. J. Partridge, *J. Org.*, **33**, 4090 (1968)
[2] F. Arndt, *Org. Syn., Coll. Vol.* **2**, 461 (1943)
[3] E. P. Kohler, M. Tishler, H. Potter, and H. T. Thompson, *Am. Soc.*, **61**, 1057 (1939); D. W. Adamson and J. Kenner, *J. Chem. Soc.*, 181 (1939)
[4] J. A. Marshall and J. J. Partridge, *Tetrahedron*, **25**, 2159 (1969)
[5] E. Mosettig and A. Burger, *Am. Soc.*, **52**, 3456 (1930); see also J. N. Bradley, G. W. Cowell, and A. Ledwith, *J. Chem. Soc.*, 4334 (1964)

Diazomethane, 1, 191–195; **2**, 102–104.

Esterification of water–soluble acids, **2**, 102, ref. 14a. The procedure has now been published.[1]

Arndt-Eistert reaction (**2**, 102–103, ref. 29a). The procedure for preparation of ethyl 1-naphthylacetate has been published.[2]

Reaction with chloromethanesulfonyl chloride, (**2**, 103–104). The procedure cited (ref. 33) has now been published.[3]

[1] E. J. Eisenbraun, R. N. Morris, and G. Adolphen, *J. Chem. Ed.*, **47**, 710 (1970)
[2] V. Lee and M. S. Newman, *Org. Syn.*, **50**, 77 (1970)
[3] L. A. Paquette and L. S. Wittenbrook, *ibid.*, **49**, 18 (1969)

2-Diazopropane, 2, 105.

Preparation. The reagent can be prepared in 70–90% yield by oxidation of acetone hydrazone with yellow mercuric oxide (British Drug Houses) catalyzed by potassium hydroxide. In the absence of base the hydrazone is not oxidized by mercuric oxide.[1] Unstable solutions are obtained by use of silver oxide in place of mercuric oxide.[2]

$$(CH_3)_2C=NNH_2 \xrightarrow[70-90\%]{HgO,\ KOH} (CH_3)_2CN_2$$

***gem*-Dimethylcyclopropanes.** Franck-Neumann[3] found that 2-diazopropane reacts at room temperature with the butenolides (1a) and (1b) to give the expected pyrazolines (2a) and (2b). On photolysis in benzene sensitized by benzophenone, these lose nitrogen to give cyclopropane derivatives (3a and 3b). In the absence of the sensitizer, 1,3-dipolar retrocycloaddition becomes a competing reaction.

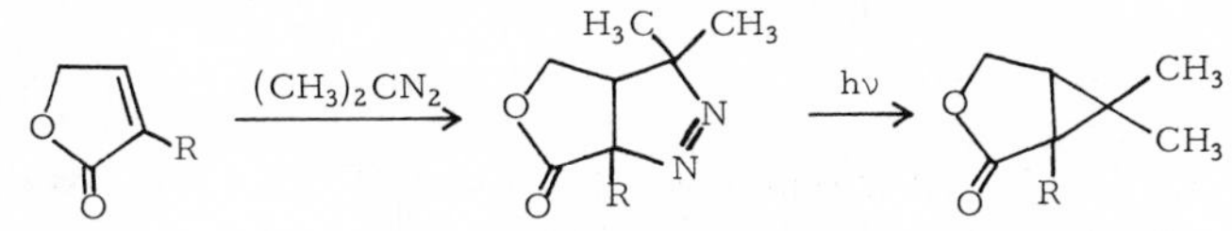

(1)   (2)   (3)

(a) R = H

(b) R = CH₃

This reaction provides a key step in the total synthesis of aristolone (7).[4] The reagent was added to the enone (4) to give the pyrazoline (5). This was converted into the cyclopropane derivative (6) by photolysis. Final steps involved bromina-

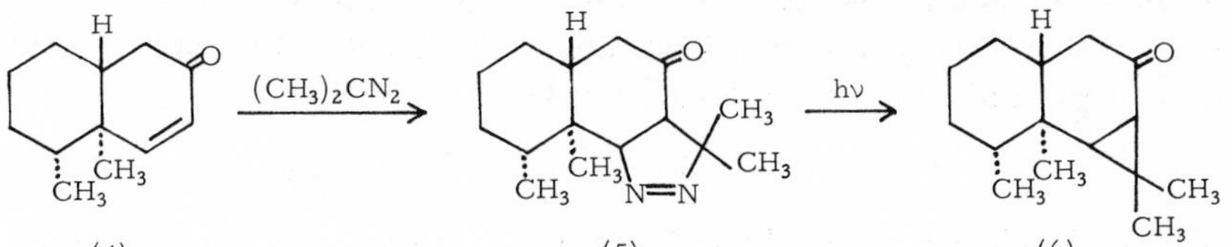

(4)   (5)   (6)

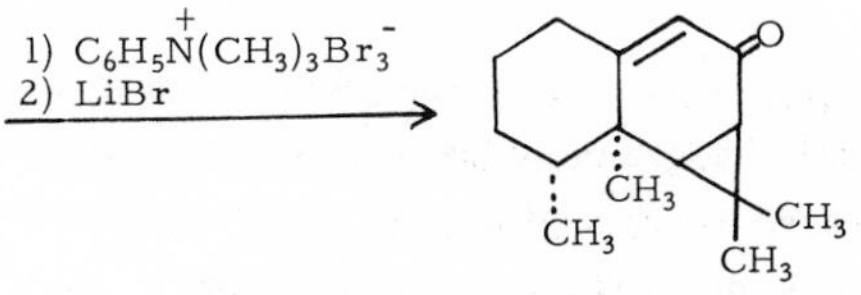

(7)

tion with phenyltrimethylammonium perbromide and the elimination of hydrogen bromide by lithium bromide in HMPT.

Reaction with tropone and tropolone.[5] The reagent reacts with tropone to give the highly reactive ketone (1); the following mechanism is proposed:

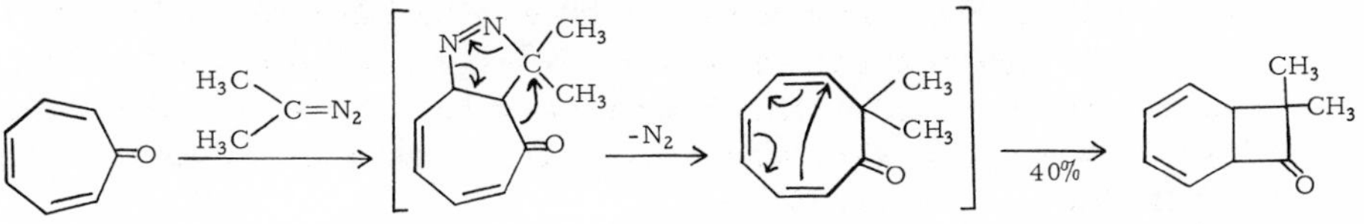

(1)

The reaction with tropolone is accompanied by O-alkylation to give the ketone (2) in 54% yield.

$$\text{(2)}$$

[1]S. D. Andrews, A. C. Day, P. Raymond, and M. C. Whiting, *Org. Syn.*, **50**, 27 (1970)
[2]D. E. Applequist and H. Babad, *J. Org.*, **27**, 288 (1962)
[3]M. Franck-Neumann, *Angew. Chem., internat. Ed.*, **7**, 65 (1968)
[4]C. Berger, M. Franck-Neumann, and G. Ourisson, *Tetrahedron Letters*, 3451 (1968)
[5]M. Franck-Neumann, *ibid.*, 2143 (1970)

Diborane, 1, 199–207; 2, 106–108.

Preparation. Diborane is generally made by treatment of sodium borohydride with boron trifluoride etherate. During a study of the reduction of organic halogen compounds with sodium borohydride, Bell *et al.*[1] noted that the stoichiometry results in generation of diborane (as monomer):

$$RX + NaBH_4 \longrightarrow RH + NaX + BH_3$$

Methyl iodide and dimethyl sulfate were chosen since they react readily and result in a gaseous product; either diglyme or THF was used as solvent. The method was shown to be satisfactory by hydroboration of some representative alkenes.

Selective reductions. Brown *et al.*[2] conducted an extensive study of reductions with diborane in THF. Most aldehydes and ketones are readily reduced; unusually high stereoselectivity was realized in the case of norcamphor, which was reduced to 98% *endo*-norbornanol and 2% *exo*-norbornanol. *p*-Benzoquinone is reduced to hydroquinone at a moderate rate, but reduction of anthraquinone is sluggish. Carboxylic acids are reduced very rapidly; indeed this group can be reduced selectively in the presence of many other substituents. Acid chlorides react much more slowly than carboxylic acids. Esters and ketones are reduced relatively slowly. Reactions with epoxides are relatively slow and complex.

Reduction of oximes. Aldoximes and ketoximes are reduced by diborane in THF at 25° to the corresponding N-monosubstituted hydroxylamines in very good yield.[3] If the reaction is carried out at 105–110° in a diglyme–THF solution, the reduction is carried to the corresponding amine. The reaction probably proceeds by way of the hydroxylamine since hydroxylamines are reduced to amines in high yield under these conditions:

Note, however, that the oximes of benzophenone and dicyclohexyl ketone are reduced only to the hydroxylamine stage.[4]

Oxime ethers are also reduced to amines, but in this case the reaction occurs at room temperature.

$$>C=NOCH_3 \xrightarrow[25^0]{B_2H_6} >CHNH_2 + CH_3OH$$

Fragmentation reactions. In extending various earlier work on solvolytic fragmentation reactions. Marshall[5,6] examined the behavior of appropriately substituted alkylboranes. The methanesulfonate (1) on treatment with a slight excess of diborane in THF followed with aqueous sodium hydroxide afforded an 85:15 mixture of hydrocarbons (2) and (3) in 90% yield. The major product (2) was

shown to be 1-methyl-*trans,trans*-1,6-cyclodecadiene, and it is considered to arise by normal hydroboration to the 9-decalylborane. Base treatment then effects a 1,3-elimination reaction to give (2). The minor product (3) presumably arises from hydroboration in the abnormal sense. Aqueous sodium hydroxide or sodium methoxide–methanol was found to give the highest yield of diene. Surprisingly, the dialkylborane derived from (1) is stable to base.

[1]H. M. Bell, C. W. Vanderslice, and A. Spehar, *J. Org.*, **34**, 3923 (1969)
[2]H. C. Brown, P. Heim, and N. M. Yoon, *Am. Soc.*, **92**, 1637 (1970)
[3]H. Feuer, B. F. Vincent, Jr., and R. S. Bartlett, *J. Org.*, **30**, 2877 (1965)
[4]H. Feuer and D. M. Braunstein, *ibid.*, **34**, 1817 (1969)
[5]J. A. Marshall, *Rec. Chem. Prog.*, **30**, 3 (1969)
[6]J. A. Marshall and G. L. Bundy, *Am. Soc.*, **88**, 4291 (1966)

Dibromofluoromethane, $CHBr_2F$. Mol. wt. 191.84, b.p. $63^\circ/739$ mm.,[1] $64.9^\circ/757$ mm.,[2] m.p. below -78°. Supplier: Dow Chemical Co. Hine *et al.*[1] purified the Dow material carefully and noted properties close to those reported by Swarts.[2]

Bromofluorocarbene. Dibromofluoromethane has been used as a precursor of bromofluorocarbene. Thus reaction of the reagent in the presence of potassium *t*-butoxide with norbornene (1) gives the three products formulated as (2), (3), and (4) in the ratio 1 : 3.2 : 4.1. The initial product is undoubtedly a cyclopropane, but, because of angle strain, isomerization to the more stable bicyclo[3.2.1]octane

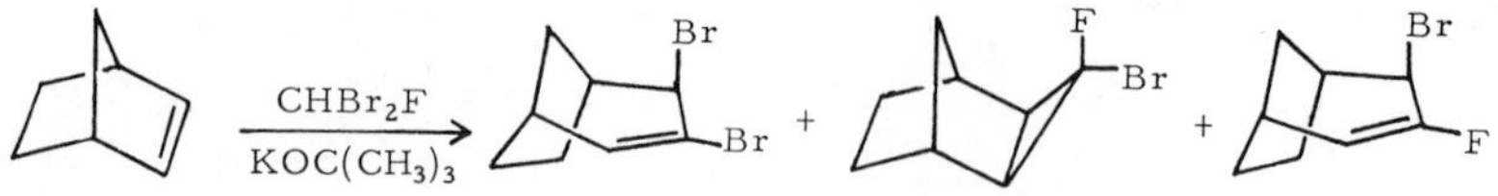

system (2) and (4) is observed. The unexpected formation of the dibromo adduct (2) may be due to an impurity or to anomalous fission of the reagent.[3]

[1]J. Hine, R. Butterworth, and P. B. Langford, *Am. Soc.*, **80**, 822 (1958)
[2]F. Swarts, *Bull. acad. roy. Belg.*, 113 (1910); [*C.A.*, **5**, 1086 (1911)]
[3]C. W. Jefford and D. T. Hill, *Tetrahedron Letters*, 1957 (1969)

3,5-Di-*t*-butyl-1,2-benzoquinone (1). Mol. wt. 220.30, m.p. 114–115° (red needles).

Preparation. The quinone is prepared in about 50% yield by oxidation of the corresponding hydroquinone, 3,5-di-*t*-butylcatechol, with lead dioxide under anhydrous conditions.[1]

Oxidation of primary amines to ketones. Corey and Achiwa[2] developed a practical method for effecting this transformation which is modeled on the biological transamination reaction involving formation and hydrolysis of Schiff bases. Normally primary amines react with quinones by conjugate addition, but in the case of 3,5-di-*t*-butyl-1,2-benzoquinone (1) this reaction is suppressed by the bulky *t*-butyl groups. Consequently the reaction of a primary amine with the quinone leads initially to the Schiff base (3), which undergoes isomerization by prototropic rearrangement to (4). Acid hydrolysis then effects cleavage to the carbonyl compound (5) and to (6). The reaction is generally carried out by allowing a solution of (1) and (2) in methanol (THF is used if necessary as cosolvent)

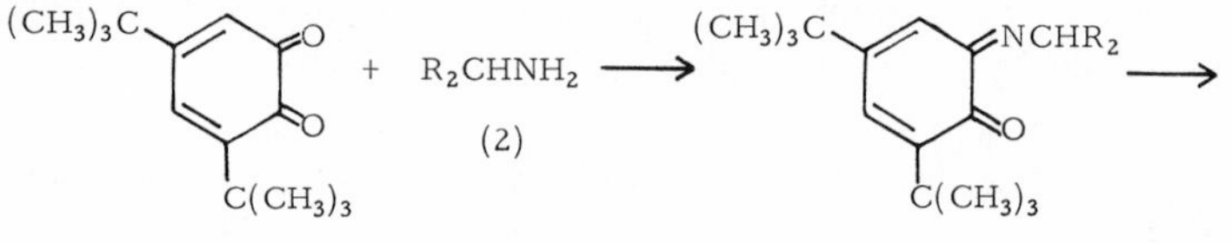

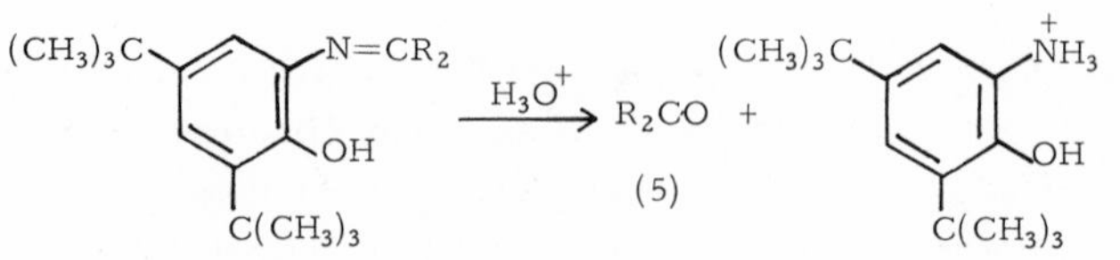

to stand for 20–30 minutes at room temperature; oxalic acid is then added to effect hydrolysis. Yields are high, usually 90% or more.

Mesitylglyoxal (7) and the 3-nitro and 3,5-dinitro derivatives have been used in the same way as the quinone. In this case an added base is needed for isomeriza-

(7)

tion of the initial Schiff base to the desired isomeric Schiff base. The tertiary base, 1,5-diazabicyclo[4.3.0]nonene-5 (**1**, 189; **2**, 98–99), is satisfactory for this purpose.

[1]W. Flaig, T. Ploetz, and H. Biergans, *Ann.*, **597**, 196 (1955)
[2]E. J. Corey and K. Achiwa, *Am. Soc.*, **91**, 1429 (1969)

Di-*n*-butylcopperlithium, $(n\text{-}C_4H_9)_2CuLi$, **2**, 152. Mol. wt. 184.70.

Preparation. Whitesides *et al.*[1] prepared the reagent by the reaction of one equivalent of tetrakis[iodo(tri-*n*-butylphosphine)copper(I)][2] with two equivalents of commercial *n*-butyllithium at −78° in THF.

$$2\ C_4H_9Li\ +\ [ICuP(C_4H_9)_3]_4\ \xrightarrow[\ -78^0\ ]{THF}\ (C_4H_9)_2CuLi$$

Oxidative coupling.[1] Treatment of the colorless or pale yellow solution prepared as above at −78° with excess oxygen leads to formation of a dark precipitate. After hydrolysis the following products were identified: octane (84%), 1-butene (14%), and 1-butanol (5%). The oxidative coupling reaction is general for copper (I) ate complexes.

$$(C_4H_9)_2CuLi\ \xrightarrow[84\%]{O_2,\ -78^0}\ C_8H_{18}$$

Cyclization of δ- or ε-haloketones. In the first paper on this reagent (**2**, 152) Corey and Posner noted that this organometallic compound is more apt than dimethylcopperlithium (**2**, 151–153) to produce halogen-copper exchange (R—X → R—Cu) at the expense of selective cross coupling. Since the reaction of dialkylcopperlithium reagents with ketones is relatively slow, the possibility presented itself of inducing a cyclization of halo ketones involving attachment of C_X to C_{CO}. To test this hypothesis, Corey and Kuwajima[3] prepared the iodoketone (1) by conventional methods and titrated it with 5 equiv. of di-*n*-butylcopperlithium in 5.7:1 pentane–ether under argon at 0° for 24 hours. The desired cyclization product (2) was obtained in more than 90% yield; only a trace of the product of cross coupling was formed. The reaction conditions, particularly the solvent and

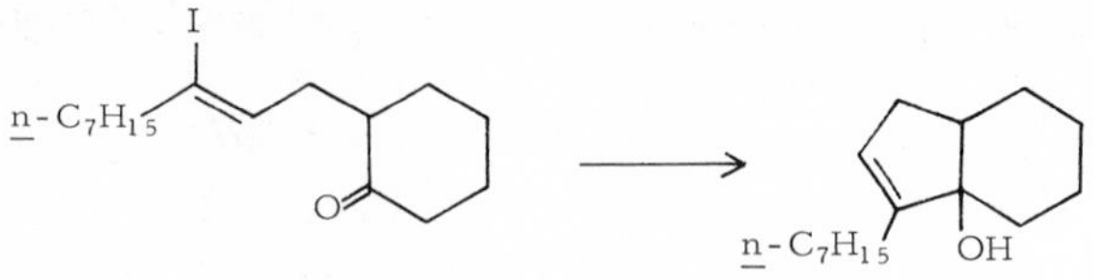

(1)   (2)

temperature, are critical for high yields. Dimethylcopperlithium and diphenyl-copperlithium were ineffective for cyclization.

This new cyclization method was used as the key step in the synthesis of a substance containing the bridged-ring system of the gibberellin family of plant hormones.[4] The tricyclic keto ketal (3) was used to supply final rings A, B, and C.

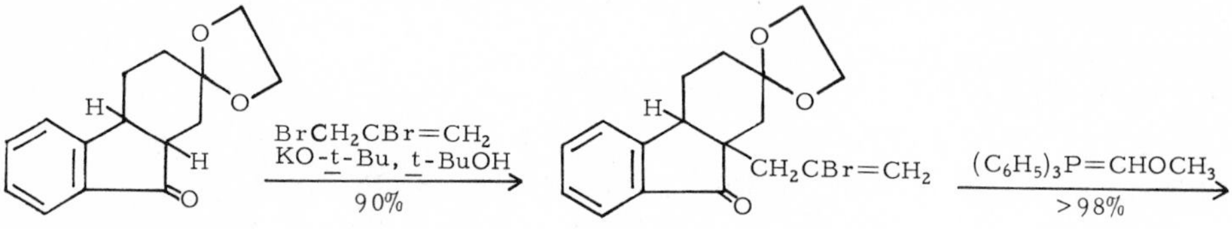

(3)   (4)

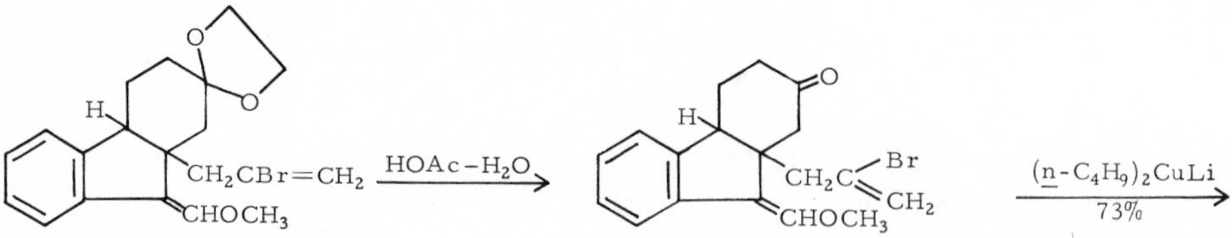

(5)   (6)

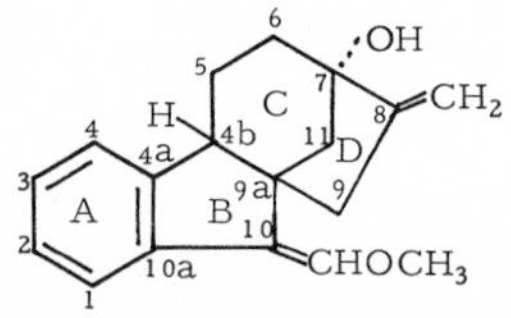

(7)

The synthesis then required construction of ring D. Alkylation of (3) with 2,3-dibromopropene gave (4) in high yield; then the free keto group was treated with methoxymethylenetriphenylphosphorane (1, 671) to give the enol ether (5) (two isomers formed about the new double bond). The protecting ketal group was then removed, and the resulting bromoketone (6) was cyclized with 6 equiv. of di-*n*-butylcopperlithium to give the desired gibbane derivative (7)[5] in good yield. Yields are high throughout. The synthesis has the advantage that it leads to functional groups at positions characteristic of gibberellins: a hydroxyl group at C_7, a 1-carbon unit at C_{10}, and the characteristic exocyclic methylene group at C_8.

Dialkylcopperlithium reagents give only low yields when used for the cyclization of saturated iodoketones of type (8) to give (9). Various anionic transition metal complexes were examined and the anion derived by treatment of nickel

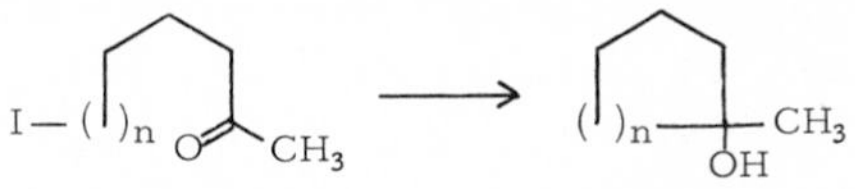

(8)   (9)

n = 1 or 2   n = 1 or 2

tetraphenylporphine (10)[6] with naphthalene–lithium was found to be most effective for this type of cyclization (60–80%). This reagent does not cyclize vinylic halo ketones of type (1).

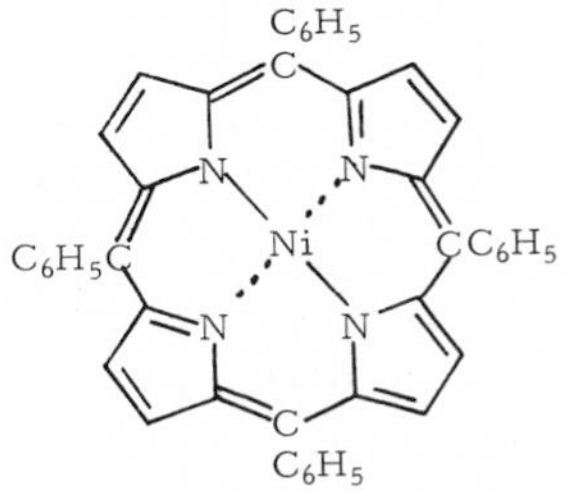

(10)

[1]G. M. Whitesides, J. San Filippo, Jr., C. P. Casey, and E. J. Panek, *Am. Soc.*, **89**, 5302 (1967)

[2]G. B. Kauffman and L. A. Teeter, *Inorg. Syn.*, **7**, 9 (1963)

[3]E. J. Corey and I. Kuwajima, *Am. Soc.*, **92**, 395 (1970)

[4]E. J. Corey, M. Narisada, T. Hiraoka, and R. A. Ellison, *ibid.*, **92**, 396 (1970)

[5]The systematic name is 7-hydroxy-10-methoxymethylene-8-methylgibba-1,3,4a(10a)-triene.

[6]A. D. Adler, F. R. Longo, J. D. Finarelli, S. Goldmacher, J. Assour, and L. Korsakoff, *J. Org.*, **32**, 476 (1967)

Di-*t*-butyl nitroxide, 1, 211.

The procedure cited for the preparation has now been published.[1]

[1]A. K. Hoffmann, A. M. Feldman, E. Gelblum, and A. Henderson, *Org. Syn.*, **48**, 62 (1968)

Dichlorobis(triphenylphosphine)palladium, $[(C_6H_5)_3P]_2PdCl_2$. Mol. wt. 702.16, m.p. 250–270° dec.

Preparation.[1] Solutions of triphenylphosphine (1 g.) in hot alcohol and of ammonium tetrachloropalladate(II) (Alfa Inorganics) (0.55 g.) in hot water are mixed and shaken. A yellow solid separates; water is added to complete precipitation of the complex metal salt, which is then crystallized from toluene; bright yellow crystals.

Carbonylation of olefins. In Reppe's[2] classical work on the carbonylation of acetylenes, nickel carbonyl was found to be an excellent catalyst, making possible relatively mild conditions (30 atm., 170°). Carbonylation of olefins with the catalyst, however, required much more vigorous conditions (200–300 atm., 250–320°). Consequently a group at Badische Anilin und Soda-Fabrik[3] undertook a search for more effective catalysts. Palladium compounds in particular were investigated, since they form complexes with olefins. Dichlorobis(triphenylphosphine)palladium was found to be particularly active, permitting carbonylation below 100° (300–700 atm.). Acids are obtained if water is present, esters if an alcohol is present. Acyl chlorides are obtained from olefins, CO, and hydrogen

$$CH_2=CH_2 \ + \ CO \ + \ C_2H_5OH \ \xrightarrow[\substack{300-700 \text{ atm.} \\ 90\%}]{60-100°} \ CH_3CH_2COOC_2H_5$$

chloride. Mild conditions are particularly important for carbonylation of Diels-Alder adducts, which often dissociate into their components at high temperatures.

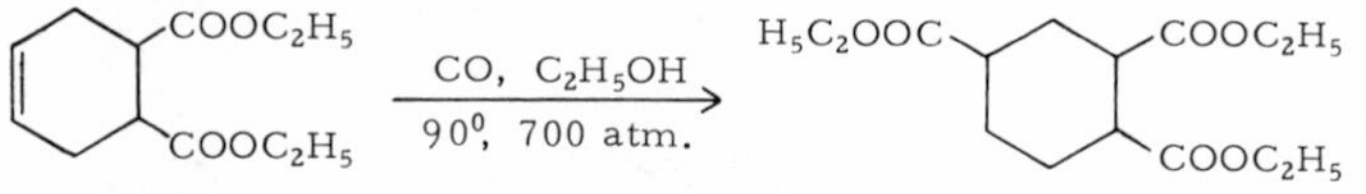

Thus the cyclohexene derivative (1) was successfully carbonylated with this catalyst. A further advantage of the catalyst is that selective carbonylation of poly-unsaturated olefins is possible. Thus it is possible to convert *trans,trans,cis-*

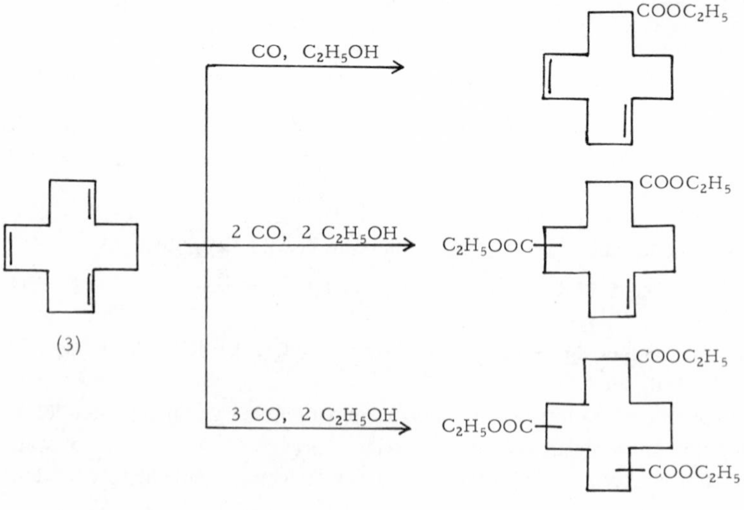

cyclodecatriene (3) into the mono-, di-, or triester. One of the *trans* double bonds reacts first, and then the *cis* double bond.

[1]J. Chatt and F. G. Mann, *J. Chem. Soc.*, 1631 (1939)
[2]W. Reppe, *Ann.*, **582**, 1 (1953)
[3]K. Bittler, N. v. Kutepow, D. Neubauer, and H. Reis, *Angew. Chem., internat. Ed.*, **7**, 329 (1968)

2,3-Dichloro-5,6-dicyano-1,4-benzoquinone (DDQ), 1, 215–219; 2, 112–117.

Dehydrogenation. Refluxed with 4,4'-dimethoxydibenzyl in purified dioxane in an oil bath at 105° for 18 hrs., the reagent effects dehydrogenation to *trans*-4,4'-dimethoxystilbene, isolated by chromatography and crystallization, in 83–85% yield.[1]

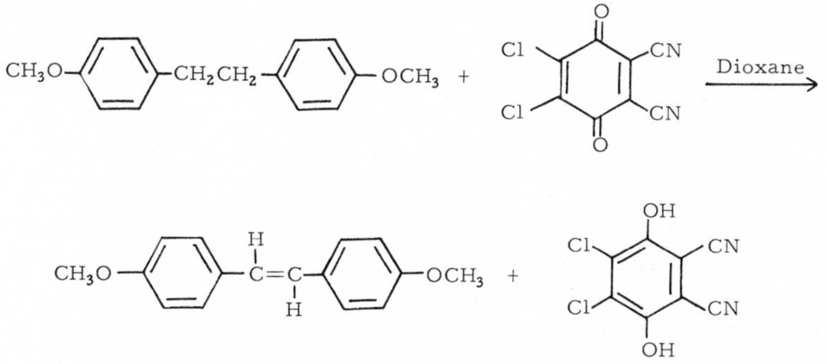

In an interesting new synthesis of a bufadienolide from cholanic acid (pregnane derivatives have been used previously), Sarel *et al.*[2] obtained as the penultimate product a β,γ-unsaturated δ-lactone (1), which required dehydrogenation to the desired α-pyrane (2). The lactone (1) was found to be relatively inert to DDQ in

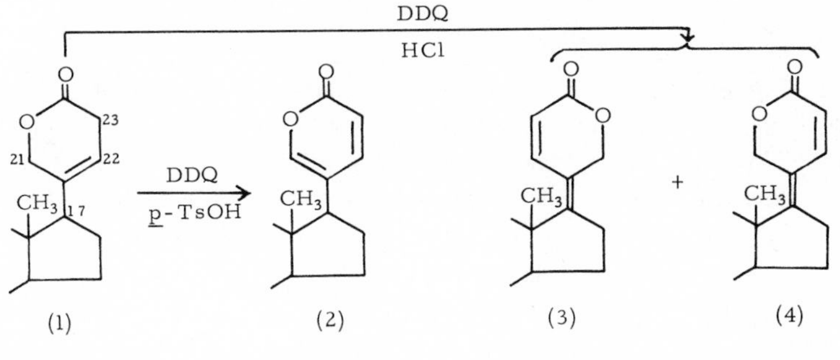

boiling *t*-butanol, benzene, or dioxane; however, reaction in the latter solvent is markedly accelerated by *p*-toluenesulfonic acid, and the product is the desired bufa-20,22-dienolide (2). Catalysis with anhydrous hydrogen chloride gives a mixture of (3) and (4) in about equal amounts, arising from removal of the C_{17} and

C_{23} hydrogens. These two products are also obtained by dehydrogenation with chloranil.

Estrone (5) is rapidly oxidized at room temperature in either dioxane or methanol to $\Delta^{9(11)}$-dehydroestrone (7, 67% yield).[3] The quinone methide (6) is the probable intermediate, since (8) is also converted into (7).

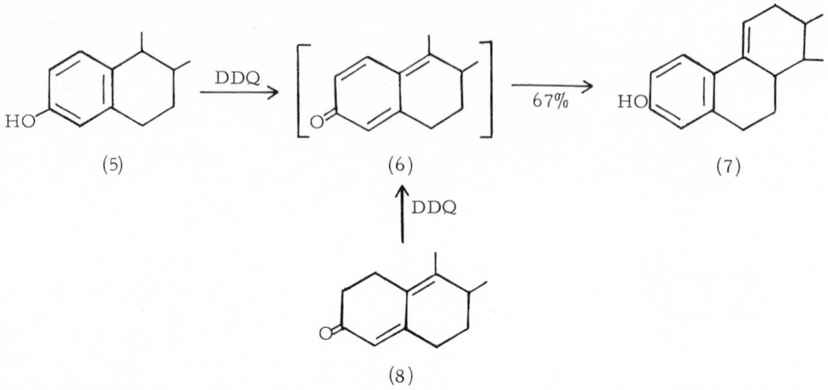

(5)   (6)   (7)

(8)

2,3,4,5,8-Pentamethyl-1,10-anthraquinone (11) has been obtained in about 50% yield by dehydrogenation of 4-hydroxy-1,2,3,5,8-pentamethylanthrone (9) as the tautomer (10) with DDQ in anhydrous ether at 0°.[4] The *ana*-anthraquinone (11) is very unstable and can be obtained only in a very dilute solution.

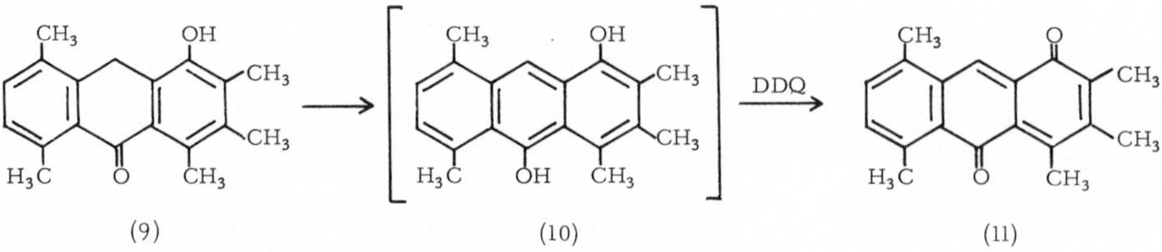

(9)   (10)   (11)

Dehydrogenation of $\Delta^{9(11)}$-estrone methyl ether. Treatment of $\Delta^{9(11)}$-estrone methyl ether (1) with DDQ in wet benzene for 2 hours gives as the major product (30% yield) the homoannular $\Delta^{8,11}$-diene (4). Further investigation showed that (4) is formed from an unstable intermediate identified as 12α-hydroxy-3-methoxy-estra-1,3,5(10),9(11)-tetraene-17-one (3). The intermediate (3) is considered to be

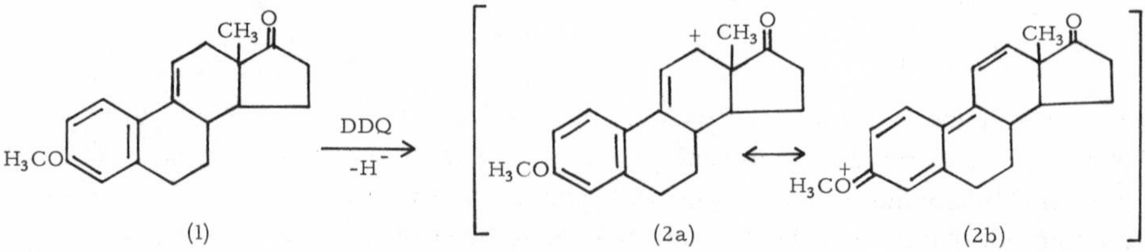

(1)   (2a)   (2b)

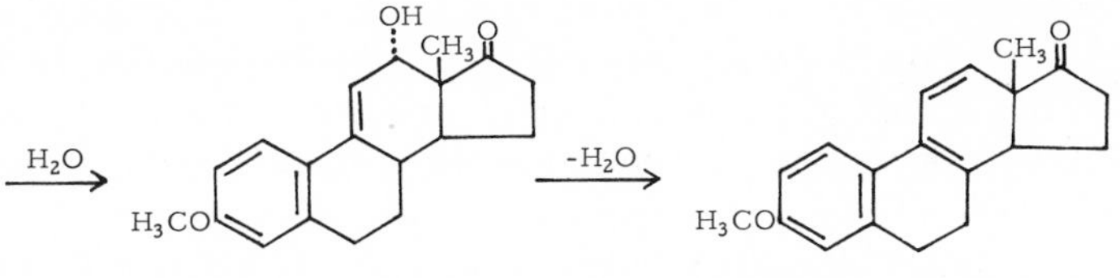

(3)   (4)

formed by abstraction of an allylic hydride ion from C_{12} to give the mesomeric cation (2a, 2b) followed by attack by water at C_{12}.[5]

Synthesis of 1,6-methano[10]annulene (5). The first synthesis by Vogel[6] involved the route shown, (1) → (5). Dichlorocarbene (CHCl_3–KO-*t*-Bu) reacts

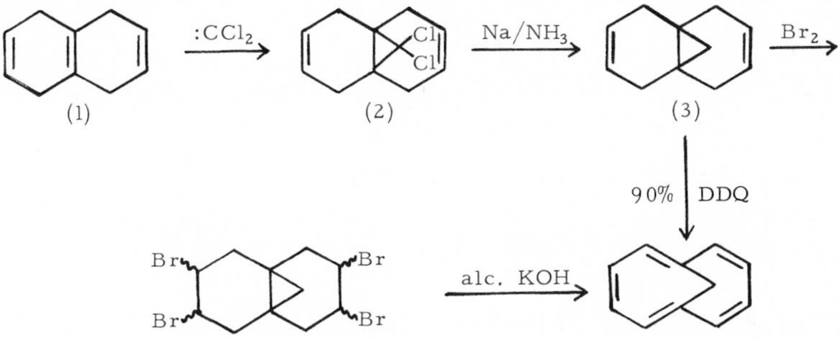

(1)   (2)   (3)

90% | DDQ

(4)   (5)

with isotetralin (1) with high selectivity to give (2); this is then reduced to (3). Dehydrogenation is effected by bromination (4) and dehydrobromination with alcoholic potassium hydroxide to give (5). No yields were reported but a "good yield" was reported for the conversion of (4) to (5).

Syntex workers[7] later found that (3) can be dehydrogenated to (5) directly by DDQ in 90% yield (2.5 moles of DDQ, refluxing dioxane, $\frac{1}{2}$–2 hrs.). This oxidation of an unactivated disubstituted olefin to a diene is rare; a related reaction to the oxidation of tetramethylethylene to 2,3-dimethylbutadiene (**2**, 117). They then found that it was possible to shorten the original synthesis even further. Reaction of (1) with a modified Simmons-Smith reagent (this volume) gave (3) directly in 50% yield. The overall yield of (5) from naphthalene is thus about 40%.

Cyclodehydrogenation (**2**, 115–116). Cardillo *et al.*[8] have extended their work on cyclodehydrogenation with this reagent to conversion of *o*-cinnamyl phenols into Δ³-flavenes. The starting materials are prepared by reaction of cinnamyl

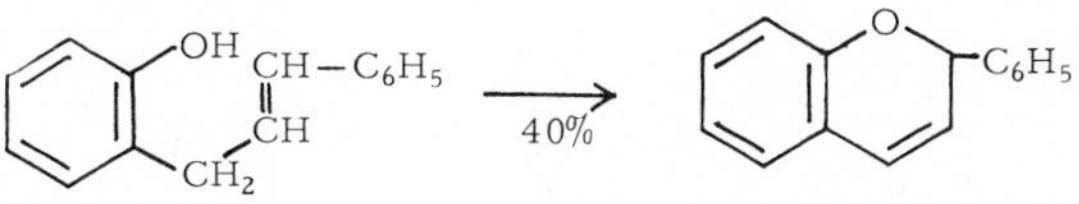

(1)   (2)

alcohol with a phenol in aqueous acetic acid (Jurd[9]). The cyclodehydration is carried out with one equivalent of DDQ in refluxing benzene for 12–20 hrs. o-Cinnamylphenols have been suggested as biogenetic precursors of flavanoids.[10]

Oxidation. In attempting the dehydrogenation of 9-isopropenyl-1,2,3,4-tetra-hydrofluorene (1) with DDQ (3 molar equiv., benzene, room temperature), Sadler and Stewart[11] obtained in addition to the expected product (2) the aldehyde (3, 45% yield). In refluxing benzene, only (3) was formed in 64% yield. The reaction

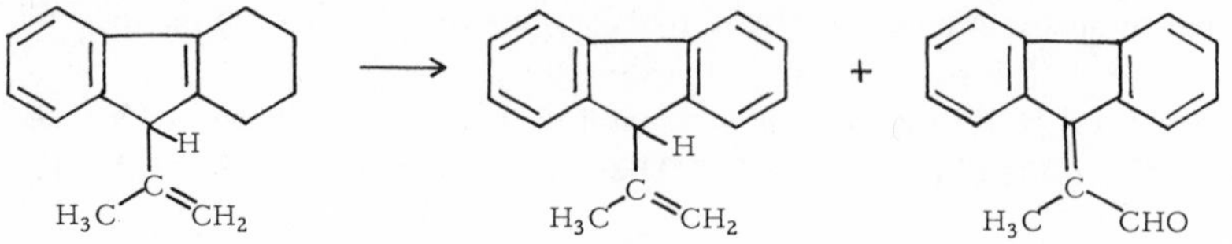

$$C_6H_5CH_2CH{=}CH_2 \longrightarrow C_6H_5CH\overset{t}{=}CHCHO \quad (50\%)$$

has been observed also with β-methylstyrene, allylbenzene, and related hydro-carbons, but indene and α-methylstyrene fail to react.

$$C_6H_5CH_2CH{=}CH_2 \longrightarrow C_6H_5CH\overset{t}{=}CHCHO \quad (50\%)$$

$$C_6H_5CH\overset{t}{=}CHCH_3 \longrightarrow C_6H_5CH\overset{t}{=}CHCHO \quad (55\%)$$

$$C_6H_5C(CH_3)\overset{c}{=}CH(CH_3) \longrightarrow C_6H_5C(CH_3){=}CHCHO \quad (20\%)$$

Oxidation of phenols (**1**, 218–219). Becker (ref. 22) also noted that oxidation of mesitol (1) results in benzylic oxidation of the p-methyl group to give (2) in 83% yield. Aldehyde formation follows successive nucleophilic attack by the

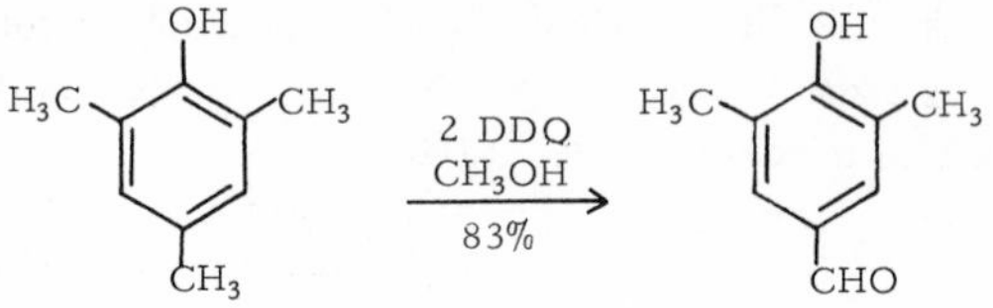

solvent upon quinone methide intermediates. In a related reaction, 6-hydroxy-tetralin has been oxidized to 6-hydroxytetral-1-one in 78% yield.[12] The methyl ether can also be used (yield of 6-methoxytetral-1-one, 70%).

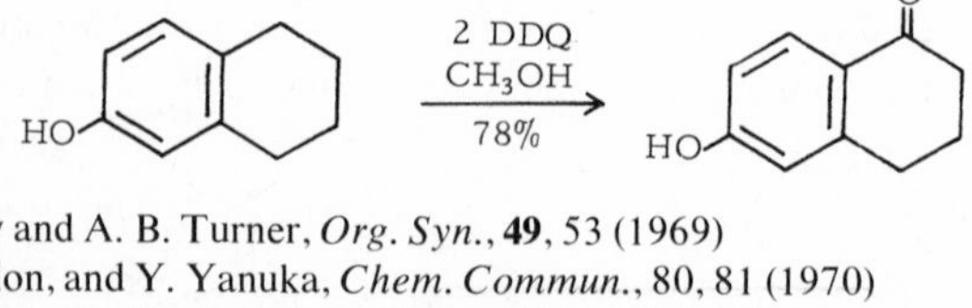

[1]J. W. A. Findlay and A. B. Turner, *Org. Syn.*, **49**, 53 (1969)
[2]S. Sarel, Y. Shalon, and Y. Yanuka, *Chem. Commun.*, 80, 81 (1970)
[3]W. Brown, J. W. A. Findlay, and A. B. Turner, *ibid.*, 10 (1968)
[4]P. Boldt and A. Topp, *Angew. Chem., internat. Ed.*, **9**, 164 (1970)

[5]J. Ackrell and J. A. Edwards, *Chem. Ind.*, 1202 (1970)

[6]E. Vogel and H. D. Roth, *Angew. Chem., internat. Ed.*, **3**, 228 (1964); V. Rautenstrauch, H.-J. Scholl, and E. Vogel, *ibid.*,**7**, 288 (1968)

[7]P. H. Nelson and K. G. Untch, *Tetrahedron Letters*, 4475 (1969)

[8]G. Cardillo, R. Cricchio, and L. Merlini,*ibid.*, 907 (1969)

[9]L. Jurd, *Experientia*, **24**, 858 (1968)

[10]M. Gregson, K. Kurosawa, W. D. Ollis, B. T. Redman, R. J. Roberts, I. O. Sutherland, A. Braga de Oliveira, W. B. Eyton, W. B. Gottlieb, and H. H. Dietrichs, *Chem. Commun.*, 1390 (1968)

[11]I. H. Sadler and J. A. G. Stewart, *ibid.*, 773 (1969)

[12]J. W. A. Findlay and A. B. Turner, *Chem. Ind.*, 158 (1970)

Dichloroketene, **1**, 221–222; **2**, 118.

Preparation. Dichloroketene has been generated *in situ* by dehydrohalogenation of dichloroacetyl chloride with trimethylamine (**1**, 222) and from trichloroacetyl bromide by dehalogenation with copper-activated zinc (**1**, 1287).

$$CCl_3COBr + Zn \longrightarrow Cl_2C{=}C{=}O + ZnBrCl$$

Cycloaddition. An early step in a new synthesis of prostaglandins by Corey *et al.*[1] involved addition of dichloroketene to *endo*-6-methoxybicyclo[3.1.0]-hexene-2 (1) (see **1**, 222, for addition of dichloroketene to cyclopentene itself).

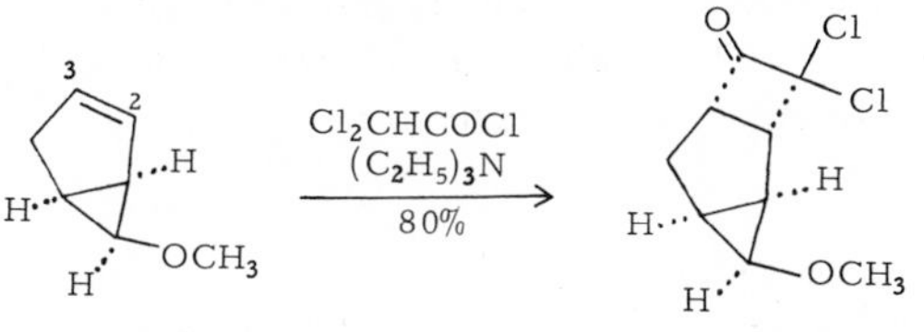

(1)   (2)

The reaction resulted in a position-specific and stereospecific addition to give the tricyclic ketone (2) in 80% yield. The *anti*-relationship of the three- and four-membered rings can be explained on steric grounds. Corey ascribed the position specificity to the fact that attachment of the carbonyl group of dichloroketene to C_3 rather than to C_2 results in a transition state stabilized by the electron-supplying cyclopropyl or *endo*-methoxy group (a or b):

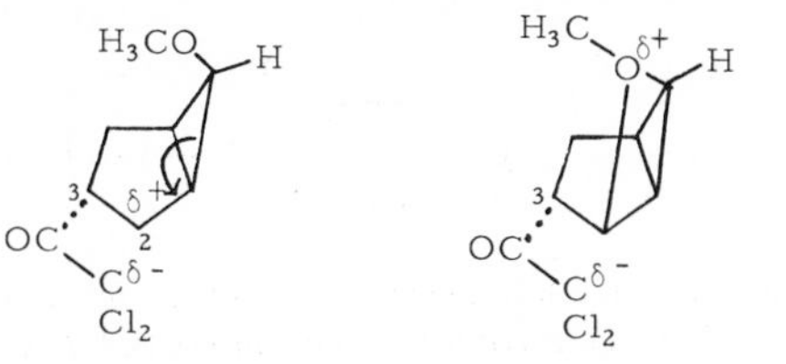

(a)   (b)

The intermediate (2) was converted by several steps into prostaglandins E_2 and $F_{2\alpha}$. However, the necessary opening of the cyclopropanol ring as one step in the

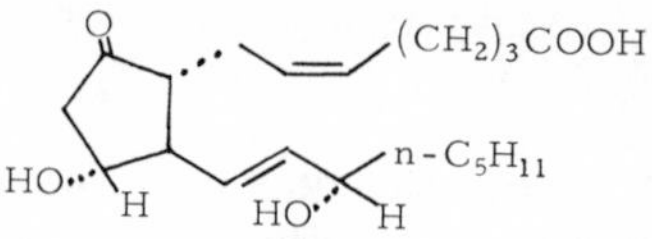

E_2 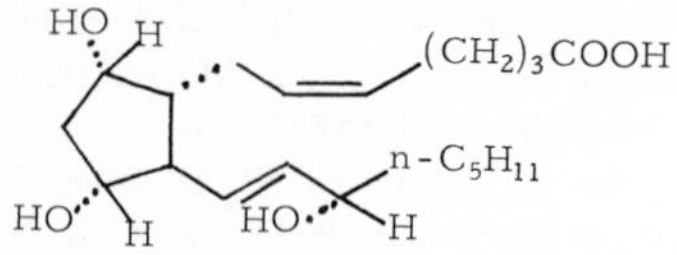 $F_{2\alpha}$

synthesis proceeded in low yield and low selectivity, and this particular approach is not considered as satisfactory as other known syntheses.

Cycloadditions of ketenes with olefins follows orbital symmetry requirements.[2]

Knoche[3] has reported the cycloaddition of dichloroketene to butyne-2 (yield 12%).

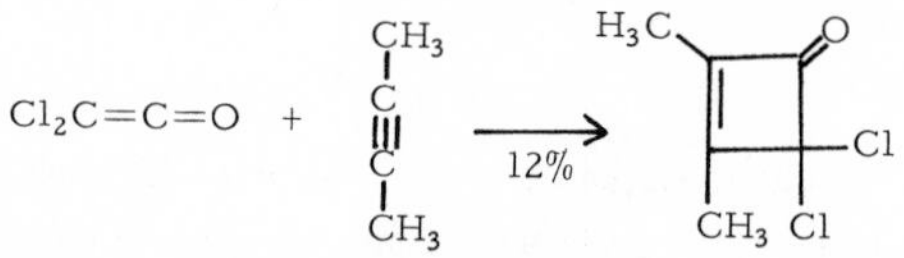

$$Cl_2C=C=O \ + $$

Synthesis of β-lactones. Dichloroketene (generated *in situ* from dichloroacetyl chloride and triethylamine) reacts with aliphatic and aromatic aldehydes in ether at low temperatures ($<20°$) to give α,α-dichloro-β-lactones.[4] Esters of α-keto-carboxylic acids also undergo the reaction.

Examples:

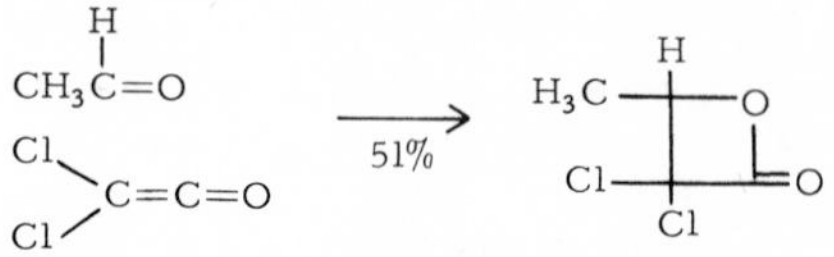

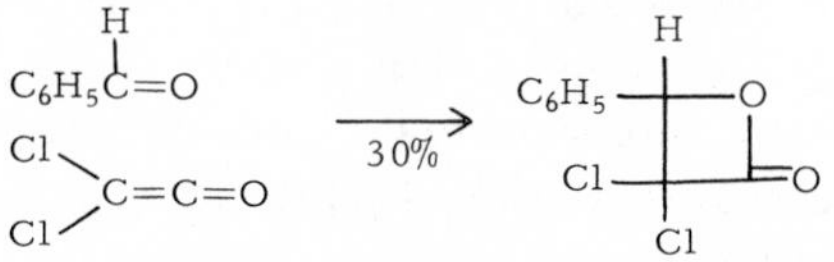

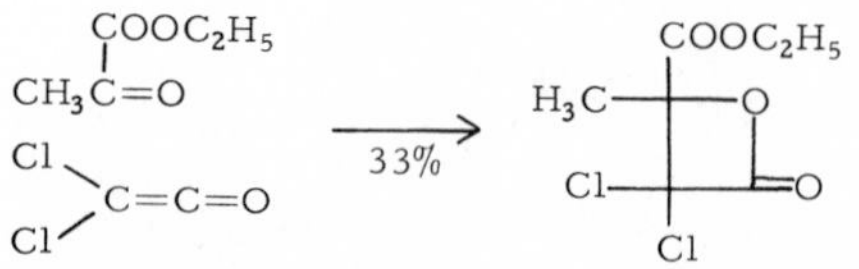

[1] E. J. Corey, Z. Arnold, and J. Hutton, *Tetrahedron Letters*, 307 (1970)
[2] R. B. Woodward and R. Hoffmann, *Angew. Chem., internat. Ed.*, **8**, 847–849 (1969)
[3] H. Knoche, *Ann.*, **722**, 232 (1969)
[4] D. Borrmann and R. Wegler, *Ber.*, **102**, 64 (1969)

Dichloromethyllithium, 1, 223–224; **2,** 119.

Reaction with carbonyl compounds. Köbrich and Werner[1] report that the reagent reacts with acetophenone (1) in THF at −75 to −110° to give the α-chloroepoxide (2). Reaction in the same solvent under reflux leads to α-chloro-α-phenylpropionaldehyde (3).

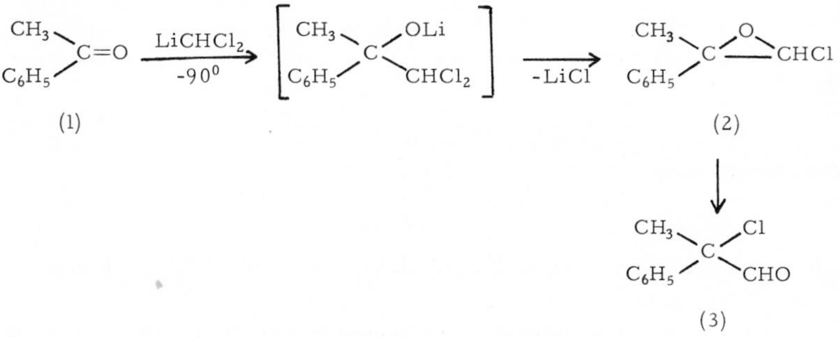

[1]G. Köbrich and W. Werner, *Tetrahedron Letters*, 2181 (1969)

Dicobalt octacarbonyl, 1, 224–228.

Ketone synthesis. Many organomercuric halides react with dicobalt octa-carbonyl in THF solution at room temperature to give symmetrical ketones:

$$2\ RHgBr \xrightarrow[\text{THF}]{Co_2(CO)_8} R\underset{\underset{O}{\|}}{C}R$$

The reaction is applicable to preparation of diaryl ketones provided that bulky groups are not attached to the *ortho*-position and that electrophilic cleavage from mercury is not hindered because of electronic factors. The procedure is also applicable to the preparation of dialkyl ketones; however, skeletal rearrangements of intermediates can lead to isomeric ketones.[1]

Examples:

$$2\ \underline{n}\text{-}C_4H_9HgBr \xrightarrow[42\%]{} (\underline{n}\text{-}C_4H_9)_2C=O$$

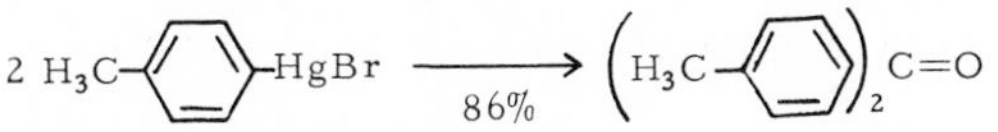

[1]D. Seyferth and R. J. Spohn, *Am. Soc.*, **91,** 3037 (1969)

Di(cobalttetracarbonyl)zinc, 2, 123–124.

Norborandiene dimerization (**2,** 123–124). Schrauzer *et al.*[1] have reported some new catalysts for the dimerization of norbornadiene to "Bisnor-S," which are nearly as effective as di(cobalttetracarbonyl)zinc but more readily accessible. These include $CoBr_2 \cdot 2\ P(C_6H_5)_3$, $CoI_2 \cdot 2\ P(C_6H_5)_3$, and $RhCl[P(C_6H_5)_3]_3$ (**2,** 448–

453; this volume). The first two are prepared by refluxing anhydrous cobalt(II) halides with triphenylphosphine in benzene. These require boron trifluoride etherate as cocatalyst. Unlike the original catalyst, the new catalysts exhibit an induction period,[2] during which they are presumably reduced to the zero valence state, probably to give true catalysts of the type $Li_n \cdot Co-Co \cdot Li_n$ where Li is a ligand.

[1] G. N. Schrauzer, R. K. Y. Ho, and G. Schlesinger, *Tetrahedron Letters*, 543 (1970)
[2] *Caution!* Once initiated, the dimerization is exothermic and external cooling may be required.

Dicyclohexylborane, $\left(\left\langle\bigcirc\right\rangle\right)_2 BH$ Mol. wt. 178.12.

This dialkylborane is prepared in the desired solvent by the reaction of cyclohexene with diborane.[1]

Dicyclohexylborane is comparable to disiamylborane (**1**, 57–59, **2**, 29) in effecting selective hydroboration of dienes[1] and of olefins containing reactive functional groups near the double bond (e.g., ethyl vinylacetate and allyl benzoate[2]).

The reagent is used in a new sequence for the preparation of alkylmercuric halides via hydroboration–mercuration.[3] A terminal olefin is converted into an organoborane by reaction with dicyclohexylborane, and the organoborane is treated with mercuric acetate. An alkyl mercuric acetate is formed, which is then converted into the more easily handled alkylmercuric halide by reaction with a sodium halide and water.

$$\left(\left\langle\bigcirc\right\rangle\right)_2 BH \;+\; R_2C{=}CH_2 \longrightarrow \left(\left\langle\bigcirc\right\rangle\right)_2 BCH_2CHR_2 \xrightarrow[\text{2) NaX}-H_2O]{\text{1) Hg(OAc)}_2}$$

$$\left(\left\langle\bigcirc\right\rangle\right)_2 BOH \;+\; R_2CHCH_2HgX$$

Typical transformations:

$$CH_3CH_2CH{=}CH_2 \longrightarrow CH_3(CH_2)_2CH_2HgCl$$

Diborane was used in early stages of the work, but in the hydroboration of styrene only 80% of the boron appears on the primary carbon atom with 20% on the secondary carbon atom.

[1]G. Zweifel, N. R. Ayyangar, and H. C. Brown, *Am. Soc.*, **85**, 2072 (1963)
[2]H. C. Brown, G. W. Kabalka, and M. W. Rathke, *ibid.*, **89**, 4530 (1967)
[3]R. C. Larock and H. C. Brown, *ibid.*, **92**, 2467 (1970)

Dicyclohexylcarbodiimide (DCC), **1**, 231–236; **2**, 126.

α-Diazoketones. Carboxylic acids (1 mol.) react with diazomethane (1.25 mol.) in the presence of DCC (1 mol.) in ether to give α-diazoketones:[1]

$$\text{RCOOH} + \text{CH}_2\text{N}_2 \xrightarrow{\text{DCC}} \text{R}\overset{\overset{\text{O}}{\|}}{\text{C}}\text{CHN}_2$$

The procedure is useful when the acid chlorides are not accessible.

Intramolecular dehydration of γ-hydroxyketones.[2] When γ-hydroxyketones are heated to about 150° with DCC, water is eliminated with formation of a cyclopropane ring. For example, 5-hydroxy-2-pentanone (1) is transformed into (3) in 80% yield. In the suggested mechanism, (2) is proposed as an intermediate. The reaction is noteworthy because dehydration is accompanied by formation of a carbon–carbon bond.

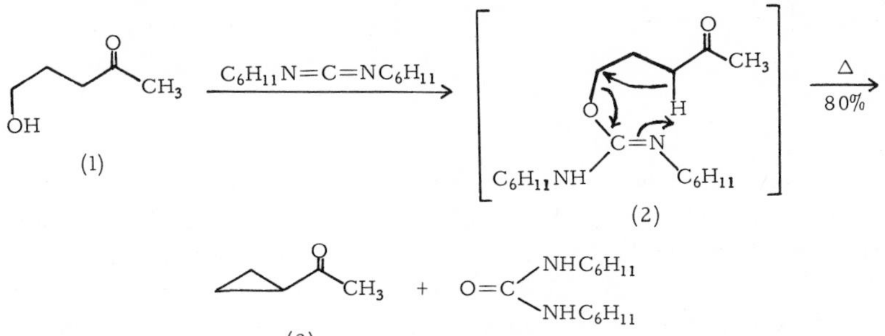

(1)   (2)   (3)

Decarbonylation. Carboxylic acids bearing an adjacent tertiary nitrogen atom undergo decarbonylation when treated with two moles each of dicyclohexylcarbodiimide and *p*-toluenesulfonic acid. Thus the tetrahydro-β-carboline carboxylic acid (I) gives the product (III) by way of (II, see arrows). The reaction

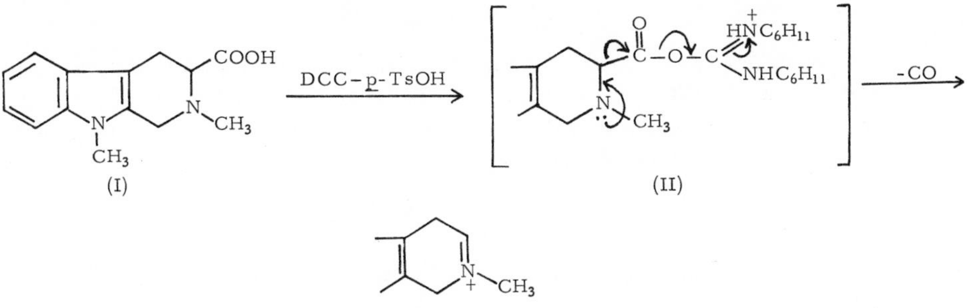

(I)   (II)   (III)

provided a key step in a new synthesis of ajmaline (4) along biogenetic lines from N-methyltryptophane.[3] One intermediate is (1), which contains a carboxyl group at C_5. On treatment with DCC and *p*-TsOH as above, the dihydro-β-carboline (2) is generated. This undergoes spontaneous cyclization to (3), *dl*-desoxyajmalal-B, which was resolved by use of *d*-camphorsulfonic acid. Completion of the synthesis of ajmaline (4) depended on certain relay operations.

(1) (2)

(3) (4)

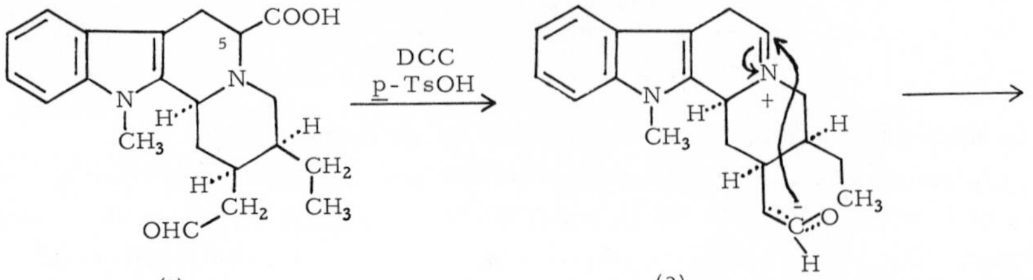

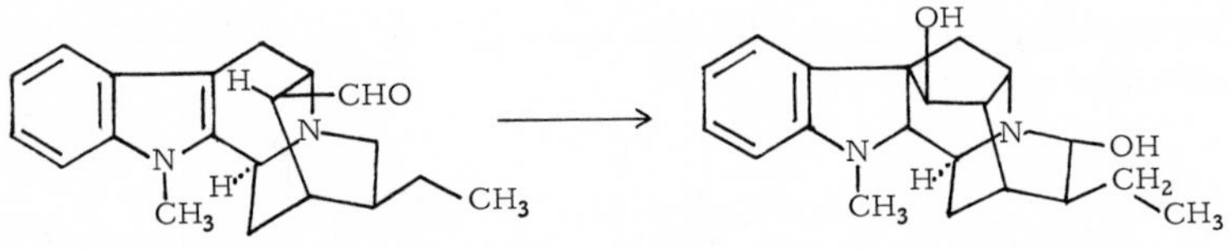

Heterocyclic compounds. DCC and carbonyl hydrazide in equimolar amounts react in refluxing DMF to give 4-cyclohexyl-3-cyclohexylamine-1,2,4-triazoline-5-one (1) in 45–50% yield. The reaction of carbonyl hydrazide (1 equivalent) and DCC (3 equivalents) in DMSO at 100° also yields (1), but the predominant product is 4-cyclohexyl-3,5-biscyclohexylamino-1,2,4-triazole (2). The reaction is applicable to other carbodiimides.[4]

(1) (2)

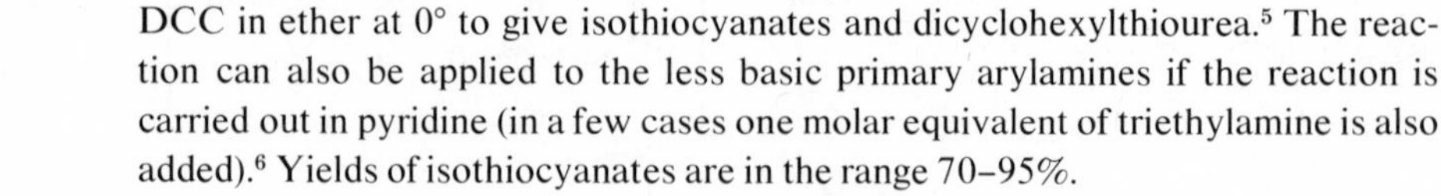

Isothiocyanates. Primary aliphatic amines react with carbon disulfide and DCC in ether at 0° to give isothiocyanates and dicyclohexylthiourea.[5] The reaction can also be applied to the less basic primary arylamines if the reaction is carried out in pyridine (in a few cases one molar equivalent of triethylamine is also added).[6] Yields of isothiocyanates are in the range 70–95%.

$$RNH_2 + CS_2 + C_6H_{11}{-}N{=}C{=}N{-}C_6H_{11} \longrightarrow R{-}N{=}C{=}S + C_6H_{11}NH\overset{S}{\overset{\|}{C}}NHC_6H_{11}$$

[1]D. Hodson, G. Holt, and D. K. Wall, *J. Chem. Soc.*, (C), 971 (1970)

[2]C. Alexandre and F. Rouessac, *Tetrahedron Letters*, 1011 (1970)

[3]E. E. van Tamelen and L. K. Oliver, *Am. Soc.*, **92**, 2136 (1970)

[4]F. Kurzer and M. Wilkinson, *J. Chem. Soc.*, (C), 19 (1970)

[5]J. C. Jochims and A. Seeliger, *Angew. Chem., internat. Ed.*, **6**, 174 (1967)

[6]J. C. Jochims, *Ber.*, **101**, 1746 (1968)

1,1-Diethoxy-2-propyne, 2, 126–127.

Synthesis of α,β-unsaturated aldehydes. 4-Keto-2-*trans*-hexenal (5), an odoriferous substance secreted by some arthropods, has been synthesized by the reaction of the Grignard derivative of 1,1-diethoxy-2-propyne (2) with propanal (1). The product (3) is then reduced specifically with sodium in liquid ammonia to

$$C_2H_5CHO + BrMgC\equiv C-CH(OC_2H_5)_2 \xrightarrow[60\%]{} C_2H_5CH-C\equiv C-CH(OC_2H_5)_2 \xrightarrow[75\%]{Na, NH_3}$$

(1)　　　　　　　　(2)　　　　　　　　　　　　OH　　　　(3)

$$C_2H_5CH-\overset{H}{C}=CCH(OC_2H_5)_2 \xrightarrow[\text{2) p-TsOH, acetone (96\%)}]{\text{1) MnO}_2 \text{ (40\%)}} C_2H_5COC\overset{H}{=}CCHO$$

OH　　H　　　　　　　　　　　　　　　　　　　　　H

(4)　　　　　　　　　　　　　　　　　　　　(5)

1,1-diethoxy-4-hydroxy-2-*trans*-hexene (4). The synthesis is completed by oxidation with active manganese dioxide and acid hydrolysis in aqueous acetone.[1]

[1]J. P. Ward and D. A. van Dorp, *Rec. trav.*, **88**, 989 (1969)

2,2-Diethoxyvinylidenetriphenylphosphorane, $(C_2H_5O)_2C\!=\!C\!=\!P(C_6H_5)_3$. Mol. wt. 376.42, m.p. 83°.

Preparation.[1]

$$(C_6H_5)_3P\!=\!CH-COOC_2H_5 \xrightarrow{[(C_2H_5)_3O]^+BF_4^-} \left[\underset{(C_6H_5)_3P}{\overset{H}{\diagup}}C\!=\!C\underset{OC_2H_5}{\overset{OC_2H_5}{\diagdown}} \right]BF_4^- \xrightarrow{NaNH_2}$$

$$(C_6H_5)_3P\!=\!C\!=\!C\underset{OC_2H_5}{\overset{OC_2H_5}{\diagup\diagdown}}$$

(1)

Synthesis of 1,3-diketo-4-pentenes.[2] This phosphorus ylide (1) adds to an aryl ketone to give a diethoxyallene (2) which immediately dimerizes (3):

$$(1) \xrightarrow{Ar_2C\!=\!O} [Ar_2C\!=\!C\!=\!C(OC_2H_5)_2] \longrightarrow$$

(2)

(3)

The reaction follows a different course with ketones which have an α-methylene group. Stable ylides (4) are formed by Michael addition and loss of ethanol. These react with aldehydes to give enol ethers of 1,3-diketo-4-pentenes (5), which are cleaved by acids to the free dicarbonyl compounds (6), stable in the reaction conditions as the enols (7). The method appears to be general, and yields are reasonable.

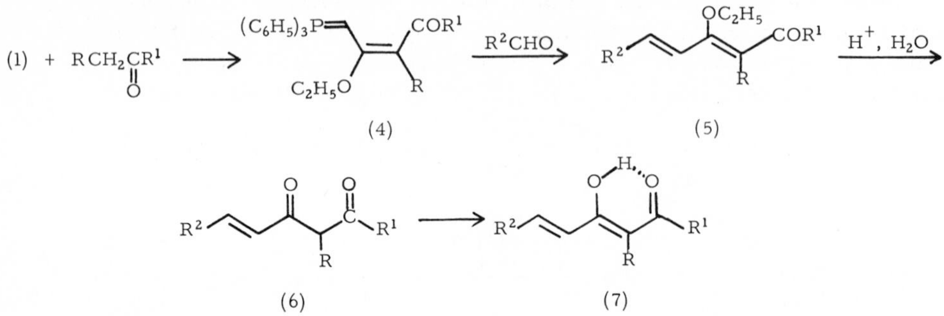

(1) + RCH_2CR^1 ⟶

(4)

R^2CHO →

(5)

H^+, H_2O →

(6)   (7)

[1]H. J. Bestmann, R. Saalfrank, and J. P. Snyder, *Angew. Chem., internat. Ed.*, **8**, 216 (1969)
[2]H.-J. Bestmann and R. W. Saalfrank, *ibid.*, **9**, 367 (1970)

Diethyl allylthiomethylphosphonate, $CH_2{=}CHCH_2SCH_2PO(OC_2H_5)_2$. Mol. wt. 224.26, b.p. 77–78°/0.11 mm.

Preparation. The reagent is prepared[1] in 67% yield by the Arbusov reaction of triethyl phosphite and allyl chloromethyl sulfide.[2]

Conversion of ketones into unsaturated aldehydes.[1] The reagent (1) is converted into the anion (2) by *sec*-butyllithium in THF at −78°. This anion reacts with a ketone to give an allyl vinyl sulfide (3) in yields generally in the range

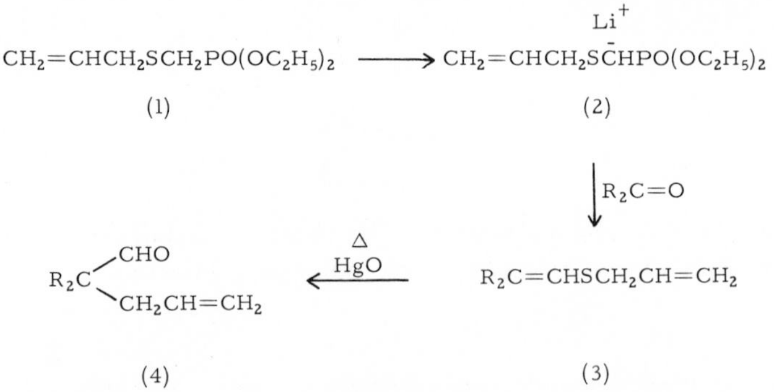

$$CH_2{=}CHCH_2SCH_2PO(OC_2H_5)_2 \longrightarrow \overset{Li^+}{CH_2{=}CHCH_2S\bar{C}HPO(OC_2H_5)_2}$$

(1)   (2)

$\downarrow R_2C{=}O$

$$R_2C\overset{CHO}{\underset{CH_2CH=CH_2}{\big\langle}} \xleftarrow[\Delta]{HgO} R_2C{=}CHSCH_2CH{=}CH_2$$

(4)   (3)

50–80%. When (3) is heated at 160–180° to effect a Claisen rearrangement, only intractable tars are obtained but, if mercuric oxide is present to convert the intermediate thioaldehyde into the corresponding aldehyde, the α-allyl aldehyde (4) is obtained in 40–83% yield. This technique should broaden the scope of the thio-Claisen rearrangement.

The products (4) derived from alicyclic ketones are useful for synthesis of spiro compounds. For example, the product (5) obtained from cyclohexanone has been transformed in several steps into the spiro enone (6) and into the spiro diketone (7).

(5)   (6)

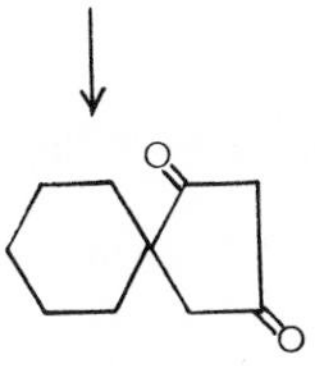

(7)

[1]E. J. Corey and J. I. Shulman, *Am. Soc.*, **92**, 5522 (1970)
[2]L. A. Walter, L. H. Goodson, and R. T. Fosbinder, *ibid.*, **67**, 655 (1945)

Diethyl carbonate, 1, 247–248; **2**, 129.

Carboethoxylation (**2**, 129; this volume, *see* **Potassium amide**).

Diethyl(2-chloro-1,1,2-trifluoroethyl)amine [N-(2-Chloro-1,1,2-trifluoroethyl)di-ethylamine], 1, 249; **2**, 130.

Fluorination. 11α-Hydroxy-19-norsteroids react with the reagent to give 11β-fluoro-19-norsteroids in yields of up to 45%.[1] The reaction is believed to involve attack on a reactive intermediate (a) by the fluoride ion eliminated in the formation of (a):

$$(C_2H_5)_2NCF_2CHClF \xrightarrow{-F^-} (C_2H_5)_2\overset{+}{N}=\overset{F}{\underset{}{C}}CHClF \xrightarrow{ROH} (C_2H_5)_2N-\overset{F}{\underset{OR}{C}}CHClF \xrightarrow{-F^-}$$

$$(C_2H_5)_2\overset{+}{N}=C-CHClF \longrightarrow RF + (C_2H_5)_2NCOCHClF$$

(a)

In that case chloride or bromide ions should compete with fluoride ion. And indeed reaction of 11α-hydroxy-19-norsteroids with the reagent in the presence of excess lithium chloride in THF gives 11β-chlorosteroids in good yield. 11β-Bromo-steroids are obtained in lower yield by use of lithium bromide in methylene chloride. Note that the reagent reacts with 11α-hydroxy-10-methylsteroids to give mainly 9(11)-enes.

The reagent was used successfully for conversion of 1-hydroxybicyclo[2.2.2]-octanes (1) into 1-fluorobicyclooctanes (2).[2] An excess of the reagent was used as solvent when the starting material was a solid. Temperatures of 60–140° were employed.

(1)   (2)

[1]E. J. Bailey, H. Fazakerley, M. E. Hill, G. E. Newall, G. H. Phillipps, L. Stephenson, and A. Tulley, *Chem. Commun.*, 106 (1970)
[2]J. Kopecký, J. Šmejkal, and M. Hudlický, *Chem. Ind.*, 271 (1969)

Diethyl cyanomethylphosphonate, 1, 250; **2**, 130–131. Replace ref. 7(b), **2**, 131, by G. R. Pettit, C. L. Herald, and J. P. Yardley, *J. Org.*, **35**, 1389 (1970)

Diethyl β-(cyclohexylimino)ethylphosphonate, 2, 131–132. Mol. wt. 261.30.

Formylolefination. The definitive paper has now been published.[1]

[1]W. Nagata and Y. Hayase, *J. Chem. Soc.*, (C), 460 (1969)

Diethyl methylenemalonate, $H_2C{=}C(COOC_2H_5)_2$. Mol. wt. 172.18, b.p. 96–98°/12 mm.

Preparation. The reagent is prepared by the reaction of paraformaldehyde and diethyl malonate in glacial acetic acid catalyzed by copper acetate and potassium acetate (yield 46%).[1] It is obtained as an oil which polymerizes at room temperature to a wax. The monomer can be recovered by distillation in vacuum.

Protection of SH groups.[2] The reagent adds readily the SH group of cysteine or glutathione to give a β,β-diethoxycarbonylethyl derivative. The derivatives are reconverted into thiols by β-elimination, preferably by a 1 N solution of potassium hydroxide in ethanol (20°, 5–10 min.).

$$\begin{array}{c}H_2CSH\\|\\H_2N{-}CHCOOH\end{array} + H_2C{=}C(COOC_2H_5)_2 \underset{KOH\,(80\%)}{\overset{74\%}{\rightleftharpoons}} \begin{array}{c}H_2C{-}SCH_2CH(COOC_2H_5)_2\\|\\H_2N\overset{|}{C}HCOOH\end{array}$$

This new protective group has been used in a synthesis of glutathione.[3]

[1]G. B. Bachman and H. A. Tanner, *J. Org.*, **4**, 493 (1939)
[2]T. Wieland and A. Sieber, *Ann.*, **722**, 222 (1969)
[3]T. Wieland and A. Sieber, *ibid.*, **727**, 121 (1969)

Diethyl methylthiomethylphosphonate, $CH_3SCH_2P(=O)(OC_2H_5)_2$. Mol. wt., 198.22; b.p. 70–72°/0.2 mm. Aldrich supplies diethyl ethylthiomethylphosphonate, $CH_3CH_2SCH_2P(=O)(OC_2H_5)_2$.

Preparation.[1] The reagent is prepared by the reaction of methylthiomethyl chloride with triethyl phosphite (110°, 6 hrs.).

Synthesis of ketones.[2] The reagent is first metalated by 1 equivalent of *n*-butyllithium in THF at −70°; this derivative is readily alkylated (2). The resulting

phosphonate is again metalated (3) and then allowed to react with an aldehyde or ketone to give an adduct (4), which when heated at 50° decomposes to the vinyl sulfide (5) with loss of diethyl phosphate anion. Previously, vinyl sulfides have been hydrolyzed to ketones (or aldehydes) by use of mercuric chloride in ethanol and 38% hydrochloric acid (steam bath, 1 hr.).[3] In the present work, hydrolysis was effected by 2 equivalents of mercuric chloride in aqueous acetonitrile (25–80°, 6–41 hrs.). The resulting ketones (6) were obtained in 75–92% yield. Overall yields in the synthesis in the two cases reported were 71% and 66%. This route provides a convenient synthesis of ketones of the type RCOR′, when R originates from the alkylating agent and R′ is derived from the aldehyde or ketone.

[1]M. Green, *J. Chem. Soc.*, 1324 (1963)
[2]E. J. Corey and J. I. Shulman, *J. Org.*, **35**, 777 (1970)
[3]J. H. S. Weiland and J. F. Arens, *Rec. trav.*, **79**, 1293 (1960)

Diethyl phenyl orthoformate, $C_6H_5OCH(OC_2H_5)_2$. Mol. wt. 196.24, b.p. 103–104.5°/10 mm. Supplier: Aldrich.

The reagent is prepared in 67% yield by acid-catalyzed ester interchange between phenol and triethyl orthoformate.

The reagent reacts with alkyl and aryl Grignard reagents under mild conditions to give the diethylacetal of an aldehyde in good yield:

$$RMgBr + C_6H_5OCH(OC_2H_5)_2 \longrightarrow RCH(OC_2H_5)_2 + C_6H_5OMgBr$$

The reaction thus effects conversion of a bromide into an aldehyde.[1]

[1]H. Stetter and E. Reske, *Ber.*, **103**, 643 (1970)

Diethyl phosphorochloridate, $(C_2H_5O)_2POCl$, **1**, 248 (formerly known as Diethyl chlorophosphite).

Olefin synthesis from α,β-unsaturated ketones. Ireland and Pfister[1] have extended the procedure of Kenner and Williams (**1**, 248, ref. 2) for deoxygenation of phenols to conversion of α,β-unsaturated ketones into olefins. For example, the α,β-unsaturated ketone (1) was reduced by lithium–ammonia to give an enolate anion which reacted with diethyl phosphorochloridate to give the phosphate ester (2) in 56% yield. This ester was reduced in high yield by lithium in a mixture of ethylamine and *t*-butanol to the olefin (3). It is noteworthy that only one olefin is formed. Actually the conversion of (1) into (3) can be carried out in 50% yield without isolation of the diethyl enol phosphate.

The sequence can also be applied to the enolate anion formed by conjugate addition of organometallic reagents to α,β-unsaturated ketones. Thus addition of diethyl phosphorochloridate to a mixture of Δ^4-cholestene-3-one (4) and dimethyl-copperlithium gives the diethyl enol phosphate (5) in 55% yield. Reduction of the ester gives the olefin (6) in high yield.

The method is related to that of Fetizon for preparation of Δ^2-steroids by lithium–ammonia reduction of diethyl enol phosphates generated from 2-bromo-3-ketones (*see* **Triethyl phosphite**, this volume).

[1]R. E. Ireland and G. Pfister, *Tetrahedron Letters*, 2145 (1969)

N,N-Diethyl-1-propynylamine, **2**, 133–134.

Oxidation of hydroxy steroids. Hydroxy steroids are oxidized to keto steroids in about 60% yield by the combination of N,N-diethyl-1-propynylamine–DMSO catalyzed by phosphoric acid. Benzene is used as solvent.[1] In later work[2] the combination of diphenylketene-*p*-tolylimine and DMSO was used with somewhat improved yields (compare Pfitzner-Moffatt reagent).

[1]R. E. Harmon, C. V. Zenarosa, and S. K. Gupta, *Chem. Commun.*, 537 (1969)
[2]R. E. Harmon, C. V. Zenarosa, and S. K. Gupta, *Chem. Ind.*, 1428 (1969)

Dihydropyrane, 1, 256–257.

Allenic alcohols. Landor *et al.*[1] have developed a new method for the preparation of allenic alcohols from prop-1-yne-3-ols, for example (1). This is converted into the tetrahydropyranyl ether (2) and then treated with ethylmagnesium

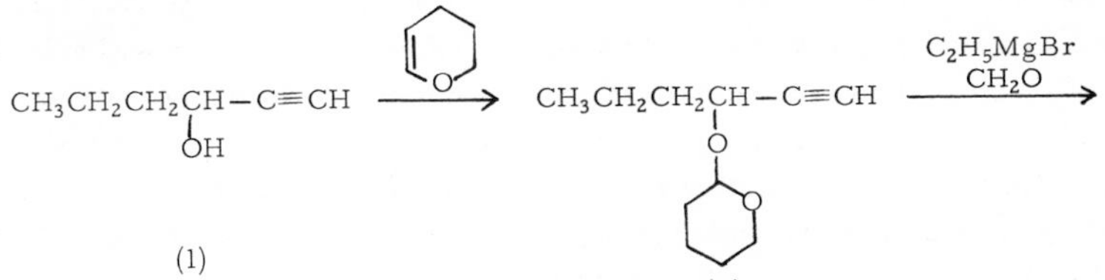

$$CH_3CH_2CH_2\underset{\overset{|}{OH}}{CH}-C\equiv CH \quad \xrightarrow{\quad \text{(dihydropyran)} \quad} \quad CH_3CH_2CH_2\underset{\overset{|}{O}}{CH}-C\equiv CH \quad \xrightarrow[\text{CH}_2\text{O}]{\text{C}_2\text{H}_5\text{MgBr}}$$

(1)

(2)

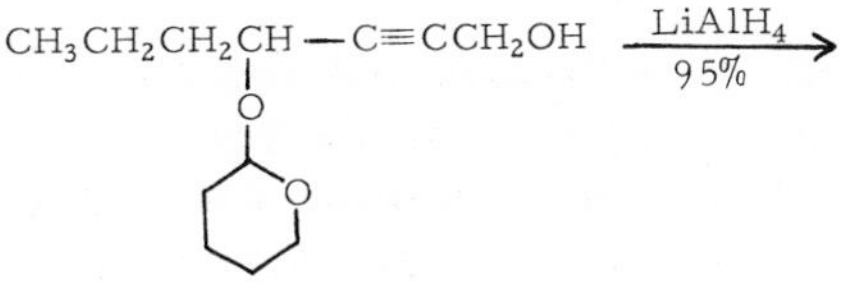

$$CH_3CH_2CH_2\underset{\overset{|}{O}}{CH}-C\equiv CCH_2OH \quad \xrightarrow[95\%]{\text{LiAlH}_4}$$

(3)

$$\left[CH_3CH_2CH_2\underset{O}{CH}-C\equiv CCH_2 \quad \underset{H}{Al}\;H \right] \quad \longrightarrow \quad CH_3CH_2CH_2C\!=\!C\!=\!CHCH_2OH$$

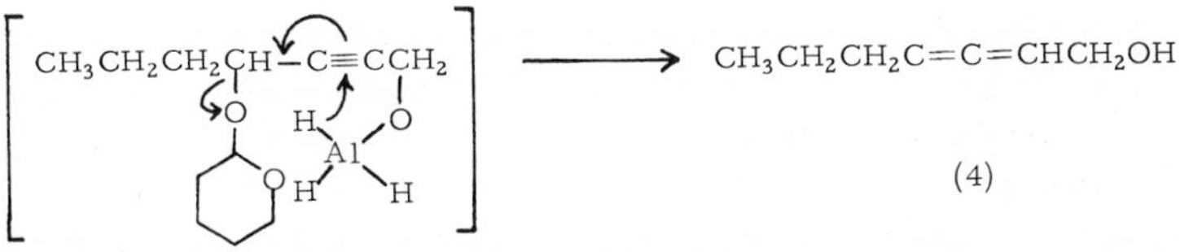

(4)

bromide and gaseous formaldehyde to give the 2-yne-1,4-diol derivative (3). Reduction of (3) with an excess of lithium aluminum hydride in ether gives the allenic alcohol (4) in high yield with elimination of the tetrahydropyranyloxy group. (Compare reduction of propargylic alcohols with LiAlH₄, 1, 593–594). The procedure is also applicable for synthesis of terminal allenic alcohols, >C(OH)CH=C=CH₂, but not for synthesis of cumulenes, which are reduced by lithium aluminum hydride to allenic and acetylenic alcohols.

[1] J. S. Cowie, P. D. Landor, and S. R. Landor, *Chem. Commun.*, 541 (1969)

Diimide, 1, 257–258; **2,** 139.

Generation in situ. Diimide has been generated from hydroxylamine-O-sulfonic acid by treatment with 0.1 N NaOH at 20°, but the decomposition is slow (1, 482). Dürckheimer[1] found that the formation of diimide occurs rapidly if a combination of hydroxylamine and hydroxylamine-O-sulfonic acid is used:

$$NH_2OH + NH_2OSO_3H \xrightarrow{2\ OH^-} NH_2NHOH + SO_4^= + 2\ H_2O$$

$$\downarrow -H_2O$$

$$NH=NH$$

Water is the best medium, but aqueous methanol can also be used. By use of this procedure fumaric acid was reduced to succinic acid in 90% yield at 50° for 2 hours. Use of hydroxylamine-O-sulfonic acid alone required 16 hrs., and the yield of reduced acid was only 42%.

Dideuteriodiimide. Berson et al.[2] generated dideuteriodiimide (N_2D_2) from potassium azodicarboxylate and acetic acid-O-*d* in the solvent methanol-O-*d*. The labeled acetic acid was prepared from acetic anhydride and deuterium oxide. The methanol-O-*d* was prepared by the method of Streitwieser et al.[3] Berson et al. used the reagent for deuteration of double bonds and found that no scrambling occurred. Generally there is some 10–15% of starting olefin which has not reacted, but the unsaturated compound can be washed out using aqueous silver nitrate solution.

Reduction of aldehydes and ketones. Although diimide is generally used for reduction of symmetrical double bonds, van Tamelen et al.[4] found that some aliphatic and aromatic ketones are reduced in low yield. Aliphatic aldehydes were found to be reduced in low yield, but benzaldehyde was reduced to benzyl alcohol in 56% yield. This reaction was found to be general by Curry et al.[5] Presumably the reduction only occurs readily when the positive charge on the carbon atom of the carbonyl group is delocalized because of the aromatic ring. Obviously hydrazine cannot be used as the source of diimide; potassium azodicarboxylate was used. *Caution!* On one occasion the salt exploded violently when exposed to strong sunlight for thirty minutes.

Reduction of allenes. Cyclic and acyclic allenes are reduced stereospecifically by diimide (generated from hydrazine, H_2O_2, $CuSO_4$) to *cis*-alkenes.[6]
Examples:

1,2-Cyclononadiene → *cis*-Cyclononene (100%)
1,2-Cyclodecadiene → *cis*-Cyclodecene (24%)
1,2-Nonadiene → *cis*-2-Nonene (17%)
3-Ethyl-1,2-pentadiene → 3-Ethyl-2-pentene (16%)

Selective reduction. Use of reagent, generated by cupric ion-catalyzed air oxidation of hydrazine, for the preparation of *cis*-cyclododecene by the selective reduction of *cis,trans,trans*-1,5,9-cyclododecatriene, the product of trimerization of butadiene,[7,8] is described by Ohno and Okamoto[9] (see also **1**, 258, ref. 10).

[1]W. Dürckheimer, *Ann.*, **721**, 240 (1969)

[2]J. A. Berson, M. S. Poonian, and W. J. Libbey, *Am. Soc.*, **91**, 5567 (1969)

[3]A. Streitwieser, Jr., L. Verbit, and P. Stang, *J. Org.*, **29**, 3706 (1964)

[4]E. E. van Tamelen, M. Davis, and M. F. Deem, *Chem. Commun.*, 71 (1965)

[5]D. C. Curry, B. C. Uff, and N. D. Ward, *J. Chem. Soc.*, (C), 1120 (1967)

[6]G. Nagendrappa and D. Devaprabhakara, *Tetrahedron Letters*, 4243 (1970)

[7]G. Wilke, *Angew. Chem.*, **75**, 10 (1963)

[8]Available from Aldrich.

[9]M. Ohno and M. Okamoto, *Org. Syn.*, **49**, 30 (1969)

Diiron nonacarbonyl (Iron ennacarbonyl), 1, 259–260; **2**, 139–140.

Trimethylenemethane. The highly unstable trimethylenemethane (1) has been obtained as the π-complex (3) by reduction of 3-chloro-(2-chloromethyl-1)-

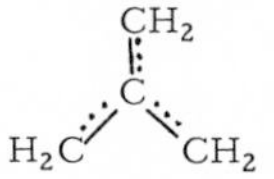

(1)

propene with excess diiron nonacarbonyl (2).[1] A convenient synthesis of such complexes involves the reaction of the more accessible methylenecyclopropanes

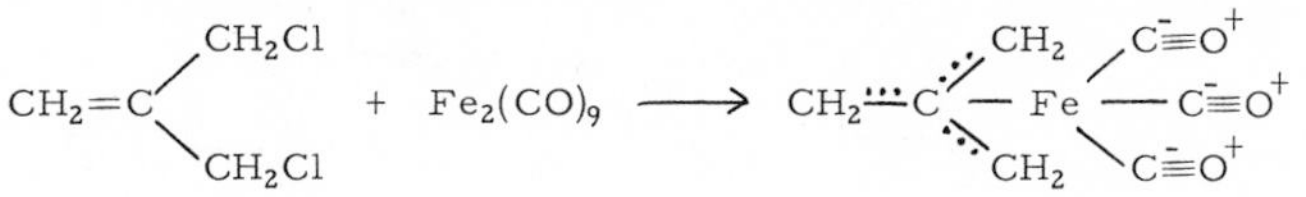

with diiron nonacarbonyl. Thus the reaction of 1-methylene-2-phenylcyclopropane (4) with an equivalent amount of diiron nonacarbonyl gives the complex (5) in 40% yield.[2]

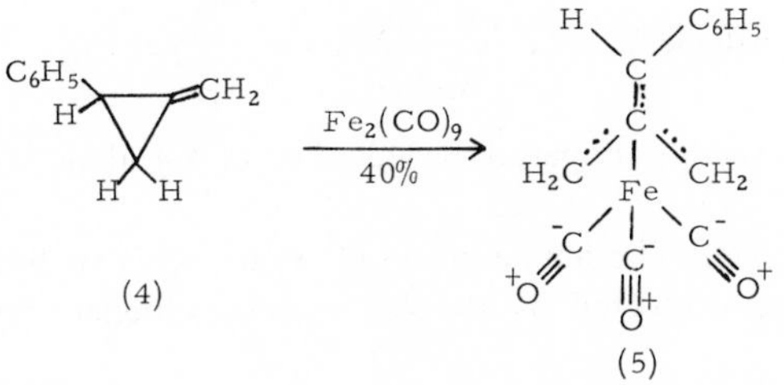

Cyclobutadieneiron tricarbonyl (**2**, 140, ref. 5). The preparation of this complex has now been published.[3]

[1]G. F. Emerson, K. Ehrlich, W. P. Giering, and P. C. Lautenber, *Am. Soc.*, **88**, 3172 (1966)

[2]R. Noyori, T. Nishimura, and H. Takaya, *Chem. Commun.*, 89 (1969)

[3]R. Pettit and J. Henery, *Org. Syn.*, **50**, 21 (1970)

Diisobutylaluminum hydride, 1, 260–262; **2**, 140–142.

Selective reduction of 2-ene-1,4-diones and 2-ene-1-ones. This metal hydride is excellent for reduction of 1,4-enediones to the corresponding unsaturated alcohols (with most reducing agents the double bond is reduced as well).[1] Thus (1) is reduced to the unsaturated compound (2) in benzene in 79% yield, and (3) is reduced to (4) in 92% yield. The cyclopentenone (5) was reduced to (6) in 98% yield. Use

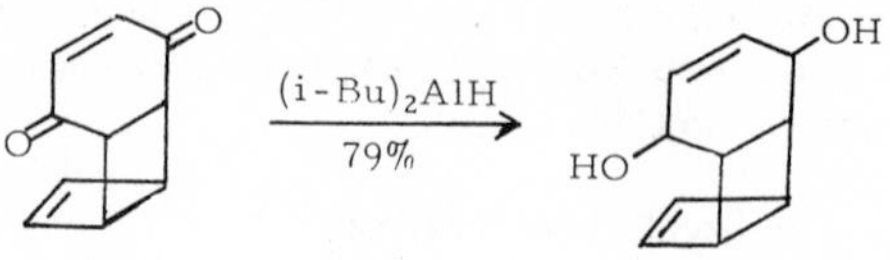

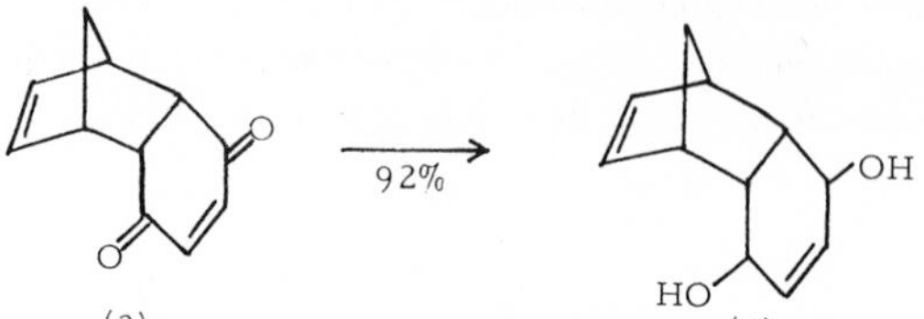

of aluminum hydride in this reduction lowered the yield of (6) to 86%.

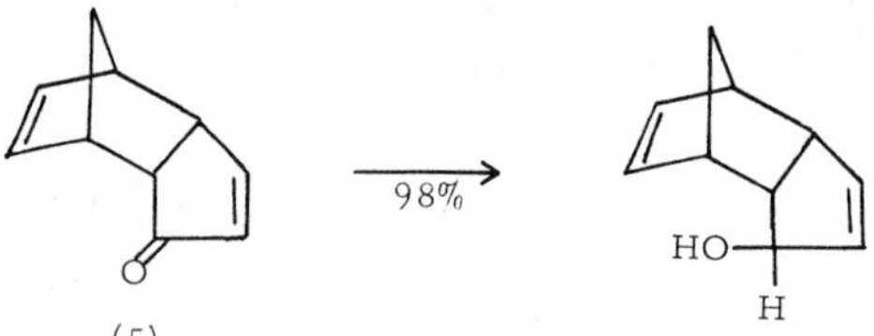

[1]K. E. Wilson, R. T. Seidner, and S. Masamune, *Chem. Commun.*, 213 (1970)

4,4′-Dimethoxybenzhydrol, 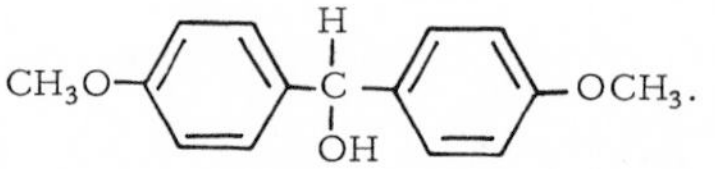   Mol. wt. 244.28, m.p. 72°.

The reagent is readily prepared by reduction of 4,4′-dimethoxybenzophenone with $NaBH_4$.

Amide protection.[1] The amide group of amino acids such as asparagine and glutamine can be protected as the 4,4′-dimethoxybenzhydryl derivative (2),

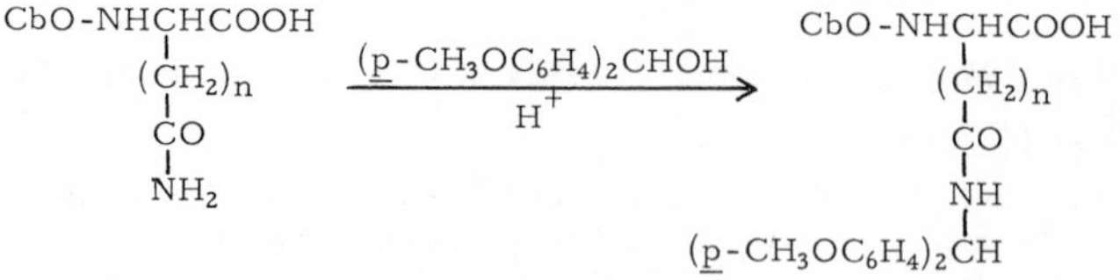

obtained by the acid-catalyzed reaction of benzoyloxycarbonylglutamine (1, $n = 1$) or -asparagine (1, $n = 2$) with the benzhydrol. The protective group is stable to catalytic hydrogenation and can be cleaved by trifluoroacetic acid/ anisole.

[1]W. König and R. Geiger, *Ber.*, **103**, 2041 (1970)

(Dimethoxymethyl)trimethoxysilane, $(CH_3O)_3SiCH(OCH_3)_2$. Mol. wt. 196.28, b.p. 82–84°/20 mm.

Preparation.[1] The reagent (1) is prepared by the reaction of hexachloro-disilane and a mixture of trimethyl orthoformate and methanol:

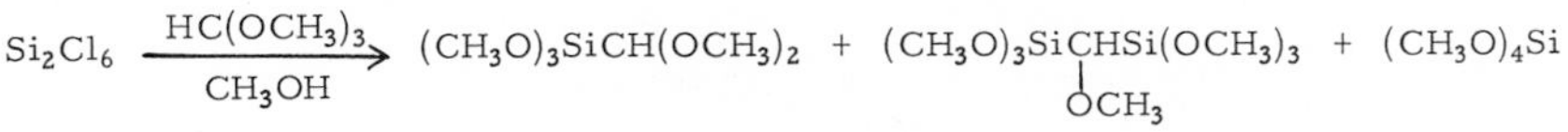

$$Si_2Cl_6 \xrightarrow[CH_3OH]{HC(OCH_3)_3} (CH_3O)_3SiCH(OCH_3)_2 + (CH_3O)_3SiCHSi(OCH_3)_3 + (CH_3O)_4Si$$

with $\underset{OCH_3}{|}$ on the second product.

$$\text{(1)} \quad 20\% \qquad\qquad\qquad 30\%$$

Methoxycarbene.[1] The reagent (1) decomposes when heated to give methoxy-carbene. Thus thermolysis in the presence of tetramethylethylene gives an essentially quantitative yield of 1-methoxy-2,2,3,3-tetramethylcyclopropane (2).

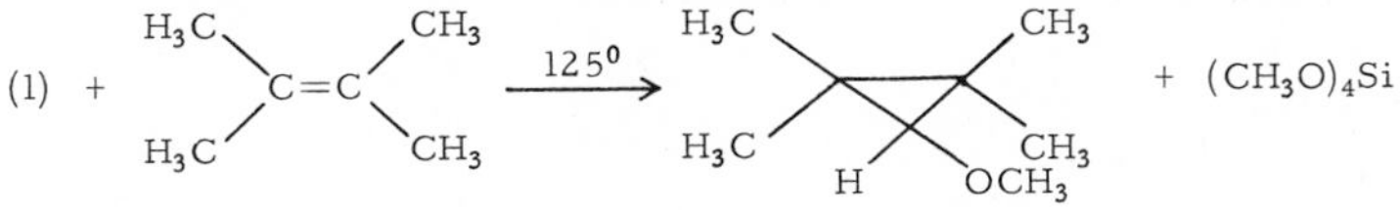

$$\text{(2)}$$

Similarly decomposition of (1) in the presence of cyclohexene gives 7-methoxy-norcarane in good yield.

[1]W. H. Atwell, D. R. Weyenberg, and J. G. Uhlmann, *Am. Soc.*, **91**, 2025 (1969)

5,5-Dimethoxy-1,2,3,4-tetrachlorocyclopentadiene, 1, 270; **2**, 143–144.

N-Ethoxycarbonylazepine (1) undergoes a [4+2] reaction with this diene, the central double bond reacting as a dienophile.[1] Use of cyclopentadiene gave only

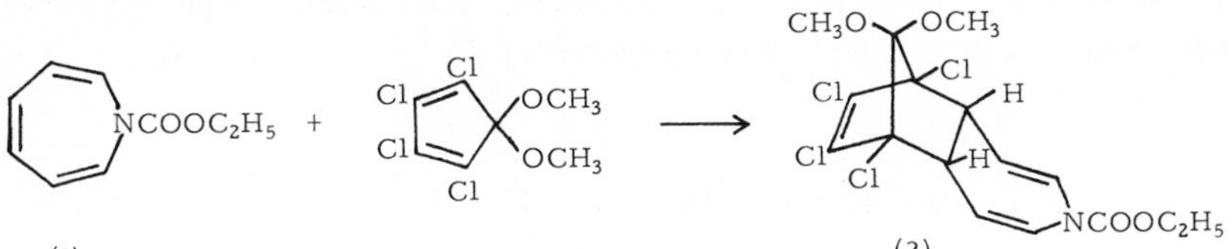

cyclopentadiene dimer and azepine dimers. Tetracyclone also failed to react.

[1]J. R. Wiseman and B. P. Chong, *Tetrahedron Letters*, 1619 (1969)

Dimethyl acetylenedicarboxylate, 1, 272–273; **2**, 145–146.

Synthesis of benzene derivatives.[1] An acetoxy-1,3-diene (conveniently gener-ated by heating an α,β-unsaturated aldehyde or ketone with isopropenyl acetate, *p*-toluenesulfonic acid catalysis) undergoes the Diels-Alder reaction with dimethyl acetylenedicarboxylate (1.5 equivalents) to give a dimethyl phthalate derivative in good yield.

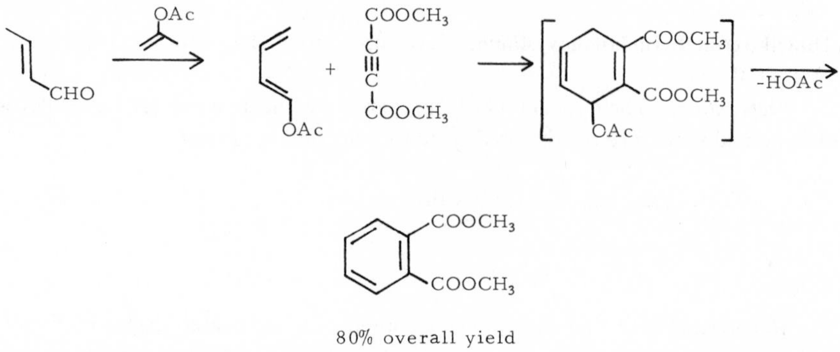

80% overall yield

With chloromaleic anhydride[2] a phthalic anhydride derivative is obtained; the intermediate adduct in this case must lose HOAc and HCl.

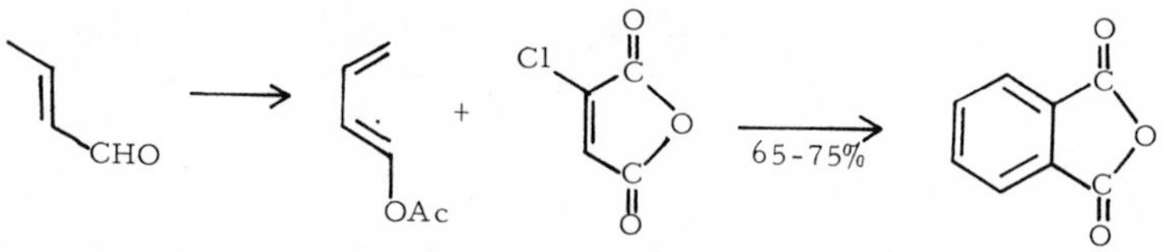

[1]J. Wolinsky and R. B. Login, *J. Org.*, **35**, 3205 (1970)
[2]Mol. wt. 132.50, n^{20}D 1.4980; supplier: Aldrich.

N,N'-Dimethyl-2-allyl-1,3,2-diazaphospholidine 2-oxide (2). Mol. wt. 174.19, b.p. 98.5–100.5°/0.45 mm.

Preparation. Allyl chloride is condensed with phosphorus trichloride in the presence of aluminum chloride to give a complex, which is decomposed by water to allylphosphoric dichloride (1).[1] This product reacts with *sym*-dimethylethyl-enediamine (triethylamine) to give the reagent (2).[2]

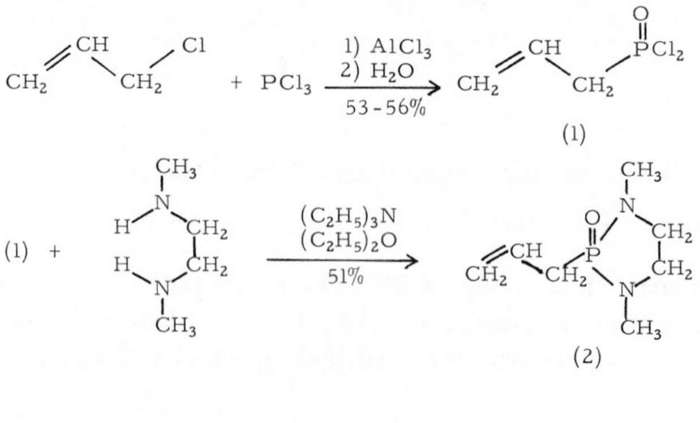

Phosphonamide synthesis of 1,3-dienes. On extension of the phosphonamide synthesis of olefins (**2**, 280–281) to 1,3-dienes, Corey[2] initially experienced some difficulties. Thus reaction of acetone with the anion of N,N,N′,N′-tetramethyl-allylphosphonodiamide (3) gave only the γ-adduct (4), which is stable to thermoly-

(3)   (4)

sis, rather than the α-adduct. Various expedients were unsuccessful in controlling the position of electrophilic attack: addition of Lewis acids, use of organozinc or organocadmium derivatives of (3). Only partial success was achieved by using the Grignard derivative of (3). Corey then achieved more successful results by reducing the bulkiness of the amide and used (2) rather than (3). Thus the reaction of (2) with acetone gave the desired α-adduct (5) in 75% yield. This modification

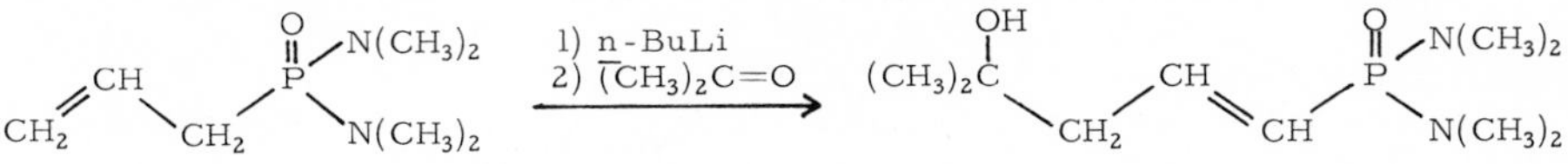

(5)   (6)

gives satisfactory yields with methyl ethyl ketone, benzaldehyde, cyclohexanone. However, if the carbonyl compound contains bulky groups (e.g., pivaldehyde), condensation takes place to a marked extent at the undesired γ-position.

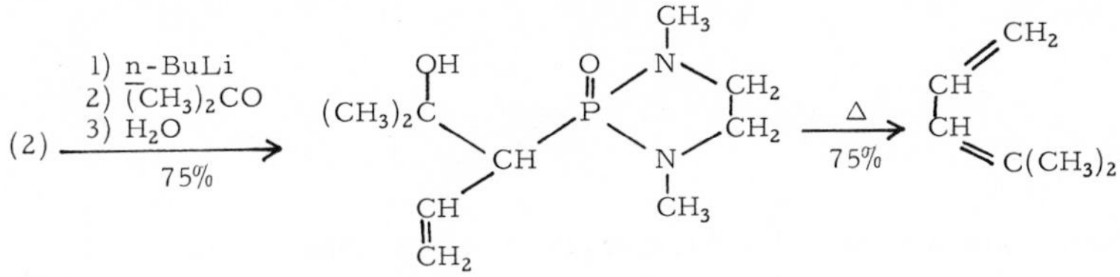

[1]A. M. Kinnear and E. A. Perren, *J. Chem. Soc.*, 3437 (1952)
[2]E. J. Corey and D. E. Cane, *J. Org.*, **34**, 3053 (1969)

(Dimethylamino)phenyloxosulfonium methylide. Mol. wt. 183.27.

$$C_6H_5-\overset{\overset{O}{\|}}{\underset{N(CH_3)_2}{S}}=CH_2 \longleftrightarrow C_6H_5-\overset{\overset{O}{\|}}{\underset{N(CH_3)_2}{S}}-CH_2^-$$

Preparation.[1] This stable ylide is prepared in solution from methyl phenyl sulfoxide[2] by the following sequence in 68% overall yield.

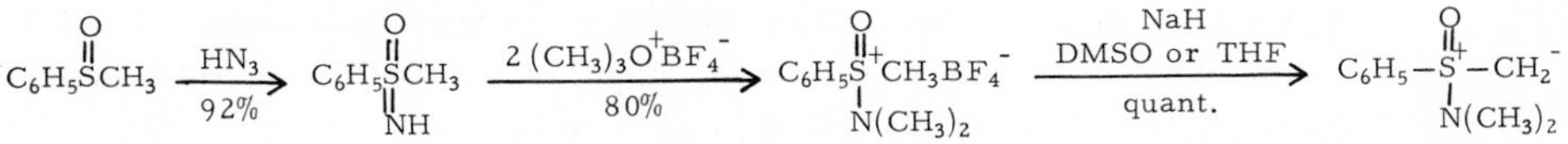

(1)

Methylene transfer (compare **Dimethyloxosulfonium methylide**, **1**, 314–315; **2**, 169–171; this volume). The ylide reacts with electrophilic olefins to yield cyclopropanes. Thus it reacts with benzalacetophenone to give an essentially quantitative yield of *trans*-1-benzoyl-2-phenylcyclopropane (2), and with dimethyl maleate to give the *trans*-adduct (3) in 75% yield.

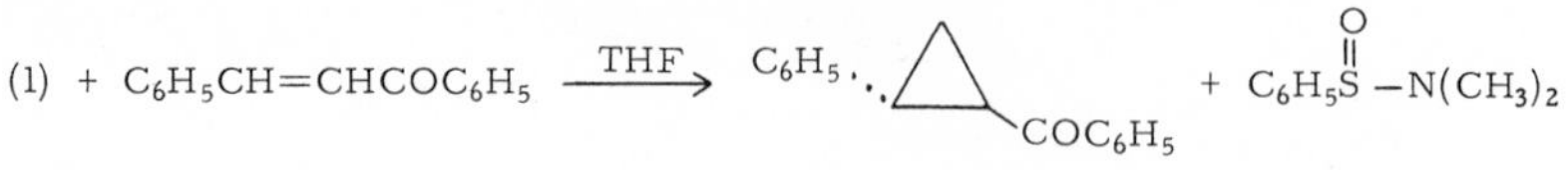

(2)

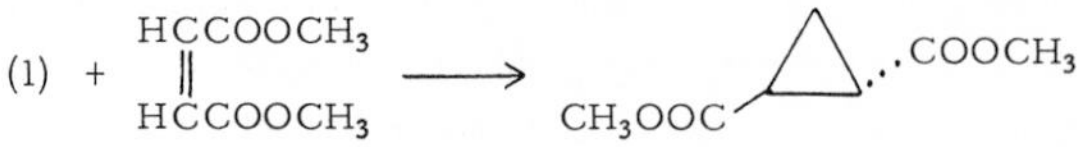

(3)

Reaction with aldehydes and ketones provides oxiranes in good yield.

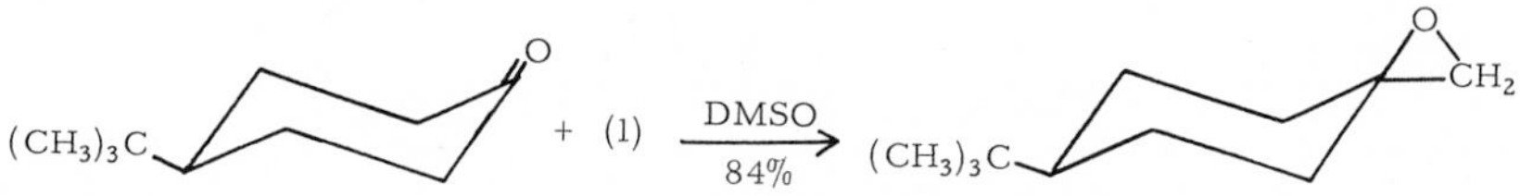

The reaction with benzalaniline gives the corresponding aziridine.

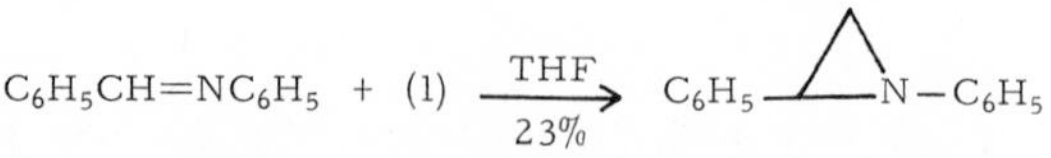

[1]C. R. Johnson, M. Haake, and C. W. Schroeck, *Am. Soc.*, **92**, 6594 (1970)
[2]C. R. Johnson and J. E. Keiser, *Org. Syn.*, **46**, 78 (1966)

Dimethylcopperlithium, **2**, 151–153.

Alkylallenes. Full details of the synthesis of alkylallenes from ethynylcarbinol acetates with organocopper reagents (**2**, 152, ref. 7) have been published: P. Rona and P. Crabbé, *Am. Soc.*, **91**, 3289 (1969).

Cross-coupling reaction. The reaction of dimethylcopperlithium and other dialkyl- or diarylcopperlithium reagents with alkyl halides proceeds without significant metal-halogen exchange to give the coupled product in yields generally in the range 70–98%.[1]

$$(CH_3)_2CuLi \ + \ RX \ \longrightarrow \ RCH_3 \ + \ CH_3Cu \ + \ LiX$$

In one case at least, the reaction of diphenylcopperlithium with (−)-(R)-2-bromobutane, predominant inversion of configuration (84–92% stereoselectivity) is observed. An S_N2-like displacement evidently is involved.

The reaction with aryl iodides shows a distinct difference in that the yield of coupled product is much higher if the reaction mixtures are oxidized with nitro-benzene or oxygen before hydrolysis. Thus the yields of coupled product without oxidation are in the range 30–50%; oxidation increases the yield to 60–70%. This result indicates that there are two reactions involved: coupling and metal-halogen exchange, formulated as in equations (1) and (2).

1. $C_6H_5I + (CH_3)_2CuLi \longrightarrow (C_6H_5)(CH_3)CuLi + CH_3I$

2. $(C_6H_5)(CH_3)CuLi + O_2 \longrightarrow C_6H_5CH_3$

Reaction with oxiranes (epoxides).[2] Dimethylcopperlithium and diphenyl-copperlithium[1] are more useful than other organometallic reagents for the nucleo-philic ring opening of epoxides. Thus dimethylcopperlithium reacts with propylene oxide (equation 1) and 1,2-epoxybutane (equation 2) to give the expected second-ary alcohols as the predominant products.

1.

$$\underset{H_3C}{\triangle}\overset{O}{} \xrightarrow[(C_2H_5)_2O,\ 0^0,\ 13.5\ hr.]{2\ (CH_3)_2CuLi} \underset{89\%}{CH_3CH_2\overset{OH}{\underset{|}{CH}}CH_3} + \underset{4\%}{\overset{H_3C}{\underset{H_3C}{>}}CHCH_2OH}$$

$$+ \ H_3C\overset{CH_3}{\underset{CH_3}{-\!\!\!\overset{|}{\underset{|}{C}}\!\!\!-}}OH \ + \ \text{unidentified products}$$

$$3\% \qquad\qquad\qquad 4\%$$

2.

$$\underset{H_3C}{}\overset{CH_2}{\diagup}\overset{O}{\triangle} \xrightarrow[(C_2H_5)_2O,\ 0^0,\ 13.5\ hr.]{2\ (CH_3)_2CuLi} CH_3CH_2CHOHCH_2CH_3 \ + \ CH_3CH_2\overset{O}{\underset{\|}{C}}CH_3$$

$$80\% \qquad\qquad\qquad 1\%$$

$$+ \ CH_3CH_2\overset{}{\underset{OH}{\overset{|}{C}}}(CH_3)_2 \ + \ \text{unidentified products}$$

$$1\% \qquad\qquad\qquad 10\%$$

Since the reagent is relatively inert toward saturated carbonyl groups, the oxide ring can be selectively opened (equations 3 and 4).

3.

$$H_3C\overset{O}{\triangle}COOC_2H_5 \xrightarrow[(C_2H_5)_2O,\ 0^0,\ 3\ hr.]{2\ (CH_3)_2CuLi} CH_3\overset{OH}{\underset{|}{CH}}-CH\overset{CH_3}{\underset{COOC_2H_5}{<}}$$

$$67\%$$

4.

$$\overset{O}{\triangleright}(CH_2)_8\overset{O}{\underset{\|}{C}}CH_3 \xrightarrow[(C_2H_5)_2O,\ -50^0,\ 0.5\ hr.]{2\ (CH_3)_2CuLi} CH_3CH_2\overset{OH}{\underset{|}{CH}}(CH_2)_8\overset{O}{\underset{\|}{C}}CH_3$$

$$68\%$$

It is also superior to Grignard reagents for conjugate addition to allylic epoxides such as 3,4-epoxy-1-butene (1).[3]

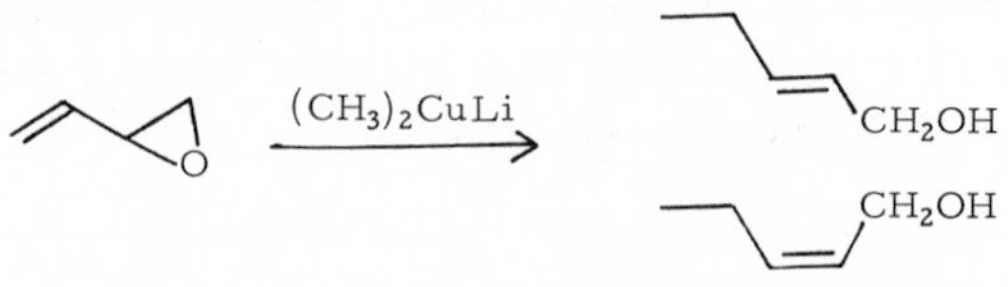

(1)   (2) cis-trans mixture

1,4-Addition to α,β-acetylenic esters. In contrast to the conjugate addition of Grignard reagents to α,β-ethylenic ketones, the related addition to acetylenic carbonyl compounds has been relatively neglected. Two laboratories[4,5] now report simultaneously that this reaction is useful for the stereospecific synthesis of tri- and tetrasubstituted olefins. Thus the reaction of dimethylcopperlithium (2) with an acetylenic ester (1) at −78 to −100° in ether or THF results in *cis*-addition to give (3) in high yield. The "enolate" (a) is apparently the intermediate. At temperatures of 0°, the *trans*-adduct corresponding to (3) predominates, presumably through equilibration of (a) to the isomeric enolate. The intermediate (a) is also useful for further reaction. Thus oxidation in the presence of excess dimethylcopperlithium furnishes (4) in fair yield. The iodide (5) is obtained by reaction with excess iodine.

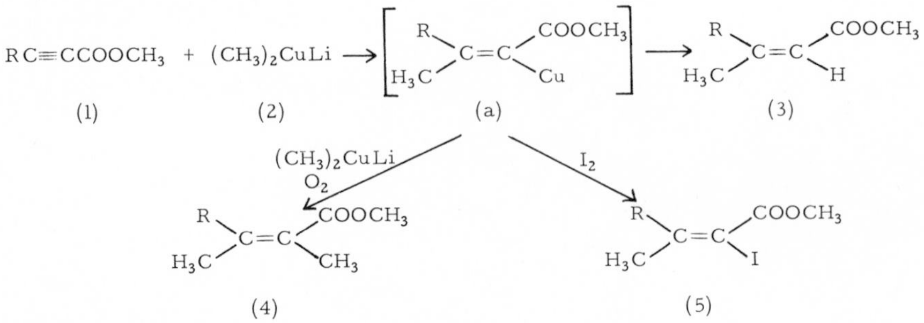

1,4-Addition to α,β-unsaturated ketones. A key step in a stereoselective synthesis of (±)-eremophil-3,11-diene (3)[6] involved introduction of an angular methyl group at C_5 of the octalone (1). Copper-catalyzed 1,4-addition of methylmagnesium bromide failed in this case presumably because of the presence of an alkyl group at C_4. However, the reaction with dimethylcopperlithium gave the desired 1,4-addition product in 77% yield. Workup included destruction of the excess organometallic reagent with dilute hydrochloric acid in order to prevent 1,2-addition to the resultant saturated ketone. The reaction has the further advantage that the product (2) has the desired *cis*-fusion characteristic of eremophilane-type sesquiterpenes. The ketone (2) was converted into (3) in high yield by preparation of

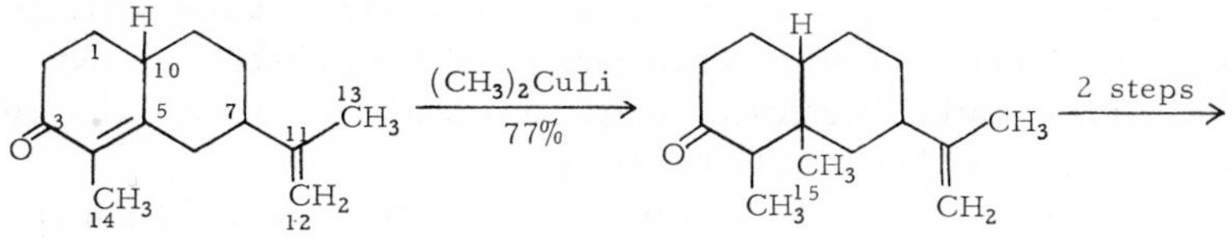

(1) $\xrightarrow[77\%]{(CH_3)_2CuLi}$ (2) $\xrightarrow{\text{2 steps}}$

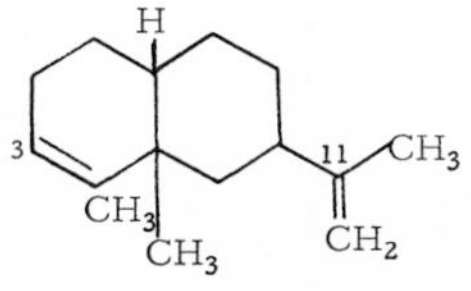

(3)

the corresponding tosylhydrazone, which was then heated with sodium ethylene glycolate (Bamford-Stevens reaction).

In a recent synthesis of the eremophilane-type fukinone (6) Marshall and Cohen[7] have also made use of the conjugate addition of dimethylcopperlithium. Reaction of the α,β-unsaturated ketone (4) with the reagent gives (5), which has the desired stereochemistry shown. The product was transformed into fukinone (6) in about ten steps.

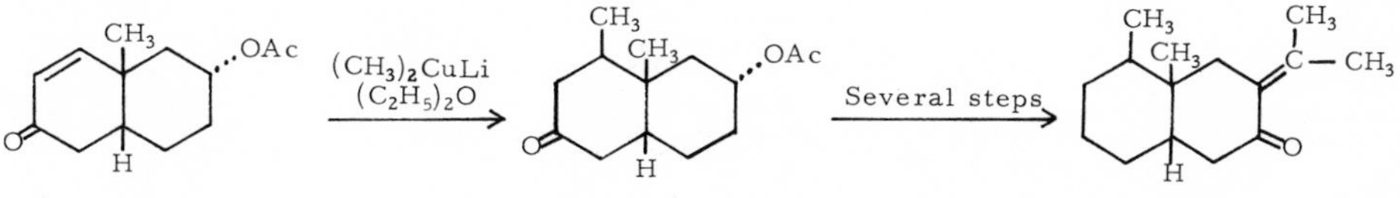

(4) $\xrightarrow[(C_2H_5)_2O]{(CH_3)_2CuLi}$ (5) $\xrightarrow{\text{Several steps}}$ (6)

Trisubstituted olefins. The reaction of a steroidal allylic acetate (1) with dimethylcopperlithium to give an alkylated trisubstituted olefin (2) (**2**, 152–153) has been extended to model compounds to gain insight into the scope and stereo-

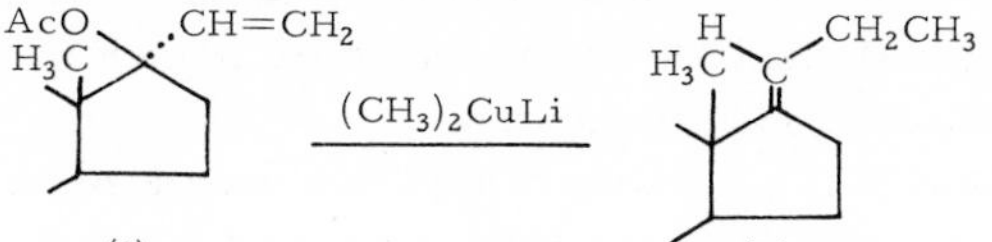

(1) $\xrightarrow{(CH_3)_2CuLi}$ (2)

chemistry of the reaction.[8] Actually two alkylation pathways are available: displacement of acetate with allylic rearrangement (a) or direct displacement (b).

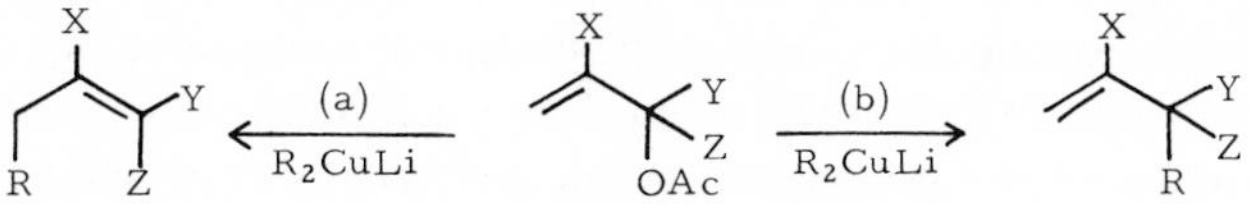

Path (a) is favored when X is equal to or smaller than the entering R group and when Z = H. Path (b) is favored when both Y and Z substituents are alkyl. Process (a) shows marked stereoselectivity, particularly when ether is used as solvent, the *trans* configuration being favored.

This trisubstituted olefin synthesis was used for the synthesis of a trienecarboxylic ester (6) having the skeleton of the C_{18}-*Crecropia* juvenile hormone, methyl *cis*-10,11-oxido-3,11-dimethyl-7-ethyltrideca-*trans,trans*-2,6-dienoate (3)

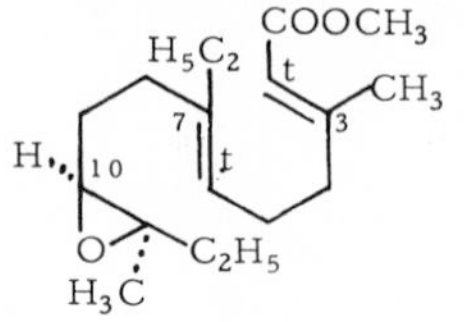

(3)

from all-*trans* ethyl farnesoate (4). This was converted into the doubly allylic acetate (5) by photosensitized oxygenation, *in situ* reduction of hydroperoxides with trimethyl phosphite followed by acetylation. Methylation of (5) with di-

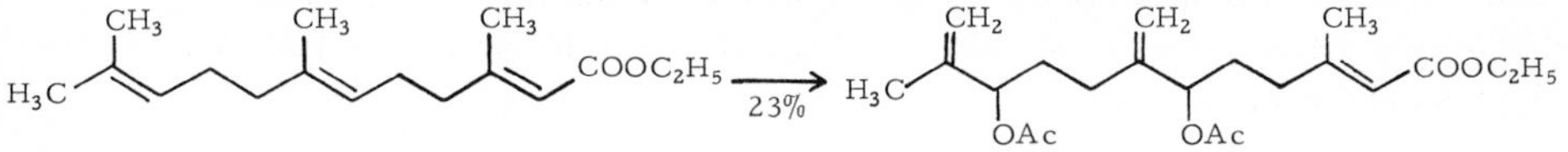

methylcopperlithium gave as the predominant product (76% yield) the undesired ethyl *trans,cis,trans*-3,11-dimethyl-7-ethyltrideca-2,6,10-trienoate (6). Thus this reaction was not suitable for a stereoselective synthesis of the juvenile hormone. Evidently a *cis*-trisubstituted double bond is formed when X (in the model above) is larger than the entering alkyl group.

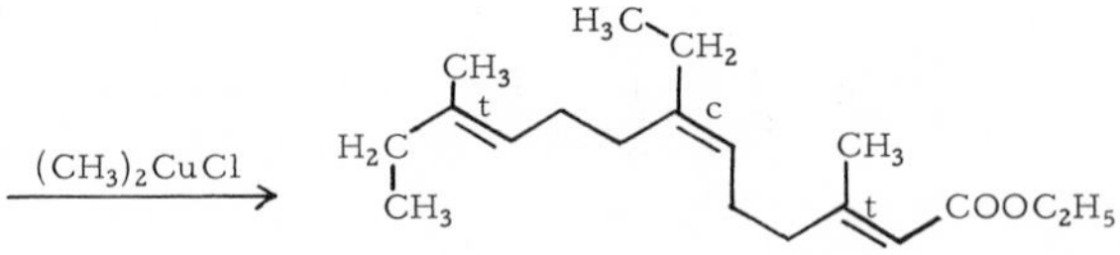

van Tamelen and McCormick[9] have used a related method for a successful synthesis of the juvenile hormone from *trans,trans*-farnesol acetate (7). This triene on treatment with *ca.* 2 equivalents of *m*-chloroperbenzoic acid underwent exclusive epoxidation at the two nonallylic centers (8). The diepoxide on treatment with lithium diethylamide (2, 247–248) undergoes a double, Hoffmann-like elimination to give (9). Alkylation of this allylic alcohol with dimethylcopperlithium as above was found to lead to the undesired configuration of the central double bond. Hence (9) was transformed into the doubly allylic chloride (10) (after protection of the primary allylic alcohol group) by reaction with tosyl

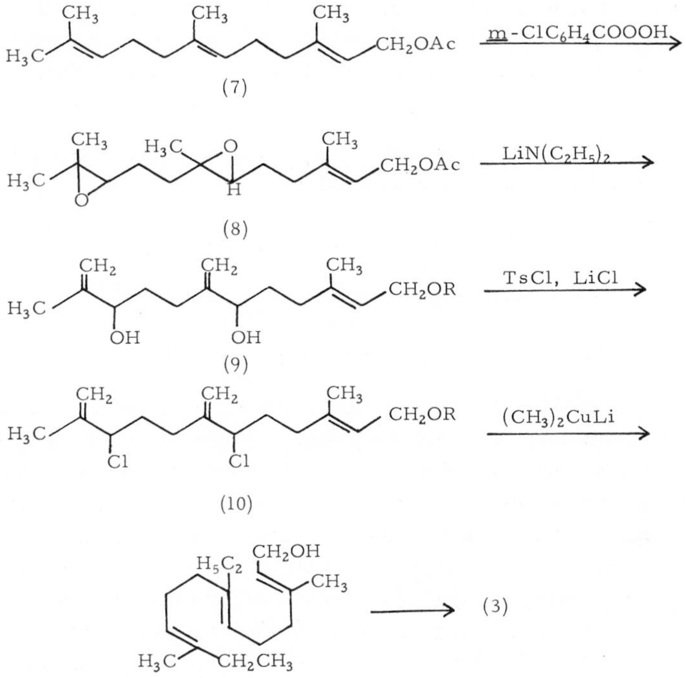

(7)

(8)

(9)

(10)

(11)

chloride–lithium chloride (this volume). Alkylation of (10) with dimethylcopper-lithium gave as one product the desired *trans,trans,cis*-triene (11) together with the *t,c,c*; *t,c,t*; *t,t,t*-isomers in the ratio 1:1:2:1. Completion of the synthesis required oxidation of the alcohol group to a carboxyl group (2, 261[10]) and terminal epoxidation (NBS–H$_2$O, followed by base).[11]

Methyl[18]annulene (3). This monosubstituted annulene has been prepared from [18]annulene (1) in two steps. Monobromo[18]annulene (2) is prepared in 44% yield by bromination with an equimolar amount of pyridinium bromide per-bromide. This reacts with a large excess of dimethylcopperlithium to give methyl-[18]annulene in 52% yield. It is imperative that no excess methyllithium be present in the reaction mixture.[12]

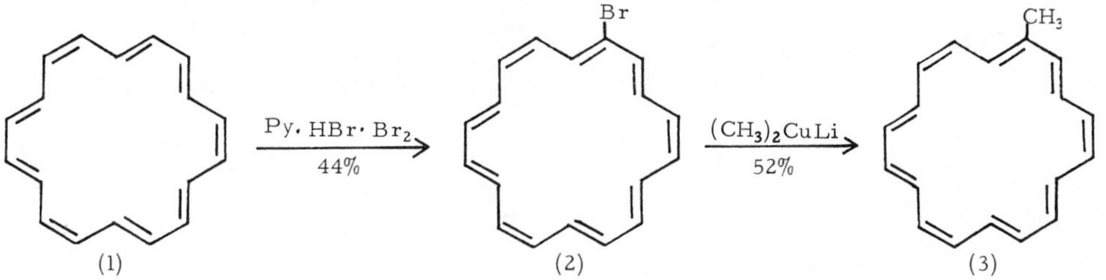

Methyl ketones from carboxylic acid chlorides.[13] The reagent, 3 moles, prepared in ether solution, reacts with acid chlorides (1 mole) at −78° in 15 minutes to give the corresponding methyl ketones in excellent yield. The acid chloride can be

$$RCOCl + (CH_3)_2CuLi \longrightarrow RCOCH_3$$

primary, secondary, tertiary, or aryl. Similar transformations were achieved with diethylcopperlithium and di-n-butylcopperlithium.

Examples:

$$\underline{n}\text{-}C_5H_{11}COCl \xrightarrow[81\%]{} \underline{n}\text{-}C_5H_{11}COCH_3$$

$$(C_6H_5)_2CHCOCl \xrightarrow[93\%]{} (C_6H_5)_2CHCOCH_3$$

$$\underline{t}\text{-}C_4H_9COCl \xrightarrow[84\%]{} \underline{t}\text{-}C_4H_9COCH_3$$

Related reagents. Trimethylmanganeselithium, $(CH_3)_3MnLi$, tan to light brown, is readily prepared from manganous bromide and three equivalents of methyllithium in ether under nitrogen.[14] Like dimethylcopperlithium, it undergoes the coupling reaction with various halides. The most efficient coupling occurs with vinylic and allylic halides: $trans$-1-iodo-1-nonene → $trans$-2-decene, 82% yield; 3-bromocyclohexene → 3-methylcyclohexene, 75%. Only moderate yields are obtained from primary iodides, aryl iodides, and cyclopropyl halides: 1-iodo-decane → n-undecane, 55% yield; iodobenzene → toluene, 50% yield; 7,7-dibromonorcarane → 7,7-dimethylnorcarane, 50% yield. The major by-products apparently result from metal-halogen exchange. Triethylmanganeselithium and tri-n-butylmanganeselithium are less effective reagents; the main reaction products result from reduction of halogen or elimination of hydrogen halide. Trimethyl-ironlithium and trimethylcobaltlithium give about the same results as trimethyl-manganeselithium. It appears therefore that reagents of the type R_2CuLi are superior to the related manganese, iron, and cobalt reagents for cross coupling.

[1]G. M. Whitesides, W. F. Fischer, Jr., J. SanFilippo, Jr., R. W. Bashe, and H. O. House, *Am. Soc.*, **91**, 4871 (1969)

[2]R. W. Herr, D. M. Wieland, and C. R. Johnson, *ibid.*, **92**, 3813 (1970)

[3]R. J. Anderson, *ibid.*, **92**, 4978 (1970); R. W. Herr and C. R. Johnson, *ibid.*, **92**, 4979 (1970)

[4]E. J. Corey and J. A. Katzenellenbogen, *ibid.*, **91**, 1851 (1969)

[5]J. B. Siddall, M. Biskup, and J. H. Fried, *ibid.*, **91**, 1853 (1969)

[6]E. Piers and R. J. Keziere, *Canad. J. Chem.*, **47**, 137 (1969)

[7]J. A. Marshall and G. M. Cohen, *Tetrahedron Letters*, 3865 (1970)

[8]R. J. Anderson, C. A. Henrick, and J. B. Siddall, *Am. Soc.*, **92**, 735 (1970)

[9]E. E. van Tamelen and J. P. McCormick, *ibid.*, **92**, 737 (1970)

[10]Double bonds are missing in formulas (10) and (11) of this citation. Correct formulas are:

[11] E. E. van Tamelen and T. J. Curphey, *Tetrahedron Letters*, 121 (1962); E. J. Corey, J. A. Katzenellenbogen, N. W. Gilman, S. A. Roman, and B. W. Erickson, *Am. Soc.*, **90**, 5618 (1968)

[12] E. P. Woo and F. Sondheimer, *Tetrahedron*, **26**, 3933 (1970)

[13] G. H. Posner and C. E. Whitten, *Tetrahedron Letters*, 4647 (1970)

[14] E. J. Corey and G. H. Posner, *ibid.*, 315 (1970)

Dimethyldiacetoxysilane, 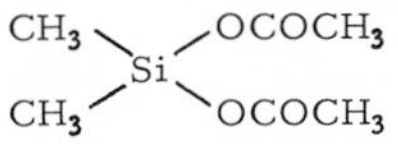
$$CH_3\underset{CH_3}{\overset{}{\diagdown}}Si\underset{OCOCH_3}{\overset{OCOCH_3}{\diagup}}$$
Mol. wt. 176.25, b.p. 164–166°, 44–45°/3 mm.

Preparation.[1] The reagent is prepared by refluxing 1 mole of dimethyldichloro-silane with two moles of acetic anhydride for four hours followed by slow distillation of the acetyl chloride produced, and then of unreacted acetic anhydride. The reagent is then isolated by distillation.

Silylation.[1] The reagent reacts with *cis*-diols and with corticosteroids in the presence of triethylamine to form siliconides (2) analogous to acetonides. It reacts with an isolated hydroxyl group to give a thermally stable silyl derivative of type

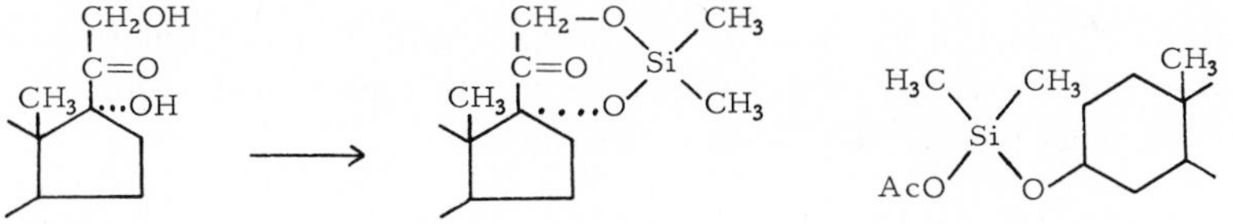

(3). These derivatives have satisfactory gas chromatographic properties.

[1] R. W. Kelly, *J. Chromatography*, **43**, 229 (1969)

Dimethyl diazomethylphosphonate, $N_2CHP(=O)(OCH_3)_2$. Mol. wt. 149.09, b.p. 59°/0.42 mm.

Preparation.[1]

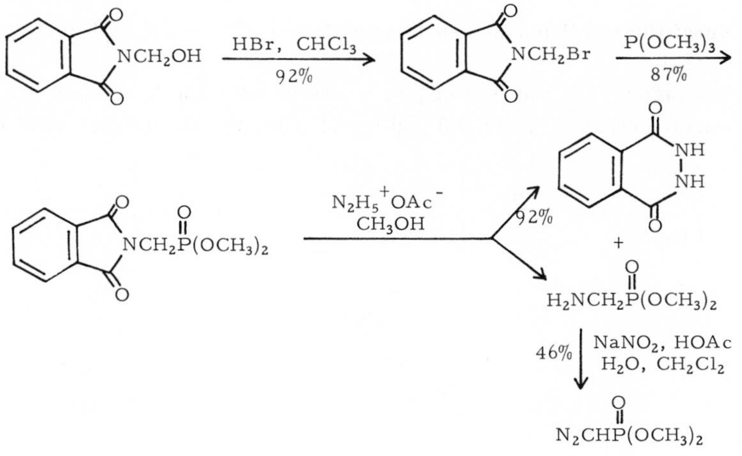

Cyclopropanation.[1] In the presence of copper powder the reagent decomposes to the carbene :$CHP(=O)(OCH_3)_2$, which adds to double bonds. Methylene chloride is the preferred solvent.

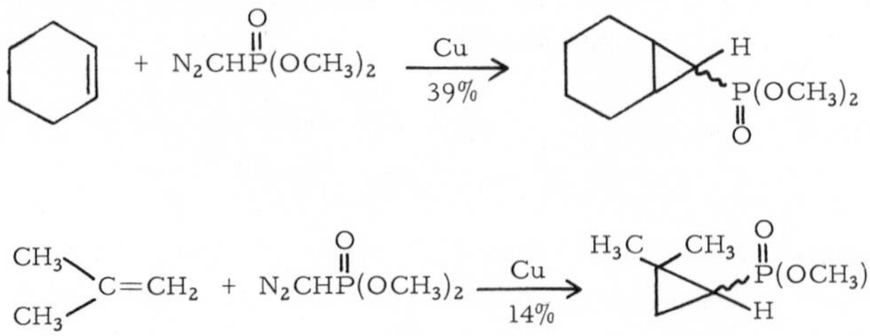

[1]D. Seyferth and R. S. Marmor, *Tetrahedron Letters*, 2493 (1970)

Dimethyldichlorosilane, $(CH_3)_2SiCl_2$. Mol. wt. 129.07, b.p. 70.3°. Supplier: Pierce Chem. Co.

Siliconides. The steroidal *cis*-diol, Δ^5-3β-acetoxy-16α,17α-dihydroxy-pregnene-20-one (1), is converted by reaction with dimethyldichlorosilane in pyridine into a cyclic silyldioxy compound (2). In analogy with acetonides, the name siliconide is suggested for such a derivative.[1]

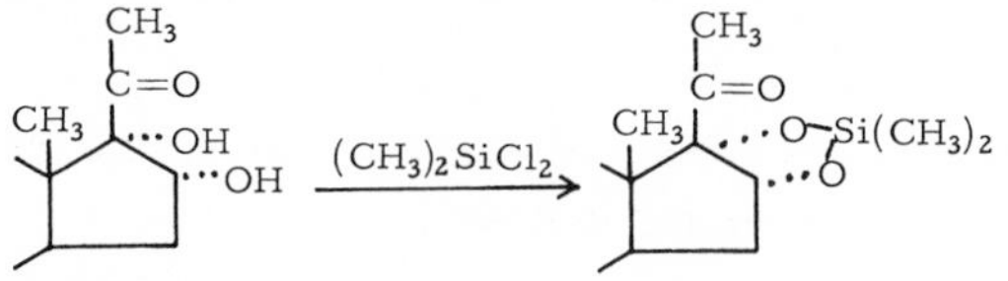

(1, 1 g.)   (2, 400 mg.)

[1]R. W. Kelly, *Tetrahedron Letters*, 967 (1969)

N,N-Dimethylformaldimmonium trifluoroacetate, $H_2C=N(CH_3)_2CF_3CO_2^-$. Mol. wt. 171.12.

Preparation.[1] The salt is prepared in methylene chloride solution from trifluoroacetic anhydride (1 mmole) and trimethylamine oxide (1 mmole) at 0°. It

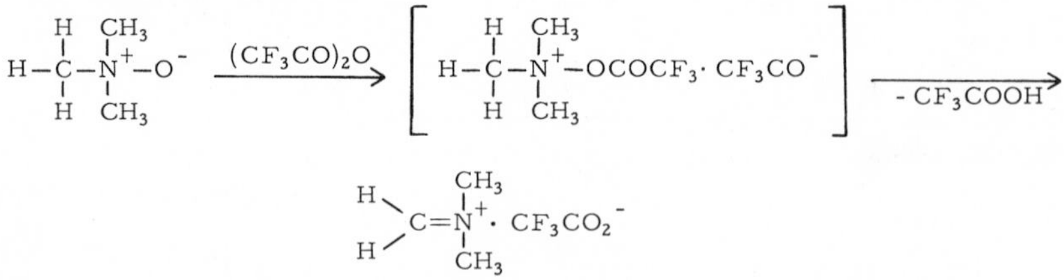

(1)

was identified by the highly characteristic proton magnetic resonance spectrum.

Mannich reaction.[2] This reaction involves the condensation of an active hydrogen compound with formaldehyde and ammonia or a primary or secondary amine (usually as the hydrochloride):

$$(CH_3)_2NH + HCHO + CH_3COCH_3 \longrightarrow (CH_3)_2N \cdot CH_2CH_2COCH_3 + H_2O$$

Since salts of the type of (1) have been considered to be the actual reagents in Mannich reactions, the French chemists[1] treated various ketosteroids with solutions of (1) and obtained the products of a Mannich reaction in higher yield than that obtained by the classical procedure.

Examples:

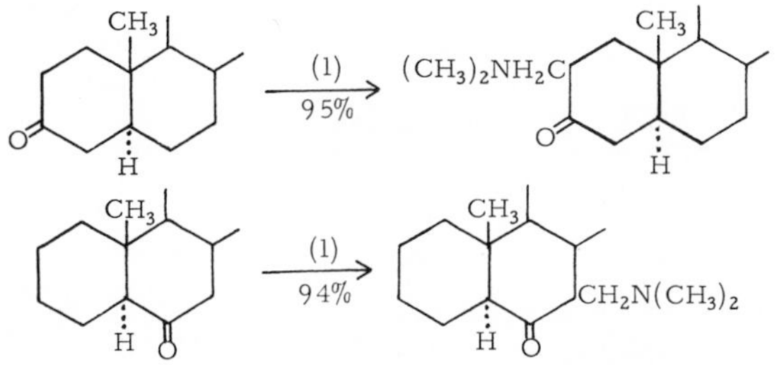

[1]A. Ahond, A. Cavé, C. Kan-Fan, H.-P. Husson, J. de Rostolan, and P. Potier, *Am. Soc.*, **90**, 5622 (1968)
[2]Review: F. F. Blicke, *Org. Reactions*, **1**, 303 (1942)

Dimethylformamide, 1, 278–281, 1110; **2**, 153–154.

Demethylation of aryl methyl ethers. Koutek and Setínek[1] reported that aryl methyl ethers can be demethylated by sodium thioethoxide (C_2H_5SNa); they used ethanol as solvent and temperatures of 190° (autoclave). However, if DMF is used as solvent, demethylation is complete at reflux temperatures in about an hour.[2] Yields are high (94–88%). Ethers of dihydric phenols give selective mono-demethylation.

[1]B. Koutek and K. Setínek, *Chem. Commun.*, **33**, 866 (1968)
[2]G. I. Feutrill and R. N. Mirrington, *Tetrahedron Letters*, 1327 (1970)

Dimethylformamide diethyl acetal, 1, 281–282; **2**, 154. Suppliers: Aldrich, Eastman, Fluka.

Dimethylformamide dimethyl acetal, $(CH_3O)_2CHN(CH_3)_2$. Mol. wt. 119.16, b.p. 102–103°/720 mm. Supplier: Aldrich.

Epoxides from vic-diols. *trans*-1,2-Dihydroxycyclohexane (1) reacts with dimethylformamide dimethyl acetal at 75° (24 hours) and then at 130° (24 hours) to give *cis*-cyclohexene oxide (2) in 88% yield. The analogous reaction with *cis*-dihydroxycyclohexane (3) yields dimethylformamide cyclohexane acetal (4).

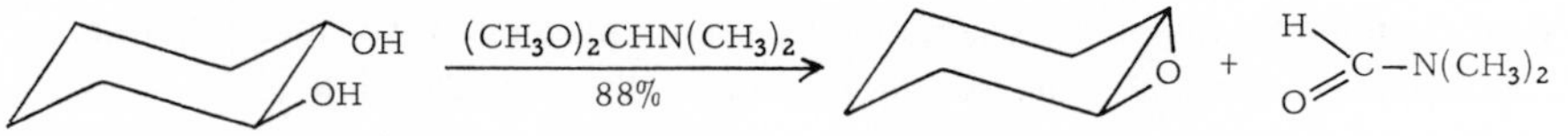

(1)   (2)

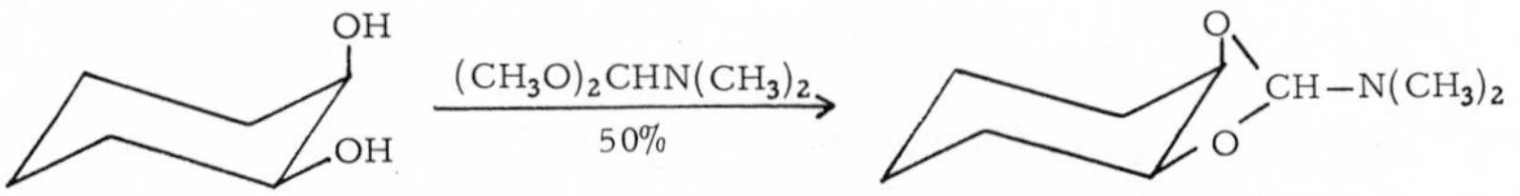

(3)   (4)

meso-Hydrobenzoin (5) is converted into *trans*-stilbene epoxide (6) by this method.

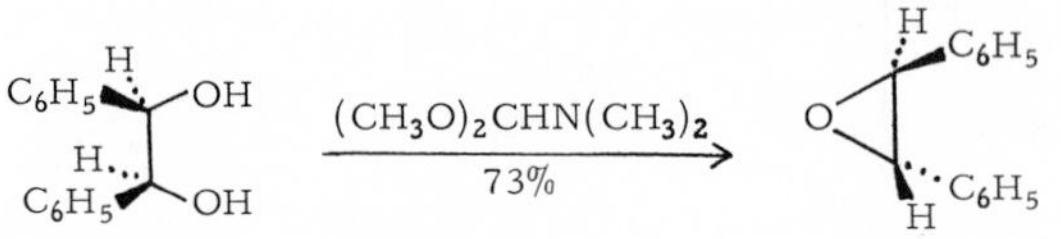

(5)   (6)

The reaction was also employed for conversion of $5\alpha,6\beta$-dihydroxycholestane into Δ^5-cholestene-α-epoxide (80% yield).

[1]H. Neumann, *Chimia*, **23**, 267 (1969)

Dimethylformamide–Thionyl chloride, 1, 286–289.

Chlorination. For chlorination of nucleosides, Dods and Roth[1] recommend that the reagent be prepared by allowing equimolar quantities of dry DMF and $SOCl_2$ to react at 27° for 30 min. The mixture is then evaporated to dryness *in vacuo* and the residue washed well with dry ether. The reagent is hygroscopic and must be used immediately. The reagent converts 2',3'-O-isopropylideneuridine into 5'-deoxy-5'-chloro-2',3'-O-isopropylideneuridine in 90% yield. Bromination is achieved in the same way by the combination of DMF and $SOBr_2$.

Squaric acid (1) is converted into dichlorocyclobutenedione (2) by reaction with thionyl chloride and a trace of DMF in 75% yield.[2] No reaction occurs with thionyl chloride alone. Use of excess thionyl chloride and a trace of DMF unexpectedly results in replacement of one of the carbonyl oxygens by two chlorine atoms to give perchlorocyclobutenone (3) in 76% yield.

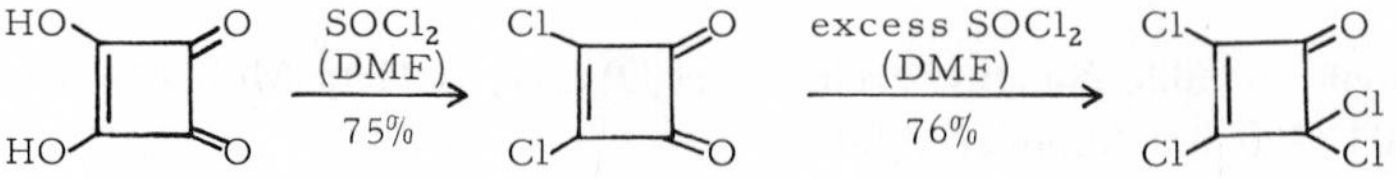

(1)   (2)   (3)

[1]R. F. Dods and J. S. Roth, *Tetrahedron Letters*, 165 (1969)
[2]R. C. DeSelms, C. J. Fox, and R. C. Riordan, *ibid.*, 781 (1970)

N,N-Dimethylhydrazine, 1, 289–290; **2**, 154–155. Additional suppliers: Aldrich, Eastman.

Acetophenone hydrazone (**2**, 154–155, ref. 4a). The procedure for preparation of acetophenone hydrazone[1] has been published.

Protection of carbonyl groups.[2] The reagent reacts with ketones under slightly alkaline conditions to form N,N-dimethylhydrazones. They can be cleaved under neutral conditions with methyl iodide in refluxing 95% ethanol. The method is

$$>C=N-N(CH_3)_2 \xrightarrow{CH_3I} [>C=N-\overset{+}{N}(CH_3)_3] \longrightarrow >C=O$$
$$I^-$$

suggested as an alternative to dioxolanes, which require acidic conditions, both for formation and hydrolysis.

[1] G. R. Newkome and D. L. Fishel, *Org. Syn.*, **50**, 102 (1970)
[2] M. Avaro, J. Levisalles, and H. Rudler, *Chem. Commun.*, 445 (1969)

Dimethyl methylphosphonate, $CH_3PO(OCH_3)_2$. Mol. wt. 124.08, $n^{20}D$ 1.4130. Supplier: Aldrich.

The corresponding anion can be prepared readily as the lithio derivative by the action of one equivalent of *n*-butyllithium in THF.[1] The anion readily reacts with benzophenone to form the adduct (1). The reaction was explored in a study of

$$\overset{OH}{\underset{|}{(C_6H_5)_2CCH_2PO(OCH_3)_2}}$$
(1)

methods of olefin synthesis, but (1) does not undergo efficient or facile cyclo-elimination to form an olefin (see **Dimethyl methylphosphonothioate, 2**, 155).

Cyclic α,β-unsaturated ketones.[2] The anion reacts with the steroidal enol lactones (1) and (3) in THF under N_2 to give the cyclic α,β-unsaturated ketones (2) and (4) directly.

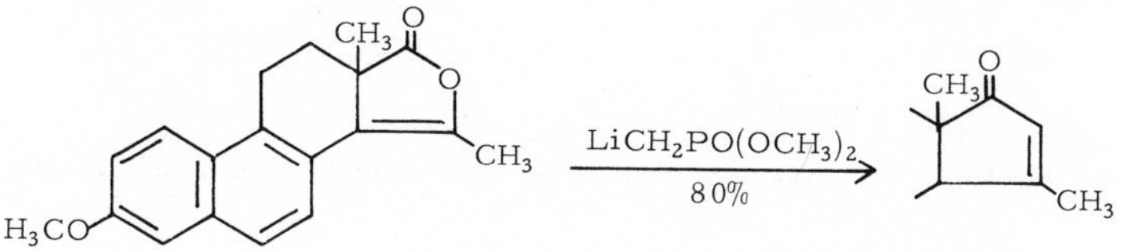

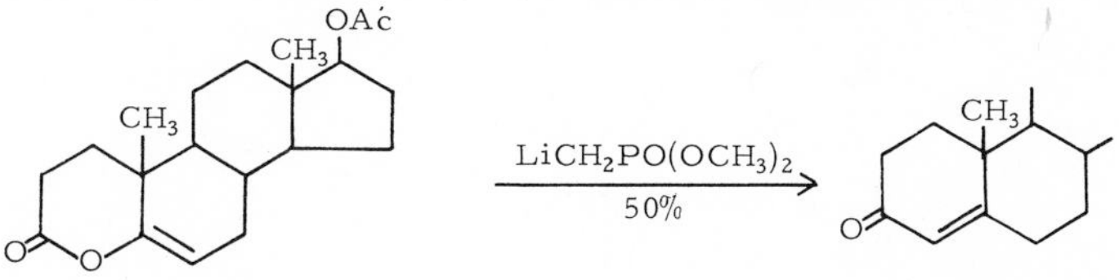

[1]E. J. Corey and G. T. Kwiatkowski, *Am. Soc.*, **88**, 5654 (1966)
[2]C. A. Henrick, E. Böhme, J. A. Edwards, and J. H. Fried, *ibid.*, **90**, 5926 (1968)

5,5-Dimethyl-N-nitroso-2-oxazolidone (3). Mol. wt. 144.13, m.p. 88.5–89.5°.

Preparation. The reagent is prepared from ethyl 3-hydroxy-3-methylbutyrate (1) by conversion into the hydrazide followed by Curtius rearrangement to 5,5-dimethyloxazolidone (2). This intermediate is then nitrosated.[1]

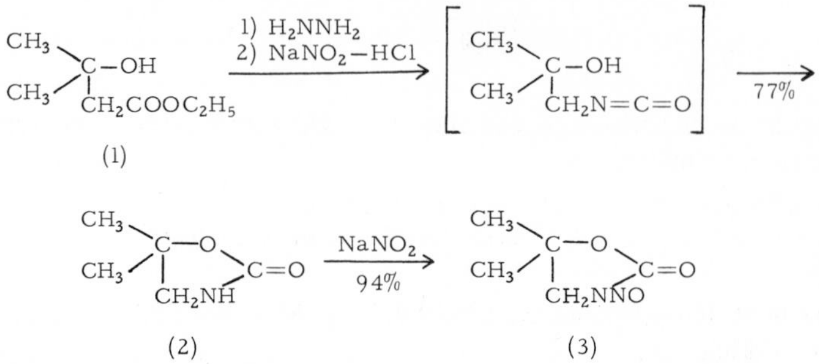

Dimethylethylidenecarbene. When treated with a base,[2] 5,5-dialkyl-N-nitrosooxazolidones decompose, probably through unsaturated carbonium ions, to give dialkylethylidenecarbenes, which can be trapped by reaction with olefins to give methylenecyclopropanes.[1] Thus 5,5-dimethyl-N-nitroso-2-oxazolidone (3) is converted into dimethylethylidenecarbene through the postulated intermediates shown. Isolated yields of cyclopropanes are 18–56%.

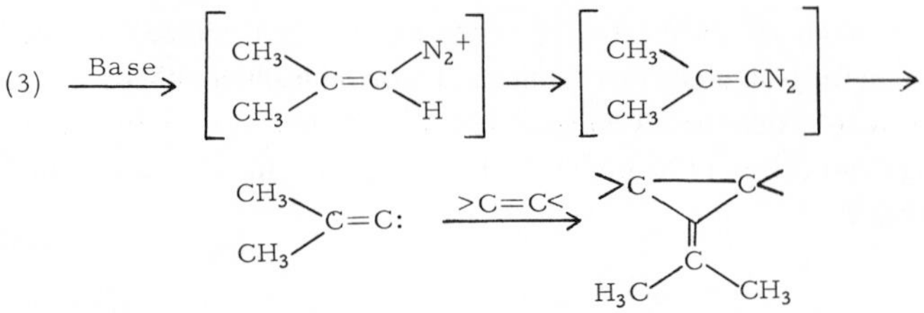

[1]M. S. Newman and T. B. Patrick, *Am. Soc.*, **91**, 6461 (1969)
[2]Lithium 2-ethoxyethoxide was used because it is soluble in the olefins used for trapping the carbene.

N,N-Dimethyl-4-pyridinamine (4-Dimethylaminopyridine). Mol. wt. 122.17, m.p. 113–114°.

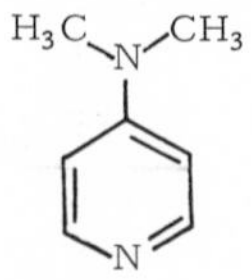

Preparation. This base is prepared by heating 4-chloropyridine with aqueous dimethylamine at 150° for 2 hrs.[1]

Acylation catalyst. This base is superior to pyridine as a catalyst for O-acylations. Even tertiary alcohols are readily acylated. Thus 1-methyl-1-cyclohexanol is converted into the acetate by treatment with acetic anhydride, trimethylamine (to combine with the acid formed), and a catalytic amount of this base in 86% yield at room temperature (14 hours). The yield of acetate was less than 5% when pyridine and/or triethylamine was used.[2]

It is also superior to pyridine as a catalyst for C-acylation, for example in the Dakin-West reaction.[3] This reaction is the conversion of α-amino acids into α-acylamino ketones by reaction with acid anhydrides catalyzed by a base (usually

$$\underset{\underset{\displaystyle RCHCOOH}{|}}{NH_2} + (CH_3CO)_2O \longrightarrow RCH(NHCOCH_3)COCH_3 + CO_2 + H_2O$$

pyridine). Actually the reaction involves several steps: azlactonization, base-catalyzed C-acylation, cleavage to a β-keto acid with decarboxylation.[4] Steglich and Höfle[2] report that N,N-dimethyl-4-pyridinamine is markedly superior to pyridine for acylation of the intermediate 5-oxazolones to the 4-acyl-5-oxaz-

olones, permitting acylation in minutes at room temperature. It also accelerates the decarboxylation step. With this catalyst Dakin-West reactions can be conducted at room temperature in 30 minutes.

[1]L. Pentimalli, *Gazz. Chim. Ital.*, **94**, 902 (1964)
[2]W. Steglich and G. Höfle, *Angew. chem., internat. Ed.*, **8**, 981 (1969)
[3]H. D. Dakin and R. West, *J. Biol. Chem.*, **78**, 91, 745 (1928)
[4]Y. Iwakura, F. Toda, and H. Suzuki, *J. Org.*, **32**, 440 (1967); W. Steglich and G. Höfle, *Tetrahedron Letters*, 1619 (1968)

Dimethyl sulfoxide, **1**, 296–310; **2**, 157–166.

Solvent Effects

Dehydrohalogenation. *t*-Alkylamines react with epichlorohydrin in methanol at 20–25° to give 1-*t*-alkylamino-3-chloro-2-propanols. On dehydrohalogenation these are converted into *t*-alkyl-2,3-epoxypropylamines. In the original procedure[1] aqueous potassium hydroxide was used for this second step; in an improved procedure[2] dehydrochlorination is effected with aqueous sodium hydroxide in the presence of DMSO, which promotes rapid and complete reaction at room temperature.

$$(CH_3)_3CNH_2 + CH_2\underset{O}{\diagup}CHCH_2Cl \longrightarrow (CH_3)_3CNHCH_2CHOHCH_2Cl \xrightarrow[51-60\%]{NaOH, DMSO}$$

$$(CH_3)_3CNHCH_2CH\underset{O}{\diagup}CH_2$$

Isopropylidene derivatives. Various workers have reported difficulties in the preparation of 1,2-O-isopropylidenemyoinositol (1), probably because this cyclitol contains three contiguous *cis*-hydroxyl groups. The reaction with acetone catalyzed by large amounts of zinc chloride is not complete after a reflux period of 50 hours.[3] Ketal exchange with 2,2-diethoxypropane catalyzed by *p*-toluene-

(1)

sulfonic acid, generally a useful method for preparation of isopropylidene derivatives of cyclitols (1, 244), is not successful in this case, probably because myoinositol is very insoluble in the reagent.[4] However, if the exchange is carried out in DMSO at 110° with removal of the methanol as formed by distillation, 1,2-O-isopropylidenemyoinositol can be obtained in 73% yield.[5]

Reduction of alkyl or aryl halides. Reduction of organic halogen compounds by sodium borohydride proceeds more rapidly in DMSO than in anhydrous diglyme.[6]

3-Substituted pentane-2,4-diones. The most generally used method for preparation of these diones, $CH_3COCHRCOCH_3$, is the reaction of sodium acetylacetonate[7] with an alkyl iodide in hydroxylic, nonpolar solvents. If DMSO is used as solvent, the reaction rate is increased to such an extent that alkyl bromides can be used as well.[8]

$$(CH_3COCHCOCH_3)\,Na \xrightarrow[30-60\%]{RBr} \underset{R}{CH_3COCHCOCH_3}$$

Williamson ether synthesis. A German patent[9] noted that ethers could be prepared from an alcohol, sodium hydroxide, and an alkyl chloride (the Williamson synthesis ordinarily involves the use of sodium metal). Canadian chemists[10] have confirmed this observation and found further that the reaction time is decreased and the yield improved by use of DMSO as solvent. For example, di-*n*-butyl ether was prepared in 95% yield from *n*-butyl alcohol and *n*-butyl chloride using sodium hydroxide in DMSO. The yield was 61% when excess alcohol was used as solvent.

Reactions

Oxidation. Oxidation of alkyl halides by DMSO requires high temperatures (100–150°), and yields are relatively low except for primary iodides (**1**, 303). Epstein and Ollinger[11] find that halides can be oxidized to carbonyl compounds by DMSO at room temperature (4–48 hours) in the presence of silver perchlorate as assisting agent. Chlorides are relatively unreactive, but bromides and iodides are oxidized relatively easily. Yields are higher with primary halides than with secondary halides. Cyclohexyl halides are oxidized to only a slight extent to cyclohexanone, the main product being cyclohexene, formed by elimination.

In the presence of *p*-toluenesulfonic acid, hydrogen chloride, or triphenylmethyl perchlorate, dimethyl sulfoxide oxidizes aliphatic and aromatic isocyanides to the corresponding isocyanates. The reaction is carried out at 50–80°, at which temperature dimethyl sulfide distils from the reaction mixture.[12]

$$\underline{n}\text{-}C_4H_9N{=}C\text{:} \ + \ (CH_3)_2S{=}O \ \xrightarrow{\ HX\ } \ \underline{n}\text{-}C_4H_9N{=}C{=}O \ + \ S(CH_3)_2$$
$$85\% \qquad\qquad 55\%$$

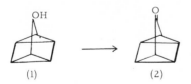

$$92\% \qquad\qquad 46\%$$

Pfitzner-Moffatt oxidation (**1**, 304–307; **2**, 162). The highly strained tetracyclic ketone quadricyclanone (**2**), 7-quadricyclo[2.2.1.0²·⁶.0³·⁵]heptanone, was first prepared[13] by oxidation of quadricyclanol (**1**) with *t*-butyl hypochlorite in pyridine–carbon tetrachloride (method of Grob and Schmid, **1**, 90, ref. 6); but the yield was only 14% and it was difficult to separate the ketone from unchanged alcohol. The oxidation later[14] was considerably improved by conversion of (1) into the tosylate

OH   O

(1)   (2)

(72% yield) and oxidation of the tosylate with DMSO (85% yield) according to the general method of Kornblum (**1**, 303).

Actually, the direct oxidation of (1) to (2) can be carried out in 81% yield with the Pfitzner-Moffatt reagent (DMSO, DCC, and H_3PO_4).[15]

Dimethyl sulfoxide–Acetic anhydride (**1**, 305; **2**, 163–165). Clement *et al.*[16] report that DMSO and acetic anhydride can be used for normal oxidation of alcohols which undergo cleavage upon oxidation with chromic acid, lead tetra-acetate, etc.

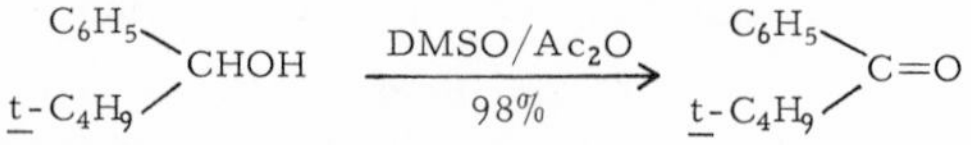

$$\underset{\underline{t}\text{-}C_4H_9}{\overset{C_6H_5}{\diagdown}}CHOH \quad \xrightarrow[98\%]{DMSO/Ac_2O} \quad \underset{\underline{t}\text{-}C_4H_9}{\overset{C_6H_5}{\diagdown}}C=O$$

Dimethyl sulfoxide–Diphenylketene-p-tolylimine (1),
Mol. wt. 283.36, m.p. 83–84°.
Preparation of the ketenimine.[17]

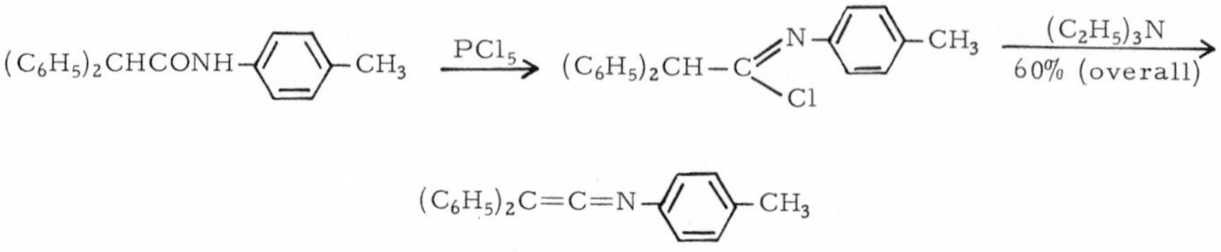

$$(C_6H_5)_2CHCONH-\!\!\!\left\langle\bigcirc\right\rangle\!\!\!-CH_3 \quad \xrightarrow{PCl_5} \quad (C_6H_5)_2CH-C\!\!\underset{Cl}{\overset{N-\left\langle\bigcirc\right\rangle-CH_3}{\diagup}} \quad \xrightarrow[60\% \text{ (overall)}]{(C_2H_5)_3N}$$

$$(C_6H_5)_2C=C=N-\!\!\!\left\langle\bigcirc\right\rangle\!\!\!-CH_3$$

(1)

Oxidation. The reagent has been used in place of DCC in Pfitzner-Moffatt type oxidations (**1**, 304–307; **2**, 162). Thus 2′,3′-O-isopropylideneadenosine (2) was oxidized to 2′,3′-O-isopropylideneadenosine-5′-aldehyde (3) in 60% yield by the ketenimine (1), DMSO, and phosphoric acid.

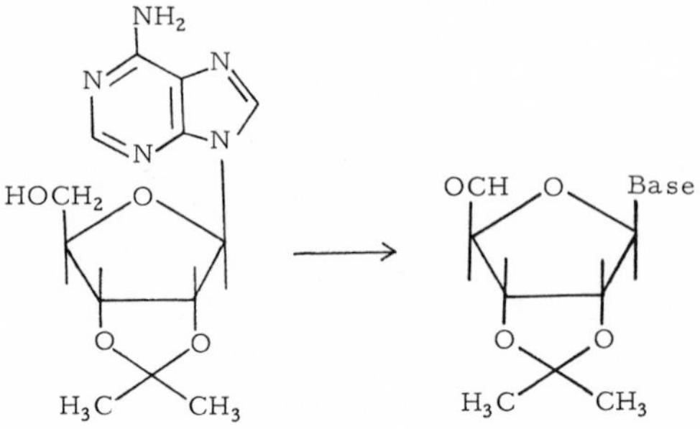

(2)   (3)

Four examples of this method of oxidation for hydroxy steroids have been reported,[19] but the yields in at least two of the cases are lower than those obtained by the original procedure.

The mechanism is believed to be similar to that suggested by Fenselau and Moffatt (**2**, 162) for DMSO–DCC oxidations.[20]

[1]V. R. Gaertner, *Tetrahedron*, **23**, 2123 (1967)
[2]*Idem, Org. Syn.*, submitted 1969
[3]S. J. Angyal, P. T. Gilham, and C. G. Macdonald, *J. Chem. Soc.*, 1417 (1957)
[4]S. J. Angyal and R. M. Hoskinson, *ibid.*, 2985 (1962).
[5]R. Gigg and C. D. Warren, *ibid.*, (C), 2367 (1969).

[6]H. M. Bell, C. W. Vanderslice, and A. Spehar, *J. Org.*, **34**, 3923 (1969)

[7]R. G. Charles, *Org. Syn.*, **39**, 61 (1959)

[8]T. M. Shepherd, *Chem. Ind.*, 567 (1970)

[9]R. Zimmerman and W. Dathe, German Pat. 1,178,840 (1965)

[10]R. G. Smith, A. Vanterpool, and H. J. Kulak, *Canad. J. Chem.*, **47**, 2015 (1969)

[11]W. W. Epstein and J. Ollinger, *Chem. Commun.*, 1338 (1970)

[12]D. Martin and A. Weise, *Angew. Chem., internat. Ed.*, **6**, 168 (1967)

[13]P. R. Story and S. R. Fahrenholtz, *Am. Soc.*, **86**, 1270 (1964)

[14]P. G. Gassman and D. S. Patton, *ibid.*, **90**, 7276 (1968). D. M. Lemal, R. A. Lovald and R. W. Harrington, *Tetrahedron Letters*, 2779 (1965), reported this transformation in a yield of 39% but with no experimental details.

[15]R. W. Hoffmann and R. Hirsch, *Ann.*, **727**, 222 (1969)

[16]W. H. Clement, T. J. Dangieri, and R. W. Tuman, *Chem. Ind.*, 755 (1969)

[17]C. L. Stevens and R. J. Gasser, *Am. Soc.*, **79**, 6057 (1957)

[18]R. E. Harmon, C. V. Zenarosa, and S. K. Gupta, *Chem. Commun.*, 327 (1969)

[19]*Idem, Chem. Ind.*, 1428 (1969); *J. Org.*, **35**, 1936 (1970)

[20]*Idem, Tetrahedron Letters*, 3781 (1969)

Dimethyl sulfoxide-derived reagent (a). Sodium methylsulfinylmethylide, 1, 310–313; 2, 166–169.

Tschugaeff reaction. In the Tschugaeff reaction[1] for the conversion of alcohols into olefins, the sodium salt of the alcohol is treated with carbon disulfide and then an alkyl halide, usually methyl iodide; the resulting xanthate is then pyrolyzed:

$$R\,CH_2CH_2OH + NaOH + CS_2 \longrightarrow R\,CH_2CH_2OCSSNa + H_2O$$

$$R\,CH_2CH_2OCSSNa + CH_3I \longrightarrow R\,CH_2CH_2OCSSCH_3 + NaI$$

$$R\,CH_2CH_2OCSSCH_3 \xrightarrow{\Delta} R\,CH{=}CH_2 + COS + CH_3SH$$

The most commonly encountered difficulty is the formation of the metal salt of the alcohol. Wynberg,[2] using standard procedures, was able to obtain the dixanthate (1) in only 5–10% yield. He then found that use of dimsylsodium as the base and

(1)

DMSO as the solvent (homogeneous reaction) gives the dixanthate in 50–60% yield. A footnote states that the xanthates of cholesterol and of *n*-octanol were obtained by this improved procedure in 90% yield.

1- and 2-Alkynes. Terminal alkynes can be prepared from the readily available terminal alkenes by bromination to 1,2-dibromoalkanes followed by dehydrobromination at room temperature for one hour with an excess (3.3 moles) of dimsylsodium in DMSO.[3] Reaction with sodamide in DMSO was only satisfactory at 65–70° and required nine hours. 1-Alkynes can be isomerized to 2-alkynes by sodamide in DMSO (30 hours, 65–70°). Use of dimsylsodium in this case leads to a mixture of 1-, 2-, and 3-alkynes.

2,3-Diphenyl-1,3-butadiene.[4] Diphenylacetylene is converted into 2,3-diphenyl-1,3-butadiene when treated with dimsyl sodium in DMSO. The yield of slightly impure product is 22–25%; recrystallization from methanol gives pure material, m.p. 47–48°, in 10.7–13.6% yield. The scope of this unusual new reaction is not known; the strongly basic conditions preclude functionality sensitive to strong base.

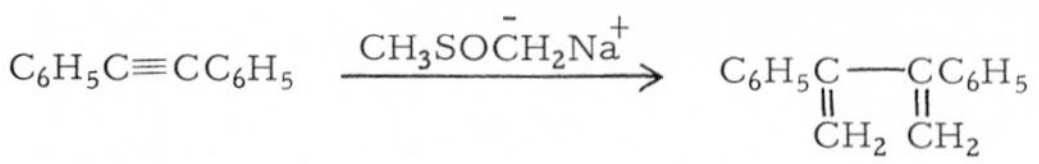

C6H5C≡CC6H5
CH3SOCH2Na+
C6H5C—CC6H5
CH2 CH2

[1]H. R. Nace, *Org. Reactions*, **12**, 57 (1962)
[2]Ae. de Groot, B. Evenhuis, and H. Wynberg, *J. Org.*, **33**, 2214 (1968)
[3]J. Klein and E. Gurfinkel, *Tetrahedron*, **26**, 2127 (1970)
[4]I. Iwai and J. Ide, *Org. Syn.*, **50**, 62 (1970)

Dimethyl sulfoxide-derived reagent (b). Dimethylsulfonium methylide, **1**, 314–315; **2**, 169–171.

Oxiranes. Canadian chemists[1] report that diaryl ketones of type (1) fail to

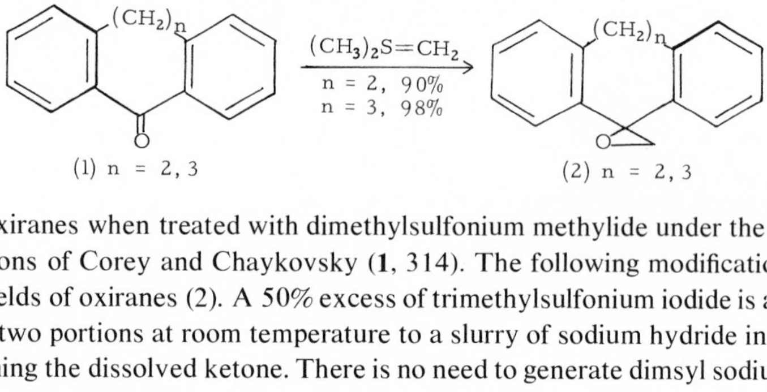

(CH2)n
(CH3)2S=CH2
n = 2, 90%
n = 3, 98%
(CH2)n
O

(1) n = 2, 3 (2) n = 2, 3

form oxiranes when treated with dimethylsulfonium methylide under the original conditions of Corey and Chaykovsky (**1**, 314). The following modification gives high yields of oxiranes (2). A 50% excess of trimethylsulfonium iodide is added in one or two portions at room temperature to a slurry of sodium hydride in DMSO containing the dissolved ketone. There is no need to generate dimsyl sodium as an ylide-forming reagent.

Indole synthesis. The reagent reacts with aromatic *o*-aminocarbonyl compounds to form indoles in 40–90% yield.[2]

O
CR
NH2
+ (CH3)2S=CH2
-H+
O-
CR
CH2S(CH3)2
N-
H
-(CH3)2S
H+
OH
CR
H
N
H
H
-H2O
R
N
H

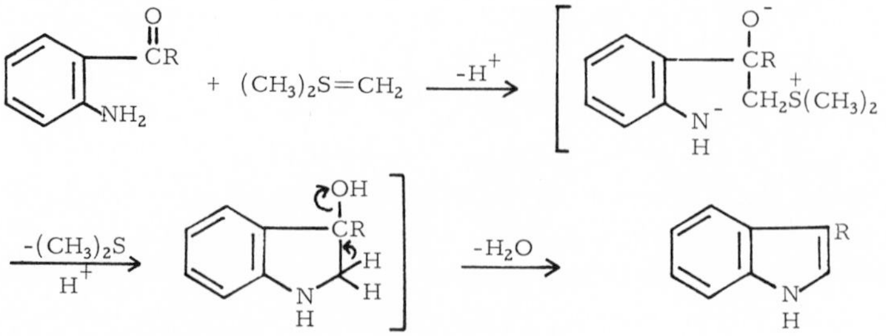

[1]K. Ackermann, J. Chapuis, D. E. Horning, G. Lacasse, and J. M. Muchowski, *Canad. J. Chem.*, **47**, 4327 (1969)
[2]P. Bravo, G. Gaudiano and A. Umani-Ronchi, *Tetrahedron Letters*, 679 (1969)

Dimethyl sulfoxide-derived reagent (c). Dimethyloxosulfonium methylide, **1**, 315–318; **2**, 171–173. The procedure cited for the preparation and use of the reagent (**2**, 171, ref. 26) has now been published.[1]

The reagent reacts stereospecifically with (−)-carvone (**1**) to give the bicyclic ketone (**2**).[2] Reduction of (**2**) followed by oxidation with Jones reagent gives an

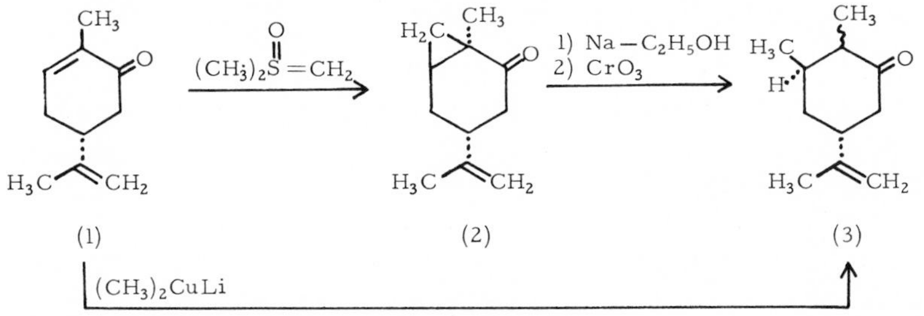

equilibrium mixture of the two methyldihydrocarvones (**3**), also obtained by the addition of dimethylcopperlithium to carvone (**1**).

Steroidal spirocyclobutanones. The reaction of 17-halo-20-ketosteroids such as (**1**) with a 2.4-fold excess of dimethyloxosulfonium methylide results in formation of the cyclobutanones (**3**) and (**4**), formed presumably by way of a spirocyclopropanone (**2**) (Favorsky rearrangement).[3]

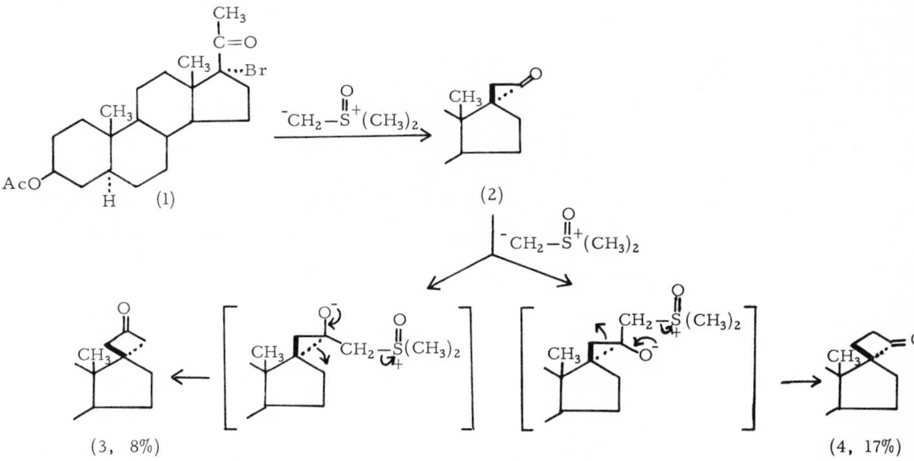

Synthesis of a luciferin. *Latia* luciferin (**4**), the specific substrate for the luciferase system of *Latia neritoides*, has been synthesized in three steps from dihydro-β-ionone (**1**). Reaction of (**1**) with dimethyloxosulfonium methylide, both

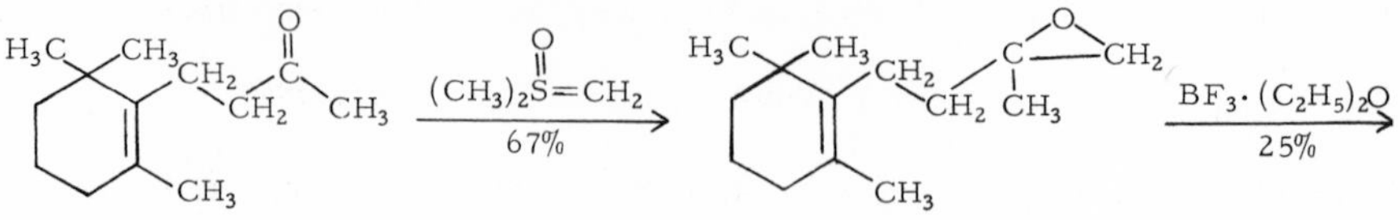

(1) → (CH₃)₂S=CH₂ 67% → (2) → BF₃·(C₂H₅)₂O 25% →

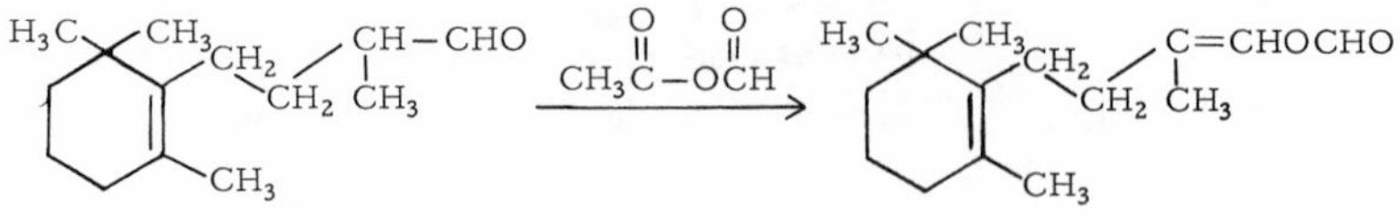

(3) → CH₃C—OCH (acetic-formic anhydride) → (4)

in DMSO, gives the epoxide (2), which is rearranged to luciferin aldehyde (3) by boron trifluoride etherate. The completing step is reaction of (3) with acetic–formic anhydride (**1**, 4; **2**, 10–12).[4]

Methylation. It was reported a few years ago that the reagent methylates acidic NH and OH groups and also some aromatic hydrocarbons (**1**, 317, ref. 22). Recently N-methylation of heterocycles has been noted.[5] Thus the reagent (3 moles) reacts with 6-benzyladenine (1) in THF to give 9-methyl-6-benzyladenine (2) in 63% yield. 6-Benzyl-9-ethyladenine (3) is obtained as a by-product (10%

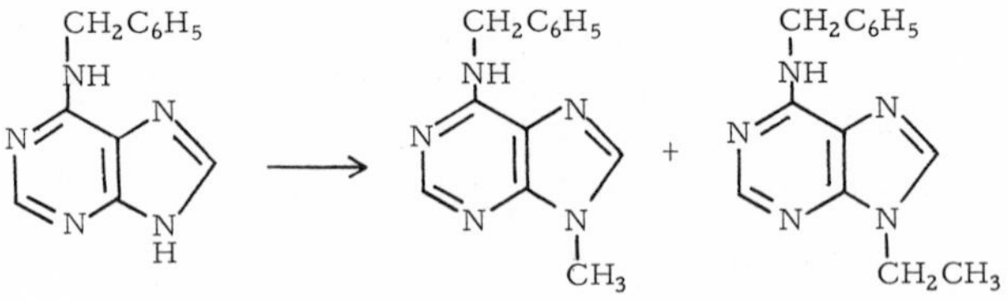

(1)   (2, 63%)   (3, 10% yield)

yield). The by-product is formed by alkylation by the rearranged ylide, dimethyl-oxosulfonium ethylide. Benzimidazole is converted in this way into the 1-methyl derivative (80% yield), and pyrrole is methylated to give 1-methylindole. 1,2,3,4-Tetrahydroharman is also methylated in the pyrrole ring to give 9-methyltetra-hydroharman (4, 90% yield). Oxindole undergoes both N- and C-alkylation.

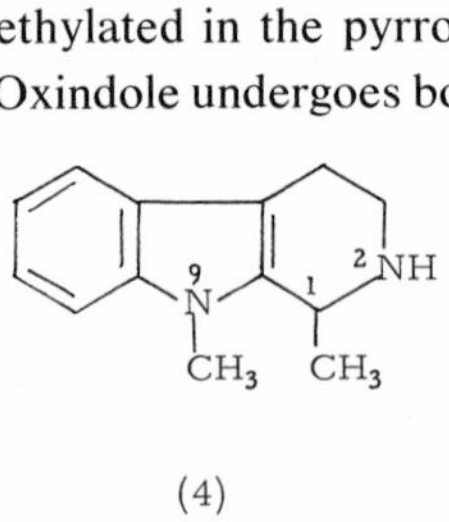

(4)

[1]E. J. Corey and M. Chaykovsky, *Org. Syn.*, **49**, 78 (1969)
[2]M. Narayanaswamy, V. M. Sathe, and A. S. Rao, *Chem. Ind.*, 921 (1969)
[3]R. Wiechert, *Angew. Chem., internat. Ed.*, **9**, 237 (1970)

[4]M. G. Fracheboud, O. Shimomura, R. K. Hill, and F. H. Johnson, *Tetrahedron Letters*, 3951 (1969)
[5]T. Kunieda and B. Witkop, *J. Org.*, **35**, 3981 (1970)

N,N-Dimethylthiocarbamoyl chloride, $(CH_3)_2NC(=S)Cl$, **2**, 173–179.

Preparation. The reagent has more recently been prepared[1] by chlorination of bis(dimethylaminothiocarbamoyl)disulfide (Thiram[2]). Compare the previously reported preparation of dimethylthiocarbamoyl chloride.[3]

$$(CH_3)_2N-\overset{\overset{S}{\|}}{C}-S-S-\overset{\overset{S}{\|}}{C}-N(CH_3)_2 \xrightarrow[80\%]{Cl_2} 2\,(CH_3)_2N-\overset{\overset{S}{\|}}{C}-Cl$$

Olefin synthesis from alcohols.[1] Primary and secondary alcohols containing a β-hydrogen can be converted into olefins by conversion into the alkoxide (sodium hydride, DMF) and then reaction of the alkoxide with N,N-dimethylthiocarbamoyl chloride to form an O-alkyl dimethylthiocarbamate. These derivatives on heating to 180–200° for 2 hours decompose to form olefins. The other product is dimethylammonium dimethylthiocarbamate (**2**). The complete sequence is formulated as follows:

$$>CH\overset{|}{\underset{|}{C}}OH \xrightarrow[75-95\%]{\begin{array}{l}1)\ NaH\\2)\ (CH_3)_2N\overset{\overset{S}{\|}}{C}Cl\end{array}} >CH\overset{|}{\underset{|}{C}}OCSN(CH_3)_2 = -\overset{|}{\underset{|}{C}}\overset{H}{\diagup}\ \overset{S}{\underset{\searrow}{C}}N(CH_3)_2 \longrightarrow \overset{C}{\underset{\wedge}{\|}} + \overset{H}{\underset{O}{\diagdown}}\overset{S}{\underset{\diagup}{C}}-N(CH_3)_2$$

$$65-90\%\qquad(1)$$

$$(1) \longrightarrow (CH_3)_2NH + COS$$

$$(1) + (CH_3)_2NH \longrightarrow (CH_3)_2NCOS^-(CH_3)_2NH_2^+$$

$$(2)$$

The method is comparable to the xanthate method[4] but has certain advantages. The preparation of xanthates requires two steps: reaction of the alkoxide with carbon disulfide followed by alkylation. Yields of olefins by the present method are at least as high as and sometimes higher than those obtained from xanthates.

Desoxy sugars.[5] Photolysis of methanolic solutions of O-alkyl dimethylthiocarbamates of sugar derivatives gives the corresponding desoxy sugar in fair yield. Thus photolysis of (1) with a 450-watt mercury arc lamp for 200 hours gives (2) in 25% yield. The original alcohol is also formed to some extent by simple

$$(1) \xrightarrow[25\%]{h\nu} (2)$$

cleavage. The alcohol can be recycled to give improved yields. The reduction process appears to involve rearrangement to the S-dimethylcarbamoyl derivative of a desoxythio sugar, which then undergoes C–S homolysis with capture of a hydrogen atom from the solvent to give the product.

Phenols → sodium arylsulfonates. The S-aryl dimethylthiocarbamates [see formula (2), **2**, 173] can also be oxidized to arylsulfonic acids, isolated as the sodium salts.[6] Hydrogen peroxide is used as oxidant and formic acid as solvent. Yields are 55–64%.

$$ArSCON(CH_3)_2 \xrightarrow{[O]} ArSO_3H$$

[1]M. S. Newman and F. W. Hetzel, *J. Org.*, **34**, 3604 (1969)

[2]Penn Salt Manufacturing Co.

[3]R. H. Goshorn, W. W. Levis, Jr., E. Jaul, and E. J. Ritter, *Org. Syn., Coll. Vol.*, **4**, 307 (1963)

[4]H. R. Nace, *Org. Reactions*, **12**, 57 (1962)

[5]R. H. Bell, D. Horton, and D. M. Williams, *Chem. Commun.*, 323 (1968)

[6]J. E. Cooper and J. M. Paul, *J. Org.*, **35**, 2046 (1970)

N,N-Dimethylthioformamide, $(CH_3)_2NC(H){=}S$. Mol. wt. 89.16, b.p. 77.0–77.6°/3.4 mm. Supplier: Eastman.

Preparation. This thioformamide is prepared[1] by reaction of dimethylformamide with phosphorus pentasulfide (56% yield).

Formamidines. The reagent reacts with primary amides to give formamidines. For example, α-naphthylamine (1) gives the base (2).[1]

Vilsmeier reaction. A British group[2] has reported several cases in which the combination of N,N-dimethylthioformamide with acetic anhydride gave definitely better yields of formyl derivatives of aromatic hydrocarbons than the usual Vilsmeier reagent (DMF–POCl₃).

[1]G. R. Pettit and L. R. Garson, *Canad. J. Chem.*, **43**, 2640 (1965)

[2]J. G. Dingwall, D. H. Reid, and K. Wade, *J. Chem. Soc.*, (C), 913 (1969)

Dimethylzinc. Mol. wt. 95.44, very pyrophoric.

Preparation.[1] A 1-l. Erlenmeyer flask with a 24/40 T-joint is charged with 600 ml. of acetic acid and 8.0 g. of Cu(OAc)₂. The mixture is brought to a boil to dissolve the acetate and cooled for a minute. Then 500 g. of 80-mesh granular zinc is added all at once. After 15 minutes, when the blue color of Cu²⁺ is dissipated, the acetic acid solution is cooled and decanted. The residue is washed with 500 ml. of acetic acid, three 300-ml. portions of ether, and finally with 500 ml. of ether. The residual zinc–copper couple is dried under a continuous flow of hydrogen of about 250 ml./min. while heating the bottom of the flask with a heat gun. When the zinc–copper couple appears dry, it is heated with the heat gun for an additional 15 minutes. To the dry zinc–copper couple, 142 g. of methyl iodide (1.0 mole) and 2.0 g. of methyl acetate as catalyst are added all at once. The flask is flushed with dry nitrogen, fitted with a reflux condenser, and heated in an oil bath at 60–65° for 80 hours, at which time refluxing appears to stop. A positive pressure of nitrogen is maintained throughout.

The flask is cooled, flushed with dry nitrogen, and the dimethylzinc is bulb-to-bulb distilled under nitrogen to a 100-ml. receiver flask cooled with a dry ice–trichloroethylene bath. Distillation is accomplished by slowly heating the still pot to 200° (2 hours). The distillate weighs 33 g. (70%) and is poured under a positive pressure of nitrogen into 300 ml. of dry, oxygen-free pentane. The molarity of the solution is taken to be $0.90\,M$. The solution is stored conveniently and safely at dry ice temperature.

Synthesis of cyclohexyl methyl ketone.[1] One hundred milliliters of dry benzene and 14.66 g. (0.10 mole) of cyclohexylcarbonyl chloride are poured into a 500-ml.

round-bottomed three-necked flask equipped with a thermometer, reflux condenser, 125-ml. pressure-equalizing addition funnel, and magnetic stirrer. After the flask has been flushed with dry nitrogen for 15 minutes, 90 ml. of $0.9\,M$ dimethylzinc in dry pentane is slowly added under nitrogen pressure to the stirred solution over a 40–45 minute period. The temperature rises to about 40°, where it is checked by ice cooling. On completion of the addition the reaction mixture is heated with stirring at 40–45° for 30 minutes and then cooled to room temperature. One hundred milliliters of saturated ammonium chloride solution is added cautiously (methane evolution). The layers are separated and the aqueous solutions extracted with two 35-ml. portions of pentane. The combined organic solution is washed with 75-ml. portions of saturated sodium bicarbonate solution and saturated sodium chloride solution and the solvent is removed by distillation through a 35-cm. vacuum-jacketed packed column (glass beads) fitted with a variable ratio reflux head to give a residue which on further distillation affords 9.20–9.65 g. (73–77%) of cyclohexyl methyl ketone, b.p. 63–64°/12 mm. The product gives a single peak on vapor phase chromatography, and its NMR spectrum has a multiplet 7.40–9.05 τ (11H) and a singlet 7.83 τ (3H).

[1]K. B. Wiberg, G. N. Taylor, and G. J. Burgmaier, *Org. Syn.*, submitted 1970

Dinitrogen tetroxide, 1, 324–329; **2**, 175–176.

1-Nitrocyclooctene (**2**, 175). The procedure of Seifert[1] for preparation of 1-nitrocyclooctene by the reaction of cyclooctene with dinitrogen tetroxide has been published.

[1]W. K. Seifert, *Org. Syn.*, **50**, 84 (1970)

Dinitrogen tetroxide–Iodine, N_2O_4–I_2.

Reaction with olefins. The addition of dinitrogen tetroxide alone is not generally a useful synthetic system because of the formation of several products. However, Stevens and Emmons[1] found that the reaction of dinitrogen tetroxide with an olefin (ether solution, N_2) in the presence of excess iodine led to β-nitroalkyl iodides, usually in good yield:

$$CH_3CH_2CH{=}CH_2 \xrightarrow[62\%]{} CH_3CH_2\underset{\underset{I}{|}}{CH}{-}CH_2NO_2$$

Furthermore the dehydrohalogenation of the product provides a convenient path to α-nitroolefins.

This sequence was used for the synthesis of *t*-butylnitroacetylene, the first representative of compounds[2] of this class. Thus the reaction of dinitrogen tetroxide and iodine with *t*-butylacetylene (1) gave a mixture of the *cis*- and *trans*-isomers (2), which was dehydrohalogenated by distillation from KOH under vacuum. The nitroacetylene is highly reactive as a dienophile, a dipolarophile and an electrophile. It is surprisingly stable to shock and has a half-life of 2–3 days at room temperature.

$$(CH_3)_3CC{\equiv}CH \xrightarrow[85\%]{N_2O_4 + I_2} (CH_3)_3CCI{=}CHNO_2 \xrightarrow[94\%]{\substack{KOH\ -100^0 \\ 0.1\ torr}} (CH_3)_3C{\equiv}CNO_2$$

$$(1) \qquad\qquad\qquad\qquad (2) \qquad\qquad\qquad\qquad (3)$$

[1]T. E. Stevens and W. D. Emmons, *Am. Soc.*, **80**, 338 (1958)
[2]V. Jäger and H. G. Viehe, *Angew. Chem., internat. Ed.*, **8**, 273 (1969)

Dioxane dibromide, 1, 333–334; **2**, 177.

Bromination. Monobromination of the highly reactive resorcinol dimethyl ether (1a) and 2-methylresorcinol dimethyl ether (1b) can be effected in 80–90% yield with dioxane dibromide.[1]

(1a, R = H)
(1b, R = CH$_3$)

(2a, R = H)
(2b, R = CH$_3$)

2,2,5-Tribromocyclopentanone ethylene ketal (2). Chapman *et al.*[2] report that attempted bromination of cyclopentanone ethylene ketal (1) with pyridinium bromide perbromide, trimethylphenylammonium perbromide, and with molecular bromine in a range of solvents all failed. Bromination was achieved success-

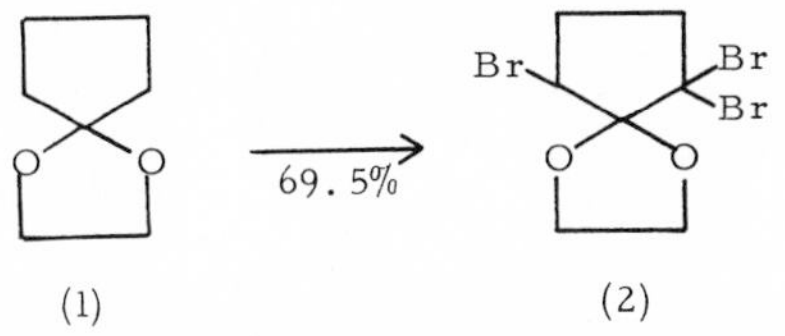

(1)   (2)

fully with dioxane dibromide. Actually it is not necessary to isolate and purify the reagent if bromine is added to the ketal in dioxane as solvent. Note, however, that Garbisch[3] reports successful bromination of (1) with bromine.

[1]D. C. Schlegel, C. D. Tipton, and K. L. Rinehart, Jr., *J. Org.*, **35**, 849 (1970)
[2]N. B. Chapman, J. M. Key, and K. J. Toyne, *ibid.*, **35**, 3860 (1970)
[3]E. W. Garbisch, Jr., *J. Org.*, **30**, 2109 (1965)

Dioxane diphosphate, 1, 335–335. Supplier: Aldrich.

Diphenylcarbodiimide, 1, 337–338; **2,** 178.
    Preparation (**2,** 178). The procedure cited has now been published.[1]

[1]P. Rajagopalan, B. G. Advani, and C. N. Talaty, *Org. Syn.*, **49,** 70 (1969)

Diphenylketene, 1, 343–345.
    2,2-Diphenylcyclobutanones. Contrary to earlier reports, diphenylketene undergoes cycloaddition with simple alkenes to give 2,2-diphenylcyclobutanones.[1] However, a long reaction time is needed for satisfactory yields. *cis*-Olefins react much faster than *trans*-olefins. Thus the reaction with *cis*-butene-2 requires 3 days, but the reaction with *trans*-butene-2 requires 3 months. With low-boiling alkenes, the reaction is carried out in a sealed tube or in an autoclave.
    Examples:

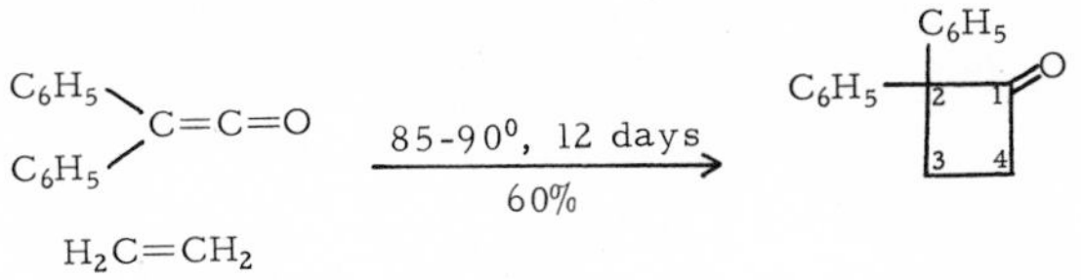

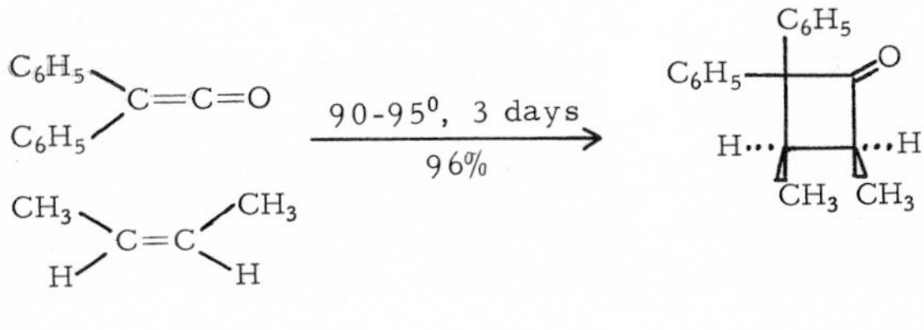

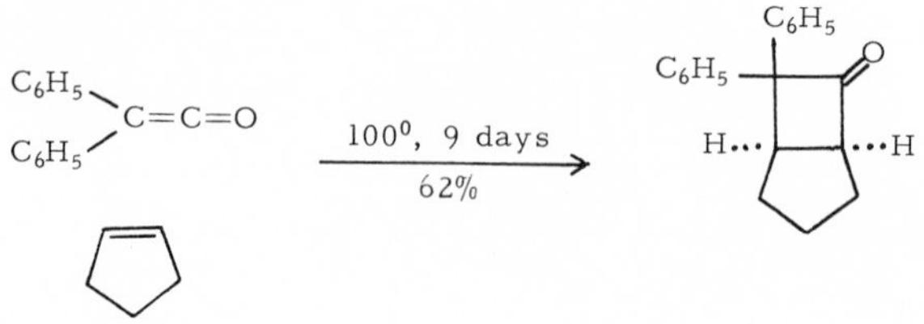

[1]R. Huisgen and L. A. Feiler, *Ber.*, **102**, 3391–3404 (1969)

1,3-Diphenylnaphtho[2.3-*c*]furane (2). Mol. wt. 320.27, m.p. 148–154°, deep red plates.

Preparation.[1] This quinonoid furane is obtained in a six-step synthesis from 2,3-dimethylnaphthalene, the last step of which involves dehydration of the lactol (1). This was accomplished in high yield by heating a suspension of the lactol in

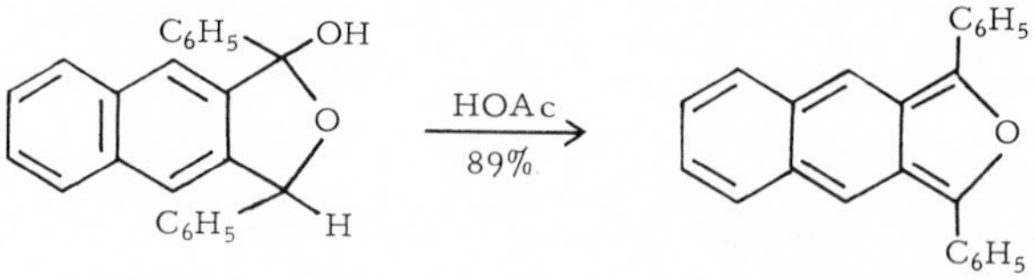

a small amount of acetic acid for a few minutes on the steam bath. Fortunately (2) was obtained directly crystalline; it decomposes on attempted recrystallization. Use of dilute hydrochloric acid for dehydration gave an amorphous product. In the absence of light and air, (2) is fairly stable in the solid state. It is described as "the first known substance with the 2,3-naphthoquinonoid structure," but this description seems to us to apply to pentacene.

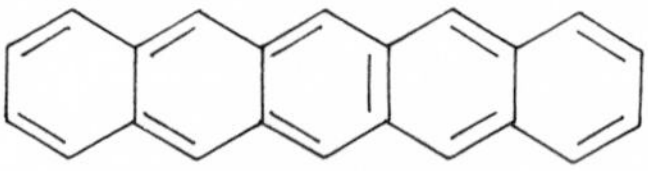

Diels-Alder reactions.[1] Like 1,3-diphenylisobenzofurane (**1**, 342–343), (2) enters into Diels-Alder reactions but it is even more reactive. Thus it reacts with dimethyl acetylenedicarboxylate and 1,4-benzoquinone at room temperature within a few minutes to give the expected adducts. It appears to be valuable for interception of particularly reactive transient dienophiles. Thus, like 1,3-diphenylisobenzofurane, this diene has been used to trap the transient 1-bromobenzo-cyclobutadiene. Treatment of *trans*-1,2-dibromobenzocyclobutene (1) with a large excess of potassium *t*-butoxide in benzene in the presence of the furane gives the Diels-Alder adduct (2) in 52% yield. The bromine atom can be eliminated by

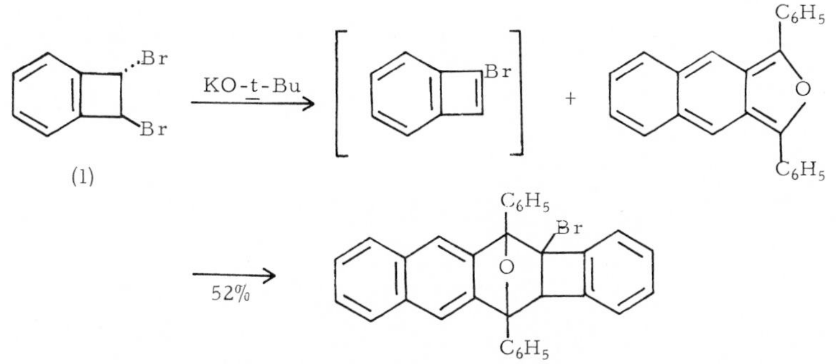

hydrogenolysis. For conversion of (2) into condensed cyclobutane aromatic systems, *see* **Phorphorus pentasulfide** (This volume).

[1]M. P. Cava and J. P. VanMeter, *J. Org.*, **34**, 538 (1969)

Diphenylphosphorochloridate, **1**, 345–346; **2**, 180.

Nitriles.[1] Aldoximes when treated with the reagent and triethylamine are converted at room temperature into nitriles in high yield. The reaction is considered to involve esterification followed by Beckmann fragmentation:

$$RCH=NOH + (C_6H_5O)_2\overset{O}{\overset{\|}{P}}Cl \xrightarrow{(C_2H_5)_3N} [RCH=NO\overset{O}{\overset{\|}{P}}(OC_6H_5)_2] \longrightarrow RC\equiv N + (C_6H_5O)_2\overset{O}{\overset{\|}{P}}OH$$

Alternatively, diphenylphosphorochloridate can be generated *in situ* by the reaction of diphenyl hydrogen phosphonate and carbon tetrachloride:

$$(C_6H_5O)_2\overset{O}{\overset{\|}{P}}H + CCl_4 \xrightarrow{(C_2H_5)_3N} (C_6H_5O)_2\overset{O}{\overset{\|}{P}}Cl + CHCl_3$$

[1]P. J. Foley, Jr., *J. Org.*, **34**, 2805 (1969)

Diphenylsulfonium cyclopropylide, $(C_6H_5)_2\overset{+}{S}\!-\!\overset{-}{\triangleleft}$ Mol. wt. 227.34.

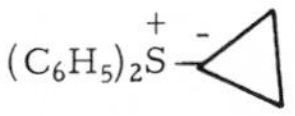

The reagent is prepared as a yellow slurry by the reaction of triphenylsulfonium fluoroborate with cyclopropyllithium:

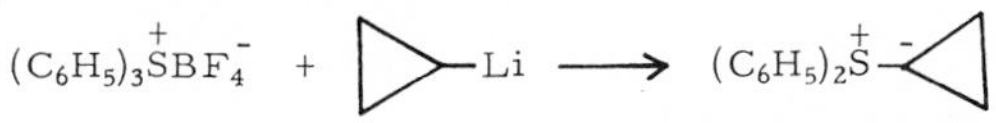

$$(C_6H_5)_3\overset{+}{S}BF_4^- + \triangleright\!-Li \longrightarrow (C_6H_5)_2\overset{+}{S}\!-\!\overset{-}{\triangleleft}$$

Spiroannelation.[1] Cyclohexanone reacts with the reagent to give the spiro-cyclobutanone (1) by way of the intermediates outlined.

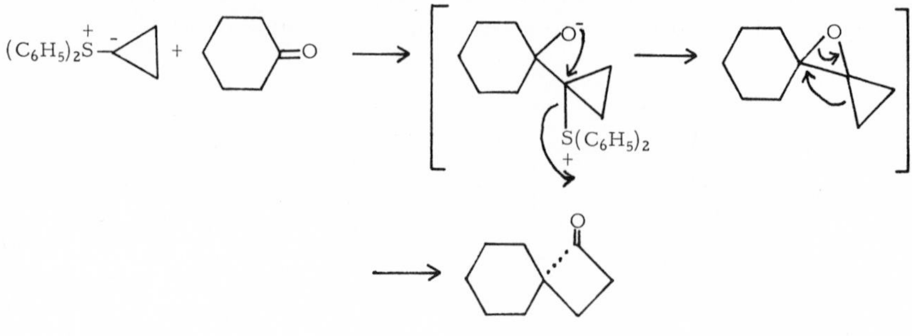

(1)

[1]B. M. Trost, R. LaRochelle, and M. J. Bogdanowicz, *Tetrahedron Letters*, 3449 (1970)

Diphenyltin dihydride, $(C_6H_5)_2SnH_2$, 1, 349. Supplier: Alfa Inorganics.

Reduction of alkyl halides. In a study of reduction of alkyl halides by organo-tin hydrides, Kuivila and Menapace[1] observed the following reactivity sequence: $(C_6H_5)_2SnH_2 \cong C_4H_9SnH_3 > (C_6H_5)_3SnH \cong (n\text{-}C_4H_9)_2SnH_2 > (n\text{-}C_4H_9)_3SnH$. Azobisisobutyronitrile is an efficient catalyst.

Diphenyltin dihydride was found to be effective for reduction of 1,4,7,7-tetrachloronorbornane (1) to 1,4-dichloronorbornane (2). Use of tri-*n*-butyltin hydride led only to reduction to 1,4,7-trichloronorbornane.[2]

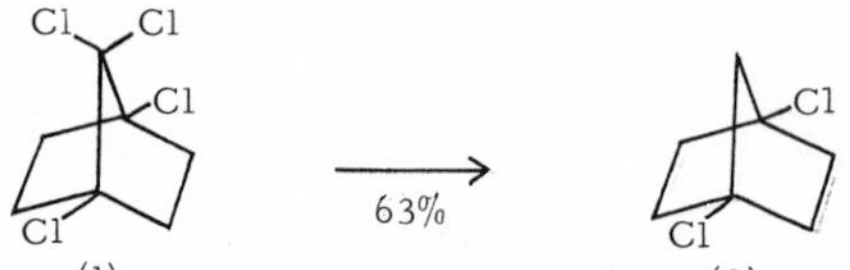

[1]H. G. Kuivila and L. W. Menapace, *J. Org.*, **28**, 2165 (1963)
[2]A. P. Marchand and W. R. Weimar, Jr., *ibid.*, **34**, 1109 (1969)

Disodium platinum tetrachloride, Na_2PtCl_4.

This salt, which is soluble in acetic acid, is recommended as a homogeneous catalyst for exchange of hydrogen in aromatic hydrocarbons for deuterium.[1] The substrate, acetic acid, heavy water, and hydrochloric acid are allowed to react in an evacuated, sealed tube at 25–120°. Aliphatics exchange only slowly by this technique. No dimerization (e.g., benzene → diphenyl) is observed. This reaction is observed with heterogeneous platinum-catalyzed exchange with heavy water.

[1]J. L. Garnett and R. J. Hodges, *Am. Soc.*, **89**, 4546 (1967); G. E. Calf and J. L. Garnett, *Chem. Commun.*, 373 (1969)

1,3-Dithiane, 2, 182–187. Supplier: Aldrich.

Preparation (**2**, 182, ref. 1). The procedure has now been published.[1]

Reaction with steroid oxides. 2-Lithio-1,3-dithiane reacts with 5α-cholestane-2α,3α-oxide (1) to give the 2β-dithianyl-3α-hydroxy derivative (2). Desulfuriza-

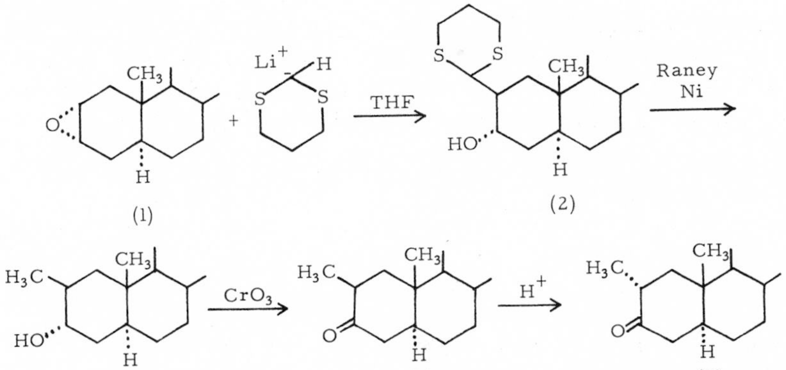

(1)   (2)

(3)   (4)   (5)

tion then gives 2β-methyl-5α-cholestane-3α-ol (3), which on oxidation with Jones reagent yields the 2β-methyl-3-ketone (4). This can be readily epimerized by ethanolic sulfuric acid to the more stable 2α-methyl-3-ketone (5), in which the alkyl group is equatorial. The same sequence of reactions when applied to 5α-cholestane-2β,3β-oxide gives 3α-methyl-5α-cholestane-2-one, epimerizable to the more stable 3β-methyl-2-ketone. The sequence of reactions is thus useful for stereospecific alkylation of steroids. In contrast, methylmagnesium iodide reacts with steroid epoxides such as (1) to give predominantly ring contraction products.[2]

The reaction has been extended to oxiranes for introduction of a corticoid-like function at C_{17}. Thus the 17-ketone (6) is converted in high yield into the 17β-oxirane (7) by reaction with dimethylsulfonium methylide. This oxide reacts with

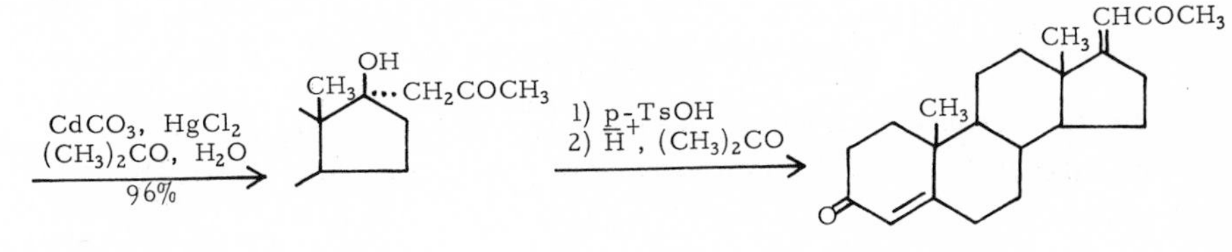

(6)   (7)   (8)

(9)   (10)

2-methyl-1,3-dithiane anion to give the 17β-hydroxydithianyl product (8) in 68% yield. Hydrolysis by the usual method gives (9) in 96% yield. The C_3-acetal group is removed with p-toluenesulfonic acid in acetone to give the Δ^4-3-ketone (78% yield), which upon more prolonged acid treatment is dehydrated to the dienedione (10) in 68% yield. The C_{17}-function in (10) is related to intermediates used in elaboration of the corticoid side chain.[3]

Aldehyde synthesis. Vedejs and Fuchs[4] report that the usual methods for hydrolysis of a 2,2-dialkyl-1,3-dithiane ($HgCl_2 + HgO$ or $CdCO_3$) to a ketone give low yields in the hydrolysis of a 2-alkyl-1,3-dithiane (1) to an aldehyde (2). They

$$RCH(OAc)_2 \xleftarrow[BF_3, \ HOAc]{Hg(OAc)_2} \quad \underset{(1)}{\text{dithiane}} \quad \xrightarrow[H_2O-THF]{HgO-BF_3} RCHO$$

(3)   (1)   (2)

report that hydrolysis can be accomplished in 60–90% yield by red mercuric oxide and boron trifluoride etherate in aqueous THF. Use of mercuric acetate and boron trifluoride etherate in acetic acid effects transacetalization to acetaldiacetates (3).

The new method was used in a synthesis of 4,4-diethoxy-2-butenal (7) as shown.

$$(C_2H_5O)_2CHCH\!-\!\!-\!CH_2 \xrightarrow[90\%]{\substack{1) \ \text{dithiane-Li} \\ 2) \ (CH_3CO)_2O}} (C_2H_5O)_2CHCHOAcCH_2\!-\!\overset{S}{\underset{H}{C}}\!-\!S$$

(4)   (5)

$$\xrightarrow[80\%]{HgO-BF_3} (C_2H_5O)_2CHCHOAcCH_2CHO \xrightarrow[77\%]{DBU} (C_2H_5O)_2CHCH=CHCHO$$

(6)   (7)

Reaction of 2-lithiodithiane and 1,1-diethoxy-2,3-epoxypropane (4) gave the expected alcohol, which was acetylated to give (5). Hydrolysis by the new procedure gave 3-acetoxy-4,4-diethoxybutanal (6). Dehydroacetoxylation was then effected with DBU (1,5-diazabicyclo[5.4.0]undec-5-ene, **2**, 101).

[1]E. J. Corey and D. Seebach, *Org. Syn.*, **50**, 72 (1970)
[2]J. B. Jones and R. Grayshan, *Chem. Commun.*, 141 (1970)
[3]*Idem, ibid.*, 741 (1970)
[4]E. Vedejs and P. L. Fuchs, *J. Org.*, **36**, 366 (1971)

Di-(2,2,2-trichloroethyl)phosphorochloridate, **2**, 187–188. Supplier: Aldrich.

Dowtherm A, **1**, 353. The preparation of 3-isoquinuclidone has now been published.[1]

[1]W. M. Pearlman, *Org. Syn.*, **49**, 75 (1969)

E

N-Ethoxycarbonyl-2-ethoxy-1,3-dihydroquinoline (EEDQ), 2, 191.

Peptide synthesis. Izumiya and Muraoka[1] examined the extent of isomerization with commonly used coupling reagents (isobutyl chloroformate, DCC, N-ethyl-5-phenylisoxazolium-3'-sulfonate, and EEDQ) and of these found that EEDQ caused the least racemization (0.2%) in the particular test chosen. The next most satisfactory was N-ethyl-5-phenylisoxazolium-3'-sulfonate (Woodward's Reagent K, 1.7–1.8%).

[1]N. Izumiya and M. Muraoka, *Am. Soc.*, **91,** 2391 (1969)

Ethoxyketene, $C_2H_5OCH=C=O$. Mol. wt. 98.10.

The reagent is generated *in situ* by dehydrochlorination of ethoxyacetyl chloride with triethylamine at $-78°$. It appears to be fairly stable at this temperature but slowly polymerizes at room temperature. It is also formed to some extent by photochemical Wolff rearrangement of ethyl diazoacetate.

Cycloaddition to olefins.[1] The reagent is generated as above in the presence of an olefin, and the reaction mixture is then sealed in a heavy glass tube and heated at 80° for 15 hours. Cyclobutanes are formed in yields of 30–56%.

Examples:

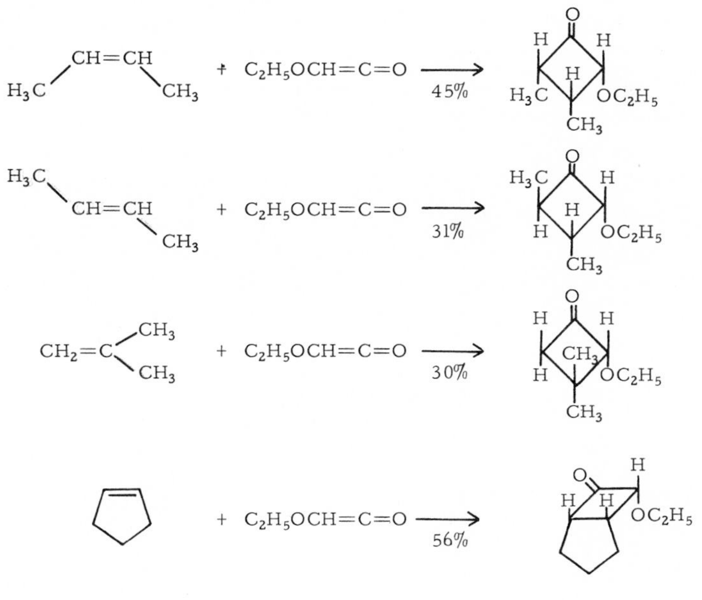

The first two examples show that the geometrical configuration of the parent olefin is maintained. Also the orientation of the ethoxy group is *syn* to the adjacent substituent in the products. The cycloadduct of an unsymmetrical olefin has a substituent vicinal to the ethoxy group (head-to-head adduct). The addition reaction is subject to steric effects; *trans*-2-butene reacts about twice as fast as *cis*-2-butene.

[1]T. DoMinh and O. P. Strausz, *Am. Soc.*, **92**, 1766 (1970)

Ethyl azidoformate, 1, 363–364; **2**, 191–192.

Ethyl azidoformate reacts with α-keto and α-ester phosphorus ylides to give 1-carboethoxy-*vic*-triazoles.[1]

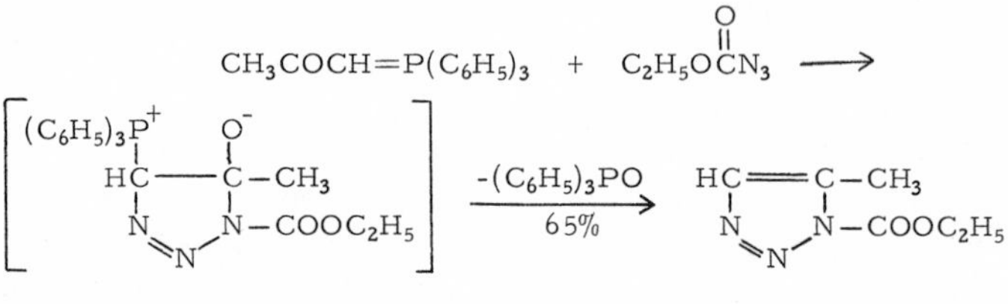

$$CH_3COCH=P(C_6H_5)_3 \quad + \quad C_2H_5O\overset{\overset{O}{\|}}{C}N_3 \quad \longrightarrow$$

[1]G. L'abbé and H. J. Bestmann, *Tetrahedron Letters*, 63 (1969)

Ethyl diazoacetate, 1, 367–370; **2**, 193–195.

Synthesis of methyl sterculate. Gensler *et al.*[1] found that the reaction of the Simmons-Smith reagent with methyl stearolate (1) to form a cyclopropene is unsuccessful. They then developed a successful method which gives sterculic acid (5) in *ca.* 30% yield. Methyl stearolate (1) was treated with ethyl diazoacetate in the presence of copper bronze to give the diester of the cyclopropene diacid; hydrolysis gave the diacid (2) in 70–90% yield. Decarbonylation of such acids has been effected previously with perchloric acid–acetic anhydride,[2] but Gensler found that the corresponding diacid chloride (3) is selectively decarbonylated by zinc chloride to give (4); esterification followed by sodium borohydride reduction gives methyl sterculate (5).

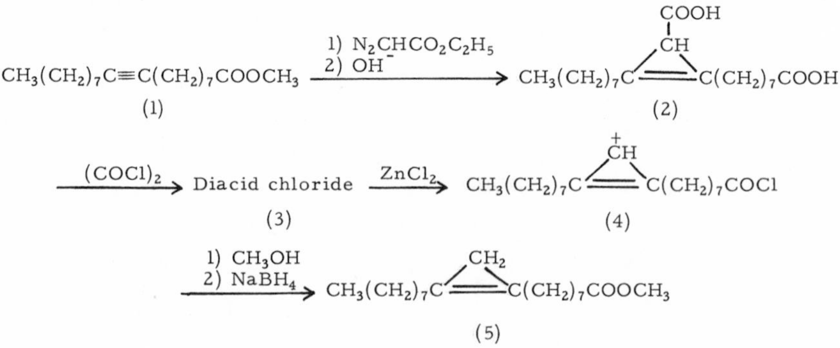

Reaction with ketones (**1**, 369–370). Mock and Hartman[3] have improved the original procedure for synthesis of β-keto esters from ketones by using triethyloxonium fluoroborate as the Lewis acid catalyst rather than boron trifluoride. Thus cyclohexanone (1) is converted into 2-carboethoxycycloheptanone (2) in 90% yield under the new conditions; the yield using BF_3 was 38%. The reaction is

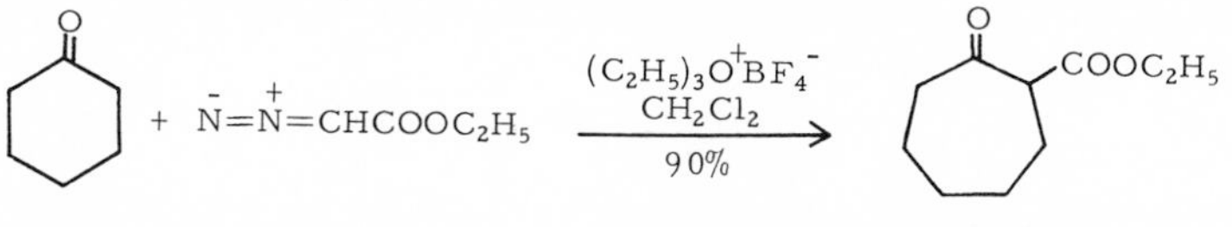

also applicable to acyclic ketones; for example, acetone is converted into ethyl 2-methylacetylacetate (3) in 78% yield.

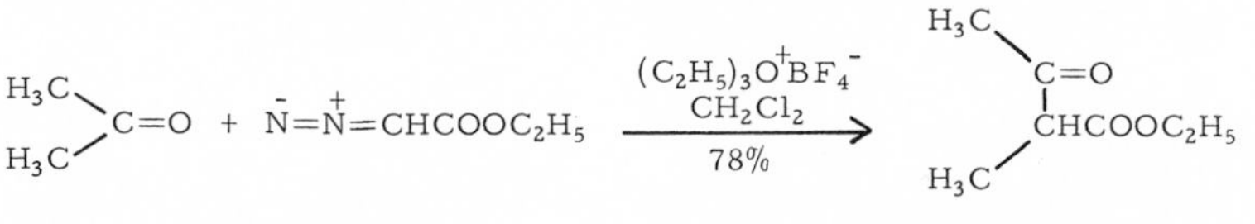

The intramolecular counterpart of this reaction is a promising synthetic route to polycyclic diketones. For example, the diazo ketone (4) on treatment with triethyloxonium fluoroborate gives the diketone (5) in high yield.

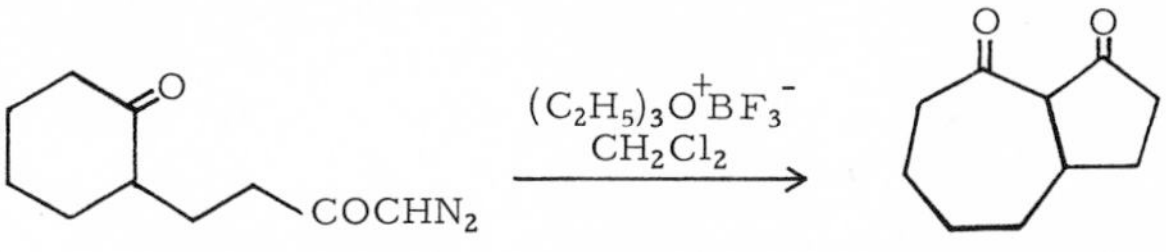

[1] W. J. Gensler, M. B. Floyd, R. Yanase, and K. Pober, *Am. Soc.*, **91**, 2397 (1969)
[2] F. L. Carter and V. L. Frampton, *Chem. Rev.*, **64**, 497 (1964); A. W. Krebs, *Angew. Chem., internat. Ed.*, **4**, 10 (1965)
[3] W. L. Mock and M. E. Hartman, *Am. Soc.*, **92**, 5767 (1970)

Ethyl diphenylphosphonite, $C_2H_5OP(C_6H_5)_2$. Mol. wt. 230.24, liq., b.p. 179°/14 mm., d_0^0 1.0896.

Benzyne. 1-Nitrosobenzotriazole (1), prepared by the reaction of benzotriazole with nitrosyl chloride, reacts vigorously with the reagent at room temperature in benzene solution with evolution of nitrogen. Reaction in the presence of tetraphenylcyclopentadienone (**1**, 1149–1150) gives the benzyne adduct, 1,2,3,4-tetraphenylnaphthalene (3), in 25% yield. Presumably the nitrene (2) is formed initially.[1]

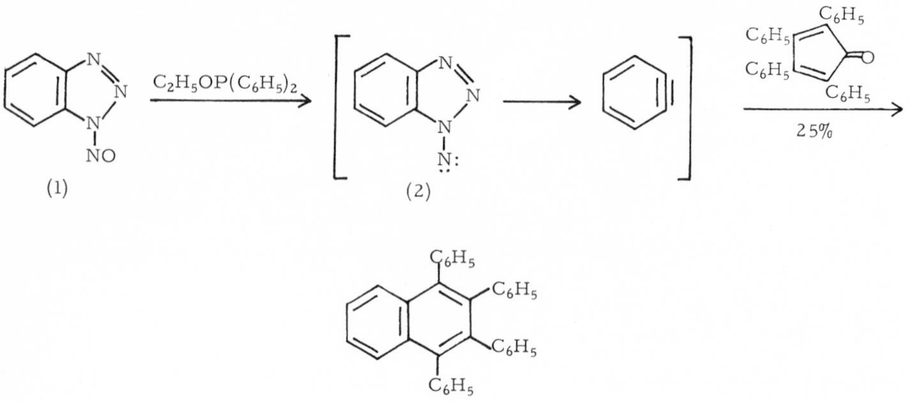

(1)   (2)   (3)

[1] J. I. G. Cadogan and J. B. Thomson, *Chem. Commun.*, 770 (1969)

Ethylene oxide, 1, 377–378; **2,** 196–197.

1,3-Dioxolanes. Carbonyl compounds react with ethylene oxide in the presence of a neutral catalyst such as tetraethylammonium bromide to form 1,3-dioxolanes, generally in yields of 70–85%. The reaction is carried out without solvent in an autoclave at 80–150°; the products are isolated by distillation.[1]

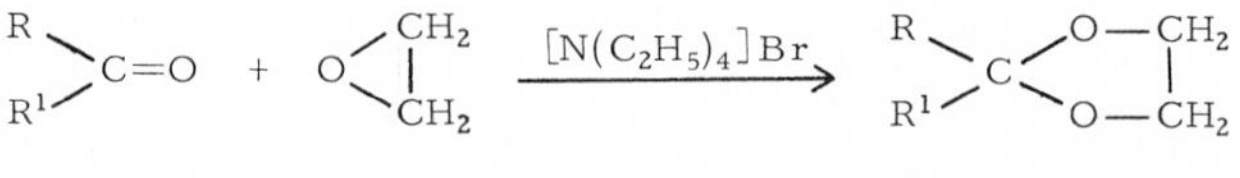

Examples:

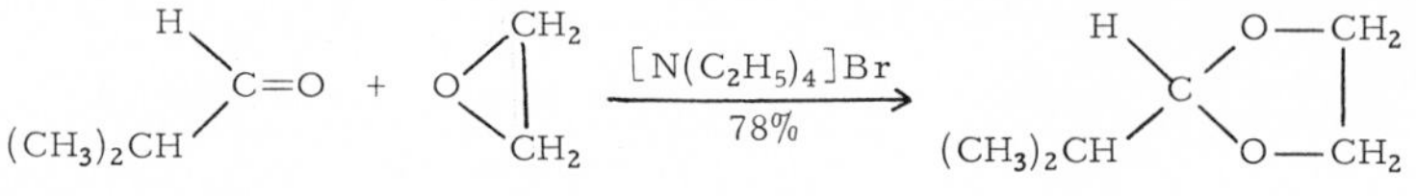

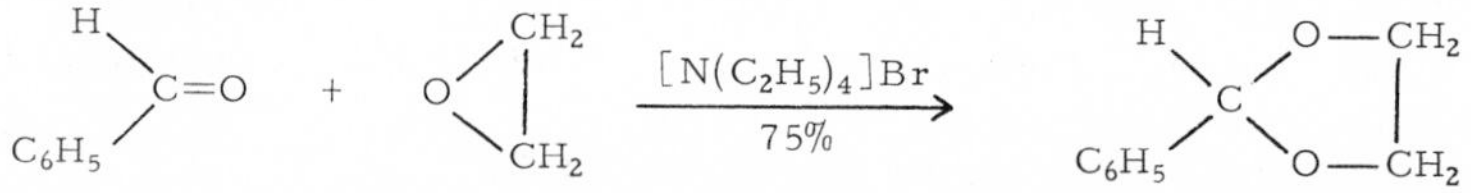

2-Haloethyl α-halocarboxylates. Aldehydes or ketones react with chloroform or bromoform and ethylene oxide in the presence of tetraethylammonium bromide (50–150°, sealed tube) to give 2-haloethyl α-halocarboxylates in 15–50% yield.[2]

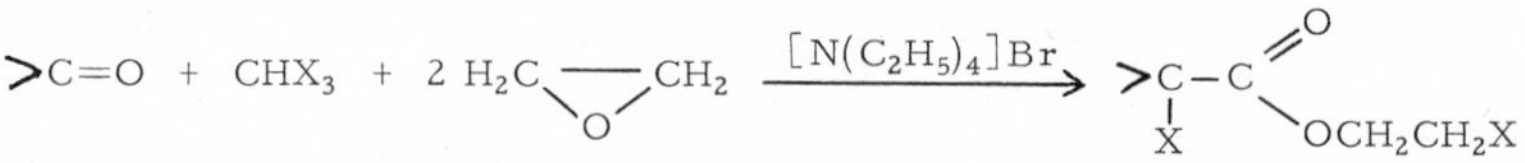

[1] F. Nerdel, J. Buddrus, G. Scherowsky, D. Klamann, and M. Fligge, *Ann.*, **710,** 85 (1968)
[2] F. Nerdel, J. Buddrus, and D. Klamann, *Ber.*, **101,** 1299 (1968)

Ethylidene iodide–Diethylzinc, CH_3CHI_2—$(C_2H_5)_2Zn$.

Methylcarbene. Simmons *et al.*[1] generated methylcarbene from ethylidene iodide and zinc–copper couple, but the yield of *exo*-7-methylnorcarane from cyclohexene was only 3.6%. Wittig and Jautelat[2] improved the yield considerably by replacing ethylidene iodide by 1-iodoethyl benzoate, $C_6H_5COOCHICH_3$. Japanese workers[3] improved the yield even more by using the combination of ethylidene iodide and diethylzinc. *cis*- and *trans*-Olefins afford derivatives whose configurations are *cis* and *trans* respectively. Olefins containing a hydroxyl group, such as

$$\xrightarrow[\substack{66\%}]{\substack{CH_3CHI_2 \\ (C_2H_5)_2Zn}}$$

endo + exo: 1.5:1

allyl alcohol, give predominantly the *anti* isomer (ratio 1.7–1). Other olefins give the *syn*-isomer predominantly.

$$CH_2=CH_2CH_2OH \xrightarrow[\substack{23\%}]{\substack{CH_3CHI_2 \\ (C_2H_5)_2Zn}}$$

[1] H. E. Simmons, E. P. Blanchard, and R. D. Smith, *Am. Soc.*, **86**, 1347 (1964)
[2] G. Wittig and M. Jautelat, *Ann.*, **702**, 24 (1967)
[3] J. Nishimura, N. Kawabata, and J. Furukawa, *Tetrahedron*, **25**, 2647 (1969)

Ethylidenetriphenylphosphorane, $(C_6H_5)_3P=CHCH_3$.

Prepared in solution by treatment of ethyltriphenylphosphonium bromide (Aldrich) with base (usually *n*-butyllithium).[1] It has also been prepared in high yield by use of dimsylsodium as base and DMSO as solvent.[2]

Modified Wittig reaction. The Wittig reaction is extremely useful for synthesis of olefins; but, unfortunately, when the products can exist as *cis*- and *trans*-isomers, both are usually formed, with some preference for the *trans*-isomer.[1] Corey[2,3] developed a stereospecific synthesis of certain trisubstituted olefins of the type (I). An aldehyde, for example heptanal, is treated with a solution of ethyl-

I

idenetriphenylphosphorane in THF at −78° for 5 minutes to form the betaine (1). This is then treated with base (*n*-butyllithium in hexane) at the same temperature to form the deep red β-oxido phosphonium ylide (2). This is allowed to warm to 0° and then treated with excess dry paraformaldehyde to generate the colorless β,β′-dioxidophosphonium derivative (3). After a suitable reaction period, workup gives 2-methyl-*cis*-2-nonene-1-ol (4) in 73% yield. Thus the reaction of (2) with an electrophile is a stereospecific reaction and gives only one betaine (3).

$$\underset{\underline{n}\text{-}C_6H_{13}}{\overset{H}{\diagdown}}C=O \xrightarrow{(C_6H_5)_3P=CHCH_3} \underset{CH_3}{\underset{|}{\underline{n}\text{-}C_6H_{13}\overset{\overset{O^-}{|}}{CH}-CH-\overset{+}{P}(C_6H_5)_3}} \xrightarrow{\underline{n}\text{-}BuLi}$$

(1)

$$\underset{CH_3}{\underset{|}{\underline{n}\text{-}C_6H_{13}\overset{\overset{O^-\overset{+}{Li}}{|}}{CH}-C=P(C_6H_5)_3}} \xrightarrow{HCHO} \underset{\underline{n}\text{-}C_6H_{13}}{\overset{O^-}{H\cdots C-C}}\overset{\overset{\overset{+}{P}(C_6H_5)_3}{\cdots CH_3}}{\underset{CH_2\overset{-}{O}\overset{+}{Li}}{}} \xrightarrow[73\%\ \text{overall}]{-(C_6H_5)_3P=O} \underset{\underline{n}\text{-}C_6H_{13}}{\overset{H}{\diagdown}}C=C\overset{CH_3}{\underset{CH_2OH}{\diagup}}$$

(2)   (3)   (4)

This sequence was used in a remarkably simple synthesis of the essential oil constituent α-santalol (6) from the aldehyde (5). This substance had been synthesized previously by the reaction of (5) with (carboethoxyethylidene)triphenyl-

(5)   →   (6)

phosphorane followed by lithium aluminum hydride reduction, but in this synthetic scheme the undesired isomer of (6) is obtained as the major product.[4]

Reaction of the β-oxidophosphonium ylide (2) with one equivalent of N-chlorosuccinimide first at −78° and then at room temperature leads to 2-chloro-cis-2-nonene (7) in about 50% yield. Surprisingly the reaction of (2) with iodobenzene dichloride leads to the isomeric 2-chloro-trans-2-nonene (8). Reactions of (2) with brominating or iodinating reagents failed to give the corresponding halo-

(7)

(2)

(8)

olefin. Since chloroolefins of the type (7) and (8) are converted into acetylenes by reaction with sodium amide in liquid ammonia, a new synthesis of acetylenes via Wittig reagents is available.[5]

[1]Review: A. Maercker, *Org. Reactions*, **14**, 270 (1965)
[2]R. Greenwald, M. Chaykovsky, and E. J. Corey, *J. Org.*, **28**, 1128 (1963).
[3]E. J. Corey and H. Yamamoto, *Am. Soc.*, **92**, 226 (1970)
[4]R. G. Lewis, D. H. Gustafson, and W. F. Erman, *Tetrahedron Letters*, 401 (1967)
[5]E. J. Corey, J. I. Shulman, and H. Yamamoto, *ibid.*, 447 (1970)

Ethyl trichloroacetate, 1, 386.

Dichlorocarbene. Tricyclo[3.2.1.0^{1,5}]octane (3) has been prepared from bicyclo[3.2.0]heptene-1(5) (1) as shown. The normal tetrahedral arrangement about the bridgehead carbons in (3) is impossible; even so, (3) is remarkably stable; the half-life at 195° is more than 20 hours.[1]

[1]K. B. Wiberg and G. J. Burgmaier, *Tetrahedron Letters*, 317 (1969)

S-Ethyl trifluorothioacetate (Ethyl thioltrifluoroacetate), $CF_3COSC_2H_5$. Mol. wt. 158.15, b.p. 90.5°/760 mm.

Preparation.[1] The reagent is prepared in 84% yield by the reaction of ethyl mercaptan with a 20–30% excess of trifluoroacetic anhydride. The reaction takes place at room temperature but is completed by heating at 100° for several hours.

N-Acylation. Weygand[2] introduced the N-trifluoroacetyl group as an N-protective group in peptide synthesis. The reagent used was the highly reactive trifluoroacetic anhydride. But this reagent has some disadvantages. Schotten-Baumann conditions cannot be used; unsymmetrical anhydrides are formed between N^α-trifluoroacetylamino acids and the generated trifluoroacetic acid; excess anhydride racemizes asymmetric centers.

The acylating properties of coenzyme A suggested to Schallenberg and Calvin[3] that S-ethyl trifluorothioacetate would be capable of acyl transfer, and in fact this reaction was readily realized under Schotten-Baumann conditions in aqueous solution at pH 8–9. The protective group is cleaved by hydrolysis at pH 10 or greater (dilute aqueous sodium hydroxide or concentrated aqueous ammonia); peptide linkages are not ruptured under these conditions. Applicability to peptide synthesis was demonstrated by synthesis of N-trifluoroacetylglycyl-DL-alanine.

Wolfrom and Conigliaro[4] found that S-ethyl trifluorothioacetate selectively acylates the amino group of 2-amino-2-desoxy-D-glucose (methanolic sodium methoxide, 24 hours, room temperature, 73% yield). They used this protective group in the synthesis of purine nucleosides. The group was removed by means of methanolic ammonia. (Note that purine nucleosides are unstable to acid.)

[1]M. Hauptschein, C. S. Stokes, and E. A. Nodiff, *Am. Soc.*, **74**, 4005 (1952).
[2]F. Weygand and E. Leising, *Ber.*, **87**, 248 (1954)
[3]E. E. Schallenberg and M. Calvin, *Am. Soc.*, **77**, 2779 (1955)
[4]M. L. Wolfrom and P. J. Conigliaro, *Carbohydrate Res.*, **11**, 63 (1969)

4-Ethyl-2,6,7-trioxa-1-phosphabicyclo[2.2.2]octane, Mol. wt. 162.13, m.p. 55–56°. Supplier: Frinton Laboratories, S. Vineland, N.J.

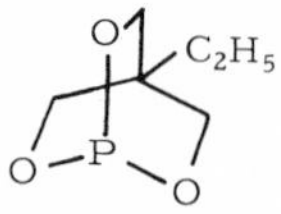

Preparation.[1] This bicyclic phosphite can be prepared in 90% yield by trans-esterification of triethyl phosphite with trimethylolpropane (triethylamine catalyst).

Singlet oxygen[2] (*see also* **Triphenyl phosphite ozonide**, this volume). Dilute methylene chloride solutions of this reagent (1) react quantitatively at −78° with one equivalent of ozone to give the 1:1 adduct (2).[2] The adduct decomposes at room temperature to the phosphate (3) with evolution of singlet oxygen as shown

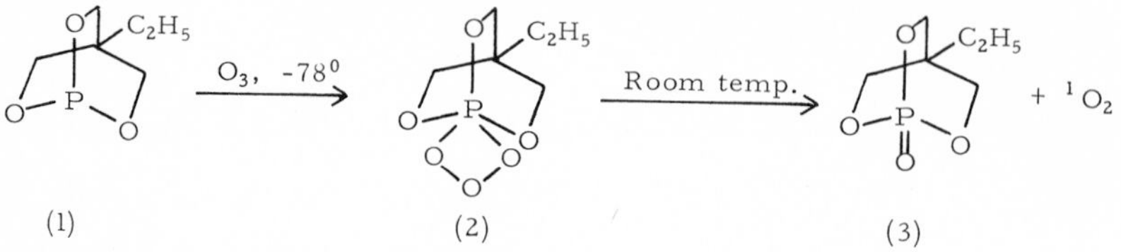

by trapping experiments (e.g., conversion of 9,10-diphenylanthracene into the *endo*-peroxide).

[1]W. S. Wadsworth, Jr. and W. D. Emmons, *Am. Soc.*, **84**, 610 (1962)
[2]M. E. Brennan, *Chem. Commun.*, 956 (1970)

F

Ferric chloride, 1, 390–392; **2,** 199.

Phenolic oxidative coupling. Oxidation of the phenolic amine (1) with 4.1 molar equivalents of ferric chloride hexahydrate gives 5-hydroxy-6-methoxyindole (2) in 10.6% yield. The amino nitrogen can also be substituted by an alkyl group.[1]

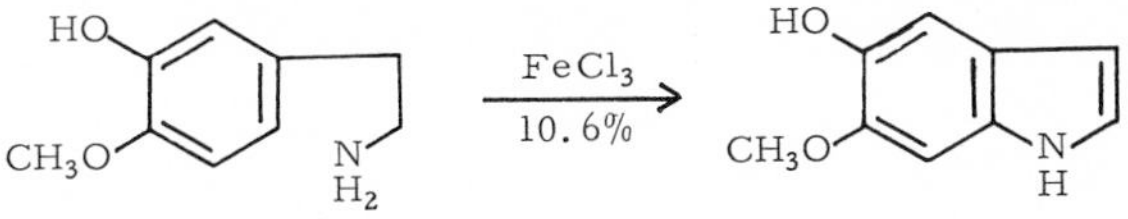

(1)   (2)

[1]T. Kametani, I. Noguchi, K. Nyu, and S. Takano, *Tetrahedron Letters*, 723 (1970)

Ferrous chloride, anhydrous, $FeCl_2$. Mol. wt. 126.76.

Preparation.[1] A mixture of 520 g. of anhydrous, sublimed ferric chloride, and 1 kg. of chlorobenzene is refluxed for $3\frac{1}{2}$ hours, during which time the trihalide is reduced with formation of ferrous chloride:

$$2\ FeCl_3\ +\ C_6H_5Cl\ \longrightarrow\ 2\ FeCl_2\ +\ C_6H_4Cl_2\ +\ HCl$$

The product is collected by suction filtration and washed with anhydrous benzene; yield: 395 g. (97%).

Ferrocene synthesis. See **Methylcyclopentadiene**, this volume.

[1]I. J. Spilners and R. J. Hartle, procedure submitted to *Org. Syn.* 1970

9-Fluorenylmethyl chloroformate (1a), 9-fluorenylmethyl azidoformate (1b).

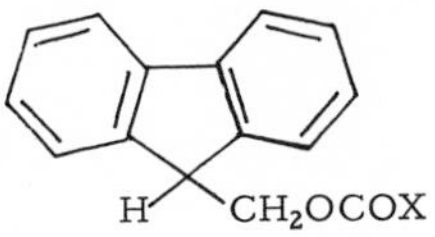

1a, X = Cl, mol. wt. 258.70, m.p. 61.5-63°

1b, X = N$_3$, mol. wt. 265.26, m.p. 83 - 85°

Reagent (1a) is prepared[1] in 86% yield by the reaction of 9-fluorenylmethanol[2] with phosgene in methylene dichloride. Reagent (1b) is obtained in 82% yield by the reaction of (1a) with sodium azide in aqueous acetone.

The 9-fluorenylmethyloxycarbonyl group (FMOC) is suggested as a new protective group for the amino function in peptide synthesis.[1] Unlike the usual protective groups, which are cleaved by acidic reagents, the protective group of FMOC-amino acids is cleaved under basic conditions, most simply by liquid ammonia. Another convenient deblocking procedure is dissolution in a simple amine such as ethanolamine or morpholine.

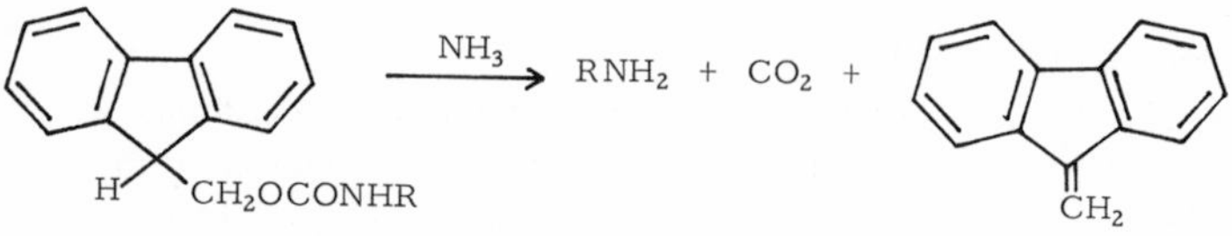

[1]L. A. Carpino and G. Y. Han, *Am. Soc.*, **92**, 5748 (1970)
[2]W. G. Brown and B. A. Bluestein, *ibid.*, **65**, 1082 (1943)

Fluoromethylenetriphenylphosphorane   (Triphenylphosphine   fluoromethylene), $(C_6H_5)_3P=CHF$. Mol. wt. 294. 30.

Preparation.

$$(C_6H_5)_3P=CH_2 \xrightarrow[\substack{80\%}]{\substack{1) \ FClO_3 \\ 2) \ NaI}} [(C_6H_5)_3\overset{+}{P}-CH_2F]I^- \xrightarrow[-LiI, \ -C_6H_6]{C_6H_5Li} (C_6H_5)_3P=CHF$$

Vinyl fluorides.[1] The Wittig reagent reacts with carbonyl compounds to give vinyl fluorides in about 50% yield. The *cis–trans* ratio is about 50 : 50.

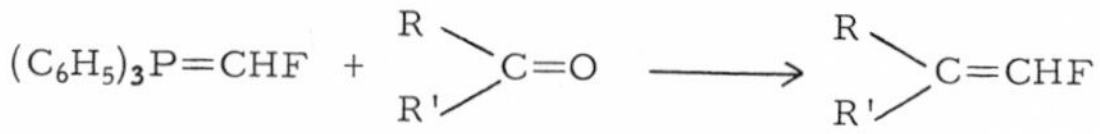

[1]M. Schlosser and M. Zimmermann, *Synthesis*, 75 (1969)

Fluoroxytrifluoromethane, 2, 200.

Addition to unsaturated steroids. The reagent adds slowly to testosterone acetate (a deactivated olefin) to give 4-fluorotestosterone. Pregnenolone acetate (1) affords an adduct (2) in modest yield, characterized by conversion into 6α-fluoroprogesterone, but the major product is a mixture of fluorinated by-products.

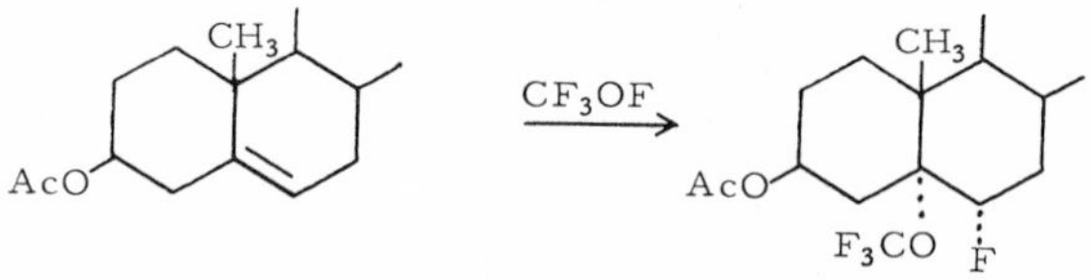

(1)   (2)

The allylic acetate (3) gives as the major product the adduct (4). Thus the reaction involves simple Markownikoff *cis*-addition without internal nucleophilic participation. However, the allylic oxygen function has a beneficient effect since in its absence a complex mixture of products is obtained.[1]

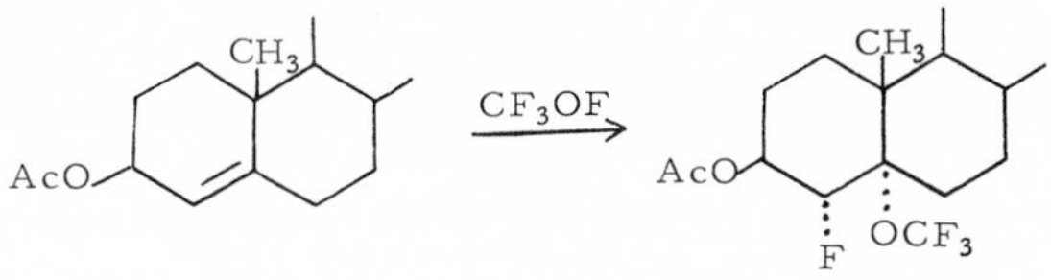

(3)   (4)

[1]D. H. R. Barton. L. J. Danks, A. K. Ganguly, R. H. Hesse, G. Tarzia, and M. M. Pechet, *Chem. Commun.*, 227 (1969)

Formamide, 1, 402–403; 2, 201–202.

Trialkyl orthoformates. Trialkyl orthoformates are prepared conveniently in one step by the reaction of formamide and an alcohol in the presence of benzoyl chloride. Yields are 45–60% for lower alcohols.[1]

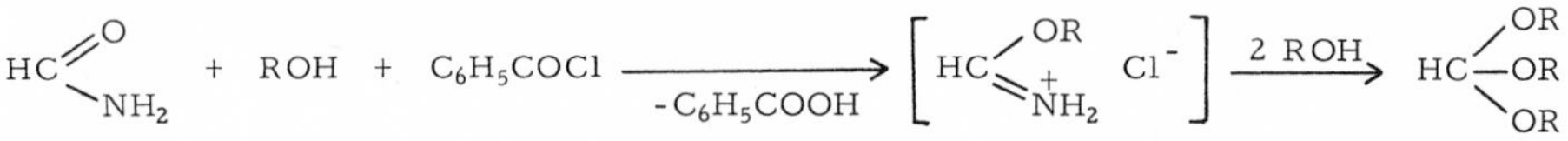

[1]R. Ohme and E. Schmitz, *Ann.*, **716**, 207 (1968)

Formamidine acetate, 1, 403–404; 2, 202.

4(5)-Methylimidazole. This imidazole can be prepared in 27.5% yield by the condensation of chloro-2-propanone and formamidine acetate in diethylene glycol. It is conveniently purified as the picrate.[1]

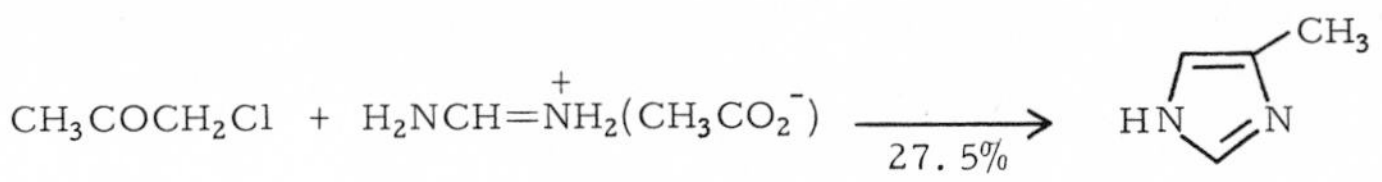

[1]C. G. Overberger and C. M. Shen, *Org. Prep. Proc.*, **1**, 1 (1969)

Formic acid, 1, 404–407; 2, 202–203.

Amides from nitriles.[1] Nitriles are converted into amides in high yield when heated at 250° (2 hours) with an equimolar amount of 100% formic acid in a pressure vessel which is preferentially cased with tantalum or silver.

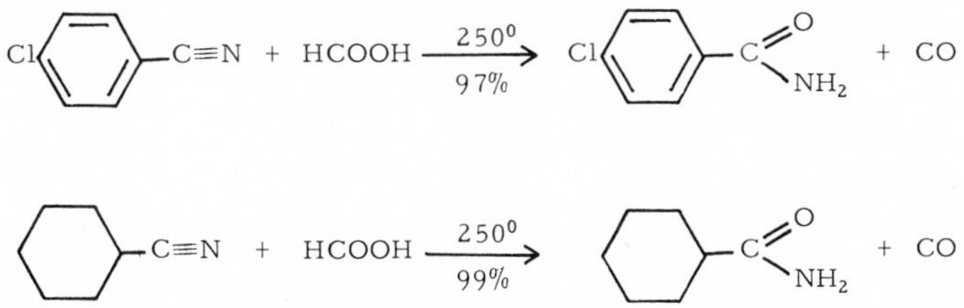

[1]F. Becke and J. Gnad, *Ann.*, **713**, 212 (1968)

N-Heptafluorobutyrylimidazole, $CF_3CF_2CF_2CO-N$ ⟨N⟩ Mol. wt. 264.11.
Supplier: Pierce Chemical Co.

Use. For glpc separation of catecholamines. Horning *et al.*[1] converted these substances first into the trimethylsilyl ethers by reaction with trimethylsilyl-imidazole in acetonitrile; acylation of primary and secondary groups was then effected by addition of N-heptafluorobutyrylimidazole. The resulting trimethyl-silyl ether–heptafluorobutyryl (TMSi–HFB) derivatives were found to have excellent glpc properties.[1]

In later work on derivatives of indole amines and indole alcohols suitable for glpc separation, Horning *et al.*[2] found that the heptafluorobutyryl (HFB) derivatives prepared by acyl transfer with N-heptafluorobutyrylimidazole are excellent for this purpose. The indole NH group as well as all amino and hydroxyl groups is acylated. No solvent is necessary; the indoles dissolve readily in the reagent at 80°. The rate of reaction varies with the substrate but is complete and quantitative after 2–3 hours at 80°. Acylation with this imidazole reagent rather than with the anhydride or acid chloride has the advantage that an acid is not liberated during the reaction.

[1] M. G. Horning, A. M. Moss, E. A. Boucher, and E. C. Horning, *Anal. Letters*, **1**, 311 (1968)
[2] J. Vessman, A. M. Moss, M. G. Horning, and E. C. Horning, *ibid.*, **2**, 81 (1969)

Hexachlorodisilane, Si_2Cl_6. Mol. wt. 268.92. Supplier: Alfa Inorganics.

Deoxygenation.[1] Phosphine oxides are reduced by this reagent in chloroform at room temperature to the corresponding phosphine in 70–90% yield. In the case of an optically active acyclic phosphine oxide the reaction takes place with complete or nearly complete inversion of configuration:

$$Si_2Cl_6 + O=P\overset{R_1}{\underset{R_3}{\cdots R_2}} \xrightarrow{\sigma} R_2\overset{R_1}{\underset{R_3}{\cdots}}P: + ''Si_2OCl_6''$$

Optically active cyclic phosphine oxides are reduced with retention of configuration. Amine oxides and sulfoxides are also reduced readily.

On the other hand, the reduction of an optically active acyclic phosphine sulfide occurs stereospecifically and with retention of configuration.[2]

$$S=P\overset{C_6H_5}{\underset{C_3H_7}{\cdots CH_3}} \xrightarrow{Si_2Cl_6} P\overset{C_6H_5}{\underset{C_3H_7}{\cdots CH_3}}$$

[1] K. Naumann, G. Zon, and K. Mislow, *Am. Soc.*, **91**, 2788, 7012 (1969)
[2] G. Zon, K. E. DeBruin, K. Naumann, and K. Mislow, *ibid.*, **91**, 7023 (1969)

Hexaethylphosphorous triamide, **1**, 425; **2**, 207.

Benzylidene dichloride.[1] The reaction of benzotrichloride with this reagent in

ether containing ethanol at room temperature rapidly affords benzylidene dichloride in practically quantitative yield. Hexamethylphosphorous triamide (**1**,

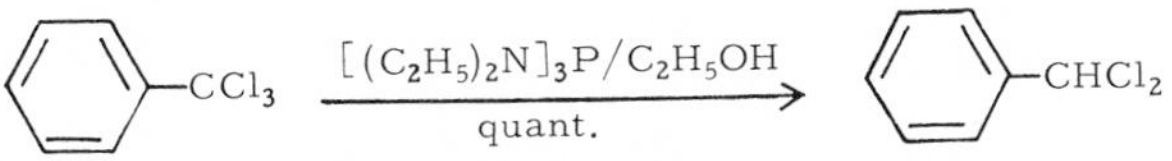

431–432; **2**, 210–211) is about equally effective, but use of triphenylphosphine,
tributylphosphine, and trioctylphosphine requires several hours at higher temperatures for a similar conversion.

Desulfurization. Alicyclic thiol sulfonates, for example 1,2-dithiolane-1,1-
dioxide (1, prepared by oxidation of the corresponding dimercaptan) are desul-

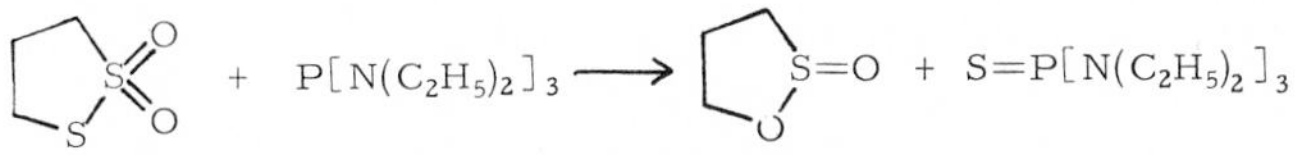

furized by the reagent to alicyclic sulfinate esters (2, 1,2-oxathiolane-2-oxide). In
the case of 1,2-dithiane-1,1-dioxide (3), two products are obtained. An open-
chain ionic intermediate is suggested, which then cyclizes through either sulfur or
oxygen, the latter being preferred.[2]

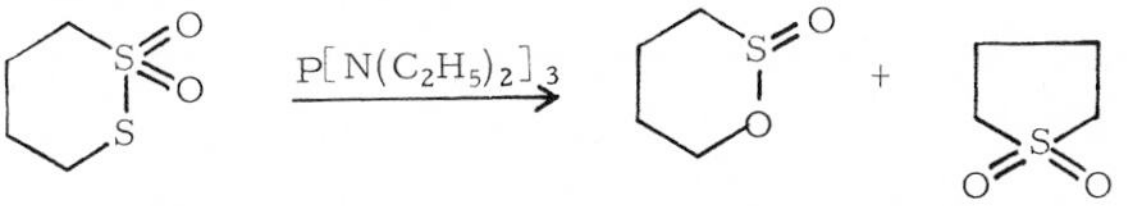

[1] I. M. Downie and J. B. Lee, *Tetrahedron Letters*, 4951 (1968)
[2] D. N. Harpp and J. G. Gleason, *ibid.*, 1447 (1969)

Hexafluoroacetone–Hydrogen peroxide   Mol. wt. 200.05.

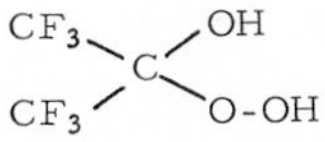

Oxidation. Hexafluoroacetone (available from Aldrich) and hydrogen peroxide form a complex which functions as a peracid. Thus it can be used in the
Baeyer-Villiger reaction and for oxidation of a primary aromatic amine to a nitro
compound. Preliminary results, however, indicate that the adduct is less powerful
than pertrifluoroacetic acid.[1]

[1] R. D. Chambers and M. Clark, *Tetrahedron Letters*, 2741 (1970)

Hexamethylphosphoric triamide (HMPT), 1, 430–431; **2**, 208–210.

C-Alkylation of ketones. Fauvarque and Fauvarque[1] found that the reaction of
alkyl Grignard reagents with ketones in HMPT results in formation of magnesium enolates with elimination of the alkane. The enolates thus formed are readily

alkylated in the same solvent at 80°. *n*-Butylmagnesium bromide was used, since

$$\underline{n}\text{-BuMgBr} + \text{R}\underset{\overset{\|}{\text{O}}}{\text{C}}\text{CH<} \xrightarrow{\text{HMPT}} \text{R}-\underset{\overset{|}{\text{OMgBr}}}{\text{C}}=\text{C<} + \text{BuH}\uparrow$$

$$\text{R}-\underset{\overset{|}{\text{OMgBr}}}{\text{C}}=\text{C<} + \text{R'X} \xrightarrow[80°]{\text{HMPT}} \text{R}-\underset{\overset{\|}{\text{O}}}{\overset{|}{\text{C}}}-\overset{|}{\underset{|}{\text{C}}}-\text{R'}$$

it is readily prepared and sufficiently soluble in HMPT. A variety of alkylating reagents has been used as such, alkyl halides, sulfates, primary tosylates; substituted halogen compounds such as $ClCH_2COOCH_3$, CH_2ClOCH_3 can be used. The method has the advantage that only C-alkylation occurs. Yields for the most part are 50–70%. A study of the site of alkylation of unsymmetrical ketones indicates that in aliphatic ketones the order is $CH_2 > CH_3 > CH$.

Oxidation of allylic alcohols. Steroidal allylic alcohols are oxidized to the corresponding α,β-unsaturated ketones in high yield by chromic anhydride in HMPT.[2] The reaction is rapid in the case of equatorial alcohols but requires some weeks in the case of axial alcohols. Nonallylic hydroxyl groups are not affected. If the alcohol is only slightly soluble in HMPT, purified acetone is used as cosolvent. The reaction, if slow, can be carried out at 50°. Benzyl alcohol is oxidized, but in low yield (30%).

Autoxidation of aromatic hydrocarbons. Normant *et al.*[3] report that HMPT is superior to DMF, DMSO, or tetramethylurea for autoxidation of aromatic hydrocarbons in the presence of alkaline metals (lithium was used).

Examples:

$$(C_6H_5)_2CH_2 \rightarrow (C_6H_5)_2C{=}O \text{ (86\% yield)}$$
Fluorene → Fluorenone (89% yield)
Anthracene → Anthraquinone (84% yield)

It was shown previously by Schriesheim (**1**, 431) that HMPT is the most satisfactory solvent for autoxidation of aromatics in the presence of bases [KOH or $(CH_3)_3COK$].

Reductions. α,β-Unsaturated acids, like α,β-unsaturated ketones (**2**, 208), are reduced by lithium in HMPT in the presence of a cosolvent. One difference is that acids form a very stable complex with HMPT; accordingly HMPT is eliminated, before acidification, by extraction with chloroform.[4]

$$C_6H_5CH{=}CHCOOH \xrightarrow[85\%]{\substack{1)\ \text{HMPT} \\ 2)\ H_2O,\ H^+}} C_6H_5CH_2CH_2COOH$$

If the reduction is carried out in the presence of an alkyl halide or a carbonyl compound, condensation at the α-position takes place as well as reduction.

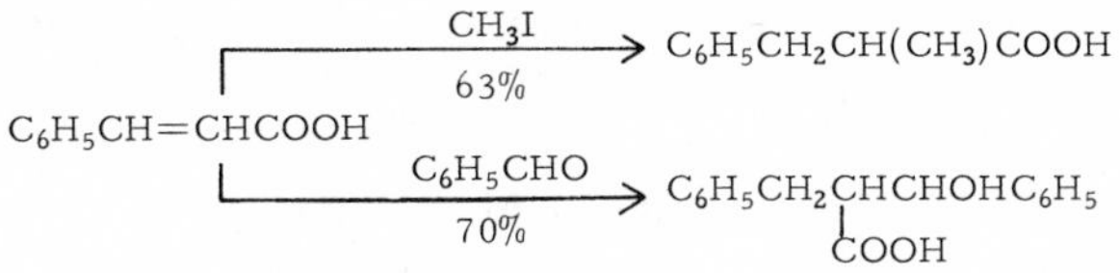

$$C_6H_5CH=CHCOOH \begin{cases} \xrightarrow[63\%]{CH_3I} C_6H_5CH_2CH(CH_3)COOH \\ \xrightarrow[70\%]{C_6H_5CHO} C_6H_5CH_2\underset{\underset{COOH}{|}}{C}HCHOHC_6H_5 \end{cases}$$

Solutions of sodium in hexamethylphosphoric triamide containing *t*-butanol are effective for reduction of alkenes.

Examples:

1-Hexene	$\longrightarrow$	n-Hexane (98% yield)
Cyclohexene	$\longrightarrow$	Cyclohexane (99% yield)
Norbornene	$\longrightarrow$	Norbornane (73% yield)
3-Hexyne	$\longrightarrow$	n-Hexane (79% yield)

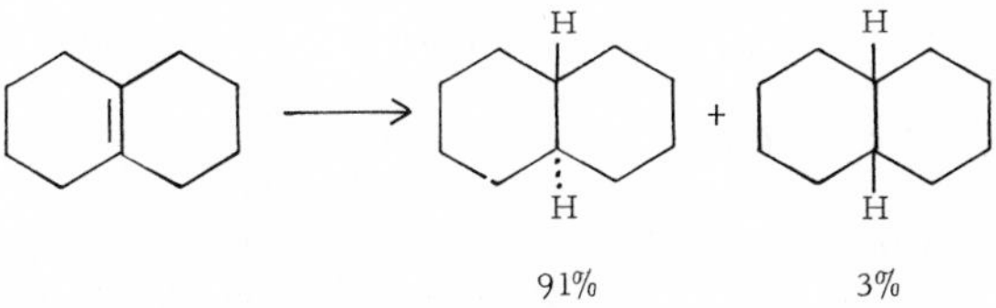

The reaction can be used for incorporation of deuterium by reduction with *t*-butanol-O-*d*.[5]

Birch reduction of α,β-unsaturated ketones. House *et al.*[6] have reported an extended study of the reduction of cyclic α,β-unsaturated ketones of the type (1) with sodium or lithium in HMPT as compared with reduction with the same metals in liquid ammonia. In general, use of HMPT somewhat favors formation of the less stable epimer, for example (3). The factors which seem to favor formation of

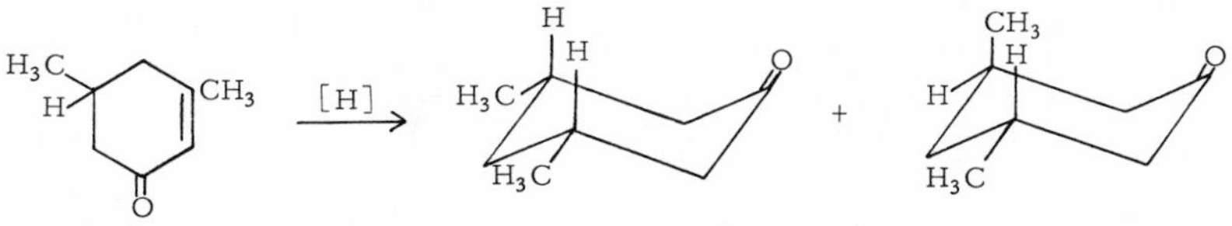

the less stable epimer are use of low temperatures (−33 to −78°), the presence of a proton donor such as *t*-butanol, use of a cosolvent such as THF, and the presence of excess metal throughout the reduction.

Conversion of allylic alcohols into chlorides and synthesis of 1,5-dienes. For synthetic purposes allylic chlorides are more useful than the corresponding, less stable, allylic alcohols. Stork[7] has described a method of converting allylic alcohols into the chlorides without rearrangement. For example, the sensitive acetal alcohol (1), with the nerol geometry, is treated at room temperature in ether–HMPT with one equivalent of methyllithium in ether, followed by addition of p-toluenesulfonyl chloride and LiCl in ether–HMPT. The reaction is allowed to stand overnight and the usual workup gives the chloride (2). By the same procedure geraniol was converted into geranyl chloride in 90% yield.

$$(CH_3O)_2CHCH_2CH_2\overset{CH_3}{\underset{|}{C}}=C\overset{H}{\underset{CH_2OH}{\diagup}} \quad\longrightarrow\quad (CH_3O)_2CHCH_2CH_2\overset{CH_3}{\underset{|}{C}}=C\overset{H}{\underset{CH_2Cl}{\diagup}}$$

$$(1) \qquad\qquad\qquad\qquad (2)$$

Stork[7] found further that the allylic chlorides couple with methallylmagnesium bromide in high yield to give 1,5-dienes provided that HMPT is used as cosolvent with THF.

$$(2) \;+\; CH_2=\overset{|}{\underset{CH_3}{C}}CH_2Cl \;\xrightarrow{95\%}\; (CH_3O)_2CHCH_2CH_2\overset{CH_3}{\underset{|}{C}}=C\overset{H}{\underset{CH_2CH_2\underset{CH_3}{\overset{|}{C}}=CH_2}{\diagup}}$$

$$(3)$$

Nucleophilic displacements. Ali and Richardson[8] state that this solvent is far superior to DMF for nucleophilic replacement reactions of sulfonate esters in the carbohydrate field. Lower temperatures (80–100°) can be used and the reaction mixture remains almost colorless.

Peptide synthesis. Based on analogy with use of onium salts derived from DMSO, Kenner and co-workers[9] developed a useful new method of peptide synthesis. Tosic anhydride[10] is allowed to react for 15 min. with excess hexamethylphosphoric triamide to form the ditosylate (1). The solution is then cooled

$$Ts_2O \;+\; 2\,[(CH_3)_2N]_3PO \;\longrightarrow\; [(CH_3)_2N]_3\overset{+}{P}-O-\overset{+}{P}[N(CH_3)_2]_3\,2\,Ts\overset{-}{O} \;\xrightarrow[\;(C_2H_5)_3NH^+\;]{RCO_2^-}$$

$$(1)$$

$$RCO-O-\overset{+}{P}[N(CH_3)_2]_3\,Ts\overset{-}{O} \;\xrightarrow[\;(C_2H_5)_3N\;]{R'NH_2}\; RCO-NHR' \;+\; [(CH_3)_2N]_3PO$$

$$(2) \qquad\qquad\qquad\qquad (3)$$

to 0° and the N-protected amino acid or peptide as the triethylammonium salt in HMPT is added to form the acyloxyphosphonium salt (2). After 5–10 min. the amino acid or peptide as the free amine is added with a further equivalent of triethylamine. The reaction is then allowed to stand at room temperature overnight. Peptides (3) are obtained in high yield (80–95%) and with high optical purity. The loss of optical activity in each of the two steps is less than 1%. Use of tosyl chloride in place of tosic anhydride results in appreciable racemization even when free glycine ester is used as the amino component.

[1]J. Fauvarque and J.-F. Fauvarque, *Bull. soc.*, 160 (1969)
[2]R. Beugelmans and M.-T. Le Goff, *ibid.*, 335 (1969)

[3]T. Cuvigny, D. Reisdorf, and H. Normant, *Compt. rend.*, **268**, 419 (1968)

[4]M. Larchevêque, *ibid.*, (C), **268**, 640 (1968)

[5]G. M. Whitesides and W. J. Ehmann, *J. Org.*, **35**, 3565 (1970)

[6]H. O. House, R. W. Giese, K. Kronberger, J. P. Kaplan, and J. F. Simeone, *Am. Soc.*, **92**, 2800 (1970)

[7]G. Stork, P. A. Grieco, and M. Gregson, *Tetrahedron Letters*, 1393 (1969)

[8]Y. Ali and A. C. Richardson, *J. Chem. Soc.*, (C), 1764 (1968)

[9]G. Gawne, G. W. Kenner, and R. C. Sheppard, *Am. Soc.*, **91**, 5669 (1969)

[10]Prepared in about 50% yield by heating tosic acid with excess phosphorus pentoxide in the presence of an inert support (Super-Cel and asbestos) to facilitate subsequent extraction; m.p. 115–127.5° [L. Field, *ibid.*, **74**, 394 (1952)].

Hexamethylphosphoric triamide–Thionyl chloride.

Acid chlorides, alkyl chlorides.[1] The combination of dimethylformamide and thionyl chloride has been known for some time to be excellent for preparation of acid chlorides and of alkyl chlorides from acids and alcohols, respectively (**1**, 286–288). The actual chlorinating agent is considered to be $[ClCH\overset{+}{=}N(CH_3)_2]Cl^-$. The combination of HMPT and thionyl chloride should be capable of forming a similar reagent:

$$
\left[\begin{array}{c} (CH_3)_2N \\ (CH_3)_2N-PCl \\ \overset{+}{(CH_3)_2N} \end{array} \right] Cl^-
$$

Indeed this combination is an excellent reagent for preparation of acid chlorides and of alkyl chlorides. HMPT as solvent is useful also because it combines with the hydrochloric acid formed; in many cases it is superior to pyridine. Reactions occur in good yield and at temperatures of about −20 to −10°. HMPT is also an excellent solvent for esterification of acyl chlorides with tertiary alcohols.

[1]J. F. Normant, J. P. Foulon, and H. Deshayes, *Compt. rend.* (C), **269**, 1325 (1969)

Hydrazine, 1, 434–445; 2, 211.

Hydrazones. A useful procedure for preparation of hydrazones free from azines is exemplified by the preparation of acetone hydrazone.[1] Acetone is first converted into acetone azine (b.p. 128–131°) by reaction with 100% hydrazine hydrate and potassium hydroxide; this product is then converted into acetone hydrazone by reaction with anhydrous hydrazine[2] and sodium hydroxide.

$$
(CH_3)_2CO \xrightarrow[86-90\%]{N_2H_4 \cdot H_2O} (CH_3)_2C{=}N{-}N{=}C(CH_3)_2 \xrightarrow[95-97\%]{N_2H_4} (CH_3)_2C{=}NNH_2
$$

[1]A. C. Day and M. C. Whiting, *Org. Syn.*, **50**, 3 (1970)

[2]Prepared by heating 100% hydrazine hydrate with an equal weight of sodium hydroxide pellets for 2 hours followed by distillation in a slow stream of nitrogen (*Caution*: Distillation in air can lead to explosion).

Hydrido(tri-*n*-butylphosphine)copper(I), $HCuPBu_3$.

The complex is obtained from tri-*n*-butylphosphine and copper(I) hydride, prepared by reduction of copper(I) bromide with diisobutylaluminum hydride at $-50°$. The complex is a useful reducing agent; it reduces iodobenzene to benzene (80% yield) and benzoyl chloride to benzaldehyde (50% yield). In addition the complex reduces primary, secondary, and tertiary alkyl-, vinyl-, and arylcopper(I) compounds to the corresponding hydrocarbons in high yields, under mild conditions and with no rearrangements.[1]

[1]G. M. Whitesides, J. S. Filippo, Jr., E. R. Stredronsky, and C. P. Casey, *Am. Soc.*, **91** 6542 (1969)

Hydrobromic acid, **1**, 450–452; **2**, 214–215.

Enol acetylation. Enol acetylation of steroidal ketones with acetic anhydride catalyzed by perchloric acid has been shown to be thermodynamically controlled (**2**, 309). However, this acetylating reagent in the case of conjugated ketones gives complex mixtures of O-acylated and C-acylated products.[1] Dienone-phenol rearrangements have been noted with Δ^4-3-ketosteroids alkylated at C_2 or C_6.[2]

Catalysis by 48% hydrobromic acid is free from these disadvantages.[3] Thus this catalyst converts testosterone acetate into a mixture of starting material and the 3,5-dienol diacetate; no 2,4-dienol diacetate was detected.

[1]A. J. Liston and P. Toft, *J. Org.*, **33**, 3109 (1968)
[2]*Idem, ibid.*, **34**, 2288 (1969)
[3]P. Toft and A. J. Liston, *Chem. Commun.*, 111 (1970)

Hydrogen peroxide, acidic, **1**, 457–465; **2**, 216.

2′-Hydroxydiphenyl-2-carboxylic acid lactone. Baeyer-Villiger oxidation of 9-fluorenone to 2′-hydroxydiphenyl-2-carboxylic acid lactone is done conveniently in even large-scale runs[1] by slow addition with cooling in acetone–dry ice of 55 ml.

$$C=O \quad \xrightarrow[\text{H}_2\text{SO}_4, \; -5^0]{90\% \; \text{H}_2\text{O}_2 - \text{Ac}_2\text{O}, \; \text{CH}_2\text{Cl}_2}$$

(2.0 moles) of 90% hydrogen peroxide[2] (FMC Corporation, Inorganic Chemicals Division, New York, N.Y. 10017) to a solution of 135 g. of concd. sulfuric acid and 350 g. of acetic anhydride (additions of not more than 3 ml. at a time). The addition is made at such a rate that the temperature does not exceed 15°.[3] To the oxidation mixture a solution of 100 g. (0.50 mole) of 9-fluorenone in 100 ml. of methylene chloride is added slowly and stirring is continued overnight at $-5°$.

Then 500 ml. of distilled water is added and the mixture boiled for 1–2 hours to destroy excess acetic anhydride and peroxides and to remove most of the methylene chloride. The solid which separates is collected after cooling and dissolved in the combined ethereal extracts (3×100 ml.) from the supernatant aqueous phase. The etheral solution is washed with 5% sodium carbonate solution, water, and brine, and dried over anhydrous sodium sulfate. Evaporation of the solvent yielded 87–96 g. (80–88.5%) of crude lactone, m.p. 87–89.5°. Two crystallizations from 95% ethanol afforded 78.5–86.2 g. (73.5–80%) of fine white crystals, m.p. 93.0–94.0°. The infrared spectrum exhibited peaks at 5.78 μ (C=O), 7.65 μ and 8.60 μ (C—O—C). The NMR spectrum $(CD_3)_2SO$ (TMS internal standard at 100 MHz) showed a characteristic aromatic multiplet at 5.4–6.6 (8 H).

Hydrogen peroxide–Acetic acid (**1**, 459–462); *oxidation of cyclobutanones to γ-lactones.* One step in a total synthesis of prostaglandins E_2 and $F_{2\alpha}$ involved oxidation of a cyclobutanone (1) to a γ-lactone (2). This reaction was accomplished in >90% yield with 30% aqueous hydrogen peroxide in glacial acetic acid at 5–10° (16 hours).[4]

(1)   (2)

[1] A. Gringauz and E. Tosk, procedure submitted to *Org. Syn.* 1970.
[2] Standing the cylinder containing hydrogen peroxide in an evaporating dish filled with water during the addition allows any drops running down to be immediately diluted.
[3] A rise in temperature even to room temperature has resulted in spontaneous and violent decomposition of the reaction mixture.
[4] E. J. Corey, Z. Arnold, and J. Hutton, *Tetrahedron Letters*, 307 (1970); E. J. Corey and R. Noyori, *ibid.*, 311 (1970)

Hydrogen peroxide, basic, 1, 466–471; 2, 216–217.

α,β-Epoxy sulfones.[1] α,β-Unsaturated sulfones, like α,β-unsaturated ketones, can be epoxidized with hydrogen peroxide under basic conditions. The reaction is stereoselective, that is, a single stereoisomer of the epoxide is obtained:

[1] B. Zwanenburg and J. ter Wiel, *Tetrahedron Letters*, 935 (1970)

Hydrogen peroxide–Phenyl isocyanate.

Epoxidation of olefins. Olefins can be epoxidized under essentially neutral conditions by this combination of reagents:

$$2\ C_6H_5NCO\ +\ H_2O_2\ +\ >\!C\!=\!C\!< \longrightarrow >\!C\!\overset{O}{\overbrace{\quad}}\!C\!< \ +\ (C_6H_5NH)_2CO\ +\ CO_2$$

The yields of epoxides are 35–70%; the yields of 1,3-diphenylurea are 92–98%. Yields of epoxides are improved by use of nonpolar solvents such as *n*-pentane or benzene. *p*-Chlorophenyl isocyanate gives somewhat higher yields than phenyl isocyanate. The active intermediate is probably an isocyanate–hydrogen peroxide complex rather than a peroxycarbamic acid $(C_6H_5NHCO_3H)$.[1]

[1] N. Matsumura, N. Sonoda, and S. Tsutsumi, *Tetrahedron Letters*, 2029 (1970)

Hydrogen peroxide–Selenium dioxide, 1, 477–478.

Epoxidation. The selenium dioxide-catalyzed hydrogen peroxide oxidation of cyclopentene (**1**, 477) and of cyclohexene[1] yields α-diols, although intermediate epoxides have been postulated. On the other hand, eight- and twelve-membered cyclic olefins do yield epoxides and not α-diols.[2]

[1] N. Sonoda and S. Tsutsumi, *Bull. Chem. Soc. Japan*, **38**, 958 (1965)
[2] J. Itakura, H. Tanaka, and H. Ito, *ibid.*, **42**, 1604 (1969)

1-Hydroxybenzotriazole, Mol. wt. 135.12, m.p. 157°. Supplier: Aldrich.

Preparation. 1-Hydroxybenzotriazole is prepared by the reaction of *o*-nitrophenylhydrazine with aqueous ammonia.[1]

Peptide synthesis. The addition of this triazole (1–2 equivalents) in the DCC-method of peptide synthesis decreases racemization, prohibits the formation of N-acylurea, and improves the yields of high-purity peptides.[2] Several substituted 1-hydroxybenzotriazoles are also effective.

[1] R. Nietzki and E. Braunschweig, *Ber.*, **27**, 3381 (1894)
[2] W. König and R. Geiger, *ibid.*, **103**, 788 (1970)

Hydroxylamine-O-sulfonic acid, 1, 481–484; **2**, 217–219.

Hydroxymethylation of quinolines.[1] Quinolines possessing free 2- and/or 4-positions react with hydroxylamine-O-sulfonic acid in methanol to give hydroxymethylquinolines (1 and/or 2). The reaction in aqueous systems gives rise to N-amination products.

[1]M. H. Palmer and P. S. McIntyre, *Tetrahedron Letters*, 2147 (1968)

N-Hydroxyphthalimide, 1, 485–486.

Preparation. Drs. W. J. van der Burg and L. J. W. M. Tax (N. V. Organon) have called our attention to a very simple preparation by Orndorff and Pratt.[1] Phthalic anhydride, hydroxylamine, and sodium carbonate are heated in water for one hour; yield 70%. Burg and Tax state that they have reproduced this procedure easily.

[1]W. R. Orndorff and D. S. Pratt, *Am. Chem. J.*, **47**, 89 (1912)

2-Hydroxypyridine (2-Pyridone)

Mol. wt. 95.10, m.p. 105–107°.

Supplier: Aldrich.

Bifunctional catalysis. Some time ago Swain and Brown[1] observed that 2-hydroxypyridine catalyzed the mutarotation of tetramethylglucose by a concerted base–acid catalysis and that it is more effective than pyridine plus phenol. A few years later Beyerman and van den Brink[2] showed that 2-hydroxypyridine and other bifunctional compounds catalyze the reaction of amines with cyanomethyl esters (reactive esters) as well as "low-energy" esters in peptide synthesis. Pyrazole and 1,2,4-triazole (**1**, 1188), were equally effective.

More recently Openshaw and Whittaker[3] in a synthesis of emetine alkaloids needed to effect condensation of an amine with a "low-energy" ester and in the particular case studied found that use of hydroxypyridine raised the yield from 2% to 87%. Twelve other mono- or polyhydroxylated heterocyclics were examined, but 2-hydroxypyridine was clearly superior.

2-Hydroxypyridine is generally applicable for the reaction of strongly basic amines with esters. It is easily recovered for further use because of its high solubility in water.

[1]C. G. Swain and J. F. Brown, Jr., *Am. Soc.*, **74**, 2538 (1952)
[2]H. C. Beyerman and W. M. van den Brink, *Proc. Chem. Soc.*, 266 (1963)
[3]H. T. Openshaw and N. Whittaker, *J. Chem. Soc.*, (C), 89 (1969)

N-Hydroxysuccinimide trifluoroacetate, 2, 219. Mol. wt. 211.10. Pierce Chemical

Co. supplies the reagent as white crystals melting at 70–72°. The material is extremely hygroscopic but is stable when protected from moisture.

Hypophosphorous acid, H_3PO_2, 1, 489–491.

Corbella *et al.*[1] used hypophosphorous acid in combination with hydrogen iodide for reduction of 2-pyrrolecarboxylic acid (1) to 3,4-dehydroproline (2).

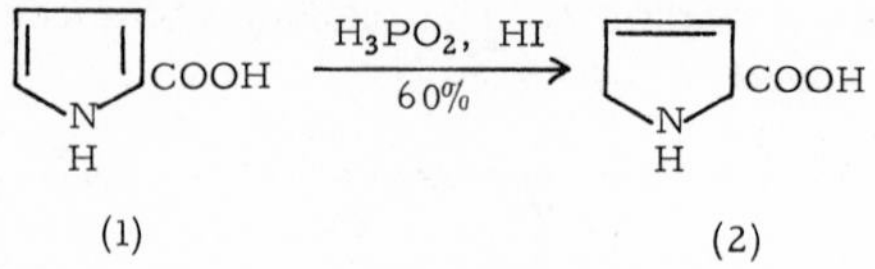

$$\text{(1)} \quad \xrightarrow[\;60\%\;]{\text{H}_3\text{PO}_2,\ \text{HI}} \quad \text{(2)}$$

Phosphonium iodide (**1**, 859–860) had been used previously for this purpose, but this reagent is not so accessible.

[1]A. Corbella, P. Gariboldi, G. Jommi, and F. Mauri, *Chem. Ind.*, 583 (1969)

I

Iodine, 1, 495–500; **2**, 220–222.

Iodination of ketones (**1**, 499). The cyclization of 1,3-dibenzoylpropane by reaction with iodine and sodium hydroxide in methanol involves iodination at the position α to one carbonyl group and elimination of the acidic hydrogen adjacent to the second carbonyl group.[1] A mixture of 35 g. (0.14 m.) of 1,3-dibenzoyl-propane and a solution of 11.2 g. (0.28 m.) of sodium hydroxide in 0.4 l. of

methanol is stirred magnetically in a round-bottomed flask to dissolve the diketone. A solution of 35 g. (0.14 m.) of iodine in 0.2 l. of methanol is then added slowly to the stirred solution via a dropping funnel at such a rate that the color of iodine is continually discharged. The resulting clear solution is stirred at room temperature for 1.5 hours, during which time the bulk of the product separates as a white solid. Collected, washed with water, and dried, *trans*-1,2-dibenzoylcyclopropane weighs 23–25 g. (66–72%), m.p. 101–103° (reported,[2] 103–104°). Removal of the solvent from the filtrate in a rotary evaporator affords a pale red solid which is treated with dilute aqueous sodium bisulfite solution to reduce unreacted iodine, collected, and dried: 8–9 g., m.p. 95–98°. Crystallization from methanol affords greater than 90% recovery of *trans*-1,2-dibenzoyl-cyclopropane, m.p. 102–103°.

Alkylmagnesium fluorides.[3] Alkylmagnesium fluorides have been prepared for the first time by reaction of alkyl fluorides (hexyl fluoride was used for the most part) with magnesium at reflux temperature in THF in the presence of various catalysts. Of these, iodine (4%) was found to be most effective; the yield of Grignard reagent was 95%, obtained in 6 days. Other less effective catalysts are cobalt chloride, bromine, ethylene dibromide, ethyl bromide.

Iodohydrins.[4] Olefins react with iodine in an aqueous solvent in the presence of an oxidizing agent such as iodic acid or oxygen catalyzed by nitrous acid to form iodohydrins in 80–90% yield. Iodohydrins readily yield epoxides on treatment with base (calcium hydroxide, sodium aluminate).

Reaction with trialkylboranes. Trialkylboranes react with iodine very sluggishly; however, in the presence of methanolic sodium hydroxide they are rapidly converted into primary iodides:

$$(R CH_2CH_2)_3B \ + \ 2 \ I_2 \ + \ 2 \ NaOH \ \longrightarrow \ 2 \ RCH_2CH_2I \ + \ 2 \ NaI \ + \ RCH_2CH_2B(OH)_2$$

The net effect is the conversion of terminal olefins into primary iodides by anti-Markownikoff hydroiodination. Note that only two alkyl groups react and the maximum yield is thus 66.7%. Yields actually obtained are close to this maximum. This difficulty can be circumvented by use of bis-3-methyl-2-butylborane (disiamylborane, **1**, 57–59; **2**, 29) rather than diborane:

$$RCH{=}CH_2 \ + \ (Sia)_2BH \ \longrightarrow \ RCH_2CH_2B(Sia)_2 \ \xrightarrow{I_2, \ NaOH} \ RCH_2CH_2I \ + \ (Sia)_2BOH \ + \ NaI$$

In this case yields of around 90% are obtained. If yields of approximately 65% are adequate, then the simpler procedure with diborane can be used.[5]

The reaction of trialkylboranes with bromine to give primary bromides is also induced with base, but in this case sodium methoxide in methanol is used as base in order to prevent an undesirable oxidative side reaction leading to alcohols.[6] Two procedures have been employed. In one, bromine (33% excess) is added to the organoborane in THF followed by a solution of sodium methoxide in methanol, all at 0°. In the other, bromine and base (10% excess) are added simultaneously at 25°. The former procedure is preferable for monosubstituted olefins, the latter is preferable for hindered olefins. Apparently, in this case, all three alkyl groups are utilized since yields of primary bromides are 30–99%.

[1] I. Colon, G. W. Griffin, and E. J. O'Connell, Jr., *Org. Syn.* submitted 1970
[2] J. B. Conant and R. E. Lutz, *Am. Soc.*, **49**, 1083 (1927)
[3] E. C. Ashby, S. H. Yu, and R. G. Beach, *ibid.*, **92**, 433 (1970)
[4] J. W. Cornforth and D. T. Green, *J. Chem. Soc.*, (C), 846 (1970)
[5] H. C. Brown, M. W. Rathke, and M. M. Rogić, *Am. Soc.*, **90**, 5038 (1968)
[6] H. C. Brown and C. F. Lane, *ibid.*, **92**, 6660 (1970)

Iodine azide, 1, 500–501; **2**, 222–223.

Aziridines.[1] β-Iodoazides, available by the addition of iodine azide to olefins, undergo reduction of the azide function followed by base-catalyzed ring closure to aziridines. The most satisfactory reagent for this purpose is lithium aluminum hydride, which can accomplish both reaction steps since it is a Lewis acid as well as a reducing agent. Competing side reactions are elimination of the elements

$$>C{=}C< \ + \ IN_3 \ \longrightarrow \ \underset{I}{>\overset{N_3}{C}{-}C<} \ \xrightarrow{LiAlH_4} \ >C\overset{\overset{H}{N}}{\underset{\qquad}{\diagup\diagdown}}C<$$

of iodine azide and hydrogenolysis of the iodo group. However, yields for the most part are satisfactory.

Iodine azide reacts stereospecifically with 3β-acetoxy-Δ[16]-androstene (1) to give 3β-acetoxy-16β-azido-17α-iodoandrostane (2) in 90% yield. Reduction with excess lithium aluminum hydride gives the 16β,17β-iminoandrostane (3).[2] The

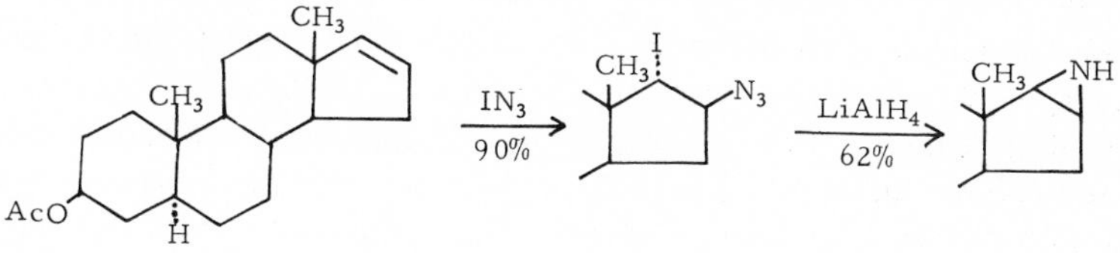

(1)   (2)   (3)

starting material is obtained readily by treatment of the 17-tosylhydrazone with methyllithium in glyme.[3]

Review.[4]

[1]A. Hassner, G. J. Matthews, and F. W. Fowler, *Am. Soc.*, **91**, 5046 (1969)
[2]G. J. Matthews and A. Hassner, *Tetrahedron Letters*, 1833 (1969)
[3]R. H. Shapiro and M. J. Heath, *Am. Soc.*, **89**, 5734 (1967)
[4]A. Hassner, *Accts. Chem. Res.*, **4**, 9 (1971)

Iodine–Dimethylsulfoxide.

Oxidation of methylpyridines to pyridine aldehydes. α-Picolines are oxidized to the corresponding aldehydes by treatment with one equivalent of iodine at room temperature to form a crystalline complex which is then dissolved in DMSO.[1] The reaction mixture is then heated to 140–160°, when dimethyl sulfide is evolved. The aldehydes are extracted after neutralization. The oxidation is considered to involve the following sequence:

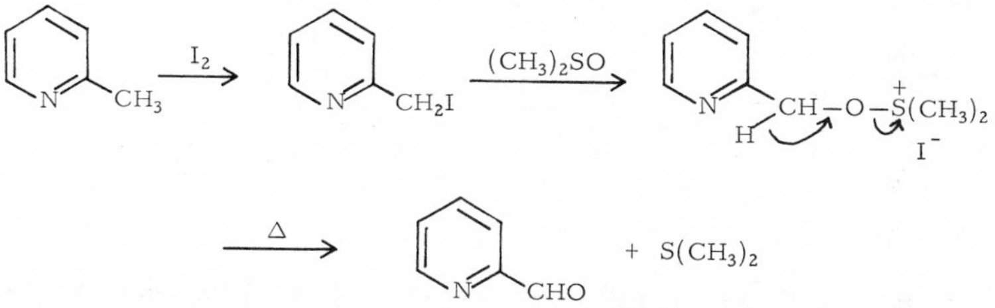

The method was suggested in a paper by Chinese chemists[2] which reported that the reaction of α-picoline N-oxide and iodine at 95–100° formed a gum which on pyrolysis decomposed to 2-pyridine aldehyde (16% yield) and α-picoline (37%).

[1]A. Markovac, C. L. Stevens, A. B. Ash, and B. E. Hackley, Jr., *J. Org.*, **35**, 841 (1970)
[2]L. Mao-Chin and C. Sae-Lee, *Acta Chim. Sinica*, **31**, 30 (1965)

Iodine isocyanate, **1**, 501; **2**, 223–224.

Addition to alkenes.[1] The addition reaction of iodine isocyanate to alkenes is much faster in methylene chloride than in ether, the commonly used solvent. The reaction is conducted by adding silver cyanate to a solution containing both the iodine and the alkene. In this manner cyclohexene is converted completely into 2-iodocyclohexyl isocyanate in about 8 minutes.

Interconversion of cis- and trans-olefins. Carlson and Lee[2] have developed the following method for interconversion of *cis*- and *trans*-olefins. Iodine isocyanate is added, for example, to a *cis*-olefin (1), and the resulting *trans*-iodocyanate (2) is converted into the carbamate (3). Hassner[3] has shown that *trans-β*-iodocarbamates on pyrolysis give rise to *cis*-fused 2-oxazolidones (4), which on hydrolysis give β-amino alcohols (5). On treatment with acid these give aziridines (6). The completing step is based on work by Clark and Helmkamp,[4] who found that deamination of aziridines under nitrosating conditions proceeds stereospecifically to the corresponding olefin with retained configuration (8). Although the sequence appears to be lengthy, yields for all the steps are good.

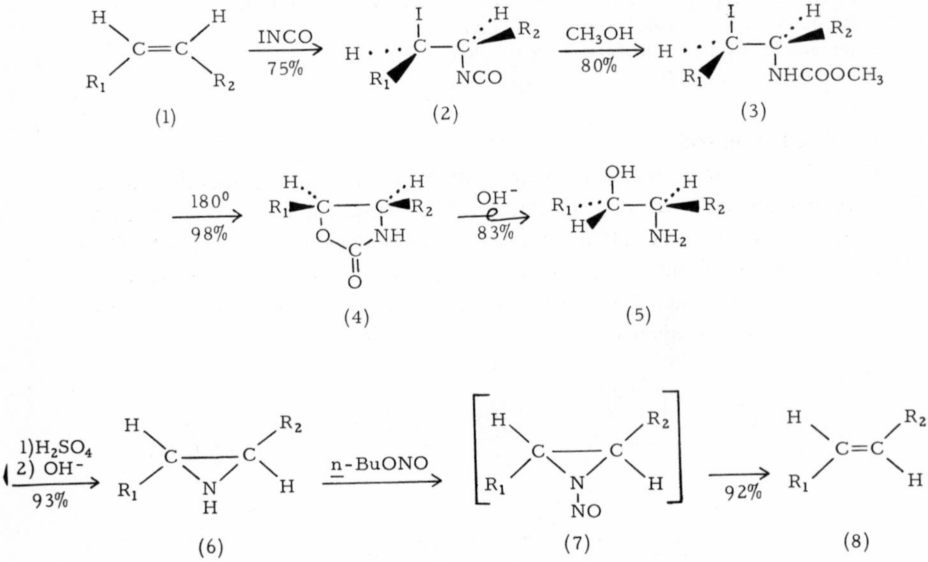

Carbamates.[5] β-Iodocarbamates can be reduced to carbamates by refluxing with powdered zinc in 1:1 acetic acid–ether for 5–24 hours. Yields are in the

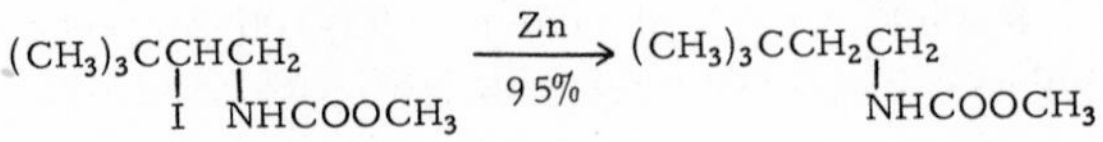

range 50–90%. However, diaxial iodocarbamates under these conditions are converted into olefins by elimination (e.g., N-carbomethoxy-2β-amino-3α-iodo-cholestane).

1H-Azepines. Paquette *et al.*[6] have developed a general synthesis of 1H-azepines which involves addition of iodine isocyanate to 1,4-dihydrobenzene derivatives, for example 1,4-dihydrobenzene (1). The *trans*-iodocarbamate (2) is obtained in 54% yield. Cyclization of (2) to the aziridine (3) is effected with

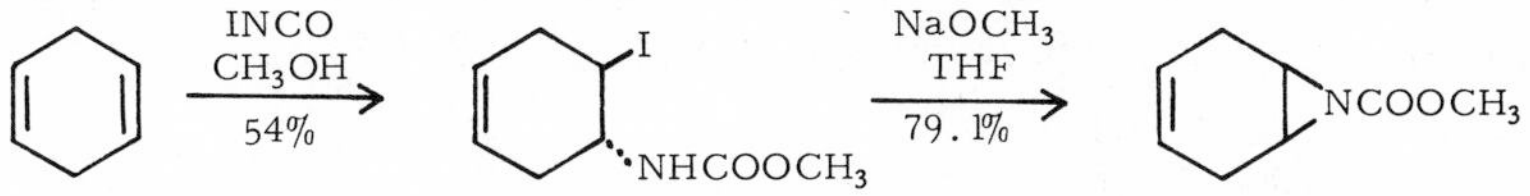

(1)   (2)   (3)

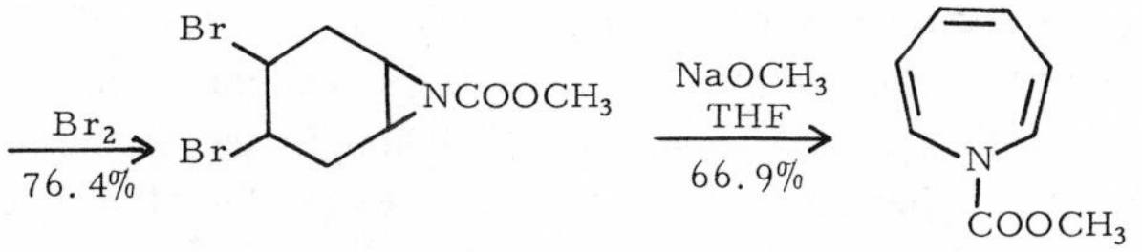

(4)   (5)

powdered sodium methoxide in THF (79.1% yield). Bromination-dehydro-bromination of (3) affords N-carbomethoxyazepine (5).

[1] C. G. Gebelein, *Chem. Ind.*, 57 (1970)
[2] R. M. Carlson and S. Y. Lee, *Tetrahedron Letters*, 4001 (1969)
[3] A. Hassner, M. E. Lorber, and C. Heathcock, *J. Org.*, **32**, 540 (1967)
[4] R. D. Clark and G. K. Helmkamp, *ibid.*, **29**, 1316 (1964)
[5] A. Hassner, R. P. Hoblitt. C. Heathcock, J. E. Kropp, and M. Lorber, *Am. Soc.*, **92**, 1326 (1970)
[6] L. A. Paquette, D. E. Kuhla, J. H. Barrett, and R. J. Haluska, *J. Org.*, **34**, 2866 (1969)

Iodine–Periodic acid dihydrate.

Iododurene has been prepared by treatment of durene either with iodine and mercuric oxide[1] or with sulfur iodide and nitric acid.[2] The present procedure,[3] by H. Suzuki of Kyoto University, Japan, is the most convenient method for preparation of monoiodo or diiodo derivatives from various polyalkylbenzenes. Shorter reaction times, higher yields, and higher purity of product are assured by use of periodic acid as oxidizing agent. The method is applicable also to the iodination of fluorobenzene.

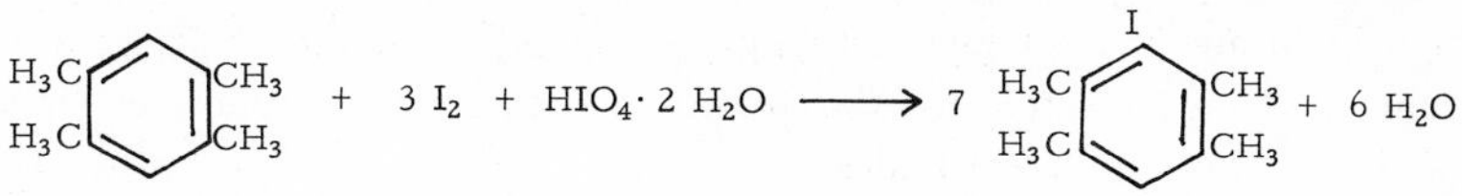

For iodination of 13.4 g. (0.1 m.) of durene, Suzuki introduced this amount of hydrocarbon into a 200-ml. two-necked flask fitted with a reflux condenser, a thermometer, and a magnetic stirrer and added 4.56 g. (0.02 m.) of periodic acid dihydrate, 10.2 g. of iodine (0.04 m.), 100 ml. of acetic acid, 20 ml. of water, and 3 ml. of concd. sulfuric acid. The mixture was heated to 65–70° with stirring until the color of iodine disappeared (about 1 hr.). The mixture was then diluted with water and the white-yellow solid that separated was collected on a Büchner funnel, washed with water, and recrystallized from 90% ethanol. The crystals were collected on a funnel and dried in air. The yield of iododurene in the form of fine, colorless needles, m.p. 78–80° was 20.8–22.6 g. (80–87%).

[1]A. Töhl, *Ber.*, **25**, 1521 (1892)
[2]A. Edinger and P. Goldberg, *ibid.*, **33**, 2875 (1900)
[3]H. Suzuki, procedure submitted to *Org. Syn.* 1969

Iodobenzene dichloride, 1, 505–506; **2**, 225–226.

Reaction with unsaturated bicyclic hydrocarbons. From a study of the chlorination of unsaturated bicyclic compounds, Masson and Thuillier[1] conclude that the reaction follows a radical addition mechanism when initiated thermally or photochemically. The stereochemistry of the addition is markedly influenced by steric effects. No Wagner-Meerwein rearrangements are observed under these conditions. An ionic mechanism is involved without initiation or in the presence of trifluoroacetic acid; in this case the usual carbonium ion rearrangements are observed.

Tanner and Gidley[2] have reached practically the same conclusions from a study of the reaction of iodobenzene dichloride with norbornene (1). The reaction follows a free-radical mechanism if initiated by irradiation (−20 to 80°) or thermally (40–80°) to give *trans*-2,3-dichloronorbornane (2) in 74% yield and *exo-cis*-2,3-dichloronorbornane (3) in 26% yield.

The course of the reaction is dramatically changed in the presence of atmospheric oxygen (0–80°), the major products now being nortricyclyl chloride (4) and *exo*-2-*syn*-7-dichloronorbornane (5). These rearrangement products evidently result from ionic chlorination.

Sulfoxides. Sulfides are converted by a stoichiometric amount of iodobenzene dichloride in aqueous pyridine at −40 to 20° to the corresponding sulfoxides in high yield and uncontaminated with corresponding sulfones.[3] There are few reagents that are selective for this purpose. The reaction applies to aliphatic, aromatic, and heterocyclic sulfides.

$$R^1SR^2 + C_6H_5ICl_2 + 2 C_5H_5N + H_2O \longrightarrow R^1SOR^2 + C_6H_5I + 2 C_5H_5N \cdot HCl$$

[1]S. Masson and A. Thuillier, *Bull. soc.*, 4368 (1969)
[2]D. D. Tanner and G. C. Gidley, *J. Org.*, **33**, 38 (1968)
[3]G. Barbieri, M. Cinquini, S. Colonna, and F. Montanari, *J. Chem. Soc.*, (C), 659 (1968)

Iodoform, CHI_3. Mol. wt. 393.79, m.p. about 120°, characteristic odor, volatile with steam.

Prepared by the action of sodium hypoiodite on ethanol, a methyl ketone, or an alcohol which on oxidation affords a methyl ketone.

Typical suppliers. Aldrich, m.p. 120–123°; Eastman, m.p. 116–121°.

Reduction to methylene dihalides.

$$CHI_3(CHBr_3) + Na_3AsO_3 + NaOH \xrightarrow[88-97\%]{} CH_2I_2(CH_2Br_2) + NaI(Br) + Na_3AsO_4$$

Iodoform   Methylene iodide

(Bromoform)   (Bromide)

Photochemical reaction with olefins. 2,2,3,3-Tetramethyliodocyclopropane:[1]

$$I_2 + 2 NaOH \longrightarrow NaI + NaOI + H_2O$$

In each of three 250-ml. round-bottomed Pyrex flasks is placed 8.4 g. (0.1 mole) of 2,3-dimethyl-2-butene (Chemical Samples Co.), 175 ml. of methylene chloride, and 50 ml. of a 5 M aqueous sodium hydroxide solution. The flasks are kept rather full in order to make efficient use of the incident light. A 1-in. Teflon-covered magnetic stirring bar is added to each flask. Three 170–190 cm. Pyrex crystallization dishes are almost filled with an ice–water mixture, each dish is placed above a Mag-Mix magnetic stirrer, and each flask is then immersed in the ice–water bath with the aid of a clamp. The three assemblies are arranged symmetrically around a Hanovia quartz immersion well (No. 19434) cooled with running tap water containing a Hanovia 450-watt medium pressure mercury lamp (No. 679A36). The edge of each flask is placed approximately 1 cm. from the wall of the well. After 2.0 g. of iodoform has been added to each flask, the mixture is irradiated until the yellow color of the iodoform disappears (the intense emission of light should be shielded to prevent damage to the eyesight of the experimenter). This process is continued until 39.4 g. (0.1 mole) of iodoform, equally distributed between the flasks, has been consumed. The reaction mixtures are combined and the organic layer separated, washed once with water, and dried over sodium sulfate. The solvent is removed with a rotatory evaporator at the water pump. After addition of 1 g. of sodium methoxide, the product is distilled under reduced pressure in an apparatus with a 5-cm. Vigreaux side arm (Ace Glass No. 9225). The receiver is cooled in an ice–water bath, and the first fraction, b.p. 45–48°/5 mm., is collected; n^{25} 1.5087. The clear distillate amounting to 14.0–15.0 g. (62.5–67%) should be stored in a refrigerator.

[1]N. C. Yang and T. A. Marolewski, *Am. Soc.*, **90**, 5644 (1968); T. A. Marolewski and N. C. Yang, procedure submitted to *Org. Syn.* 1969

Iodosobenzene diacetate, 1, 508–509.

Sulfoxides. This reagent was found to be the most suitable for oxidation of thiaxanthone (1) to the sulfoxide (2). A variety of other oxidizing reagents gave mainly the sulfone.[1]

(1)   78%   (2)

Oxidative cyclization. The reaction of 2,2'-diaminodiphenylmethanes (1) with iodosobenzene diacetate in benzene at room temperature for several days affords dibenzo-[c,f]-[1,2]diazepines (2) in 30–60% yield.[2] The reaction is not applicable to 2,2'-diaminobenzophenones.

(1) $C_6H_5\overset{+}{I}OAc(AcO^-)$ (2)

[1]J. P. A. Castrillón and H. H. Szmant, *J. Org.*, **32**, 976 (1967)
[2]R. J. Dubois and F. D. Popp, *J. Heterocyclic Chem.*, **6**, 113 (1969)

Iridium tetrachloride, 2, 228–229. Additional supplier: Alfa Inorganics.

The procedure for the reduction of 4-*t*-butylcyclohexanone to the axial alcohol, *cis*-4-*t*-butylcyclohexanol, (**2**, 229, ref. 1), has now been published.[1]

Reduction of cyclohexanones. Henbest and Mitchell[2] have described in detail their method for reduction of substituted cyclohexanones predominantly to axial alcohols using a soluble iridium–phosphite catalyst prepared *in situ* from iridium tetrachloride and phosphorous acid (or an easily hydrolyzed ester of this acid). Often 96% or more of the axial alcohol is obtained in this way.

[1]E. L. Eliel, T. W. Doyle, R. O. Hutchins, and E. C. Gilbert, *Org. Syn.*, **50**, 13 (1970)
[2]H. B. Henbest and T. R. B. Mitchell, *J. Chem. Soc.*, (C), 785 (1970)

Iron, 1, 519; **2,** 229.

Coupling of allylic halides. Hall and Hurley[1] have confirmed patent claims that allylic halides can be coupled by iron powder; the nature of the solvent is highly important. The dipolar aprotic solvent DMF was found most satisfactory. Added inorganic bromide or iodide salts exert a marked catalytic effect. In favorable cases 90% yields of nonconjugated dienes can be obtained. In recent times nickel carbonyl has been used extensively for this coupling reaction (**1,** 722–723; **2,** 290–292); however, nickel carbonyl is highly toxic and flammable.

[1]D. W. Hall and E. Hurley, Jr., *Canad. J. Chem.,* **47,** 1238 (1969)

Iron pentacarbonyl, 1, 519–520; **2,** 229–230.

Carbodiimides. Carbodiimides are obtained in about 50% yield by the iron pentacarbonyl-catalyzed reaction of azides with isocyanides.[1]

$$RN_3 \ + \ R'N \stackrel{\_}{\equiv} C: \ \xrightarrow{\ Fe(CO)_5\ } \ R-N=C=N-R' \ + \ N_2$$

Sulfides. Sulfoxides are deoxygenated to give sulfides when heated with iron pentacarbonyl in diglyme or di-*n*-butyl ether in fair to excellent yields (50–95%).[2]

$$R_2SO \ + \ Fe(CO)_5 \ \xrightarrow{\ 130-135^0\ } \ R_2S$$

[1]T. Saegusa, Y. Ito, and T. Shimizu, *J. Org.,* **35,** 3995 (1970)
[2]H. Alper and E. C. H. Keung, *Tetrahedron Letters,* 53 (1970)

L

Lead dioxide, 1, 533–536; **2,** 233–234.

Oxidation of phenols. Oxidation of phenols in apolar solvents with lead dioxide gives mainly polymeric ethers.[1] However, oxidation in polar solvents (acetic acid and formic acid) yields diphenoquinones and p-benzoquinones almost exclusively.[2]

[1] H. M. van Dort, C. R. H. I. de Jonge and W. J. Mijs, *J. Pol. Sci.*, (C), **22,** 431 (1968)
[2] C. R. H. I. de Jonge, H. M. van Dort, and L. Vollbracht, *Tetrahedron Letters*, 1881 (1970)

Lead tetraacetate, 1, 537–563; **2,** 234–238.

Oxidative decarboxylation (**1,** 554–557; **2,** 235–237). Cupric acetate, $Cu(OAc)_2.H_2O$, markedly catalyzes decarboxylation of primary and secondary acids, and in this case alkenes, rather than alkanes, are formed in good to excellent yield, for example:

$$CH_3(CH_2)_3COOH + Pb(OAc)_4 \xrightarrow{Cu(OAc)_2} CH_3CH_2CH=CH_2 + CO_2 + Pb(OAc)_2 + 2 HOAc$$

In addition, the reaction can be further catalyzed by a variety of bases such as lithium acetate and pyridine. A typical decarboxylation is carried out by dissolving the catalyst (0.535 mmole), pyridine (1.39 mmoles), and the alkanoic acid (21.2 mmoles) in a solvent (benzene, chlorobenzene). Lead tetraacetate (9.86 mmoles) in the same solvent is added, and the resulting mixture is stirred in the dark for one hour. Although alkenes have been reported to react with lead tetraacetate, under these experimental conditions the alkene is not attacked.[1]

Further examples:

$$C_6H_5CH_2CH_2CH_2CH_2COOH \xrightarrow[\substack{Cu(OAc)_2 \\ 40-65\%}]{Pb(OAc)_4} C_6H_5CH_2CH_2CH=CH_2$$

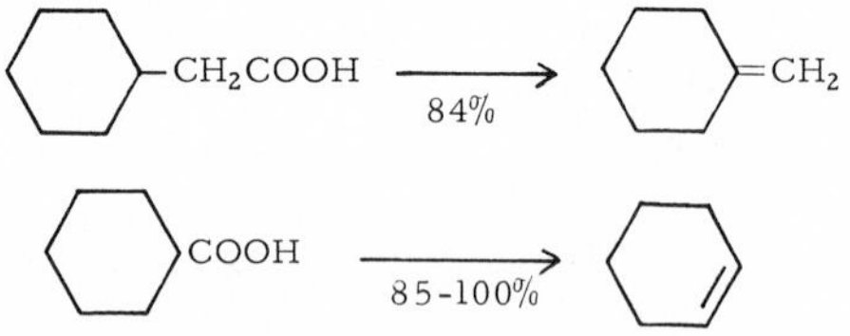

The method was shown to be applicable to the preparation of ω-alkenoic acids from dibasic acids. Thus suberic acid was converted into the half ester; oxidative decarboxylation of this gave 6-heptenoic acid in good yield.

$$HOOC(CH_2)_6COOH \longrightarrow HOOC(CH_2)_6COOC_2H_5 \longrightarrow CH_2=CH(CH_2)_4COOH$$

The oxidative bisdecarboxylation of α,β-dicarboxylic acids is usually carried out with lead tetraacetate and pyridine, in benzene or acetonitrile as solvent, at 50–60° [procedure of Grob (**1**, 554–555)]. However, these conditions are reported[2] to be unsuccessful in some cases, for example with (**1**) (1,4-dimethoxycarbonyl-bicyclo[2.2.2]octane-2,3-dicarboxylic acid). The decarboxylation, however, is successful in refluxing benzene (45% yield)[3] or can be conducted at room tempera-ture in comparable yield if dimethyl sulfoxide or dioxane is used as solvent.[4]

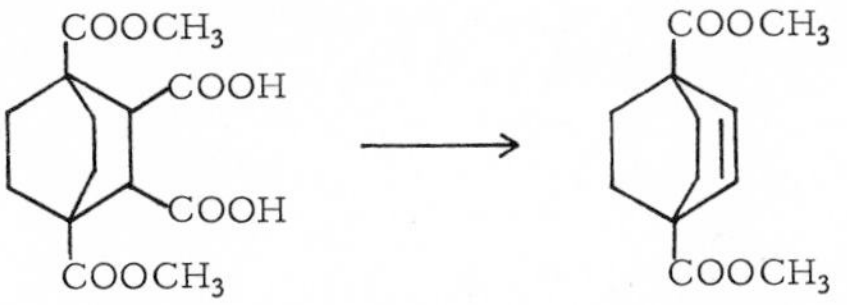

In a new and convenient degradation of the side chain of eburicoic acid (**1**) to a methyl ketone group (**5**), the first step involved oxidative decarboxylation, using the procedure of Kochi (**2**, 235). The derived triene (**2**) was obtained in about 50% yield. Remaining steps involved anionic conjugation (**3**); ozonolysis

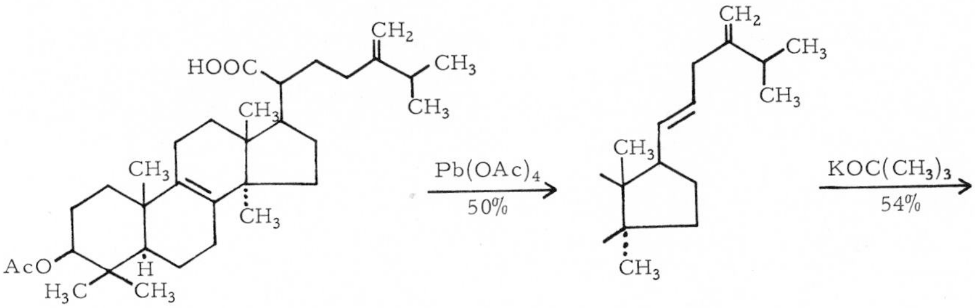

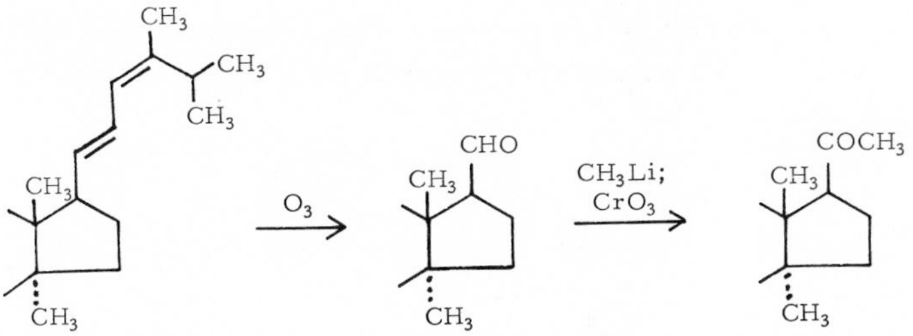

(**4**), and reaction with methyllithium followed by oxidation of the resulting alcohol (**5**).[5]

Oxidation of hydrazones. The hydrazone of 3β-hydroxy-Δ^{24}-lanostene-7-one (1) is oxidized in CH_2Cl_2 at 0–5° by lead tetraacetate to give 3β-hydroxy-7α-acetoxy-Δ^{24}-lanostene (2) as the major product. The $\Delta^{7,24}$-diene (3) is a minor product.[6] The reaction is useful, since 7α-hydroxy derivatives have been used to

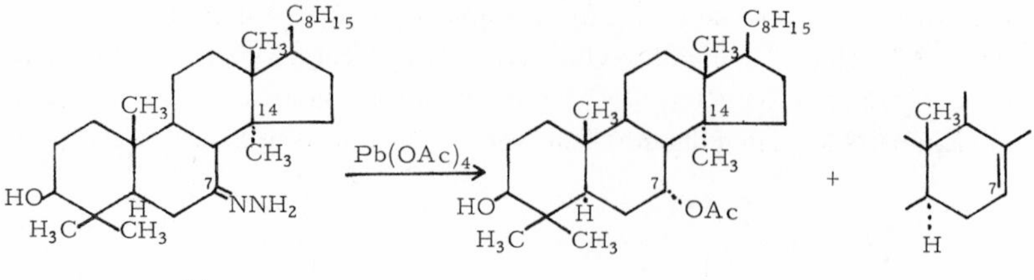

(1)   (2, 68%)   (3, 20%)

functionalize the angular methyl group at C_{14}. 7-Ketones in the lanosterol series have been reduced to the 7α-ols by acid-catalyzed catalytic hydrogenation, but this method is not suitable when the side chain is unsaturated.

Aroylhydrazines are oxidized by lead tetraacetate at room temperature to aroic acids by way of an aroyldiimide.[7]

$$ArCONHNH_2 \xrightarrow{Pb(OAc)_4} [ArCON=NH] \xrightarrow[80-90\%]{} ArCOOH$$

Synthesis of aziridines.[8] Lead tetraacetate oxidation of certain 1,1-disubstituted hydrazones, for example, (1)–(3), in the presence of nucleophilic olefins commonly used as carbene traps, gives aziridines in fair to good yields. Presumably singlet

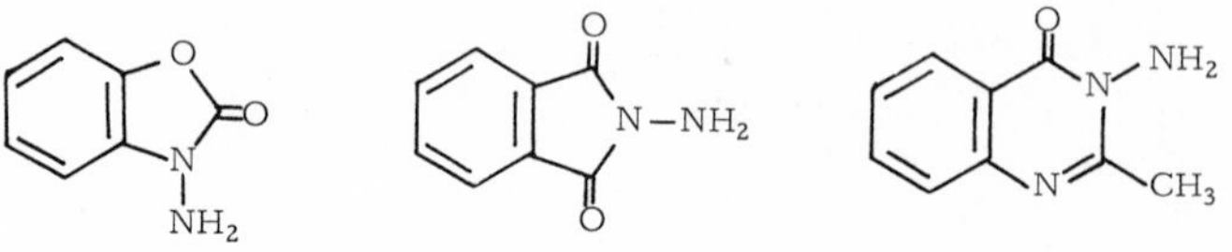

(1)   (2)   (3)

aminonitrenes are formed, which react stereospecifically with the olefin. Unfortunately no simple method was found for removing the heteroaromatic groups from the products.

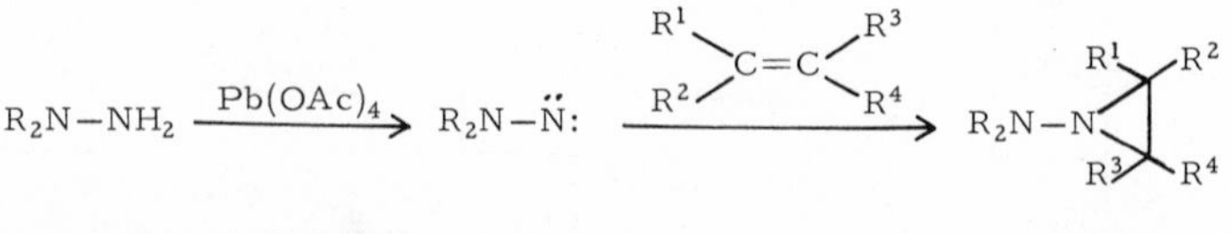

Benzynequinone. Oxidation of the quinone (1) in methylene chloride with lead tetraacetate in the presence of tetracyclone leads to 6,7,8,9-tetraphenyl-1,4-naphthoquinone (2) in about 40% yield. Benzynequinone is evidently formed and is trapped by tetracyclone with decarbonylation of the original adduct.[9]

(1)

(2)

Allylic oxidation. Δ^5-Steroids (cholesteryl acetate and diosgenyl acetate) undergo allylic acetoxylation at C_7 when oxidized with lead tetraacetate in benzene or glacial acetic acid solution. 7β-Acetoxy and 7α-acetoxy derivatives are formed in a ratio of $3:2$.[10]

Acetoxylation of steroidal δ-lactones. When isoandrolactone acetate (1) is treated in boiling benzene with 4 moles of lead tetraacetate, a slow reaction occurs and after 150 hours the α-acetoxylated lactone (2) is obtained in nearly 70% yield. In the Δ^5-unsaturated series allylic acetoxylation at C_7 takes place as well. The

reaction is useful since some natural products differ only in the state of oxidation of the lactone ring and direct introduction of an oxygen function in the lactone ring has presented difficulties.[11]

[1] J. D. Bacha and J. K. Kochi, *Tetrahedron*, **24**, 2215 (1968)

[2] G. Smith, C. L. Warren, and W. R. Vaughan, *J. Org.*, **28**, 3323 (1963); L. G. Humber, G. Myers, L. Hawkins, C. Schmidt, and M. Boulerice, *Canad. J. Chem.*, **42**, 2852 (1964)

[3] J. C. Kauer, B. E. Benson, and G. W. Parshall, *J. Org.*, **30**, 1431 (1965)

[4] N. B. Chapman, S. Sotheeswaran, and K. J. Toyne, *Chem. Commun.*, 214 (1965)

[5] D. H. R. Barton, D. Giacopello, P. Manitto, and D. L. Struble, *J. Chem. Soc.*, (C), 1047 (1969)

[6] D. H. R. Barton, P. L. Batten, and J. F. McGhie, *Chem. Commun.*, 450 (1969); *J. Chem. Soc.*, (C), 1033 (1970)

[7] J. B. Aylward and R. O. C. Norman, *ibid.*, (C), 2399 (1968)

[8] D. J. Anderson, T. L. Gilchrist, D. C. Horwell, and C. W. Rees, *ibid.*, (C), 576 (1970)

[9] C. W. Rees and D. E. West, *Chem. Commun.*, 647 (1969)

[10] M. Stefanovic, A. Jokić, Z. Maksimović, Lj. Lorenc, and M. Lj. Mihailovic, *Helv.*, **53**, 1895 (1970)

[11] M. Stefanović, Z. Djarmati, and M. Gašić, *Tetrahedron Letters*, 2769 (1970)

Lindlar catalyst, 1, 566–567. Supplier: Fluka.

The detailed preparation has now been published.[1]

Ozonization of norbornadiene (1) followed by reduction of the ozonization intermediate with Lindlar's catalyst affords 4-cyclopentene-*cis*-1,3-dialdehyde (2) in 35% yield. The method represents a new and selective method for reduction

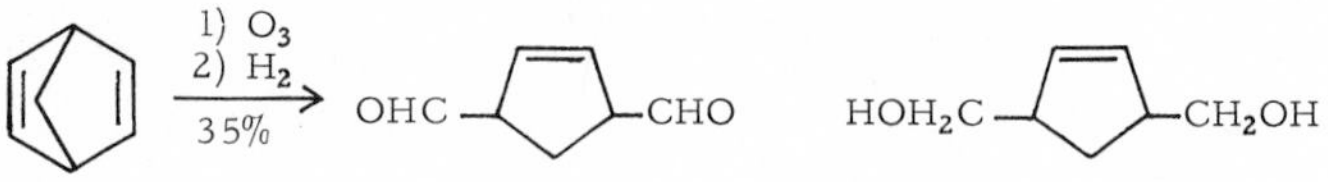

(1)   (2)   (3)

of unsaturated ozonization intermediates. Reduction with zinc in acetic acid does not lead to an isolable product. Reduction with sodium borohydride leads to *cis*-1,3-bishydroxymethylcyclopent-4-ene (3).[2]

[1]H. Lindlar and R. Dubuis, *Org. Syn.*, **46**, 89 (1966)
[2]C. A. Grob and H. R. Pfaendler, *Helv.*, **53**, 2156 (1970)

Lithio ethyl acetate, $LiCH_2COOC_2H_5$. Mol. wt. 94.04.

Preparation.[1] Lithio ethyl acetate is prepared in essentially quantitative yield by the reaction of ethyl acetate with lithium bis(trimethylsilyl)amide[2] in THF at $-78°$. It has been prepared also but in about 15% yield by the addition of ethyl

$$LiN[Si(CH_3)_3]_2 + CH_3COOC_2H_5 \longrightarrow HN[Si(CH_3)_3]_2 + LiCH_2COOC_2H_5$$

bromoacetate to butyllithium in hexane–ether at $-75°$.[3] Solutions of lithio ethyl acetate are stable indefinitely at $-78°$; this anion is more stable than sodio ethyl acetate.

β-Hydroxy esters.[1] The reagent reacts rapidly with carbonyl compounds to give β-hydroxy esters in high yield after hydrolysis:

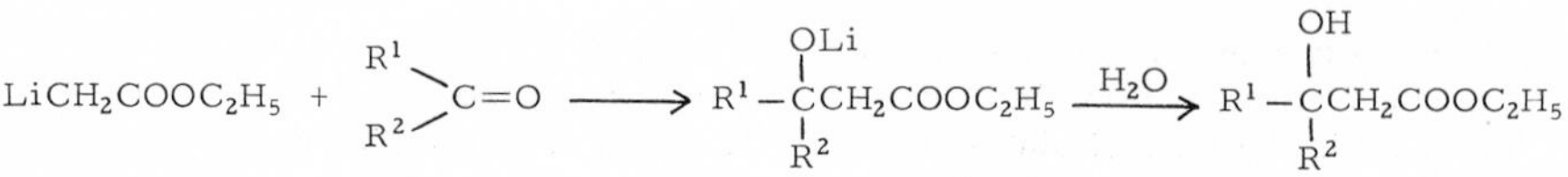

$$LiCH_2COOC_2H_5 + \underset{R^2}{\overset{R^1}{>}}C=O \longrightarrow R^1-\underset{R^2}{\overset{OLi}{\underset{|}{C}}}CH_2COOC_2H_5 \xrightarrow{H_2O} R^1-\underset{R^2}{\overset{OH}{\underset{|}{C}}}CH_2COOC_2H_5$$

Examples:

$$\underset{CH_3}{\overset{CH_3}{>}}C=O \xrightarrow[90\%]{} (CH_3)_2\overset{OH}{\underset{|}{C}}CH_2COOC_2H_5$$

$$C_6H_5CHO \xrightarrow[80\%]{} C_6H_5\overset{OH}{\underset{|}{CH}}CH_2COOC_2H_5$$

$$C_6H_5CH=CHCHO \xrightarrow[94\%]{} C_6H_5CH=CH\overset{OH}{\underset{|}{CH}}COOC_2H_5$$

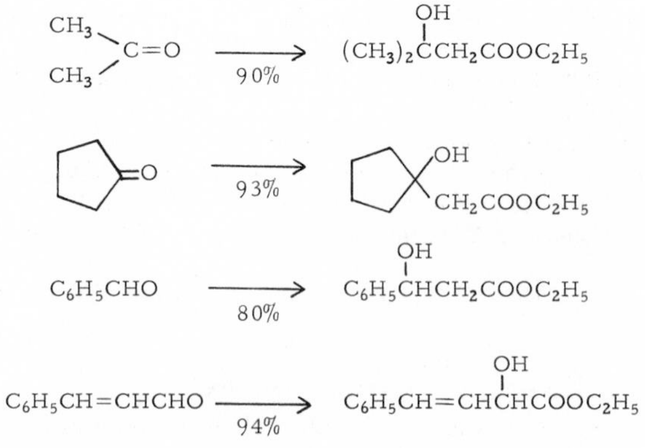

The last example shows that the reaction is successful even with α,β-unsaturated carbonyl compounds.

[1]M. W. Rathke, *Am. Soc.*, **92**, 3222 (1970)
[2]Prepared by the reaction of hexamethyldisilazane and *n*-butyllithium in hexane; the hexane is then evaporated and replaced with THF [E. H. Amonoo-Neizer, R. A. Shaw, D. O. Skovlin, and B. C. Smith, *J. Chem. Soc.*, 2997 (1965)]
[3]W. M. Jones and R. S. Pyron, *Tetrahedron Letters*, 479 (1965)

3-Lithio-1-trimethylsilylpropyne, 2, 239–241.

Synthesis of acetylenes (**2**, 240). Ireland *et al.*[1] applied the lithio-1-trimethylsilylpropyne procedure of Corey and Kirst[2] to a synthesis of the enediyne (3), desired as an intermediate in a triterpenoid synthesis. But application of this coupling procedure to the *trans*-dibromide (1) afforded less than 5% of the desired disilyl derivative (2). They then explored the coupling of (1) with propargyl

Grignard reagent with subsequent trimethylsilylation of the crude product in order to effect separation of acetylenic and allenic material. In this way they obtained the crystalline bistrimethylsilylenediyne (2) in 51% yield. The protecting groups were removed efficiently by the procedure of Schmidt and Arens (**2**, 437).

Ireland *et al.* then compared the two methods and found that both procedures were equally effective with simple allylic halides and benzylic halides. However, the Grignard coupling is better suited to reactive, base-sensitive allylic halides such as (1), whereas the lithiotrimethylsilylpropyne reagent is preferable for less reactive, saturated alkyl halides.

Total synthesis of α-*santalol.*[2] α-Santalol (7) has been synthesized from (−)-π-bromotricyclene (1) by essentially the same procedure used previously by Corey and Kirst for the synthesis of *trans,trans*-farnesol (**2**, 240–241). The starting material was converted into the terminal acetylene (2) by reaction with lithio-1-trimethylsilylpropyne followed by desilylation (**2**, 239–240). This was converted into the propargylic alcohol (3) by way of the lithium derivative by reaction with paraformaldehyde. *trans*-Hydroalumination was then effected by treatment with *n*-butyllithium followed by diisobutylaluminum hydride. Treatment with iodine

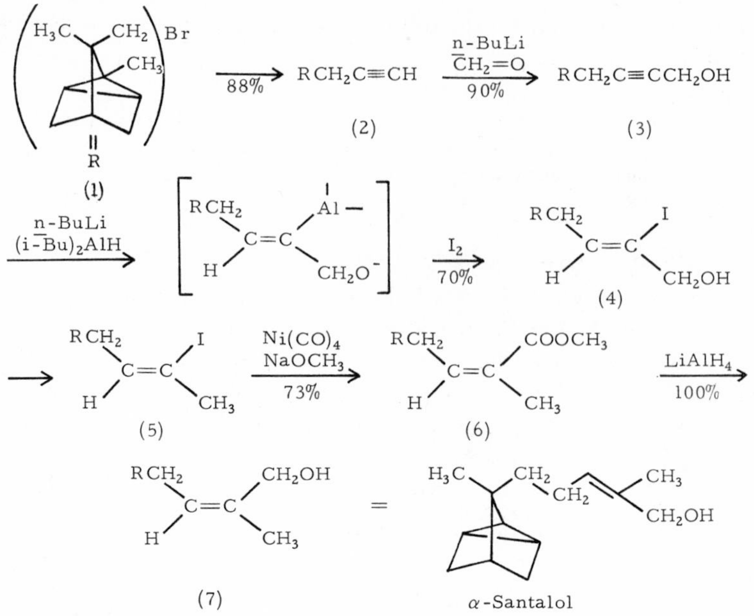

(2)

(3)

(1)

(4)

(5)

(6)

(7)   α-Santalol

gave the iodo allylic alcohol (4). The next step involved transformation of the hydroxymethyl into a methyl group by conversion through the mesylate to the bromide and reaction of this with sodium borohydride in DMSO. The next steps in the synthesis were conversion of the vinylic iodide (5) into the methyl ester by reaction with nickel carbonyl–sodium methoxide in methanol (*see* **Nickel carbonyl,** This volume). Reduction with lithium aluminum hydride completed the synthesis of α-santalol (7).

[1]R. E. Ireland, M. I. Dawson, and C. A. Lipinski, *Tetrahedron Letters*, 2247 (1970)
[2]E. J. Corey, H. A. Kirst, and J. A. Katzenellenbogen, *Am. Soc.*, **92**, 6314 (1970)

Lithium, 1, 570–573.

Synthesis of cyclopropanols.[1] A new and convenient synthesis of cyclopropanols involves the following sequence:

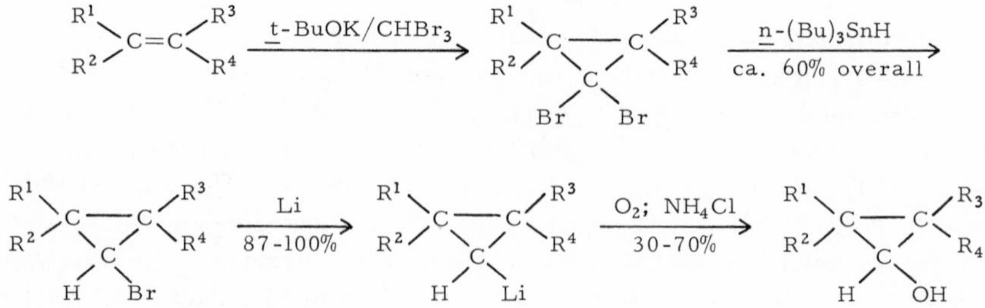

The first two steps are well known and proceed in about 60% yield. Conversion of the cyclopropyl bromide into the cyclopropyllithium is carried out with lithium in diethyl ether according to the procedure of Seyferth and Cohen.[2] The yield is nearly quantitative. The final step, oxygenation, is performed with CO_2-free oxygen at $-71°$, followed by hydrolysis with saturated ammonium chloride. Yields are generally 50–70%.

An interesting feature is that oxygenation of a cyclopropyllithium derived from a *syn/anti* mixture of bromides leads to the more stable cyclopropanol.

[1] D. T. Longone and W. D. Wright, *Tetrahedron Letters*, 2859 (1969)
[2] D. Seyferth and H. M. Cohen, *J. Organometal. Chem.*, **1**, 15 (1963/64)

Lithium–Alkylamine reduction, **1**, 574–581; **2**, 241–242.

Reduction of naphthalene (**2**, 241–242, ref. 6a). The reduction of naphthalene with lithium and a mixture of ethylamine and dimethylamine has now been published.[1]

Reduction of carboxylic acids. Burgstahler *et al.*[2] a few years ago reported briefly that carboxylic acids can be reduced to aldehydes by lithium and ethylamine; however, the yields for the most part were low except in the case of fairly high molecular weight acids. Bedenbaugh *et al.*[3] report that the reaction is actually of considerable value. They used methylamine rather than ethylamine and maintained a basic medium. Under these conditions an intermediate imine can be isolated. This intermediate is hydrolyzed rapidly by aqueous acids to an aldehyde; or it can be reduced to the corresponding amine either catalytically or with lithium in methylamine. The conversions are illustrated for pentanoic acid as starting material.

$$CH_3(CH_2)_3COOH \xrightarrow{\text{Li, } CH_3NH_2} CH_3(CH_2)_3CH=NCH_3 \xrightarrow[60\%]{H_3O^+} CH_3(CH_2)_3CHO$$

$$52\% \downarrow \begin{array}{l} H_2/\text{cat. or} \\ \text{Li, } CH_3NH_2 \end{array}$$

$$CH_3(CH_2)_3CH_2NHCH_3$$

One disadvantage is that aromatic, olefinic, and other unsaturated centers are reduced as well; however, benzoic acid can be reduced to benzaldehyde in about 25% yield if ammonium nitrate is added to the reduction medium.

[1] E. M. Kaiser and R. A. Benkeser, *Org. Syn.*, **50**, 88 (1970)
[2] A. W. Burgstahler, L. R. Worden, and T. B. Lewis, *J. Org.*, **28**, 2918 (1963)
[3] A. O. Bedenbaugh, J. H. Bedenbaugh, W. A. Bergin, and J. D. Adkins, *Am. Soc.*, **92**, 5774 (1970)

Lithium aluminum hydride, 1, 581–595; **2,** 242.

Reduction of allylic alcohols (**1,** 592, ref. 32). The preparation of 1,1-diphenyl-cyclopropane has now been published.[1]

Reduction of aryl halides. Aryl halides have generally been considered to be resistant to hydrogenolysis by lithium aluminum hydride. However, Karabatsos and Shone[2] found that certain aromatic halides are reduced by the reagent in diglyme at 100°. The order of reactivity is I > Br > Cl > F. These chemists concluded, however, that triphenyltin hydride is superior for this purpose.

More recently Brown and Krishnamurthy[3] reported that aryl iodides and bromides are reduced in high yield by lithium aluminum hydride in tetrahydrofurane; the former are reduced at a reasonable rate even at room temperature, the latter require higher temperatures (reflux). Optimum rates are obtained with a ratio of at least 2 moles of reducing reagent per mole of aryl halide. Actually a 4 : 1 molar ratio was adopted for routine use. Brown and Krishnamurthy suggest that the difference in the results from the two laboratories may be due to the use in the earlier work of "slurries" of lithium aluminum hydride. Since the reagent is ordinarily readily soluble in THF or diglyme, the reagent in the earlier work may have been defective. Aryl chlorides are reduced so slowly that iodo and bromo substituents can be selectively reduced in the presence of chloro substituents. The rate of reduction is enhanced by electron-withdrawing substituents and is decreased by electron-releasing groups.

[1]M. J. Jorgenson and A. F. Thatcher, *Org. Syn.,* **48,** 75 (1968)
[2]G. J. Karabatsos and R. L. Shone, *J. Org.,* **33,** 619 (1968)
[3]H. C. Brown and S. Krishnamurthy, *ibid.,* **34,** 3918 (1969)

Lithium aluminum hydride–Aluminum chloride, 1, 595–599; **2,** 243.

Tetraphenylcyclopentadiene. This useful diene (2) is not easily available, although tetraphenylcyclopentadienone (**1,** tetracyclone) is prepared readily (**1,** 1149). An early study[1] of conversion of (1) into (2) by reduction was not particularly promising, the best yield (18%) being obtained by inverse lithium

(1)   (2)

aluminum hydride reduction. Cava[2] has now found that reduction of (1) (1 mmole) in ether with lithium aluminum hydride (2 mmoles) and aluminum chloride (2 mmoles) gives the diene (2) in 93% yield. Reduction with excess sodium borohydride gives a mixture of alcohols which can be dehydrated to (2) with iodine in refluxing benzene in an overall yield of 55%.

[1]N. O. V. Sonntag, S. Linder, E. I. Becker, and P. E. Spoerri, *Am. Soc.*, **75**, 2283 (1953)
[2]M. P. Cava and K. Narasimhan, *J. Org.*, **34**, 3641 (1969)

Lithium amalgam, Li(Hg)$_x$. *Prepared* from lithium metal and mercury.

Wurtz reaction. Wiberg and Connor[1] noted that lithium amalgam is far more satisfactory than either sodium or lithium for conversion of 3-bromomethyl-cyclobutyl bromide (1) into bicyclo[1.1.1]pentane (2). When sodium metal was

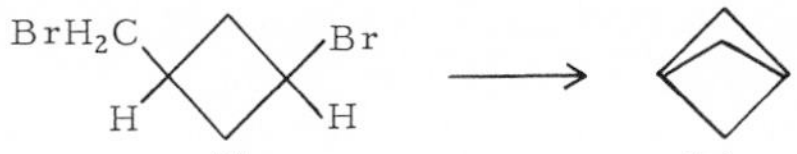

used, the yield of (2) was 0.5%; use of lithium amalgam in refluxing dioxane raised this yield to 4.2%.

The most convenient synthesis of 1,3-dimethylbicyclo[1.1.0]butane (4) employs the following two-step process.[2] Hydrogen bromide is added to allene and/or methylacetylene to give a mixture of *cis*- and *trans*-1,3-dibromo-1,3-dimethyl-cyclobutanes (3). This mixture is stirred with a suspension of 2% lithium amalgam in dioxane for 48 hours at room temperature; (4) is obtained in 85% yield.

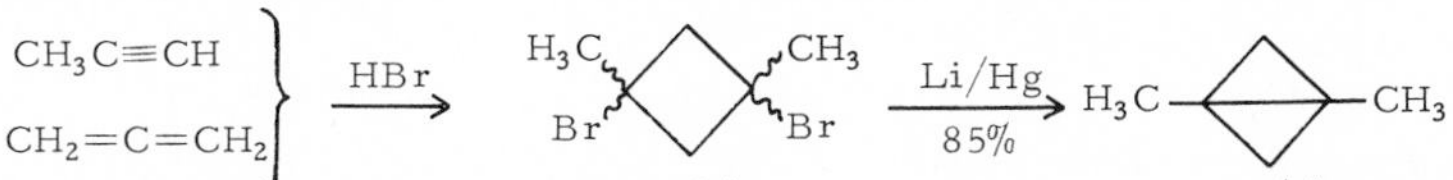

The most convenient synthesis of cyclobutane is the reaction of 1,4-dibromo-butane with lithium amalgam in refluxing dioxane; yield 70%.[3]

Dehalogenation. Two laboratories have reported cases where two different reactive intermediates are liberated from two different starting materials by the same reagent (lithium amalgam) and then allowed to react together. In one case,[4] 5,6-benzo-1,2,3,4-tetramethylbicyclo[2.2.0]hexa-2,5-diene (3) has been prepared by treating an ether solution of *o*-bromofluorobenzene (1) and of 3,4-dichloro-1,2,3,4-tetramethylcyclobutene (2) with lithium amalgam. The reaction can be viewed formally as a Diels-Alder reaction between benzyne and tetramethyl-cyclobutadiene.

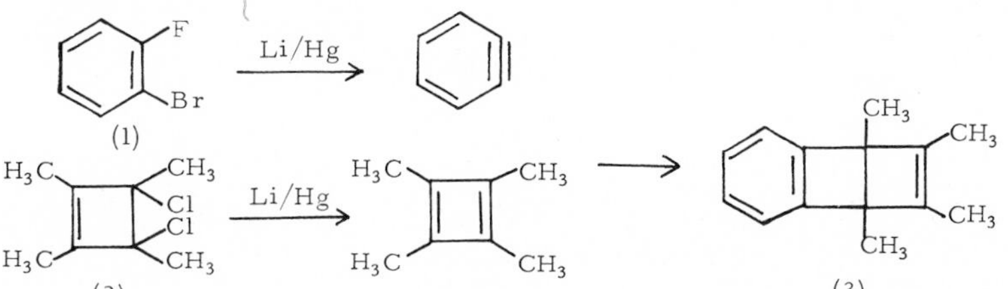

In the second case,[5] 1,6,7,8-tetrachloro-2,3,4,5-tetramethylbicyclo[4.2.0.0$^{2.5}$]

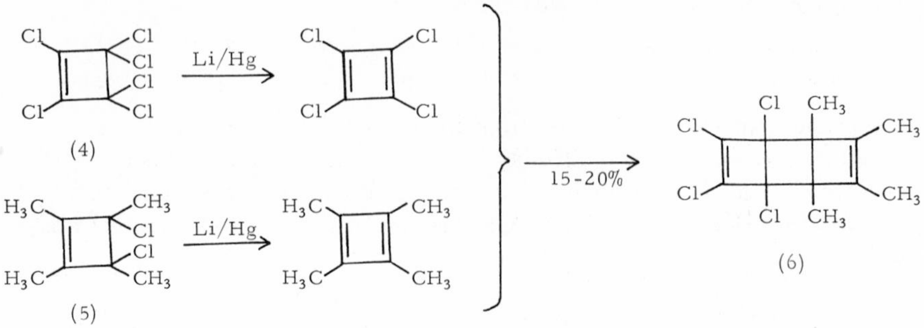

octadiene (6) has been obtained from treatment of a mixture of hexachlorocyclo-butene (4) and of 3,4-dichloro-1,2,3,4-tetramethylcyclobutene (5) in boiling ether with lithium amalgam.

[1]K. B. Wiberg and D. S. Connor, *Am. Soc.*, **88**, 4437 (1966)
[2]K. Griesbaum and P. E. Butler, *Angew. Chem., internat. Ed.*, **6**, 444 (1967)
[3]D. S. Connor and E. R. Wilson, *Tetrahedron Letters*, 4925 (1967)
[4]D. T. Carty, *ibid.*, 4753 (1969)
[5]R. Criegee and R. Huber, *Angew. Chem., internat. Ed.*, **8**, 759 (1969)

Lithium amide, 1, 600–601.

3-Methylcyclopropene. 3-Methylcyclopropene (2) was originally prepared in about 2% yield by the reaction of *trans*-1-chloro-2-butene (1, Aldrich) with sodium amide in THF.[1] German chemists[2] have increased the yield to about 30% by use of lithium amide in boiling dioxane. A particular advantage of the new procedure is that (2) is formed in high purity (98%).

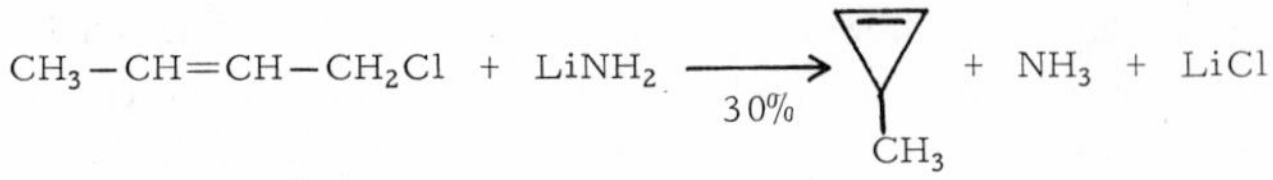

Lithium amide[2] is also superior to sodium amide[3] for dehydrochlorination of methallyl chloride (3) to give 1-methylcyclopropene (4).

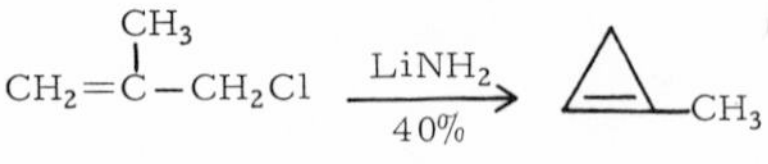

[1]S. Wawzonek, B. J. Studnicka, and A. R. Zigman, *J. Org.*, **34**, 1316 (1969)
[2]R. Köster, S. Arora, and P. Binger, *Angew. Chem., internat. Ed.*, **9**, 810 (1970)
[3]F. Fisher and D. E. Applequist, *J. Org.*, **30**, 2089 (1965)

Lithium–Ammonia, 1, 601–603; **2,** 205.

Reductive alkylation and carbomethoxylation. Stork[1] has now published details of his method of generating the less stable enol of an α,β-unsaturated ketone and trapping this intermediate by alkylation or carboxylation before equilibration can occur (**1,** 601–602). For example, reduction of $\Delta^{1(9)}$-2-octalone (1) with lithium–ammonia followed by addition of methyl iodide instead of the usual proton source gives in about 50% yield a mixture of 1-methyl-*trans*-2-decalone (2, 83%) and the product of reduction, *trans*-2-decalone (3, 17%). Direct alkylation of *trans*-2-

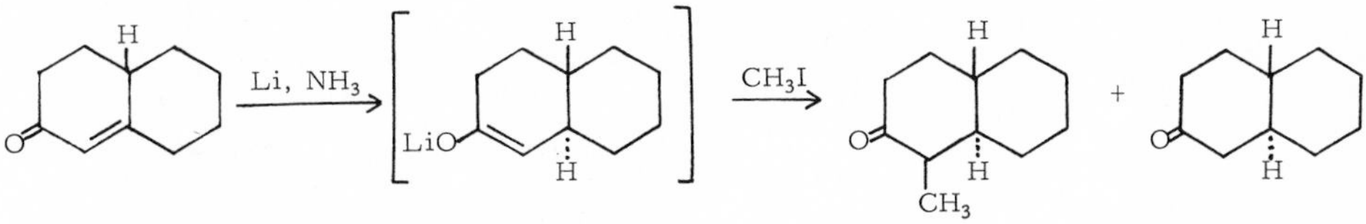

(1)   (2)   (3)

decalone leads to substitution at the 3-position. The trapping procedure was applied to construction of hydrophenanthrene systems. Thus the sequence was applied to 10-methyl-$\Delta^{1(9)}$-2-octalone (4), and 1,3-dichloro-2-butene was used as the alkylating agent; the crude alkylation product (5) was cyclized with concentrated sulfuric acid to the dodecahydrophenanthrone (6).

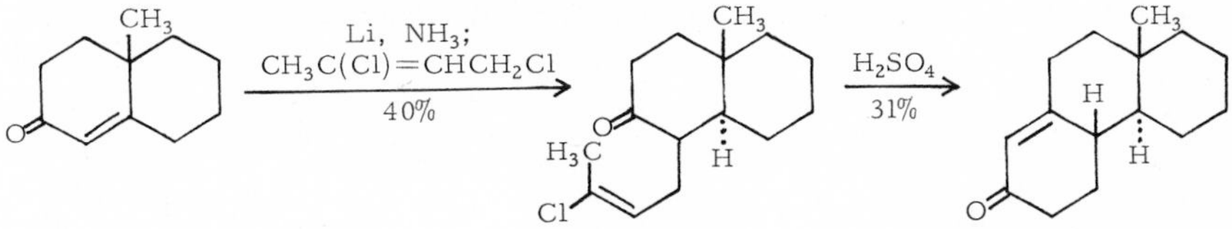

(4)   (5)   (6)

The sequence is equally useful for alkylation at a position which already carries an alkyl group. In this case formation of the simple reduction product (such as 3, above) becomes a serious matter. However, if the liquid ammonia is replaced before alkylation by THF, the desired alkylation occurs in reasonable yield. Thus 1,10-dimethyl-$\Delta^{1(9)}$-2-octalone (7) was alkylated under these conditions to the desired 1,1,10-trimethyl-*trans*-2-decalone (8) in satisfactory yield.

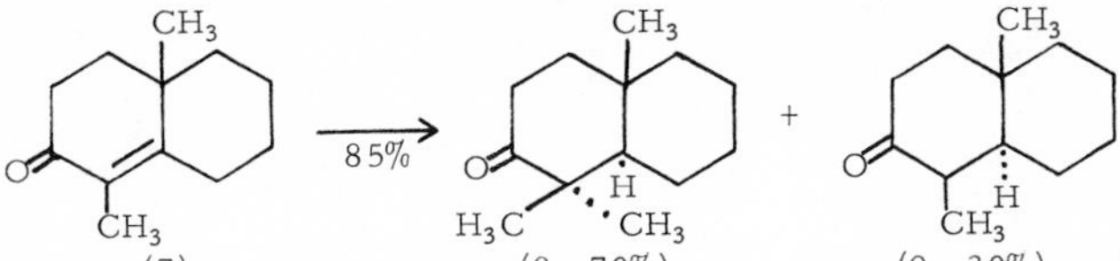

(7)   (8, 70%)   (9, 30%)

The THF conditions were also used for alkylation of $\Delta^{9(10)}$-1-octalone (10). Some alkylation occurred at the 2-position, but the main products (62%) were *cis*- and *trans*-9-methyl-1-decalone, (11) and (12), in the approximate ratio 3 : 1.

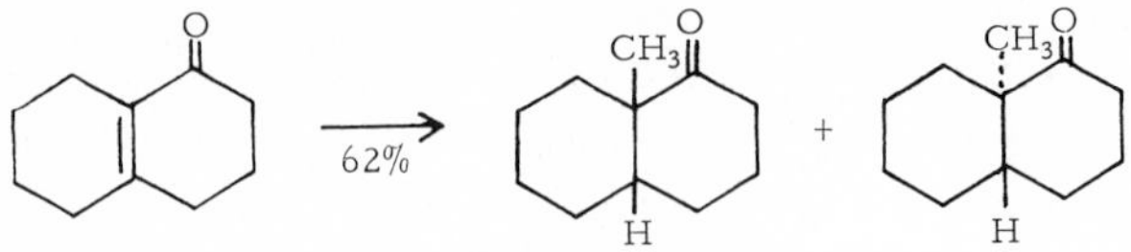

(10)   (11)   3:1   (12)

Stork also reported two instances in which the enolate ion was trapped with carbon dioxide. Thus addition of lithium in ammonia to (4), followed by replace-

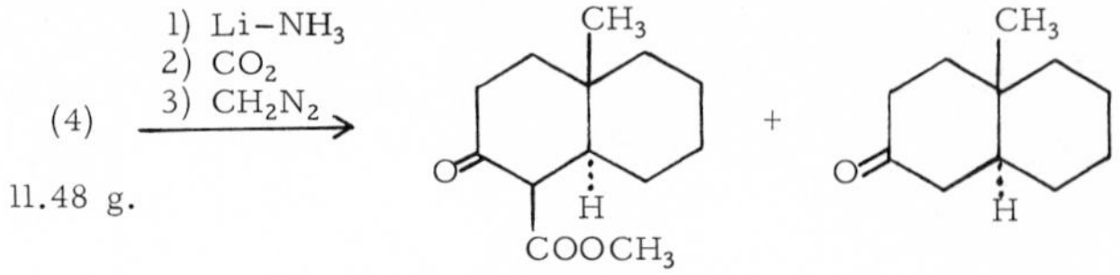

(13, 5.2 g.)   (14, 3.55 g.)

ment of the ammonia by anhydrous ether and addition of solid carbon dioxide gave a β-keto acid, isolated as the methyl ester (13).

This sequence has obvious application to synthesis of resin acids and was used in fact as a key step in a synthesis of methyl desisopropyldehydroabietate (20) by Spencer et al.[2] Thus reductive carbomethoxylation of the tetrahydropyranyl ether (15)[3] gave the desired product in 68% yield (simple reduction also occurs).

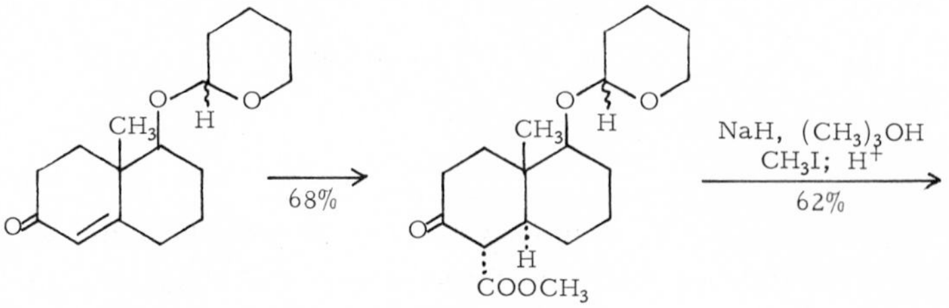

(15)   (16)

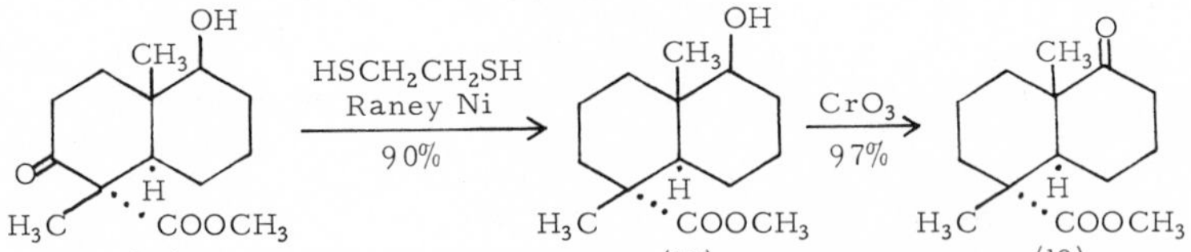

(17)   (18)   (19)

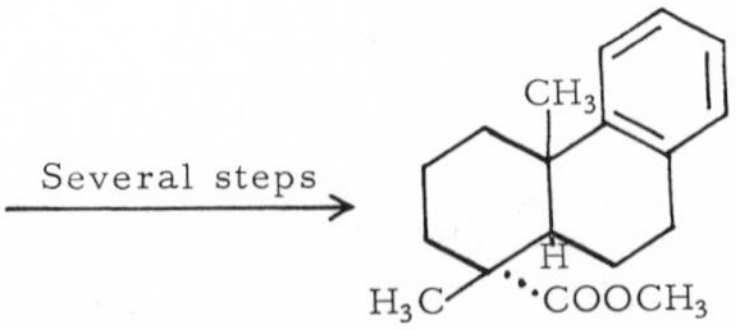

(20)

Alkylation of (16) gives predominantly (17), the product of electronically controlled axial methylation. Thus in two steps the desired, *trans* ring fusion is
obtained and also the desired stereochemistry in the abietic series. The remaining
steps to (19) proceeded in high yield, and this product was transformed by
obvious methods into methyl desisopropyldehydroabietate (20).

The Stork procedure was also used as a key step in the synthesis of (−)-sanda-
racopimaric acid (24) from testosterone acetate (21).[4]

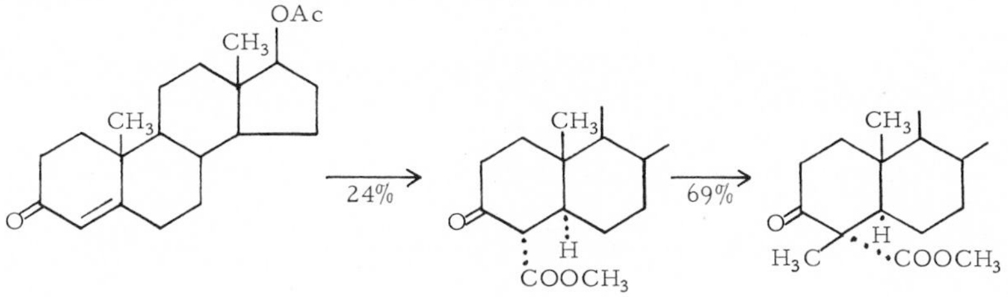

(21)   (22)   (23)

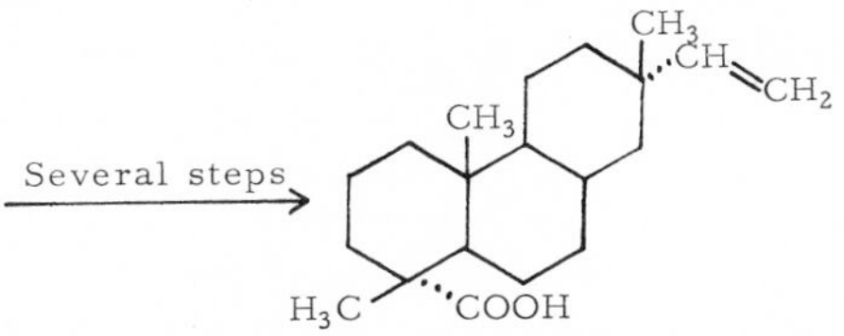

(24)

Note, however, that reductive carbomethoxylation of the α,β-unsaturated
ketone (25) gave (26) in only low yield.[5]

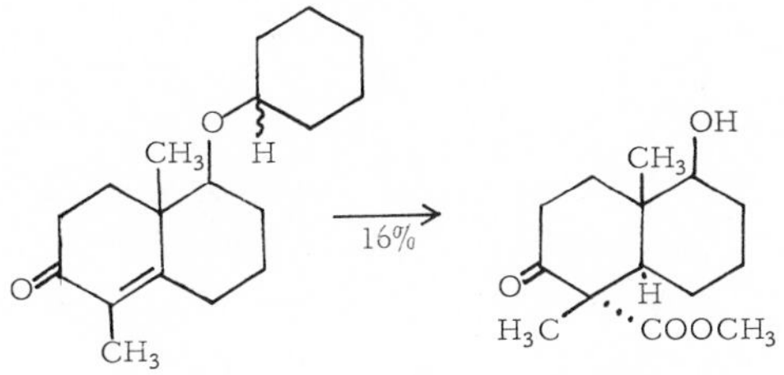

(25)   (26)

vic-Cyclopropanediols.[6] Reduction of a dilute THF-ammonia solution of the
cyclic β-diketone (1) with an excess of lithium gives the *vic*-cyclopropanediol (2)
in 65–95% yield (not entirely pure). Reduction with a limited amount of lithium
gives (3) in 70% yield.

(1)   (2)   (3)

[1]G. Stork, P. Rosen, N. Goldman, R. V. Coombs, and J. Tsuji, *Am. Soc.*, **87**, 275 (1965)
[2]T. A. Spencer, T. D. Weaver, R. M. Villarca, R. J. Friary, J. Posler, and M. A. Schwartz, *J. Org.*, **33**, 712 (1968)
[3]Use of the free alcohol led to considerable undesired carbomethoxylation at C_2 (steroid numbering).
[4]A. Afonso, *Am. Soc.*, **90**, 7375 (1968)
[5]T. A. Spencer, R. J. Friary, W. W. Schmiegel, J. F. Simeone, and D. S. Watt, *J. Org.*, **33**, 719 (1968)
[6]W. Reusch and D. B. Priddy, *Am. Soc.*, **91**, 3677 (1969)

Lithium bisbenzenesulfenimide, $LiN(SC_6H_5)_2$. Mol. wt. 239.38.

The reagent is prepared by the reaction of *n*-butyllithium with bisbenzenesulfenimide.[1] The sulfenimide is obtained in 60% yield by the reaction of benzenesulfenyl chloride with ammonia:

$$2\ C_6H_5SCl\ +\ 3\ NH_3\ \xrightarrow[60\%]{}\ (C_6H_5)_2NH\ +\ 2\ NH_4Cl$$

M.p. 129-130° dec.

Primary alkylamines.[2] The reagent (1) reacts with alkyl bromides or alkyl *p*-toluenesulfonates to form an N-alkyl bisbenzenesulfenimide (2), which is not isolated, but hydrolyzed with 3 N hydrochloric acid. A primary alkylamine (3) and benzenesulfenyl chloride are formed. THF and dimethoxyethane are the most suitable solvents. Yields for the most part are in the range 40–80%.

$$(C_6H_5S)_2NH\ \xrightarrow{n-C_4H_9Li}\ [(C_6H_5S)_2NLi]\ \xrightarrow[ROTs]{RBr\ or}\ [(C_6H_5S)_2NR]\ \xrightarrow{HCl}\ RNH_2\ +\ 2\ C_6H_5SCl$$

(1)   (2)   (·3)

Water-soluble amines, such as β-methoxyethylamine, are obtained successfully by treating the N-alkyl bisbenzenesulfenimide (2) with benzenethiol in the absence of water:

$$(2)\ +\ 2\ C_6H_5SH\ \longrightarrow\ RNH_2\ +\ 2\ C_6H_5SSC_6H_5$$

(3)

The reaction conditions are mild, and the method is suitable for preparation of amines containing nitrile, ester, and amide groups; these groups, under the conditions of the Gabriel synthesis, are hydrolyzed to carboxylic acid groups.

[1]H. Lecher, *Ber.*, **58**, 409 (1925)
[2]T. Mukaiyama and T. Taguchi, *Tetrahedron Letters*, 3411 (1970)

Lithium *t*-butoxide, $LiOC(CH_3)_3$. Mol. wt. 80.05.

The reagent is prepared by the reaction of *t*-butanol with *n*-butyllithium in hexane (exothermic reaction).

***t*-Butyl esters.** Esters of carboxylic acids[1] and sulfonic acids[2] have been prepared by reaction of acid chlorides with alkoxides. The method is particularly useful for preparation of highly hindered esters. The method is recommended for the preparation of *t*-butyl *p*-toluate.[3]

[1]M. S. Newman and T. Fukunaga, *Am. Soc.*, **85**, 1176 (1963)
[2]P. J. Hamrick, Jr., and C. R. Hauser, *J. Org.*, **26**, 4199 (1961)
[3]G. P. Crowther, E. M. Kaiser, R. A. Woodruff, and C. R. Hauser, *Org. Syn.* submitted 1969; E. M. Kaiser and R. A. Woodruff, *J. Org.*, **35**, 1198 (1970)

Lithium carbonate (1, 606).

Cyclohepta-2,4,6-trienones. 2,2,7-Tribromocycloheptanones are dehydrobrominated with lithium carbonate in boiling DMF to give cycloheptatrienones in about 50% yield.[1] A double-bond migration is probably involved.[2]

[1]G. Jones, *J. Chem. Soc.*, (C), 1230 (1970)
[2]E. W. Collington and G. Jones, *ibid.*, (C), 2656 (1969)

Lithium cyanohydridoborate, $LiBH_3CN$. Mol. wt. 45.90. Supplier: Alfa Inorganics.

Preparation. The reagent is prepared in 90–95% yield by the reaction of lithium borohydride with hydrocyanic acid.[1]

Reduction. An early report[2] stated that aldehydes and α-hydroxy ketones are reduced by this reagent but that ketones, nitro compounds, carboxylic acids, and esters are not reduced. Borch and Durst,[3] however, found that some ketones are reduced, but more slowly than with sodium borohydride. One advantage over the latter reagent is that lithium cyanohydridoborate is stable in acid up to pH ~3 and hence can be used for reduction of groups sensitive to high pH (e.g., thiamine).

Probably the most useful application of the reagent is for reductive amination of aldehydes and ketones. The carbonyl compound is allowed to react with ammonia or primary or secondary amines at pH <7 in absolute methanol in the presence of the reagent. The corresponding amine is obtained in 40–80% yield (isolated).

$$C_6H_5CHO \ + \ CH_3CH_2NH_2 \xrightarrow[80\%]{} C_6H_5CH_2NHC_2H_5$$

$$C_6H_5CH_2\overset{\text{O}}{\underset{\|}{C}}COONa \xrightarrow[49\%]{NH_3} C_6H_5CH_2CH(NH_2)COOH$$

[1]G. Wittig and P. Raff, *Ann.*, **573**, 195 (1951)
[2]G. Drefahl and E. Keil, *J. Prakt. Chem.*, **6**, 80 (1958)
[3]R. F. Borch and H. D. Durst, *Am. Soc.*, **91**, 3996 (1969)

Lithium diisopropylamide, 1, 611; **2,** 249.

Alkylation of dialkylacetic acids.[1a,b] A dialkylacetic acid, for example isobutyric acid, is converted into the sodium salt by treatment with sodium hydride in THF. Treatment with lithium diisopropylamide then forms the metalated derivative (1), which is then alkylated to give an alkyldimethylacetic acid (2). In the original

$$(CH_3)_2CHCOONa \ + \ LiN[CH(CH_3)_2]_2 \longrightarrow [(CH_3)_2CCO_2]^=Li^+Na^+$$

$$(1)$$

$$\xrightarrow[60-90\%]{RX} RC(CH_3)_2COOH$$

$$(2)$$

procedure[1a] the free acid was employed, but in this case two moles of reagent are required. This sequence is an alternative to the Haller-Bauer sequence.[2]

The original method was found to give low yields (29–44%) in the case of straight-chain aliphatic acids. However, if HMPT is used as cosolvent with THF, yields of about 90% are obtained. On the other hand, use of HMPT–THF for the alkylation of branched-chain aliphatic acid decreases the yield.[3]

Monoalkylation of alkylacetic acids. Essentially the same procedure can be used for monoalkylation of alkylacetic acids to give dialkylacetic acids.[4] These have generally been prepared by the malonic ester synthesis,[5] but this process has the disadvantage that if two different alkyl groups are to be introduced, dialkylation by the first of the two alkylating agents can be a side reaction. Also, the use of a malonic ester derivative requires removal of one of the activating groups at the end of the sequence. The new method gives yields of 50–88% of dialkylacetic acids:

$$RCH_2COONa \xrightarrow{LiN[CH(CH_3)_2]_2} [RCHCO_2]^=Li^+Na^+ \xrightarrow{R'X} R'RCHCOOH$$

The order of introduction of the alkyl group does not appear to be important, and the steric bulk of the R group does not seem to be significant. The principal limitation is that salts of lower carboxylic acids have poor solubility in the solvent system (THF–heptane) generally employed.

Monoalkylation of toluic acids. When toluic acid or dimethylbenzoic acids are treated with two equivalents of lithium diisopropylamide in THF–heptane at 0°, a deep red homogeneous solution of the metalated acid is formed. This is readily alkylated by reaction with an alkyl halide. Yields are in the range 30–70%. A

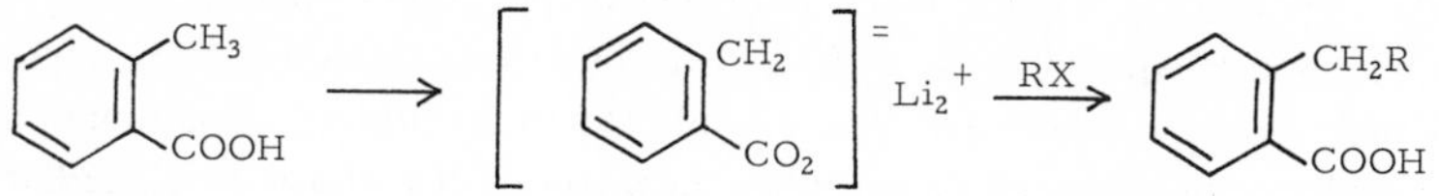

second methyl group can also be present and one methyl group can be preferentially alkylated, the order of reactivity being $o > p > m$.[6]

Synthesis of aldehydes and nitroparaffins. A straight-chain carboxylic acid can be converted into the α-anion by this base in the presence of hexamethylphosphoric triamide (**1**, 430–431; **2**, 208–210; This volume) as cosolvent in THF. Reaction of this anion with ethyl formate followed by neutralization with 10% hydrochloric acid furnishes an aldehyde:

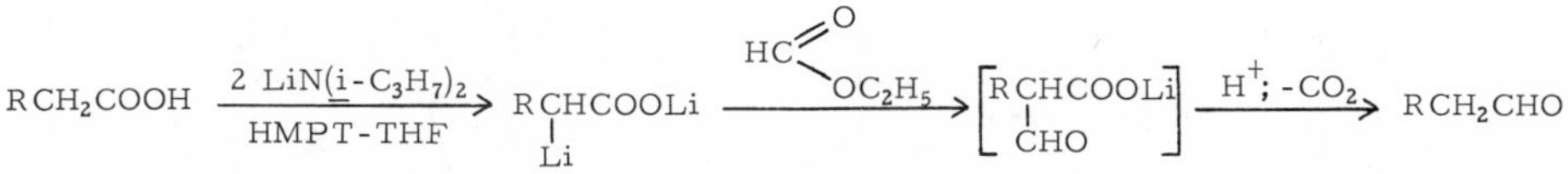

Yields are moderate in the case of straight-chain acids, but are low with α-branched chains. The method is applicable to unsaturated acids and in this case is superior to Rosenmund reduction.

Nitroparaffins can be prepared similarly by reaction of the α-anion with an alkyl nitrate (*n*-propyl nitrate was used):

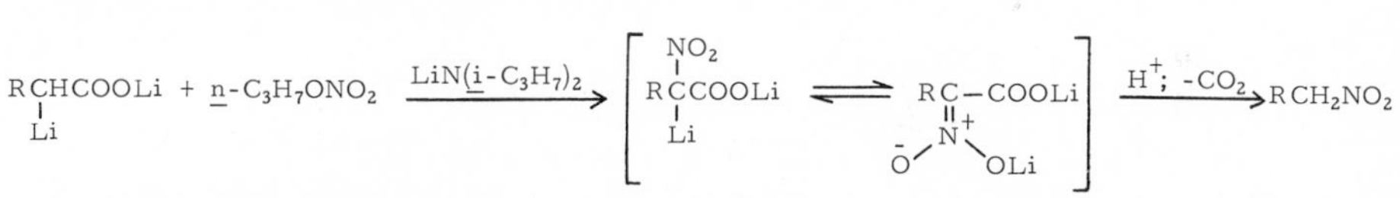

Yields are in the range 45–68%.[7]

Directed aldol condensation (**2**, 249). The procedure for the directed aldol condensation of acetaldehyde and benzophenone to give β-phenylcinnamaldehyde (**2**, 249, ref. 2) has now been published.[8]

[1](a) P. L. Creger, *Am. Soc.*, **89**, 2500 (1967); (b) *idem*, *Org. Syn.*, **50**, 58 (1970)

[2]K. E. Hamlin and A. W. Weston, *Org. Reactions*, **9**, 1 (1957)

[3]P. E. Pfeffer and L. S. Silbert, *J. Org.*, **35**, 262 (1970)

[4]P. L. Creger, *Am. Soc.*, **92**, 1397 (1970)

[5]A. C. Cope, H. L. Holmes, and H. O. House, *Org. Reactions*, **9**, 107 (1957); H. O. House, "Modern Synthetic Reactions, 163–184, Benjamin, New York, 1965

[6]P. L. Creger, *Am. Soc.*, **92**, 1396 (1970)

[7]P. E. Pfeffer and L. S. Silbert, *Tetrahedron Letters*, 699 (1970)

[8]G. Wittig and A. Hesse, *Org. Syn.*, **50**, 66 (1970)

Lithium–Ethylenediamine, **1**, 614–615; **2**, 250.

Reduction of epoxides. Hallsworth and Henbest (**1**, 579, refs. 12 and 13) found that some steroidal epoxides, which were unreactive to lithium aluminum hydride, are easily reduced with a large excess of lithium in ethylamine. However, some olefin is also formed in some cases. Brown *et al.*[1] now report that the combination of lithium and ethylenediamine at 50° is excellent for reduction of labile epoxides of bicyclic ketones, which are reduced only slowly by lithium aluminum hydride and usually with some extensive rearrangement. They chose ethylenediamine rather than ethylamine because the reduction is less vigorous in ethylenediamine than in ethylamine and thus easier to control. Also isolation of the alcohol is simplified because ethylenediamine is very soluble in water and only slightly soluble in ether, whereas ethylamine is miscible both in water and in ether. By this procedure, norbornene oxide (1) is reduced to pure *exo*-norbornanol (2) in 87% yield (isolation). Analysis by glpc indicated that two rearranged alcohols (3, 4) are formed to a minor extent and that (2) is formed in 99.3% yield.

(1)   (2, 99. 3%)   (3, 0. 2%)   (4, 0. 5%)

Even the very labile 2-methylene-7,7-dimethylnorbornane epoxide (5) is reduced in this way to the tertiary alcohol (6) in 89% yield. Reduction with lithium aluminum hydride gives chiefly a primary alcohol.

(5)   (6)

[1] H. C. Brown, S. Ikegami, and J. H. Kawakami, *J. Org.*, **35**, 3243 (1970)

Lithium 2-methoxyethoxide, $CH_3OCH_2CH_2OLi$. Mol. wt. 82.03.

Vinyl halides. Treatment of 5,5-dialkyl-N-nitrosooxazolidones (1) with lithium 2-methoxyethoxide in 2-methoxyethanol saturated with sodium iodide, lithium bromide, or lithium chloride leads to high yields of vinyl halides (3).[1] The reaction is considered to proceed through an unsaturated carbonium ion (2).

(1)   (2)   (3) X = I, Br, Cl

Vinylsilanes have also been prepared from the same precursors. Thus treatment of (1), R = CH$_3$, with lithium ethoxide (1, 612–613) in benzene containing triethylsilane gives triethyl (2-methyl-1-propenyl)silane (5) in 61% yield. Since carbenes are known to insert into the SiH bond, an unsaturated carbene (4) is apparently involved.[2]

$$(1),\ R = CH_3 \xrightarrow{R'O^-Li^+} [(CH_3)_2C=C:] \xrightarrow[61\%]{HSi(C_2H_5)_3} (CH_3)_2C=C\underset{Si(C_2H_5)_3}{\overset{H}{\diagup}}$$

(4)　　　　　　　　　　　　　　　(5)

[1]M. S. Newman and C. D. Beard, *Am. Soc.*, **92**, 4309 (1970)
[2]M. S. Newman and T. B. Patrick, *ibid.*, **92**, 4312 (1970)

Lithium perhydro-9b-boraphenalylhydride (3). Mol. wt. 184.06.

Preparation. The reaction of *trans,trans,trans*-cyclododecatriene (1) with triethylamine borane at 140° in perhydrocumene gives *cis,cis,trans*-perhydro-9b-boraphenalene (2) in 73% yield.[1,2]

Reaction of (2) with lithium hydride in THF gives a solution of the addition compound (3).[3]

(1)　　　　　　　　　　　　(2)　　　　　　　　　　　(3)

Reduction of cyclic ketones.[3] This unusual derivative of lithium borohydride is not only an active reducing agent for cyclic ketones, but also exhibits high stereoselectivity. Thus 2-methylcyclopentane is reduced to *cis*-2-methylcyclopentanol in 94% yield, and norcamphor gives *endo*-norborneol (99% yield). In contrast to the usually predominant attack of lithium aluminum hydride from the axial direction, with the new reagent attack from the equatorial position is greatly favored. Previously this steric control of reduction of cyclic ketones was achieved by use of dialkylboranes in which the alkyl groups are bulky, e.g., disiamylborane (1, 57–59; 2, 29) and diisopinocamphenylborane (1, 262–263). However, such reagents reduce ketones very slowly, whereas reductions with (3) are complete within 30 minutes at 0°.

[1]G. W. Rotermund and R. Köster, *Ann.*, **686**, 153 (1965)
[2]The original configuration assigned in ref. 1 (*cis,cis,cis*) has been revised to *cis,cis,trans* by H. C. Brown and W. C. Dickason, *Am. Soc.*, **91**, 1226 (1969)
[3]*Idem, ibid.*, **92**, 709 (1970)

Lithium *n*-propyl mercaptide, $LiSCH_2CH_2CH_3$. Mol. wt. 82.09.

Preparation.[1] The reagent is prepared by addition of freshly distilled *n*-propyl mercaptan to a suspension of lithium hydride in hexamethylphosphoric triamide (HMPT). The reagent is sensitive to oxygen and should be stored under argon at 0°.

Cleavage of methyl esters.[1] The reagent (in HMPT) is very effective for cleavage of methyl esters under mild conditions. Thus the very hindered methyl mesitoate is cleaved to the acid in quantitative yield at 25° (1.25 hours). Methyl O-methylpodocarpate (1) is readily cleaved to O-methylpodocarpic acid in quantitative yield at 25° (1.5 hours). This reaction had been effected previously with potassium *t*-butoxide in DMSO, but required 2 hours at 52°.[2] Bartlett and Johnson also examined the selective ester cleavage of methyl 3β-acetoxy-Δ⁵-etienate (2).

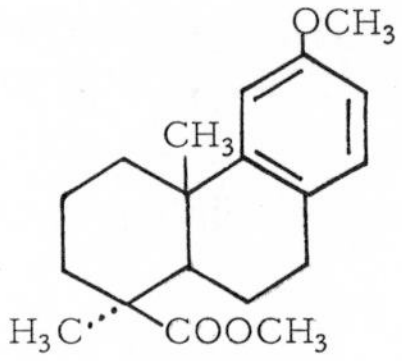

(1) 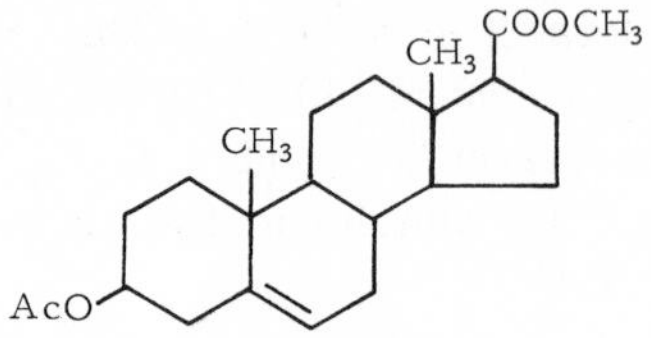   (2)

This reaction had been effected previously with lithium iodide in refluxing 2,6-lutidine (**1**, 615–616). This combination afforded 25–28% of starting material, 49–51% of the desired acetoxy acid, and 5–10% of the hydroxy acid. Use of the mercaptide reagent converted the acetoxy ester into 3β-acetoxy-Δ⁵-etianic acid in 92% yield (25°, 24 hours).

[1]P. A. Bartlett and W. S. Johnson, *Tetrahedron Letters*, 4459 (1970)
[2]F. C. Chang and N. F. Wood, *ibid.*, 2969 (1964)

Lithium tri-*t*-butoxyaluminum hydride, 1, 620–625; **2**, 251–252.

The reduction of pregnenolone acetate to 3β-acetoxy-20β-hydroxy-5-pregnene (**1**, 622, ref. 11) has been published.[1]

[1]K. Heusler, P. Wieland, and Ch. Meystre, *Org. Syn.*, **45**, 57 (1965)

M

Magnesium, 1, 627–629; **2**, 254.

Methylene magnesium halides. Methylene bromide and iodide react with magnesium turnings, or better with magnesium amalgam, at room temperature in ether–benzene to give a clear, stable solution of a methylene magnesium halide. Addition of a crystal of iodine facilitates initiation of the reaction. These new *gem*-dimetallic compounds methylenate carbonyl compounds in 45–80% yield. The reagents react with α,β-unsaturated carbonyl compounds by 1,2- or 1,4-addition.

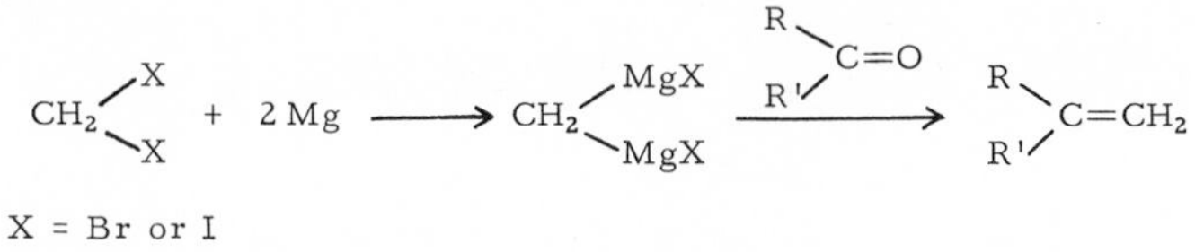

$$CH_2 \begin{smallmatrix} X \\ X \end{smallmatrix} + 2\,Mg \longrightarrow CH_2 \begin{smallmatrix} MgX \\ MgX \end{smallmatrix} \xrightarrow{\ \substack{R \\ R'}C=O\ } \begin{smallmatrix} R \\ R' \end{smallmatrix}C=CH_2$$

X = Br or I

Other alkylidene halides do not show any evidence of formation of a *gem*-dimetallic species, although a reaction does occur.[1]

[1]F. Bertini, P. Grasselli, G. Zubiani, and G. Cainelli, *Tetrahedron*, **26**, 1281 (1970)

Magnesium methoxide, 2, 255–256.

Synthesis of decalin-1,8-dione systems. Magnesium methoxide has been used as the base in a one-step synthesis of the above system from cyclohexene-3-one and esters of 2-carboxyglutaric acid. The reaction involves a combined Michael addition and Dieckmann condensation.[1] The diketone exists as the *cis*-enol

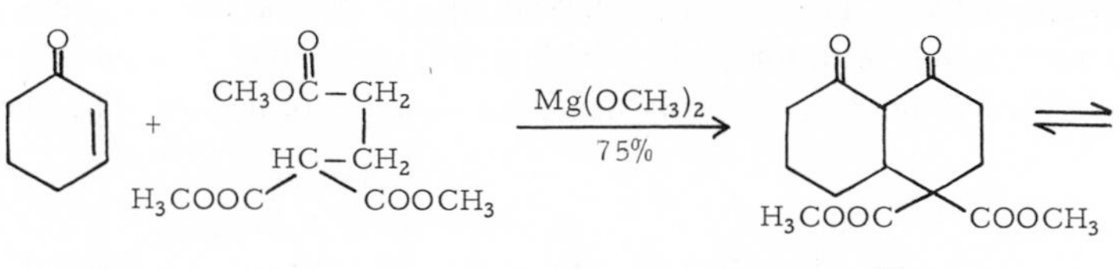

(2)

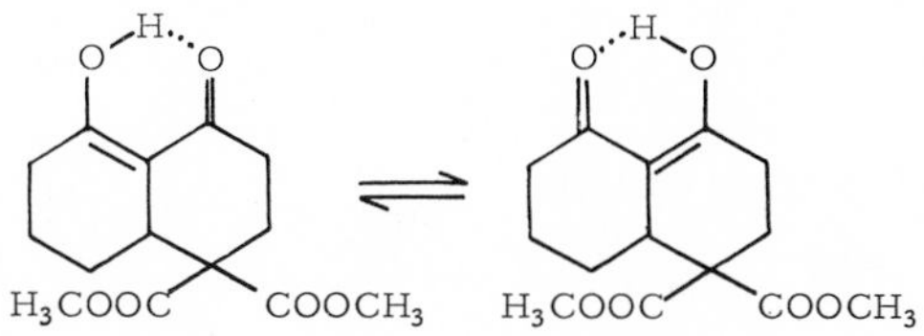

(3a)   (3b)

chelate. Sodium methoxide has also been used as base, but yields are lower than with magnesium methoxide.

[1] K. Schank and N. Moell, *Ber.*, **102**, 71 (1969)

Magnesium methyl carbonate, MMC, **1**, 631–633; **2**, 256.

α-Methylenebutyrolactones. This grouping characterizes a number of sesquiterpenes and other natural products, but has not been readily available by synthesis. In a new two-step synthesis[1] a butyrolactone (1) is carboxylated in high yield with magnesium methyl carbonate to give (2), and this is then transformed into the α-methylenebutyrolactone (4) by a variation of a method used by van Tamelen and Bach.[2] The acid is allowed to react with aqueous formaldehyde and

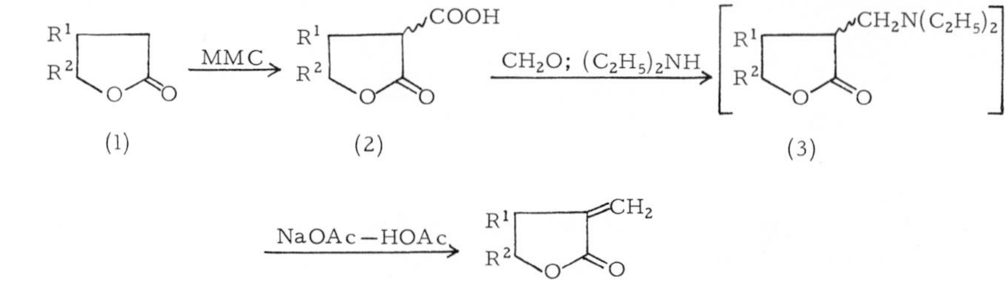

(1)   (2)   (3)

(4)

diethylamine to give the Mannich product (3), which is not isolated but treated with sodium acetate in acetic acid. The yields in this second sequence are about 50%.

The method is apparently not applicable to six-membered ring lactones; these do not give acidic products with MMC.

In an alternative synthesis,[3] a methoxycarbonyl lactone, for example (5), is treated successively with formaldehyde–dimethylamine hydrochloride (Mannich reaction), methyl iodide, and finally heated in dimethylformamide. The final step

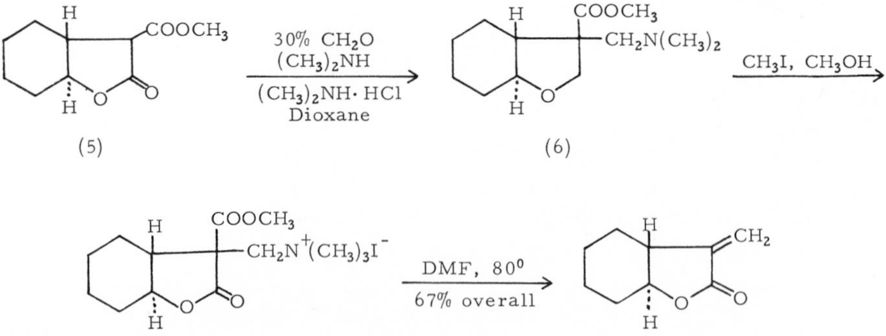

(5)   (6)

(7)   (8)

is considered to proceed by displacement of the carboxylate anion by iodide followed by decarboxylation and β-elimination of trimethylamine. The overall yield of (8) is 67%.

Carboxylation of resorcinols. The acidity of the aromatic hydrogens in resorcinols prompted Israeli chemists[4] to investigate the possibility of carboxylation with magnesium methyl carbonate. The reaction was successful. Thus resorcinol gave β-resorcylic acid (45%) and 2,4-dihydroxyisophthalic acid (15%), together with starting material (43%).

One interesting example of this procedure is the carboxylation of cannabidiol (1a), a component of hashish and marihuana, to give cannabidiolic acid (1b). Some starting material is not consumed but is readily recoverable from the product, a carboxylic acid.

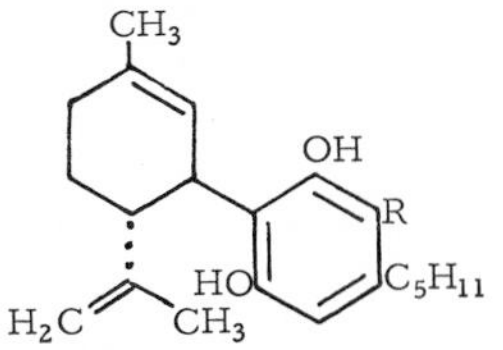

    1a   R = H
    1b   R = COOH

α-Carboxylation of ketones. The reagent has been widely used for α-carboxylation of cyclic ketones; in synthetic work in the pyrethrin series, Crombie *et al.*[5] found that MMC is extremely useful for carboxylation of a methyl ketone to give a β-keto acid; for example:

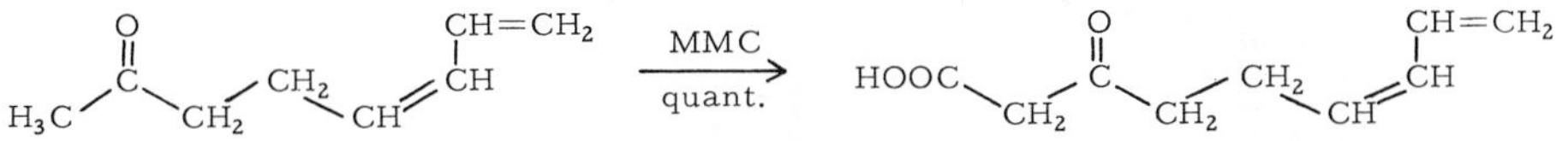

This procedure provides a considerable improvement over use of diethyl carbonate, generally used for this purpose.

[1]J. Martin, P. C. Watts, and F. Johnson, *Chem. Commun.*, 27 (1970)
[2]E. E. van Tamelen and S. R. Bach, *Am. Soc.*, **77**, 4683 (1955); *idem, ibid.*, **80**, 3079 (1958)
[3]E. S. Behare and R. B. Miller, *Chem. Commun.*, 402 (1970)
[4]R. Mechoulam and Z. Ben-Zvi, *ibid.*, 343 (1969)
[5]L. Crombie, P. Hemesley, and G. Pattenden, *Tetrahedron Letters*, 3021 (1968)

Manganese dioxide, 1, 637–643; **2,** 257–263.

New preparation. Carpino[1] reports that an active form of manganese dioxide can be prepared by treating aqueous permanganate solutions with decolorizing carbon.[2] A brown-black powder is precipitated on the excess carbon. If this powder is dried in an oven at 105–110° for 8–24 hours, it becomes comparable to Attenburrow material in activity.

Oxidative aromatization. Activated manganese dioxide (excess) oxidatively dehydrogenates certain cyclohexene aldehydes, ketones, and Schiff bases, but not esters, to the corresponding aromatic derivative. For example, 2-methyl-cyclohexene-3-carboxaldehyde (3 and 4), obtained by the Diels-Alder reaction of pentadiene-1,3 (1) with acrolein (2) is oxidized to *o*-tolualdehyde (5) in 69% yield. 4-Acetylcyclohexene is oxidized to acetophenone in 71% yield.[3]

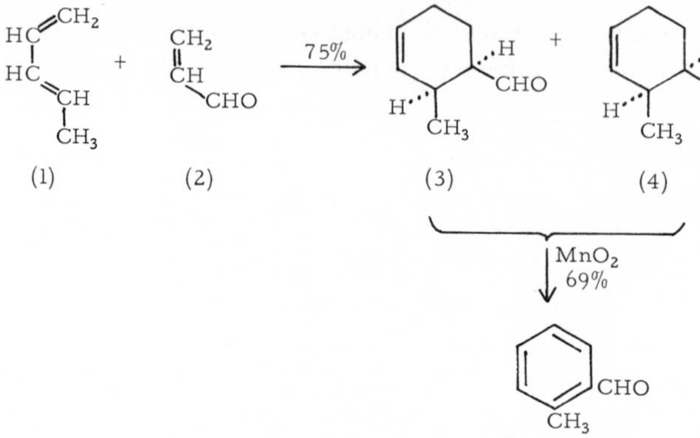

(1)   (2)   (3)   (4)

(5)

Oxidative condensation (**2**, 258). Manganese dioxide effects the oxidative condensation of the norbelladine derivative (1) to the dienone (2) in 10–13% yield.[4] Polymeric material is also formed, but (2) is essentially the only monomeric oxidation product. Use of potassium ferricyanide gives an array of products. The dienone (2) undergoes acid-catalyzed rearrangement to (3), which has the ring system of the amaryllis alkaloid nivalidine.

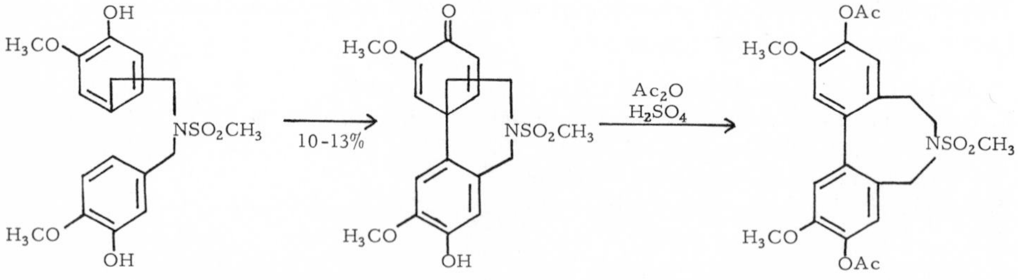

(1)   (2)   (3)

Oxidation of diosphenols. Oxidation of 15,15′-dehydro-β-carotene-3,4-dione (1) with manganese dioxide in acetone at 20° gives the purple cyclopentanedione (2) in about 30% yield.[5] The reaction presumably involves initial oxidation to the

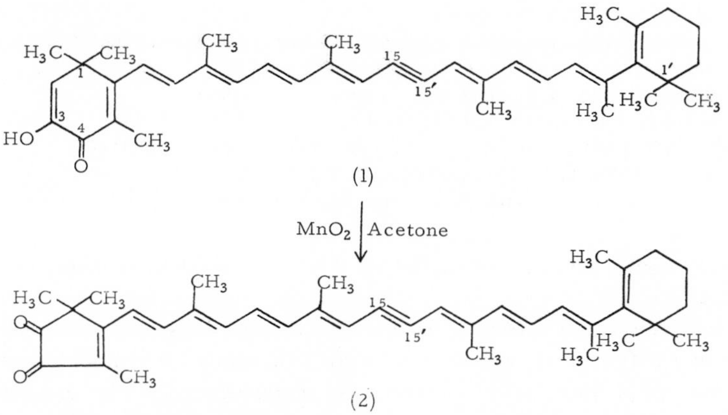

(1)

MnO$_2$ | Acetone

(2)

2,3,4-trione (a), benzylic acid rearrangement to a hydroxy acid (b), and further oxidation to the observed product (2). Carotenoids with end groups corresponding to those of (a) have been isolated.

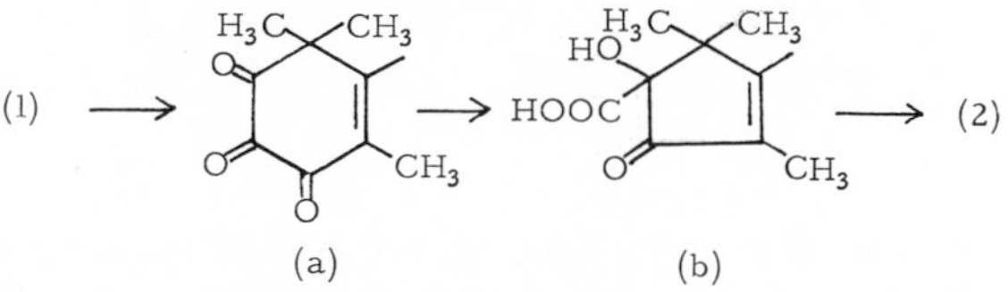

(1) $\longrightarrow$ (a) $\longrightarrow$ HOOC (b) $\longrightarrow$ (2)

(a)   (b)

Oxidation of α,β-unsaturated aldehydes (**2**, 261). Woo and Sondheimer[6] pre-pared methyl [18]annulenecarboxylate (2) by the method of Corey *et al.* (**2**, 263, ref. 42) by treating [18]annulenecarboxaldehyde (1) in THF and methanol with hydrogen cyanide and manganese dioxide (44% yield). In this case the classical method of oxidation by Jones reagent failed owing to complete destruction of the annulene system.

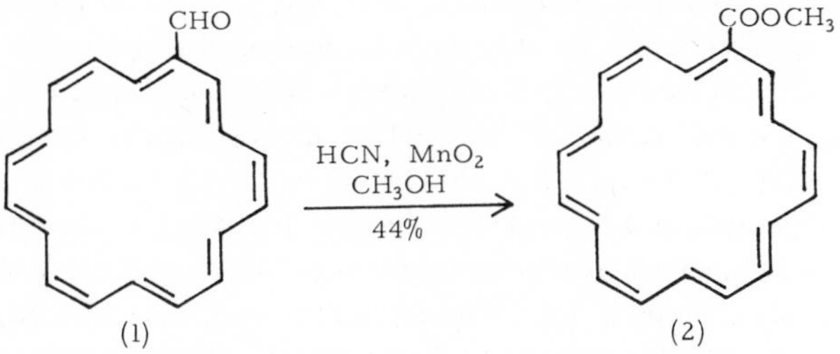

CHO   COOCH$_3$

HCN, MnO$_2$
CH$_3$OH
44%

(1)   (2)

Oxidation of nitrogen compounds. The manganese dioxide oxidation of organic nitrogen compounds has been reviewed in depth by Meth-Cohn and Suschitzky.[7] These authors state that samples prepared by various methods seem to have about the same activity.

[1]L. A. Carpino, *J. Org.*, **35**, 3971 (1970)
[2]Not all carbon samples are equally effective. The best results were obtained with Nuclear C-190N; less effective are Darco G-60 and Mallinckrodt USP.
[3]J. C. Leffingwell and H. J. Bluhm, *Chem. Commun.*, 1151 (1969)
[4]B. Frank and H. J. Lubs. *Ann.*, **720**, 131 (1968)
[5]R. Holzel, A. P. Leftwick, and B. C. L. Weedon, *Chem. Commun.*, 128 (1969)
[6]E. P. Woo and F. Sondheimer, *Tetrahedron*, **26**, 3933 (1970)
[7]O. Meth-Cohn and H. Suschitzky, *Chem. Ind.*, 443 (1969)

Manganic acetylacetonate, $Mn(C_5H_7O_2)_2$. Mol. wt. 243.06. Supplier: Alfa Inorganics.

Catalyzed peroxide oxidation.[1] Amides and lactams are selectively oxidized to imides by *t*-butyl hydroperoxide or peracetic acid catalyzed by trace amounts of transition metal ions, such as manganic acetylacetonate. For example, 2-piperidone was oxidized to glutarimide (2) in 72% yield by peracetic acid[2] catalyzed by the reagent.

(1)   (2)

The procedure can also be used to oxidize certain aromatic hydrocarbons (diphenylmethane → benzophenone, 48% yield), sulfides and sulfoxides to sulfones (dimethyl sulfide → dimethyl sulfone, 100% yield).

[1]A. R. Doumaux, Jr., J. E. McKeon, and D. J. Trecker, *Am. Soc.*, **91**, 3992 (1969); A. R. Doumaux, Jr., and D. J. Trecker, *J. Org.*, **35**, 2121 (1970)
[2]A 25% solution in ethyl acetate, available from Union Carbide Corp., was used.

Mercuric acetate, **1**, 644–652; **2**, 264–267.

Markownikoff hydration of olefins (**2**, 265–267). Brown and Geoghegan[1] have reported a detailed study of the oxymercuration–demercuration procedure. Yields of alcohols in the case of terminal olefins, $RCH=CH_2$ and $R_2C=CH_2$, and disubstituted internal olefins, $RCH=CHR'$, are practically quantitative. Trisubstituted olefins of the type $R_2C=CHR'$ show a wide variation in reactivity. 1-Phenylcyclopentene and 1-phenylcyclohexene, for example, are unreactive.

The procedure has been applied to unsaturated alcohols; if the hydroxyl group is capable of participation to form a tetrahydrofurane or tetrahydropyrane, these cyclic ethers are formed in good yield. Otherwise the expected diols are obtained. The procedure is also applicable to non-conjugated and conjugated dienes. In each case diols are obtained in good yield and the structures are consistent with two consecutive 1,2-Markownikoff additions.[2]

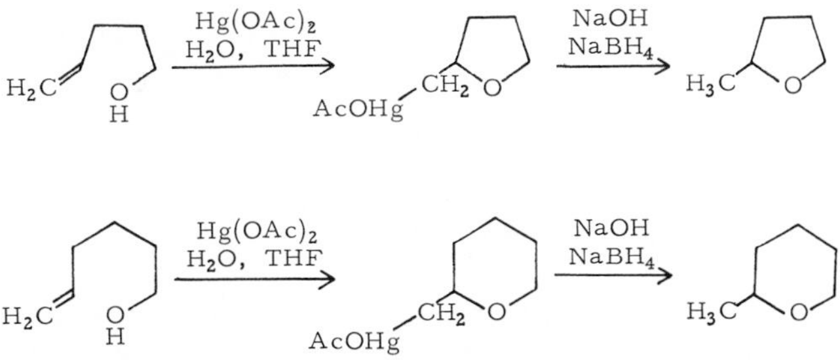

Solvomercuration. Brown and Geoghegan (**2**, 265–267) developed a convenient method for Markownikoff hydration of olefins involving oxymercuration followed by *in situ* reduction of the oxymercurial with sodium borohydride. Brown and Rei[3] report that, if the reaction is carried out in the presence of alcohols rather than water (solvomercuration), ethers are obtained in high yield except when *t*-butanol is utilized as the nucleophilic reagent. In this case, yields are improved by using mercuric trifluoroacetate rather than mercuric acetate. This salt (m.p. 167–169°) is obtained in high yield by the reaction of mercuric oxide with trifluoroacetic anhydride in trifluoroacetic acid. Actually mercuric trifluoroacetate can be used to advantage for preparation of all the ethers.

Replacement of alcohols by acetonitrile gives N-alkylacetamides, hydrolysis of which leads to the corresponding amines. In this case anhydrous mercuric nitrate, which contains a weakly nucleophilic anion, proved highly satisfactory.[4]

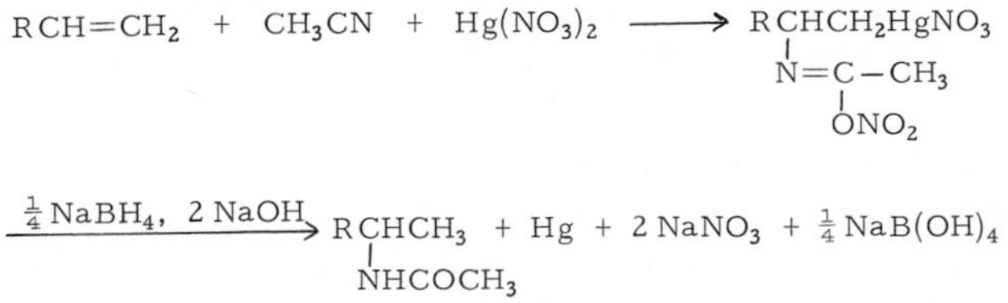

Transannular cyclization. In a recent total synthesis of indole and dihydroindole alkaloids,[5] a key step is a transannular cyclization reaction with mercuric acetate. For example, the reaction of 4β-dihydrocleavamine (1) with mercuric acetate yields the indolenine (2), which was isolated as the more stable reduction product (3).

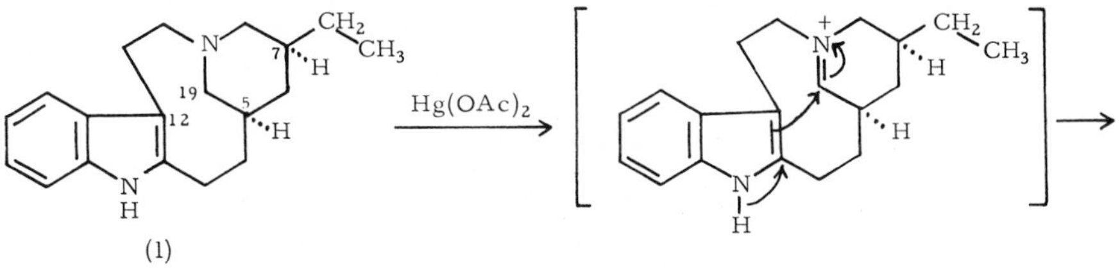

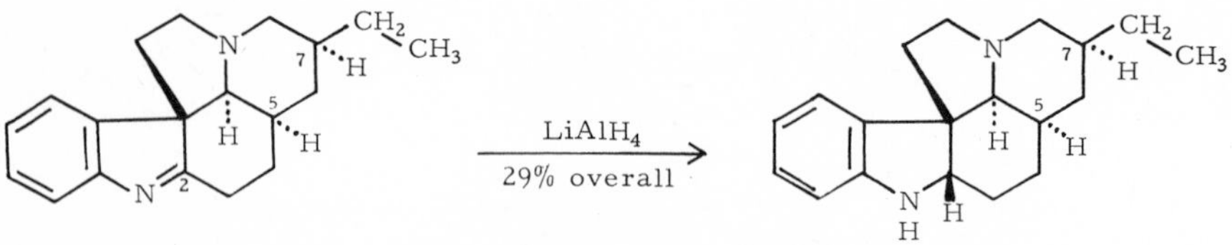

(2)   (3)

The reaction was also applied to the *Aspidosperma* alkaloids. Thus oxidation of (−)-quebrachamine (4) with mercuric acetate followed by reduction with lithium aluminum hydride gives (+)-aspidospermidine (5).

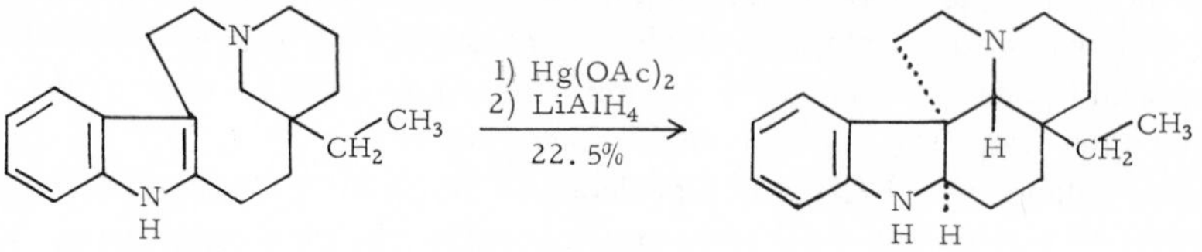

(4)   (5)

[1]H. C. Brown and P. J. Geoghegan, Jr., *J. Org.*, **35**, 1844 (1970)
[2]H. C. Brown, P. J. Geoghegan, Jr., J. T. Kurek, and G. J. Lynch, *Organometal. Chem. Syn.*, **1**, 7 (1970)
[3]H. C. Brown and M.-H. Rei, *Am. Soc.*, **91**, 5646 (1969)
[4]H. C. Brown and J. T. Kurek, *ibid.*, **91**, 5647 (1969)
[5]J. P. Kutney, E. Piers, and R. T. Brown, *ibid.*, **92**, 1700 (1970); J. P. Kutney, R. T. Brown, E. Piers, and J. R. Hadfield, *ibid.*, **92**, 1708 (1970)

Mercuric azide, $Hg(N_3)_2$. Mol. wt. 284.66.

Solid mercuric azide is sensitive to light and shock. The reagent can be generated safely *in situ* by the reaction of mercuric acetate and sodium azide in 50% aqueous THF.[1]

Synthesis of alkyl azides.[1] Terminal alkenes and strained cyclic alkenes react with the reagent to give a mercurial intermediate, which on reduction with sodium borohydride gives an azide. Internal olefins do not react. The method is an extension of the hydroxymercuration reaction of Brown (**2**, 265–267).

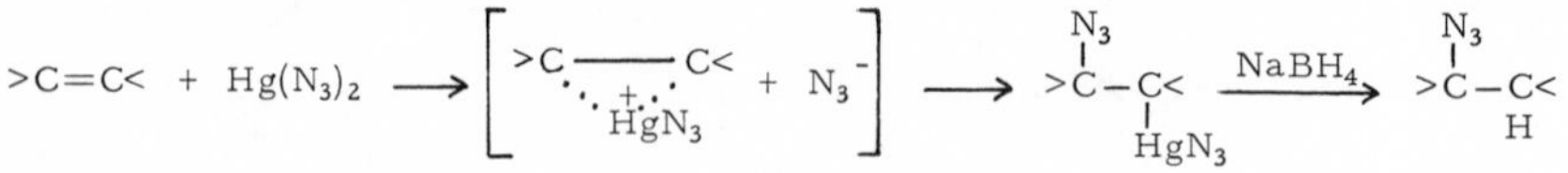

1-Heptene ⟶ 2-Heptyl azide (83% yield)

Methylenecyclohexane ⟶ 1-Methylcyclohexyl azide (60% yield)

[1]C. H. Heathcock, *Angew. Chem., internat. Ed.*, **8**, 134 (1969)

Mercuric cyanide, 1, 655.

Pyrimidine nucleosides. Nucleosides can be obtained directly by heating a suspension of the pyrimidine base with acetobromoglucose and mercuric cyanide in an inert solvent, toluene or acetonitrile.[1]

[1]G. T. Rogers and T. L. V. Ulbricht, *Chem. Commun.*, 508 (1969); *idem., J. Chem. Soc.,* (C), 2450 (1969)

Mercuric nitrate, $Hg(NO_3)_2$. Mol. wt. 324.62. Suppliers: Merck. Alfa Inorganics.

Amides.[1] The solvomercuration–demercuration of terminal olefins or cyclic olefins with acetonitrile and mercury(II) nitrate followed by reduction of the intermediate organomercury compound affords amides, hydrolyzable to amines. Neither mercuric acetate nor mercuric trifluoroacetate is satisfactory in this reaction. Preliminary attempts to use a tertiary olefin failed.

$$RCH{=}CH_2 \ + \ CH_3C{\equiv}N \ + \ Hg(NO_3)_2 \ \longrightarrow \ \underset{\underset{\underset{ONO_2}{|}}{\underset{N{=}CCH_3}{|}}}{RCHCH_2HgNO_2} \ \xrightarrow{\ 1/4\ NaBH_4,\ 2\ NaOH\ }$$

$$\underset{NHCOCH_3}{\underset{|}{RCHCH_3}} \ + \ Hg \ + \ 2\,NaNO_3 \ + \ 1/4\ NaB(OH)_4$$

[1]H. C. Brown and J. T. Kurek, *Am. Soc.*, **91**, 5647 (1969)

p-Methoxybenzyloxycarbonyl azide (*p*-Methoxybenzylazidoformate), 1, 668–669.

In a new simplified preparation,[1] the reagent is obtained directly and in high yield by the reaction of *p*-methoxybenzyl chloroformate with sodium azide in the presence of pyridine (compare preparation of *t*-butyl azidoformate, **2,** 48).

[1]H. Yajima and Y. Kiso, *Chem. Pharm. Bull. Japan,* **17,** 1962 (1969)

4-Methoxy-5,6-dihydro-2*H*-pyrane, 2, 271.

Preparation. Owen and Reese[1] have developed a convenient method suitable for the large-scale preparation of tetrahydro-4*H*-pyrane-4-one (3), the precursor of 4-methoxy-5,6-dihydro-2*H*-pyrane (**2,** 271). Ethylene is acylated with 3-chloro-propionyl chloride (1) in virtually quantitative yield to give 1,5-dichloropentane-3-one (2). This is then hydrolyzed with sodium hydrogen phosphate at 90°.

$$\text{(1)} \qquad\qquad\qquad\qquad \text{(2)} \qquad\qquad\qquad\qquad \text{(3)}$$

Protection of OH groups (2, 271). Preparation of 2'-O-methoxytetrahydro-pyranyl-uridine and -adenosine and of 5'-O-methoxytetrahydropyranyl-thymidine has been described. Yields are 75–85%.[2] The protective group has also been used in the preparation of ribonucleotide 2',5'-bisketals.

[1]G. R. Owen and C. B. Reese, *J. Chem. Soc.*, (C), 2401 (1970)
[2]C. B. Reese, R. Saffhill, and J. E. Sulston, *Tetrahedron*, **26**, 1023 (1970); D. P. L. Green, T. Ravindranathan, C. B. Reese, and R. Saffhill, *ibid.*, **26**, 1031 (1970)

α-Methoxymethylenetriphenylphosphorane, **1**, 669–670; **2**, 271. Replace ref. 2 (**2**, 271) by G. R. Pettit, B. Green, G. L. Dunn, and P. Sunder-Plassmann, *J. Org.*, **35**, 1385 (1970)

Methoxymethylsulfonate, $CH_3OCH_2OSO_2CH_3$. Mol. wt. 140.16; b.p. 73–75°/10^{-2} mm.

Preparation.[1] The reagent is prepared in essentially quantitative yield from acetyl methanesulfonate and dimethoxymethane:

$$CH_3COOSO_2CH_3 + CH_3OCH_2OCH_3 \longrightarrow CH_3OCH_2OSO_2CH_3 + CH_3COOCH_3$$

Oxyalkylation.[1] The reagent reacts with alcohols (primary and secondary) and with tertiary amines as shown:

$$ROH + CH_3OCH_2OSO_2CH_3 \longrightarrow ROCH_2OCH_3 \text{ (ca. 85% yield)}$$

$$C_6H_5CH_2N(CH_3)_2 \xrightarrow[96\%]{} C_6H_5CH_2\overset{+}{N}(CH_3)_2(CH_2OCH_3)CH_3SO_3^-$$

With primary and secondary amines the reaction is more complex, leading eventually to the respective aminal and the amine sulfonate salt. Benzene reacts to give diphenylmethane in 94% yield. Toluene gives an isomeric mixture of ditolyl-methanes. The sulfonate cleaves a number of ethers.

[1]M. H. Karger and Y. Mazur, *Am. Soc.*, **91**, 5663 (1969)

Methoxymethyl vinyl ketone, $CH_2{=}CHCOCH_2OCH_3$. Mol. wt. 100.11.

Preparation. Wenkert and Berges[1] generated the reagent *in situ* from 1,4-dimethoxy-2-butanone[2] by reaction with ethanolic potassium hydroxide. Ireland *et al.*[3] prepared the reagent in crude form by the pyrolysis of 1,4-dimethoxy-2-butanone with sodium benzoate.[4] In a typical reaction 10.06 g. of the precursor on pyrolysis with 12.5 g. of sodium benzoate afforded 4.38 g. of distillate that contained 2.7–2.8 g. of methoxymethyl vinyl ketone (assayed by vpc and NMR). This crude distillate can be used directly for annelation.[5]

Annelation. Both Wenkert[1] and Ireland[3] have used this vinyl ketone in Robinson annelation reactions, for example:

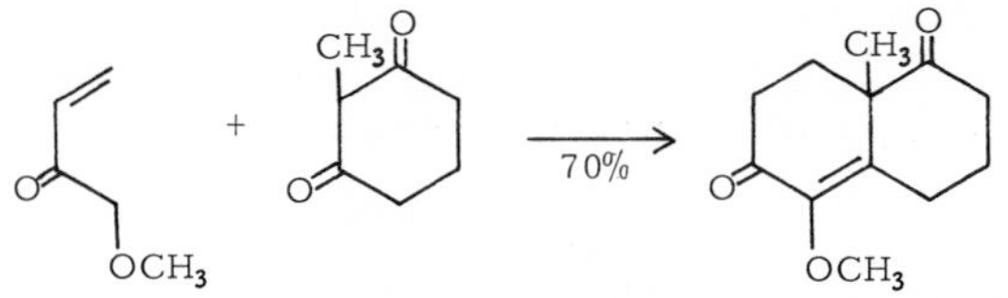

The enol ethers which are obtained in this way have been converted into methoxycyclopropanes and then into angular methylated decalones (*see* Simmons-Smith reagent, this volume).

[1]E. Wenkert and D. A. Berges, *Am. Soc.*, **89**, 2507 (1967)
[2]G. F. Hennion and F. P. Kupiecki, *J. Org.*, **18**, 1601 (1953)
[3]R. E. Ireland, D. R. Marshall, and J. W. Tilley, *Am. Soc.*, **92**, 4754 (1970)
[4]Method of S. Archer, W. B. Dickinson, and M. J. Unser, *J. Org.*, **22**, 92 (1957)
[5]Personal communication from Dr. Ireland.

d- and *l-α*-Methylbenzylamine, 2, 271–273.

Asymmetric synthesis. Following the procedure of Cope *et al.* (**2**, 272), Chamberlain and McKervey[1] prepared the optically active complex (3) of humulene (1) by treatment with (−)-*trans*-dichloro(ethylene)-α-methylbenzylamine platinum-(II), (2). Epoxidation of (3) with perlauric acid gives optically inactive humulene (1) and humulene 1,2-epoxide with $\alpha_D = -3.2°$. Since the natural epoxide has an

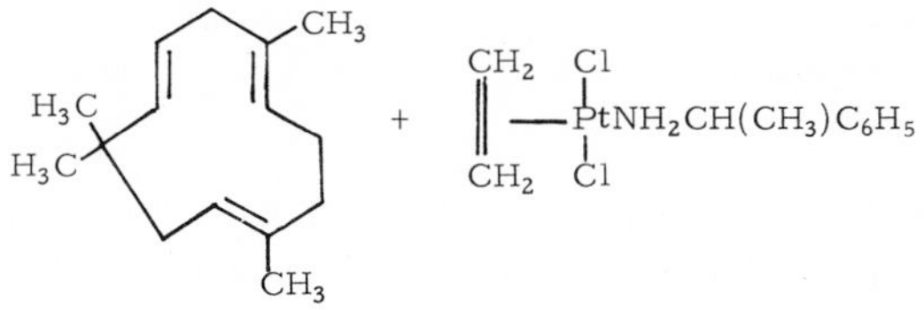

(1)   (2)

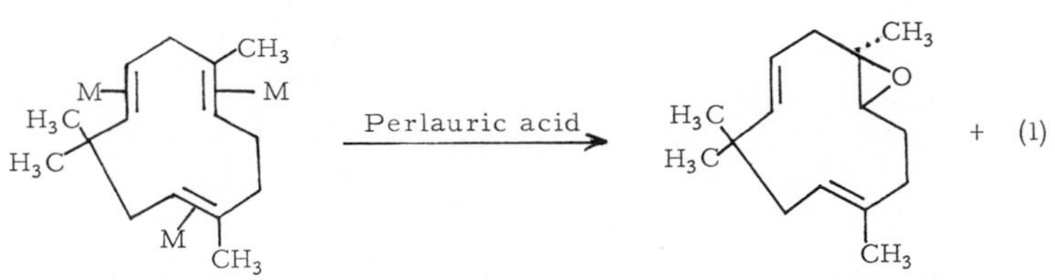

(3) M = Pt(Cl$_2$)NH$_2 \cdot$ CH(CH$_3$)C$_6$H$_5$
α_D -15.4^0

(4) α_D -3.2^0

α_D value of −31.2°, the optical yield is 10.3%. It was not possible to obtain optically active humulene from the platinum complex, presumably because of conformational mobility of the ring system.

Resolution of 1,2-cyclononadiene. Cope *et al.*[2] examined the resolution of the cyclic allene 1,2-cyclononadiene (1) by way of the diastereomeric platinum complex containing this amine (compare resolution of cycloalkenes (2, 272–273)). A

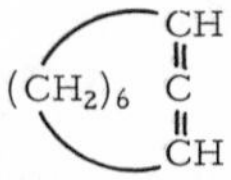

$$(CH_2)_6 \quad \begin{array}{c} CH \\ \| \\ C \\ \| \\ CH \end{array}$$

(1)

yellow, crystalline complex was obtained, but fractional crystallization resulted in only partial resolution (*ca.* 44%) owing to mutarotation. Actually (+)- and (−)- (1) of high optical purity are prepared more readily by synthesis from (−)- and (+)-*trans*-cyclooctene, respectively, by the method shown:

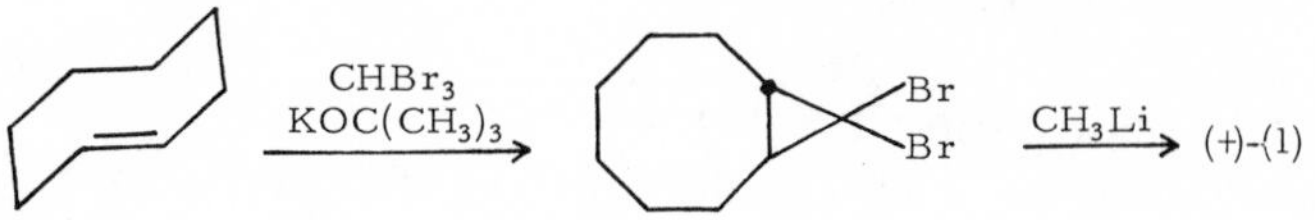

(−) $\qquad\qquad$ $\alpha_D + 44.2^0$ $\qquad\qquad$ $\alpha_D + 138^0$

[1]T. R. Chamberlain and M. A. McKervey, *Chem. Commun.*, 366 (1969)
[2]A. C. Cope, W. R. Moore, R. D. Bach, and H. J. S. Winkler, *Am. Soc.*, **92**, 1243 (1970)

Methylcyclopentadiene, 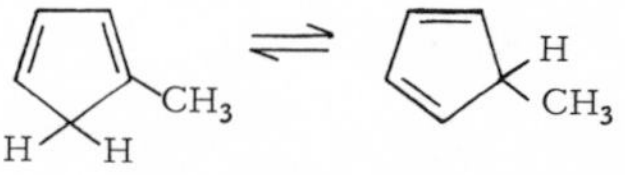 Mol. wt. 78.11, b.p. 72°.

Preparation.[1] The dimer (available from K and K Laboratories, Inc.) is cracked by dropwise addition to a distilling flask at 180° in the presence of a trace of iron.

Synthesis of 1,1'-dimethylferrocene.[1] In a 5-l. three-necked flask fitted with

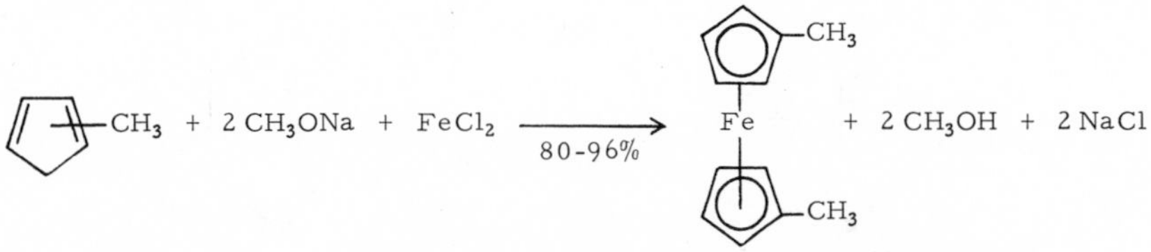

a mechanical stirrer, a thermometer, a drying tube, and an inlet tube to maintain a nitrogen atmosphere throughout the reaction, 2.5 l. of absolute methanol is introduced and to it is added with stirring 305 g. (13.3 atoms) of freshly cut sodium at such a rate that a moderate reflux is maintained. The flask is heated by means of a heating mantle until all the sodium has reacted. The temperature of the sodium methoxide slurry in methanol is adjusted to 40–50°, and 395 g. (3.1 moles) of anhydrous ferrous chloride is added portionwise. This is followed by addition of 1 g. of electrolytic iron powder (Fisher) to retard dimerization of methylcyclopentadiene. The mixture warms up slightly while it is being stirred for about 30 minutes.

Freshly distilled methylcyclopentadiene is withdrawn in 50-ml. portions from its cold storage and added dropwise through an addition funnel to the brown reaction mixture at about 60°. After completion of the addition of 468 g. (5.85 moles) of methylcyclopentadiene, stirring of the mixture is continued for 5 hours at about 55° and finally for 10 hours at room temperature.

The reaction mixture is poured into a suitable number of separatory funnels containing a total of 6800 ml. of 12% sulfuric acid and 3000 ml. of petroleum ether. After stirring, the layers are separated. The blue aqueous layer is treated with aqueous 5% sodium thiosulfate until colorless and extracted with petroleum ether. The combined petroleum ether extracts are washed repeatedly with water until neutral and dried over Drierite. After filtering, the petroleum ether is evaporated on a rotary evaporator and the mixture is dried at 80° (2 mm.). The dried residue is 600 g. or 2.8 moles (96%) of brown-orange 1,1'-dimethylferrocene, m.p. 38°. The product may be identified or analyzed for purity by gas chromatography, mass spectrometry, and infrared spectroscopy.

[1]I. J. Spilners and R. J. Hartle, procedure submitted to *Org. Syn.* 1970

Methylene bromide–Lithium.

Epoxides. The reaction of carbonyl compounds with methylene bromide (iodide) and magnesium amalgam gives methylenic olefins (**2**, 274):

$$CH_2(MgX)_2 \; + \; R_2C{=}O \; \longrightarrow \; R_2C{=}CH_2$$

Cainelli[1] reports that, if magnesium amalgam is replaced by lithium or lithium amalgam, oxiranes are formed in relatively good yields:

$$R_2C{=}O \; + \; CH_2Br_2 \; \xrightarrow{\;2\,Li\;} \; R_2C\overset{\displaystyle\diagdown}{\underset{O}{\diagup}}CH_2 \; + \; 2\,LiBr$$

α,β-Unsaturated ketones are unreactive, at least under conditions where saturated ketones react. Probably the carbenoid species α-bromomethyllithium is involved.

The lithium can be replaced by *n*-butyllithium, but yields are considerably lowered.

[1]F. Bertini, P. Grasselli, G. Zubiani, and G. Cainelli, *Chem. Commun.*, 1047 (1969)

Methyl fluoride–Antimony pentafluoride.

When methyl fluoride is dissolved in SbF_5–SO_2 solution at $-78°$ a clear, colorless solution is obtained. NMR, PMR and Raman spectra indicate that a 1:1 donor–acceptor complex (I) is formed.[1] This complex is a powerful methylating

$$\delta^+ \; CH_3 \cdots \underset{F}{\overset{F}{\diamond}} \cdots \underset{F}{\overset{F}{\underset{F}{\overset{|}{Sb}}}} \overset{F}{\underset{F}{\diamond}}$$

(I)

agent. Thus dimethyl ether is converted into trimethyloxonium ion (1) and dimethyl sulfide into trimethylsulfonium ion (2). Aromatic hydrocarbons (3) and even alkanes (4) are methylated. In neat SbF_5, methyl fluoride undergoes a self-condensation reaction.

(1) $CH_3OCH_3 \; + \; CH_3F - SbF_5 \; \longrightarrow \; (CH_3)_3O^+ SbF_6^-$

(2) $CH_3SCH_3 \; + \; CH_3F - SbF_5 \; \longrightarrow \; (CH_3)_3S^+ SbF_6^-$

(3) $ArH \; + \; CH_3F - SbF_5 \; \longrightarrow \; ArCH_3 \; + \; HF \; + \; SbF_5$

(4) $RH \; + \; CH_3F - SbF_5 \; \longrightarrow \; RCH_3 \; + \; HF \; + \; SbF_5$

Alkyl halides (other than fluorides) when dissolved in SbF_5–SO_2 form dialkylhalonium ions:[2]

$$2\,RX \; + \; SbF_5 - SO_2 \; \longrightarrow \; RXR^+ SbF_5X^-$$

[1]G. A. Olah, J. R. DeMember, and R. H. Schlosberg, *Am. Soc.*, **91**, 2112 (1969)
[2]G. A. Olah, and J. R. DeMember, *ibid.*, **91**, 2113 (1969)

Methyl fluorosulfonate, CH_3OSO_2F. Mol. wt. 114.10, b.p. 92–94°, m.p. ~ −95°.[1]
The reagent is prepared by mixing dimethyl sulfate and fluorosulfonic acid in a 1.2 : 1 ratio and distilling fairly slowly.

Alkylation. Methyl fluorosulfonate is a convenient and powerful alkylating agent. It alkylates amines, amides, nitriles, and ethers.

[1]M. G. Ahmed, R. W. Adler, G. H. James, M. L. Sinnott, and M. C. Whiting, *Chem. Commun.*, 1533 (1968)

Methyl iodide, 1, 682–685; **2**, 274.
N-Methylamino acid derivatives.[1] CbO- and Boc-amino acids are converted into the corresponding N-methylamino acid derivatives in excellent yield by methyl iodide and silver oxide in DMF. An unprotected carboxyl group is also converted into the methyl ester by this procedure.

[1]R. K. Olsen, *J. Org.*, **35**, 1912 (1970)

Methyllithium, 1, 686–689; **2**, 274–278.
Allene synthesis. The preparation of 1,2-cyclononadiene from *cis*-cyclooctene (**1**, 689, ref. 5) has now been published.[1]
Spiropentanes. In studying the formation of allenes from *gem*-dibromocyclo-

propanes and alkyllithium reagents (**1**, 686–688; **2**, 274–276), Moore and Ward[2] carried out the reaction in the presence of cyclohexene as a trapping agent for intermediate carbenes. Only allenes and no trapped products were formed in the case of *gem*-dibromopropanes derived from acylic olefins. However, the reaction of 7,7-dibromonorcarane (1) with methyllithium in ether–cyclohexene at −80° gave a spiropentane (2) in 10% yield together with the bicyclopropylidine derivative (3, 30% yield). Moore and Ward suggested that the intermediate involved in the formation of (2) and (3) might be the carbene (4), the highly strained allene

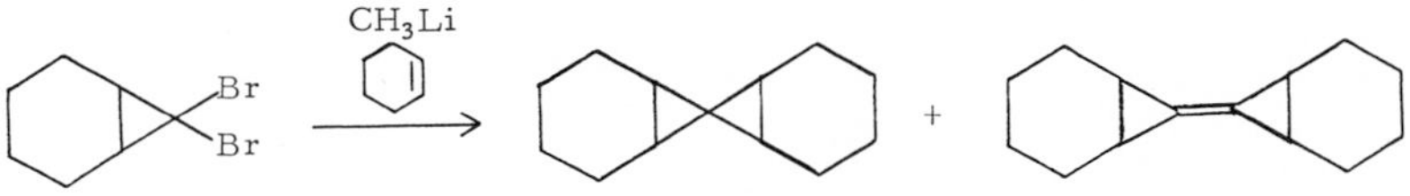

(1)   (2)   (3)

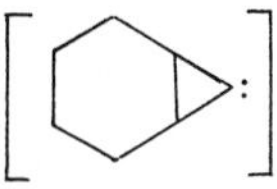

(4) 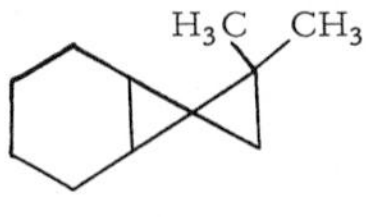   (5)

1,2-cycloheptadiene, or a species intermediate between the carbene and allene. The formation of spiropentanes may be general since the reaction of (1) with isobutylene (methyllithium) gives (5) in unspecified yield.

The synthesis has since been extended to Δ³-7,7-dibromonorcarene (6) to give spiropentanes in about 30% yield.[3] Moreover, the reaction is stereospecific, (7) and (8) being formed exclusively from *cis*- and *trans*-2-butene, respectively. The

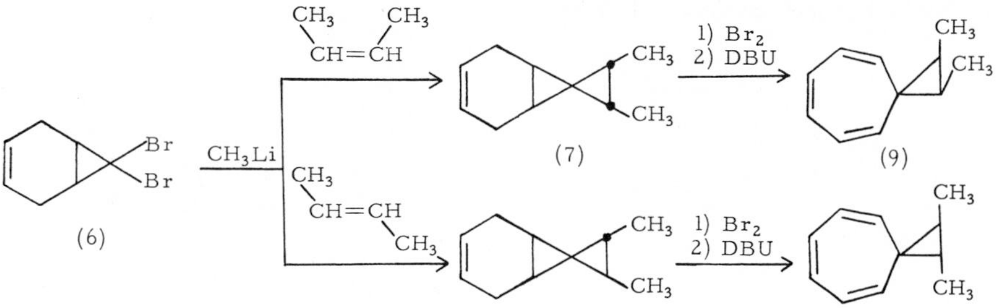

(6)   (7)   (9)   (8)   (10)

two compounds were prepared as part of a study of the norcaradiene–cyclo-heptatriene valence tautomerism.[4] Bromination of (7) and (8) gave uncharacterized dibromides, which were dehydrobrominated with 1,5-diazabicyclo[5.4.0]undec-5-ene (DBU, **2**, 101) to give (9) and (10) in 15% yield. NMR spectroscopy indicated that not much, if any, of the norcaradiene form is present in (9) and (10).

[1]L. Skattebøl and S. Solomon, *Org. Syn.*, **49**, 35 (1969)
[2]W. R. Moore and H. R. Ward, *J. Org.*, **25**, 2073 (1960)
[3]M. Jones, Jr., and E. W. Petrillo, Jr., *Tetrahedron Letters*, 3953 (1969)
[4]Review: G. Maier, *Angew. Chem., internat. Ed.*, **6**, 402 (1967)

Methylmagnesium iodide, **1**, 689–690; **2**, 278.

Demethylation (**1**, 689–690). The procedure of Wilds and McCormack was found to be particularly useful in a recent total synthesis of the steroid hormone equilin which requires as one step demethylation of (1).[1] Equilin and derivatives

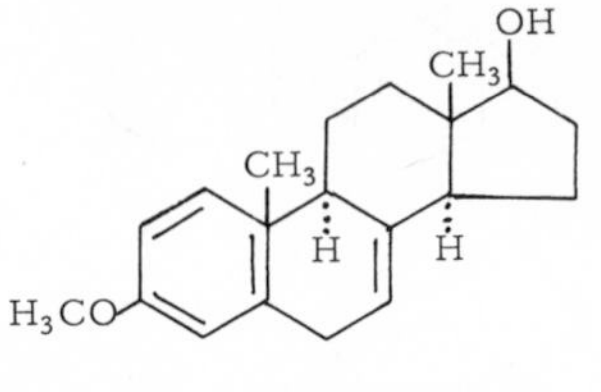

(1)

are markedly prone to acid-catalyzed isomerization, and hence the usual acidic reagents could not be used. However, (1) can be efficiently demethylated by fusion with methylmagnesium iodide at 165–170° (83% yield).

[1]R. P. Stein, G. C. Buzby, Jr., and H. Smith, *Tetrahedron*, **26**, 1917 (1970)

N-Methylmorpholine oxide–Hydrogen peroxide, **1**, 690.

The Schneider-Hanze reagent (**1**, 690, ref. 1) was used to transform the tetraene (1) into the ketol (2) in 46% yield[1]. A diol (3) was also formed in 11% yield, and a minor by-product was a 17-ketone (4) resulting from oxidative cleavage of the C_{17} side chain. Somewhat lower yields were obtained in analogous oxidations.

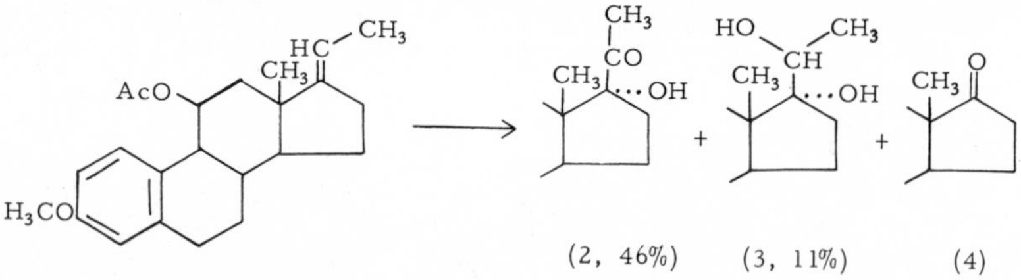

(1)   (2, 46%)   (3, 11%)   (4)

[1]B. Gadsby, M. R. G. Leeming, G. Greenspan, and H. Smith, *J. Chem. Soc.*, (C), 2647 (1968)

Methylphenyl-N-*p*-toluenesulfonylsulfoximine (2), $C_6H_5\overset{O}{\underset{NSO_2C_6H_4CH_3\text{-}\underline{p}}{\overset{\|}{\underset{\|}{S}}}}-CH_3$ Mol. wt. 277.33, m.p. 107–108°.

Preparation.[1] This N-*p*-toluenesulfonylsulfoximine (2) is prepared in 90% yield by the reaction of the sulfoxide (1) with tosyl azide catalyzed by copper. It is converted into its anion salt (3) by reaction with sodium hydride in DMSO. The anion salt is stable at room temperature.

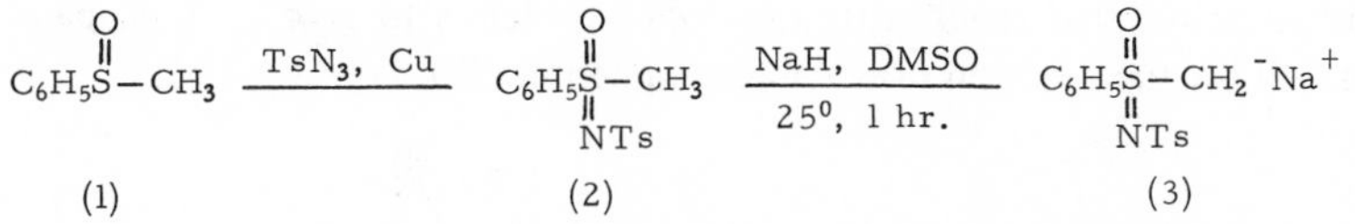

$$C_6H_5\overset{O}{\underset{}{\overset{\|}{S}}}-CH_3 \xrightarrow{\text{TsN}_3,\text{ Cu}} C_6H_5\overset{O}{\underset{NTs}{\overset{\|}{\underset{\|}{S}}}}-CH_3 \xrightarrow[25°,\ 1\ \text{hr.}]{\text{NaH, DMSO}} C_6H_5\overset{O}{\underset{NTs}{\overset{\|}{\underset{\|}{S}}}}-CH_2^-\ Na^+$$

$$\qquad\quad (1)\qquad\qquad\qquad\qquad (2)\qquad\qquad\qquad\qquad\qquad (3)$$

Oxiranes.[1] The anion salt (3) reacts with aldehydes or ketones to give oxiranes in good yield. The reaction with acetophenone is formulated. The product (4) is isolated by quenching with water and extraction with an appropriate solvent. The

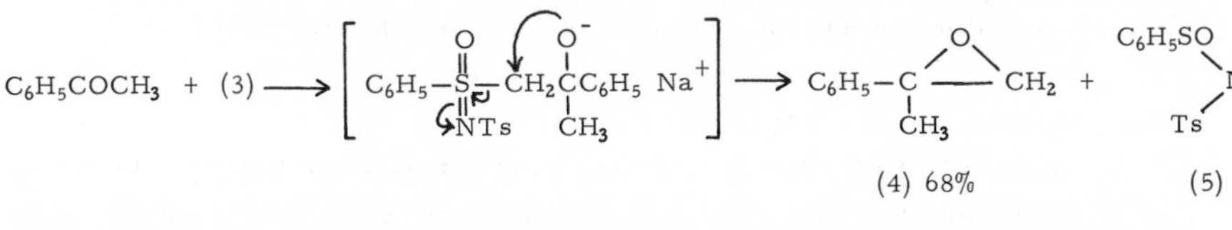

$$C_6H_5COCH_3 + (3) \longrightarrow \left[C_6H_5-\overset{O}{\underset{\underset{NTs}{\|}}{\overset{\|}{S}}}-CH_2\overset{O^-}{\underset{CH_3}{\overset{|}{C}}}C_6H_5\ Na^+ \right] \longrightarrow C_6H_5-\overset{O}{\underset{CH_3}{\overset{}{C}}}-CH_2 + \overset{C_6H_5SO}{\underset{Ts}{\overset{}{N^-}}} Na^+$$

$$\qquad\qquad\qquad\qquad\qquad\qquad\qquad\qquad\qquad\qquad (4)\ 68\%\qquad\qquad (5)$$

sodium salt of N-benzenesulfinyl-*p*-toluenesulfonamide (5) remains in the aqueous layer. Stereoselective synthesis of oxiranes can be achieved with this reagent; for example:

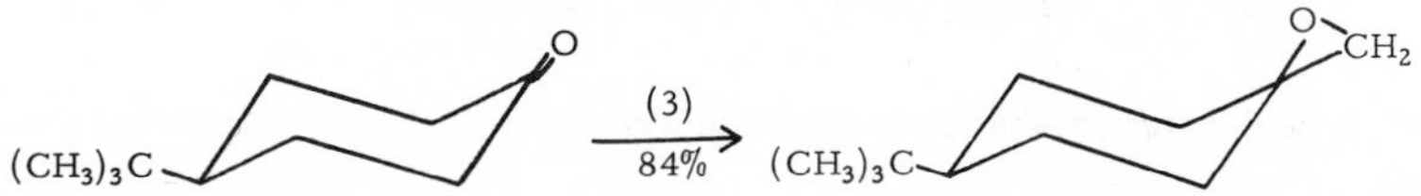

The anion (6) derived from dimethyl N-*p*-toluenesulfonylsulfoximine[2] can be used in the same way:

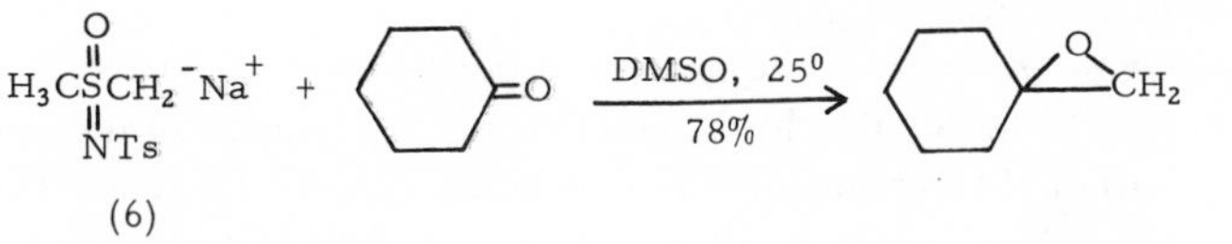

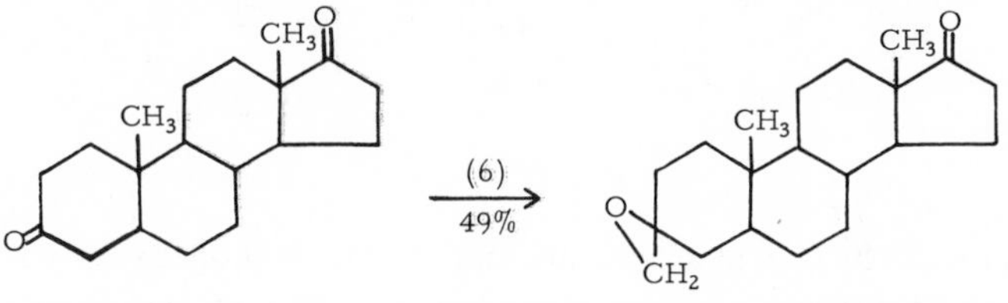

The nucleophilic methylene transfer reaction can be realized with substances other than aldehydes and ketones. Thus the reaction of benzalaniline (7) with either (3) or (6) gives the aziridine (8) in 86% yield.

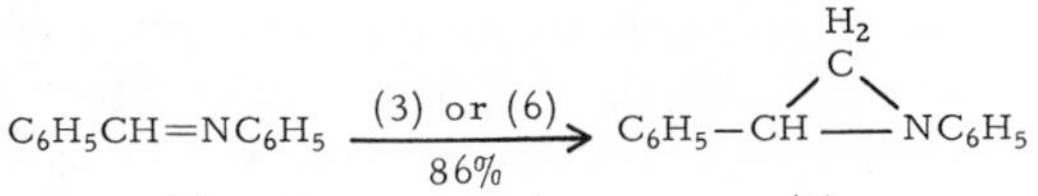

$$C_6H_5CH{=}NC_6H_5 \xrightarrow[86\%]{(3)\ or\ (6)} C_6H_5-\overset{\displaystyle H_2}{\overset{\displaystyle C}{\underset{}{CH}}}{-\!\!-\!\!-}NC_6H_5$$

(7)   (8)

The reagents also react with electrophilic olefins to give cyclopropanes, as illustrated for the reaction of benzalacetophenone (9):

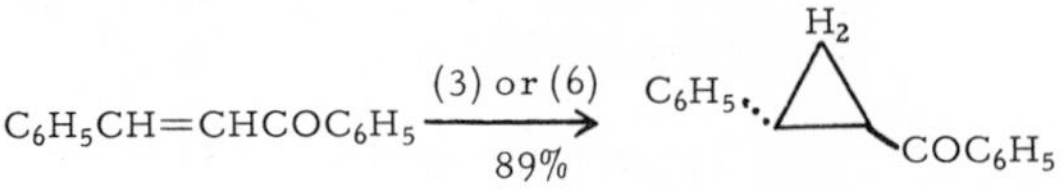

$$C_6H_5CH{=}CHCOC_6H_5 \xrightarrow[89\%]{(3)\ or\ (6)}$$

(9)   (10)

[1] C. R. Johnson and G. F. Katekar, *Am. Soc.*, **92**, 5753 (1970)
[2] Available from Columbia Organic Chemical Co., Inc., Columbia, S.C. 29205

Molecular sieves, 1, 703–705; **2,** 286–287.

Ketimines. The formation of ketimines from aldehydes or ketones and amines is reversible and, in general, it is necessary to remove the water formed. Azeotropic distillation has been usually used. Kyba[1] recommends use of molecular

$$>\!\!C{=}O\ +\ H_2NR \rightleftarrows \left[>\!\!\overset{\displaystyle OH}{\underset{}{C}}\!-\!NHR \right] \rightleftarrows\ >\!\!C{=}NR\ +\ H_2O$$

sieves, type 4A, for this purpose, particularly when the ketone or amine is readily volatile.

[1] E. P. Kyba, *Org. Prep. Proc.*, **2,** 149 (1970)

Molybdenum hexacarbonyl, 2, 287.

Oxidation of enol ethers. The procedure of Sheng and Zajacek (**2,** 287) when applied to 5,6,7,8-tetrahydrochromane (1) leads to cleavage of the double bond to give a dicarbonyl compound, 6-ketononanolide (2), in 50% yield.[1] The reaction

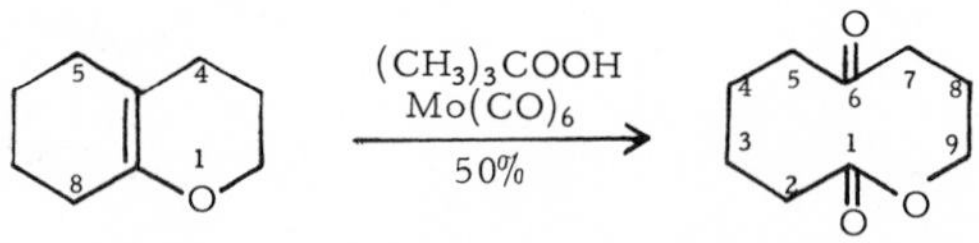

$$\xrightarrow[50\%]{\underset{Mo(CO)_6}{(CH_3)_3COOH}}$$

(1)   (2)

may well proceed by way of an epoxide. Both *t*-butyl hydroperoxide and cumene hydroperoxide have been used, but the isolation of the product is simpler with the former reagent. Oxidation of dihydropyrane (3) gives 4-formyloxybutanal (4) in 25% yield together with another unidentified product.

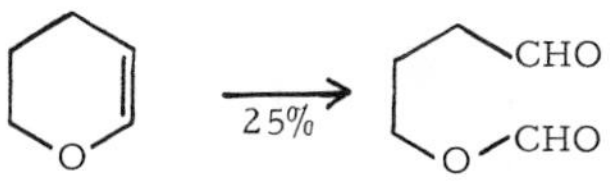

(3)   (4)

Disulfides. The reaction of sulfonyl chlorides (alkyl or aryl) with molybdenum hexacarbonyl in anhydrous tetramethylurea at 70° (N_2) provides a convenient synthesis of disulfides. The metal carbonyl does not function as a catalyst;

$$2\,RSO_2Cl \xrightarrow[55-75\%]{Mo(CO)_6} RSSR$$

RSO_2Cl and $Mo(CO)_6$ are used in a molar ratio of 1 : 1.1 to 1.3.[2]

Nitrosamines → *amines.* Aromatic nitrosamines are converted into secondary amines when refluxed in 1,2-dimethoxyethane for 15–18 hours with a slight molar

$$\underset{R}{ArN}-N{=}O \xrightarrow[60-70\%]{Mo(CO)_6} \underset{R}{ArNH}$$

excess of molybdenum hexacarbonyl. This metal carbonyl gives higher yields than either chromium or tungsten carbonyl but is somewhat less effective than iron pentacarbonyl. Treatment of non-aromatic nitrosamines with these metal carbonyls gives amine metal carbonyl complexes.[3]

[1]R. D. Rapp and I. J. Borowitz, *Chem. Commun.*, 1202 (1969)
[2]H. Alper, *Angew. Chem., internat. Ed.*, **8**, 677 (1969)
[3]*Idem, Organometal. Chem. Syn.*, **1**, 69 (1970)

N

Naphthalene–Lithium, 2, 288–289.

Acetylenic diols. Ref. 4, **2**, 289: Definitive paper, S. Watanabe, K. Suga, and T. Suzuki, *Canad. J. Chem.*, **47**, 2343 (1969). Naphthalene–lithium was found to be superior to other metalating agents (naphthalene–sodium, *n*-butylmagnesium bromide) in this synthesis.

Synthesis of β-hydroxy acids. β-Hydroxy acids are usually prepared by the Reformatsky reaction.[1] Two laboratories[2,3] have recently reported a general one-step synthesis by the reaction of a carbonyl compound and an acid in the presence of 2 equivalents of an ion-radical of the type $ArH^{\dot{-}}M^+$. The best results were obtained with naphthalene–lithium and THF as solvent:

$$CH_3COOH \xrightarrow{2[C_{10}H_8]^{\dot{-}}Li^+} \bar{C}H_2{-}\overset{O}{\overset{\|}{C}}{-}O^- + {>}C{=}O \longrightarrow {>}\overset{OH}{\overset{|}{C}}CH_2COOH$$

Thus acetic acid condenses with cyclohexanone to give (1′-hydroxycyclohexane-1′-yl)acetic acid in 65% yield.

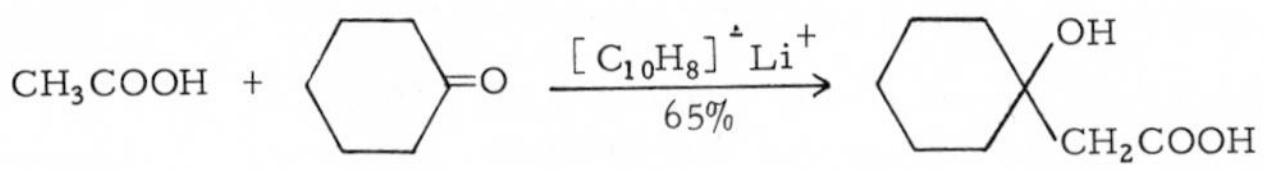

Further examples:

$$CH_3COOH + C_6H_5COC_6H_5 \xrightarrow[58\%]{} (C_6H_5)_2\overset{OH}{\overset{|}{C}}{-}CH_2COOH$$

$$CH_3COOH + CH_3(CH_2)_6CHO \xrightarrow[80\%]{} CH_3(CH_2)_6CHOHCH_2COOH$$

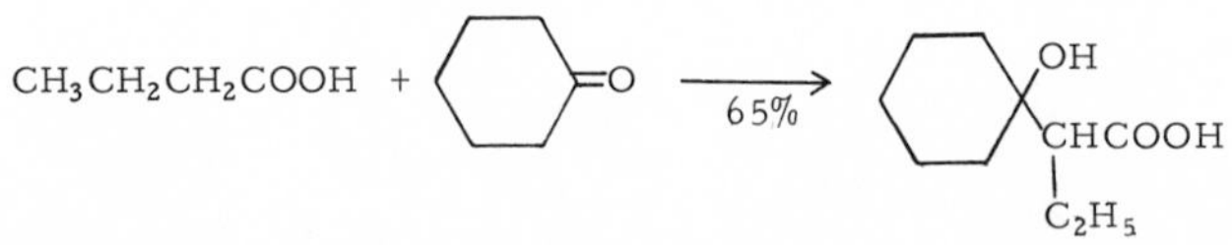

[1]D. G. M. Diaper and A. Kuksis, *Chem. Rev.*, **59**, 89 (1959)
[2]S. Watanabe, K. Suga, T. Fujita, and K. Fujiyoshi, *Chem. Ind.*, 1811 (1969)
[3]B. Angelo, *Bull. soc.*, 1848 (1970)

Nickel boride, 1, 720; **2,** 289–290.

Hydrogenation catalyst. C. A. Brown[1] has compared P-1 nickel boride with W-2 Raney nickel as hydrogenation catalysts and finds that the former is somewhat more active and produces less double-bond migration. In addition it is not pyrophoric. Almost all olefins can be hydrogenated, but tri- and tetrasubstituted olefins are somewhat sluggish. Brown also reported two instances of selective hydrogenation: 2-methyl-1,5-hexadiene (1) is converted into 2-methyl-1-hexene (93% purity), and 4-vinylcyclohexene (2) is converted into 4-ethylcyclohexene (98% pure). Brown notes that the structure of P-1 nickel boride is not certain;

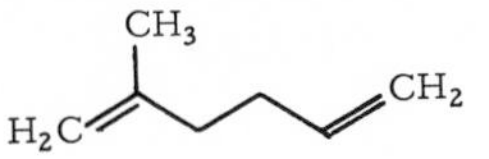

(1) 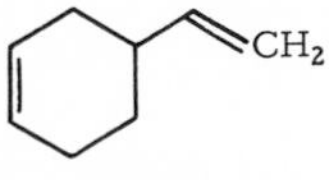   (2)

chemical analysis does not agree with Ni_2B, the original postulated structure.

P-2 Nickel boride shows a remarkable ability to selectively hydrogenate the strained double bond of norbornene. Thus hydrogenation of 5-methylenenorbornene (1) and of the *endo*-dicyclopentadiene (3) give essentially quantitative yields of the dihydro derivatives (2) and (4), respectively.[2]

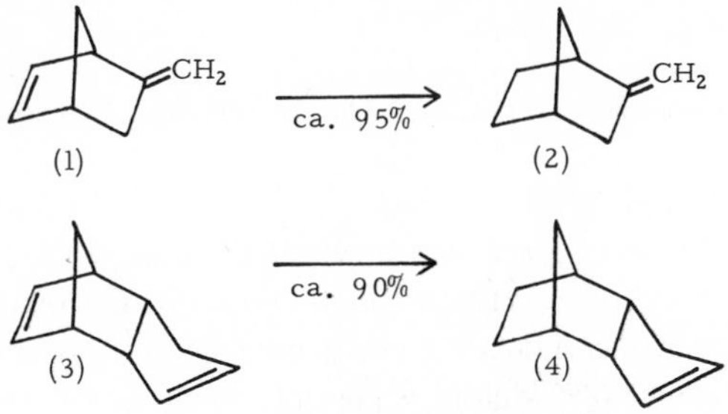

(1)   ca. 95%   (2)

(3)   ca. 90%   (4)

Selective hydrogenation of C≡C. Corey has used P-1 type nickel boride catalyst for selective hydrogenation of C–C triple bonds in the presence of C–C double bonds. Thus in a total synthesis of the sesquiterpene sesquicarene (3) Corey and Achiwa[3] hydrogenated (1) to (2) in 90% yield.

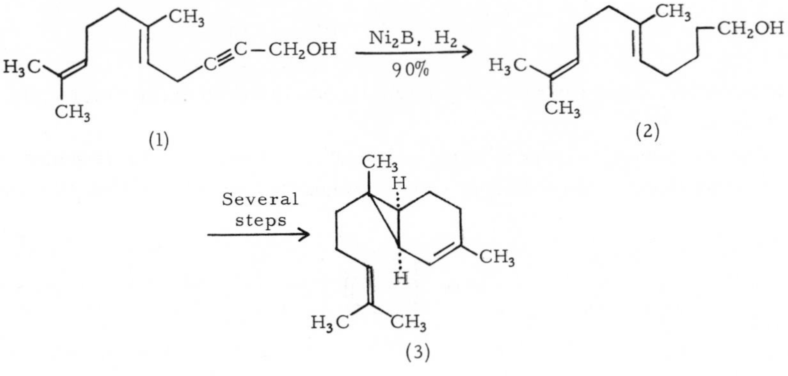

(1) Ni_2B, H_2 (2)
           90%

Several steps

(3)

The reaction was also used in a synthesis of *dl*-sirenin (6)[4] for the hydrogenation of (4) to (5) in 80% yield.

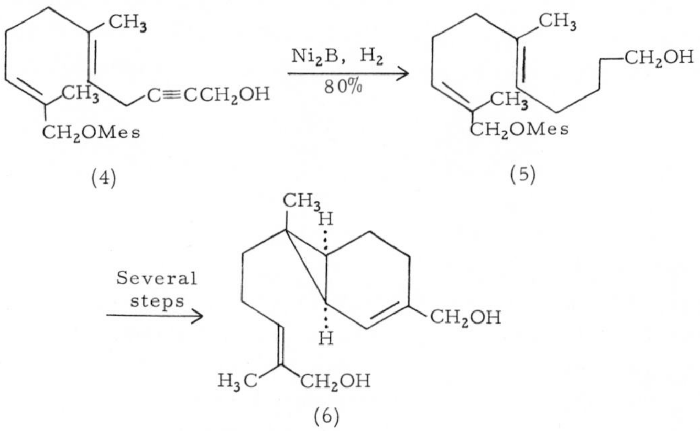

(4)

(5)

Several
steps
$\longrightarrow$

(6)

[1]C. A. Brown, *J. Org.*, **35**, 1900 (1970)
[2]*Idem*, *Chem. Commun.*, 952 (1969)
[3]E. J. Corey and K. Achiwa, *Tetrahedron Letters*, 1837 (1969)
[4]E. J. Corey, K. Achiwa, and J. A. Katzenellenbogen, *Am. Soc.*, **91**, 4318 (1969)

Nickel carbonyl, **1**, 720–723; **2**, 290–293.

Reaction of ethyl diazoacetate with π-allylnickel bromide (**2**, 291). Ethyl diazoacetate reacts with π-allylnickel bromide (0°, ether) with evolution of nitrogen to give ethyl *trans*-β-vinylacrylate (69% yield), ethyl *cis*-β-vinylacrylate (19% yield), and ethyl Δ⁴-pentenoate (8% yield).[1] A possible mechanism is suggested based on

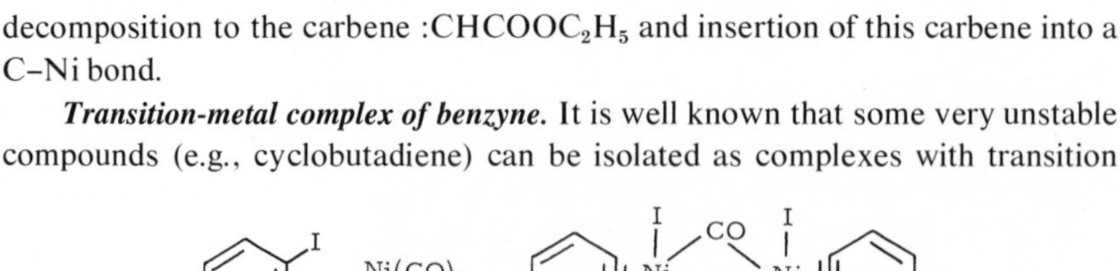

decomposition to the carbene :CHCOOC₂H₅ and insertion of this carbene into a C–Ni bond.

Transition-metal complex of benzyne. It is well known that some very unstable compounds (e.g., cyclobutadiene) can be isolated as complexes with transition

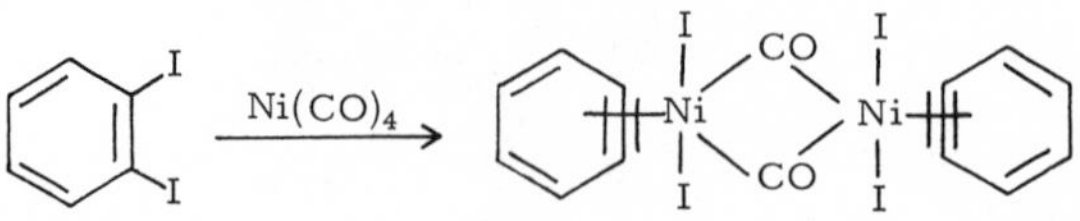

(1)

metals. The first such complex of benzyne has been prepared by heating *o*-diiodobenzene with nickel carbonyl in pentane or cyclohexane to 70° in a sealed tube.[2] The black crystalline product is π-benzyne diiodo-μ-carbonylnickel dimer (1).

Conjugate addition of acyl groups to α,β-unsaturated carbonyl compounds.[3] Nickel carbonyl forms unstable complexes with organolithium compounds (ether, $-50°$), which react smoothly at $-50°$ with a variety of conjugated enones to give 1,4-diketones. For example, the reagent (1) prepared from *n*-butyllithium and nickel carbonyl reacts with mesityl oxide (2) to give 4,4-dimethyl-2,5-nonanedione

$$\underline{n}\text{-}C_4H_9Li \xrightarrow{Ni(CO)_4} [\underline{n}\text{-}C_4H_9CONi(CO)_3]^-Li^+ + (CH_3)_2C{=}CHCOCH_3 \longrightarrow (CH_3)_2CCH_2COCH_3$$

(1)   (2)

with the (CH₃)₂CCH₂COCH₃ having CO and $\underline{n}\text{-}C_4H_9$ substituents

(3)

(3) in 89% yield. The original communication should be consulted for three possible mechanisms for this reaction.

Intramolecular allylic coupling (**1**, 722–723; **2**, 290–292). The key step in a total synthesis of the sesquiterpene elemol (3) involves reaction of the dibromo ester (1) with nickel carbonyl (7 equiv.) in N-methylpyrrolidone to form the monocyclic ester (2). Several other products are obtained, separable by chromatography. Significantly none of the *cis*-adduct is formed. Reaction of (2) with

(1)   (2)   (3)

methylmagesnium bromide affords *dl*-elemol.[4]

Carboxylation of organic halides.[5] Vinyl halides when treated with several equivalents of nickel carbonyl in methanol containing 2–3 equiv. of sodium or potassium methoxide are converted into the corresponding methyl esters. The

$$C_6H_5CH{\overset{t}{=}}CHBr \xrightarrow[\substack{95\%}]{\substack{Ni(CO)_4,\ CH_3OH,\ CH_3ONa \\ 2\ hr.,\ 25°}} C_6H_5CH{\overset{t}{=}}CHCOOCH_3$$

usual sequence is observed: I > Br > Cl; in fact, chlorides are generally un-reactive. Alkyl halides do not undergo this reaction even under more forcing conditions than those used for vinyl halides. However, the combination of nickel carbonyl and potassium *t*-butoxide in *t*-butanol reacts with both vinyl halides and alkyl iodides to give *t*-butyl esters of carboxylic acids; for example:

$$\text{1-Iodoheptane} \xrightarrow[\substack{24 \text{ hrs. } 50^0}]{\substack{\text{Ni(CO)}_4 \\ \text{t-BuOH, t-BuOK}}} \underline{\text{t}}\text{-Butyl octanoate (66\% yield)}$$

Yields are somewhat lower with this reagent because dehydrohalogenation is a serious side reaction.

Aminocarbonylation can be achieved by use of a mixture of nickel carbonyl and an amine, as shown in a typical example.

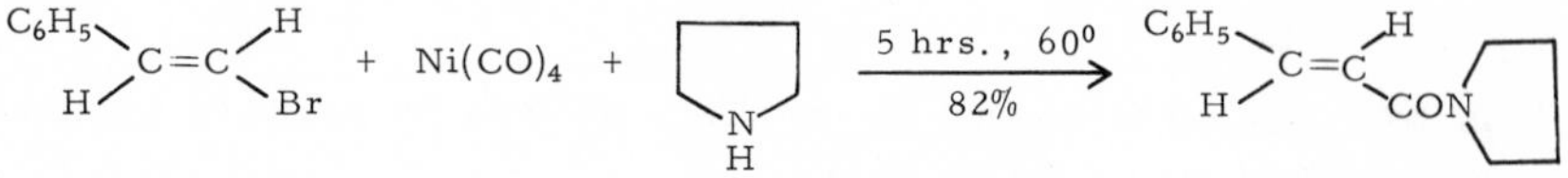

A related reaction is the conversion of vinyl bromides into vinyl cyanides by reaction with potassium hexacyanodinickelate,[6] as shown for the reaction of *trans*-1-bromo-2-phenylethylene.

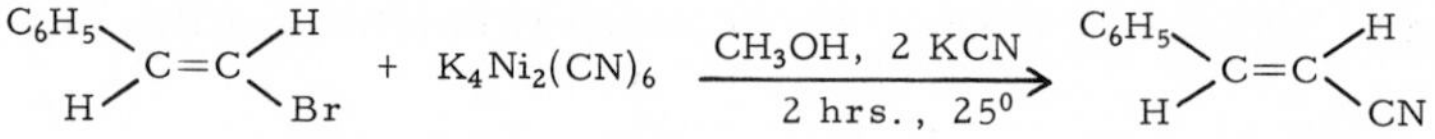

78%

[1]I. Moritani, Y. Yamamoto, and H. Konishi, *Chem. Commun.*, 1457 (1969)
[2]E. W. Gowling, S. F. A. Kettle, and G. M. Sharples, *ibid.*, 21 (1968)
[3]E. J. Corey and L. S. Hegedus, *Am. Soc.*, **91**, 4926 (1969)
[4]E. J. Corey and E. A. Broger, *Tetrahedron Letters*, 1779 (1969)
[5]E. J. Corey and L. S. Hegedus, *Am. Soc.*, **91**, 1233 (1969)
[6]Preparation: W. M. Burgess and J. W. Eastes, *Inorg. Syn.*, **5**, 197 (1957); I. Hashimoto, N. Tsuruta, M. Ryang, and S. Tsutsumi, *J. Org.*, **35**, 3748 (1970)

Nitric acid, 1, 733–735.

Oxidation:[1] In a 1-l. beaker mounted in an ice bath are placed 15.4 g. (0.14 mole) of resorcinol, 400 ml. of water, and 75 g. of crushed ice. Then 16 ml. (0.3 mole) of concd. sulfuric acid is added with stirring, followed by an ice-cold solution of 20.52 g. (0.2975 mole) of sodium nitrite in 150 ml. of water in 1–2 ml.

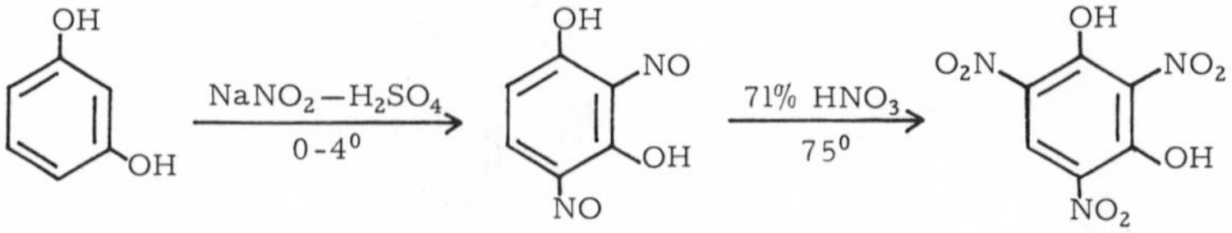

portions over a period of 45 minutes, while maintaining a reaction temperature of 0–4°. The mixture is allowed to stand in the ice bath for 3 hours to complete the reaction. The product, heavily contaminated with sodium sulfate, is collected on a Whatman No. 4 paper and allowed to dry overnight.

The next step is done in a hood owing to evolution of nitric oxide. A 500-ml. beaker is charged with 65 ml. of 71% ("conc'd.") nitric acid, and the pulverized crude 2,4-dinitrosoresorcinol is added in small amounts with stirring in 60–70 minutes, keeping the temperature at 4–8°. The ice bath is removed to allow the mixture to warm to room temperature for 30 minutes. Then the mixture is heated slowly to 75° for 45 minutes to complete the reaction. The mixture is cooled to room temperature and the product is filtered off with Whatman No. 4 paper, then washed with five 40-ml. portions of water. Dilute sodium carbonate solution (25 ml.) is added and let stand 1 hour before collecting and drying the product, 2,4,6-trinitroresorcinol. Yield 28.7 g. (84%), m.p. 168–170°. Purity and stability may be improved by crystallization from dil. sodium carbonate solution.

[1]T. Urbanski, procedure submitted to *Org. Syn.* 1970

p-**Nitrophenylphosphorodichloridate**, **1**, 744–745. Supplier: Aldrich.

2-Nitroprop-2-yl hydroperoxide (1). Mol. wt. 121.09.

This hydroperoxide (1) is formed by autoxidation of 2-nitropropane catalyzed by cuprous chloride:

$$(CH_3)_2CH{-}NO_2 \ + \ O_2 \ \xrightarrow{\ CuCl\ } \ (CH_3)_2C\underset{NO_2}{\overset{OOH}{<}}$$

(1)

Tert-Amines → *sec-amines.*[1] When a tertiary amine in pyridine is shaken with 2-nitropropane and CuCl under oxygen, it is converted into a nitrosamine (5), which can be isolated in yields of 15–65% and then reduced to a secondary amine. The reaction proceeds through oxidation to an amine oxide (2) with conversion of (1) into 2-nitro-2-propanol (3). This decomposes into acetone and nitrous acid. The nitrous acid traps the secondary amine formed from the amine oxide (2) after rearrangement to the carbinolamine (4). It is noteworthy that even

$$\underset{R}{\overset{R}{>}}NCH_2R' \ + \ (1) \ \longrightarrow \ \underset{R}{\overset{R}{>}}\overset{\overset{O}{\|}}{N}CH_2R' \ + \ (CH_3)_2C\underset{NO_2}{\overset{OH}{<}} \ \xrightarrow{-(CH_3)_2C=O}$$

(2)   (3)

$$\underset{R}{\overset{R}{>}}N\overset{\overset{OH}{|}}{C}HR' \ + \ HNO_2 \ \xrightarrow{-H_2O} \ \underset{R}{\overset{R}{>}}NNO \ + \ R'CHO$$

(4)   (5)

very hindered amines are readily dealkylated under the mild conditions employed.

[1]B. Franck, J. Conrad, and P. Misbach, *Angew. Chem., internat. Ed.*, **9**, 892 (1970)

(S)-1-Nitroso-2-methylindoline-2-carboxylic acid, (S)-16,

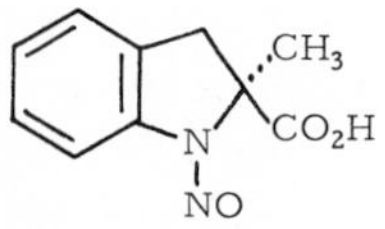

(S)-16

This is a chiral reagent analogous to (S)-1-amino-2-hydroxymethylindoline (*which see*) developed by Corey *et al.*[1] for effecting the resolution of an α-amino acid with regeneration of the chiral reagent.

[1]E. J. Corey, R. J. McCaully, and H. S. Sachdev, *Am. Soc.*, **92**, 2476 (1970)

Nitrosyl chloride, **1**, 748–755; **2**, 298–299.

ω-Cyanoaldehydes. ω-Cyanoaldehydes (4) can be prepared in a four-step synthesis from cycloalkenes (1). The first step involves addition of nitrosyl chloride (H^+ or $h\nu$) to give a cyclic α-chlorooxime (2), followed by reaction with a sodium alkoxide in the corresponding alcohol or THF to form an α-alkoxycyclo-alkanone oxime (3). The final steps involve reaction with phosphorus penta-chloride and hydrolysis. Overall yields are in the range 45–80%.[1]

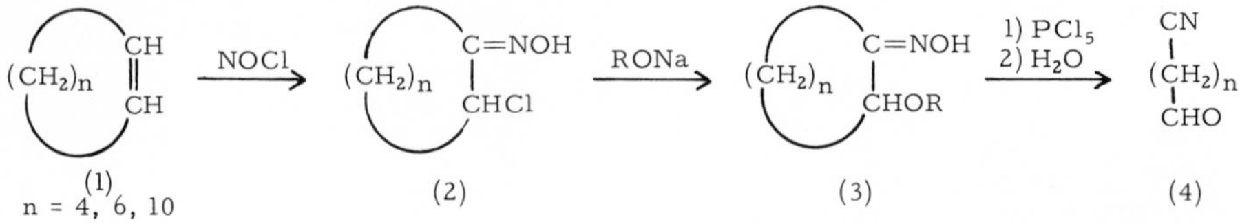

Chlorination. Sulfoxides having an α-methylene group yield the correspond-ing α-chlorosulfoxides on treatment with nitrosyl chloride in pyridine–chloro-form.[2]

[1]M. Ohno, N. Naruse, S. Torimitsu, I. Terasawa, *Am. Soc.*, **88**, 3168 (1966); M. Ohno, N. Naruse, and I. Terasawa, *Org. Syn.*, **49**, 27 (1969)
[2]R. N. Loeppky and D. C. K. Chang, *Tetrahedron Letters*, 5415 (1968)

Nitrosyl fluoride, **1**, 755; **2**, 299.

Review.[1]

Reaction with $\Delta^{9(11)}$-*steroids.* Nitrosyl fluoride adds to $\Delta^{9(11)}$-steroids in methylene chloride or 1,2-dichloroethane (10 days) to give, in moderate yield, the corresponding 11-nitrimino-9α-fluoro derivatives, which can be hydrolyzed

(Al_2O_3) to the medically important 9α-fluoro-11-ketosteroids. The nitrimino group can be reduced to the nitramine or the 11-amine.[2]

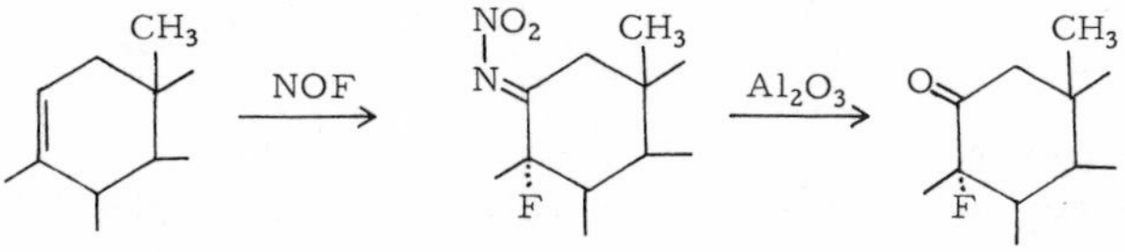

The reaction of nitrosyl fluoride with a 9(11)-unsaturated steroid (1) in ethyl acetate at 50° is unusual in that it gives the corresponding 12-keto-$\Delta^{9(11)}$-steroid (2) in about 20% yield.[3] The nature of the oxidant in this reaction is not known.

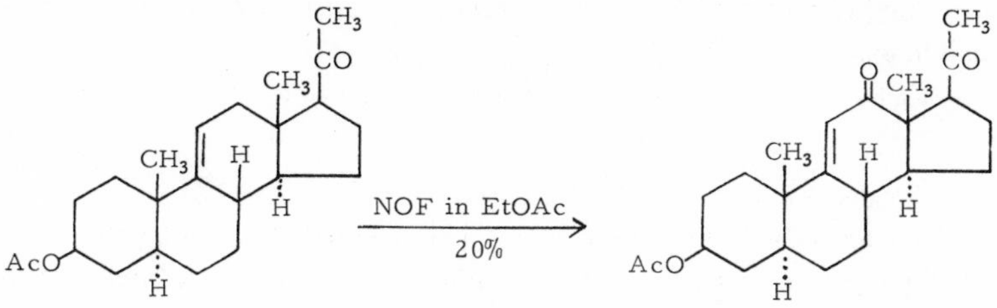

(1)   (2)

[1] R. Schmutzler, *Angew. Chem., internat. Ed.*, **7**, 440 (1968)
[2] J. P. Gratz and D. Rosenthal, *Steroids*, **14**, 729 (1969)
[3] D. Rosenthal and J. P. Gratz, *J. Org.*, **34**, 409 (1969)

Nitryl iodide, 1, 757; **2**, 300.

Addition of olefins. The work of Hassner *et al.* (**1**, 757) on addition of nitryl iodide to olefins has now been published.[1] The reaction probably proceeds by a free-radical addition involving $NO_2 \cdot$.

Ref. 2, **2**, 300: Definitive paper: W. A. Szarek, D. G. Lance, and R. L. Beach, *Carbohydrate Res.*, **13**, 75 (1970)

[1] A. Hassner, J. E. Kropp, and G. J. Kent, *J. Org.*, **34**, 2628 (1969)

O

Osmium-on-carbon, Os—C.

Reduced osmium on carbon is an excellent catalyst for selective hydrogenation of α,β-unsaturated aldehydes to unsaturated alcohols.[1] Cinnamaldehyde → cinnamyl alcohol (95% yield). Reduced rates are observed with alumina as the support. This selective reduction is not applicable to α,β-unsaturated ketones; thus hydrogenation of mesityl oxide afforded methyl isobutyl ketone.

The same selective reduction has been achieved previously by platinum or ruthenium catalysts inhibited by additional metals.

[1]P. N. Rylander and D. R. Steele, *Tetrahedron Letters*, 1579 (1969)

Oxalyl chloride, 1, 767–772; **2,** 301–302.

Chlorocarbonylation. The reaction of adamantane (1) with one equivalent of oxalyl chloride under free-radical conditions (benzoyl peroxide) followed by methanolysis gives a mixture of methyl adamantane-1- and -2-carboxylate, (2 and 3) easily separable by fractional distillation. Chlorocarbonylation with a fivefold excess of reagent followed by methanolysis gives a mixture of methyl adamantane-

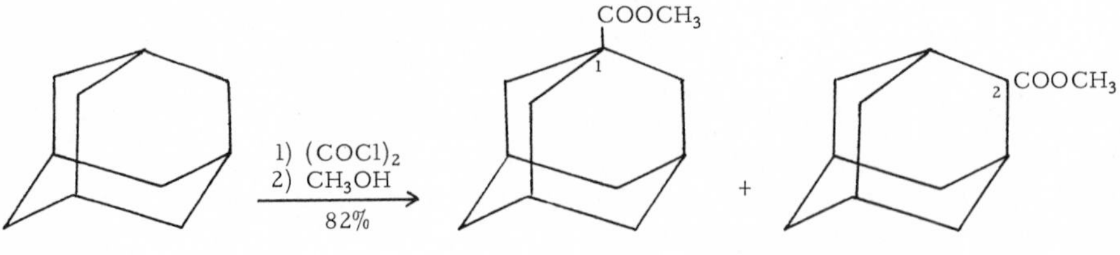

dicarboxylates, the major component being the 1,3-isomer.[1]

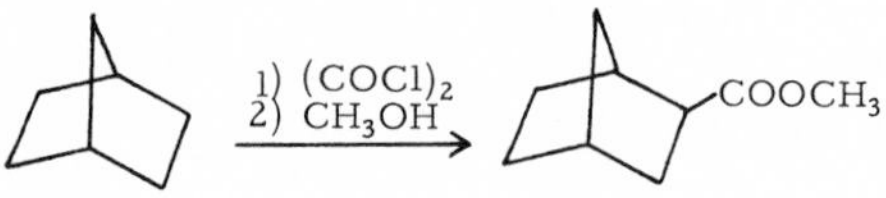

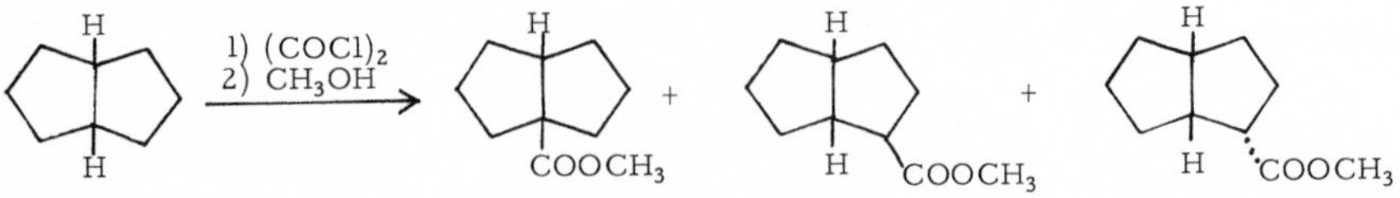

Chlorocarbonylation of norbornane (4) under the same conditions shows high stereoselectivity in that methyl *exo*-[2.2.1]bicycloheptane-2-carboxylate (5) is essentially the only product. The reaction with *cis*-[3.3.0]bicyclooctane (6) also shows stereoselectivity.[2]

[1] I. Tabushi, J. Hamuro, and R. Oda, *J. Org.*, **33**, 2108 (1968)
[2] I. Tabushi, T. Okada, and R. Oda, *Tetrahedron Letters*, 1605 (1969)

P

Palladium hydroxide on barium sulfate, 1, 782; 2, 305.

In 1955 Kuhn and Haas[1] published a procedure for the preparation of a hydrogenation catalyst consisting of palladium hydroxide catalyst supported on barium sulfate; the catalyst was prepared from $PdCl_2$, H_2SO_4, and $Ba(OH)_2$. Later work[2] indicated that the activity of the catalyst was variable and not reproducible, and the trouble was shown to be due to traces of barium carbonate, which lowers the activity. In a new procedure, barium hydroxide is replaced by barium acetate. Thus barium acetate is warmed in distilled water with sodium sulfate and the suspension of barium sulfate is treated with palladium chloride in dilute hydrochloric acid. After neutralization with sodium hydroxide the light brown catalyst is washed with water and dried at 105° for 2 hrs. The catalyst was used for the hydrogenation of various acid anhydrides and carbonyl compounds and shown to have reproducible activity.

[1]R. Kuhn and H. J. Haas, *Angew. Chem.*, **67**, 785 (1955)
[2]R. Kuhn and I. Betula, *Ann.*, **718**, 50 (1968)

trans-3-Pentene-2-one, 2, 306–307.

Preparation. Odom and Pinder[1] recommend preparation of the reagent by Friedel-Crafts acylation of propylene with acetyl chloride to give 2-chloropropyl methyl ketone, which is then dehydrochlorinated to give the reagent. The pro-

$$CH_3CH{=}CH_2 \ + \ CH_3COCl \xrightarrow{\text{AlCl}_3} CH_3\underset{\underset{Cl}{|}}{C}HCH_2COCH_3 \xrightarrow[25-27\%]{K_2CO_3} \underset{H}{\overset{H_3C}{>}}C{=}C\underset{COCH_3}{\overset{H}{<}}$$

cedure is based on that of Jones and Taylor,[2] who, however, did not report a yield.

Robinson annelation. Robinson annelation of substituted cyclohexanones with the reagent offers an entry into eremophilone-type sesquiterpenes. Thus the total synthesis of (±)-nootkatone (2) has been achieved by the annelation of (1) with *trans*-3-pentene-2-one in the presence of sodium hydride.[3] In this case only one isomer was isolated. However, Coates and Shaw[4] noted that the steric outcome of

(1)   (2)

the condensation was markedly affected by the reaction conditions. Thus condensation of the pyrrolidine enamine of 2-methylcyclohexane-1,3-dione (3) with the reagent in benzene–acetic acid–sodium acetate gave predominantly a product (4) with the *trans*-orientation of the methyl groups. If the benzene is replaced by formamide, a mixture of approximately equal amounts of (4) and (5) is obtained.

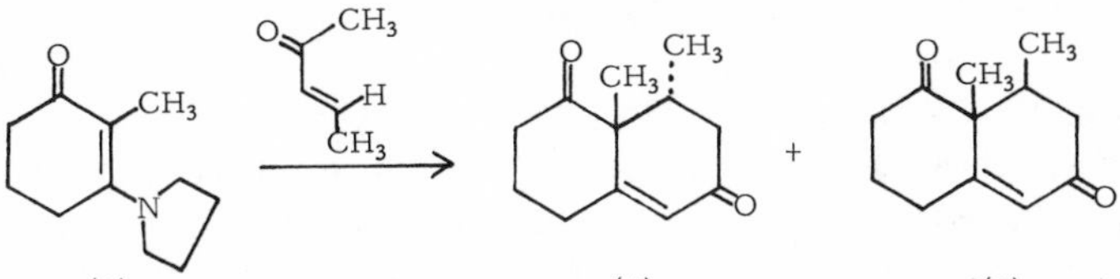

(3)　　　　　　　　(4)　　　　　　　(5)

[1]H. C. Odom and A. R. Pinder, procedure submitted to *Org. Syn.* 1969
[2]N. Jones and H. T. Taylor, *J. Chem. Soc.*, 1345 (1961)
[3]H. C. Odom and A. R. Pinder, *Chem. Commun.*, 26 (1969)
[4]R. M. Coates and J. E. Shaw, *ibid.*, 47 (1968)

Peracetic acid, 1, 785–791; **2,** 307–309.

Epoxidation of olefins. The reaction of buffered peracetic acid (**2,** 307–309) with 3,3,6,6-tetramethyl-1,4-cyclohexadiene (1) gives the monoepoxide (2) and the *cis*-diepoxide (3) as the only volatile products. The low yields reflect isolation by preparative gas chromatography.[1]

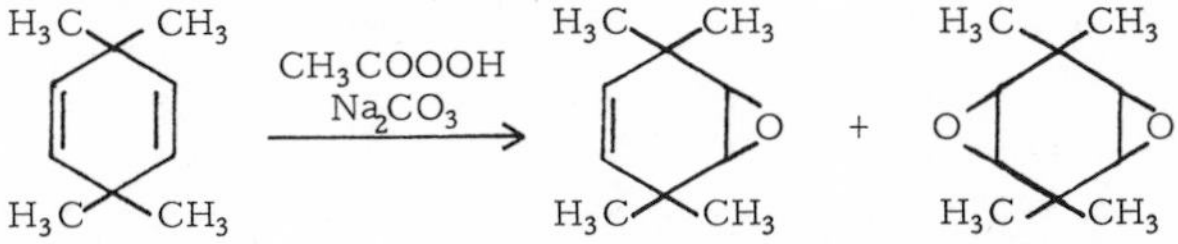

(1)　　　　　　(2, 30%)　　　　(3, 11%)

[1]R. W. Gleason and J. T. Snow, *J. Org.*, **34,** 1963 (1969)

Perbenzoic acid, 1, 791–796.

Episulfoxides. Episulfides are oxidized to episulfoxides by perbenzoic acid in methylene chloride (−20 to −30°; 40–75% yield). The simultaneously formed benzoic acid is converted into the ammonium salt by reaction with dry ammonia and removed by filtration.[1] Sodium metaperiodate had been used previously for this oxidation, but yields were lower.[2]

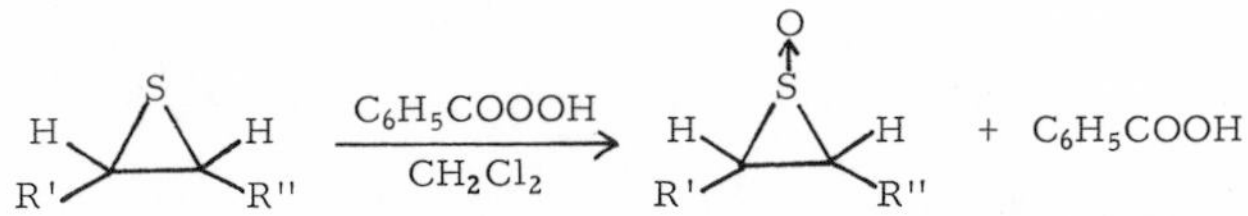

[1]K. Kondo, A. Negishi, and M. Fukuyama, *Tetrahedron Letters*, 2461 (1969)
[2]G. E. Hartzell and J. N. Paige, *Am. Soc.*, **88,** 2616 (1966); *idem, J. Org.*, **32,** 459 (1967)

Perchloric acid, **1**, 796–802; **2**, 309–310.

Steroidal 17,20-acetonides. *p*-Toluenesulfonic acid is a satisfactory catalyst for conversion of steroidal 20,21-glycols and glycerols into acetonides. Acetonation of 17,20-glycols proceeds very slowly with this catalyst. However, use of perchloric acid as catalyst provides 17,20-acetonides in 15 minutes or less.[1]

[1]M. L. Lewbart and J. J. Schneider, *J. Org.*, **34**, 3505 (1969)

Periodic acid, **1**, 815–819; **2**, 313–315.

Oxidation of a phenylhydrazine to a phenyl azo compound. Oxidation of the hydrazine (1a) or (1b) with two equivalents of the reagent gives the corresponding azo compound (2a) or (2b) in 90% yield.[1] This reaction had been carried out pre-

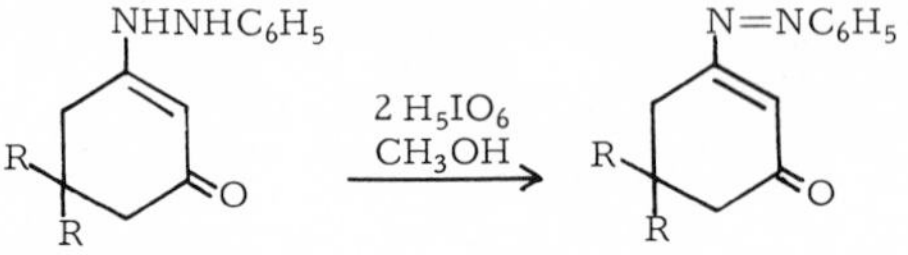

(1a, R = H)   (2a, R = H)
(1b, R = CH$_3$)   (2b, R = CH$_3$)

viously with NBS in boiling carbon tetrachloride, but the yield in this case was 46%.[2]

[1]A. J. Fatiadi, *J. Org.*, **35**, 831 (1970)
[2]B. Eistert, G. Kilpper, and J. Göring, *Ber.*, **102**, 1379 (1969)

Perlauric acid, **2**, 315–316.

Meakins *et al.*[1] state that perlauric acid is more suitable than perbenzoic acid for quantitative study of peroxidation of 3-substituted Δ^5-cholestenes. The reagent converts lumisterol (1) into the 5β,6β-epoxide in 80% yield.

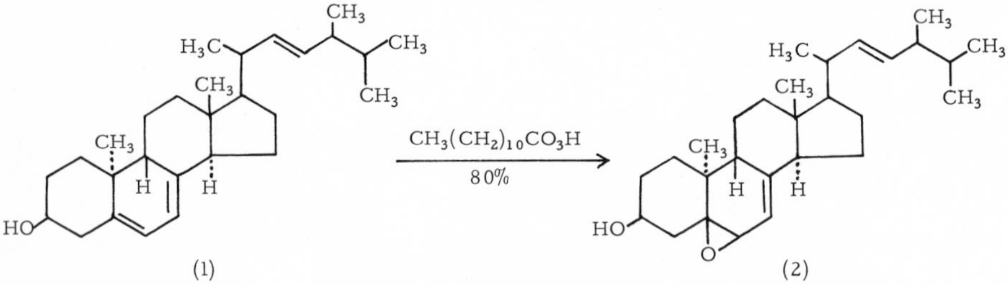

(1)   (2)

[1]K. D. Bingham, G. D. Meakins, and J. Wicha, *J. Chem. Soc.*, (C), 510 (1969)

Perphthalic acid, **1**, 819–820.

Bayer-Villiger oxidation. The oxidation of 3-methoxy-16β-acetyl-D-nor-1,3,5(10)-estratriene (1) to the acetate (2) presented difficulties; the usual per-

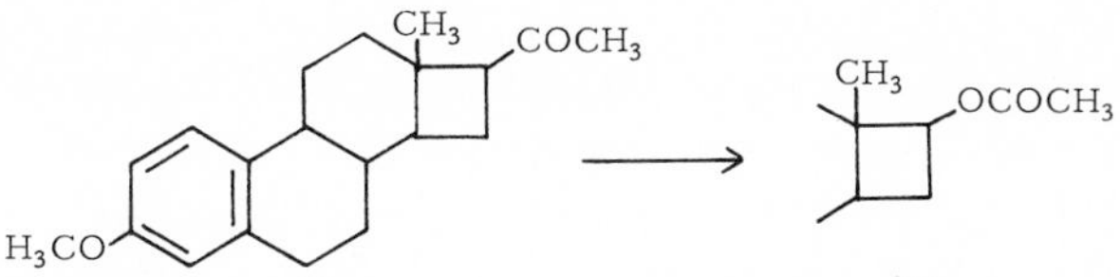

(1) (2)

acids (perbenzoic acid, *m*-chloroperbenzoic acid, and peracetic acid) either gave back starting material or products in which ring A was destroyed. The transformation was accomplished in ether solution with perphthalic acid (large excess) in 85% yield by reaction at –3° for a month.[1]

[1]J. Meinwald and J.-L. Ripoll, *Am. Soc.*, **89**, 7075 (1967)

Pertrifluoroacetic acid, 1, 821–827; **2,** 316.

The procedure for oxidation of 2,6-dichloroaniline to 2,6-dichloronitrobenzene has been published.[1]

[1]A. S. Pagano and W. D. Emmons, *Org. Syn.*, **49**, 47 (1969)

Phenacylsulfonyl chloride (ω-Acetophenonesulfonyl chloride), $C_6H_5COCH_2SO_2Cl$. Mol. wt. 218.66, m.p. 88°.

Preparation.[1] The reagent is prepared by the reaction of acetophenone with dioxane sulfotrioxide[2] followed by conversion of the resulting sulfonic acid into the chloride with phosphorus trichloride.

Secondary amines. The use of toluenesulfonamides in the monoalkylation of amines suffers from the disadvantage that vigorous conditions are required for removal of the protective group. Hendrickson and Bergeron[3] have used the phenacylsulfonyl group, removable by reduction with zinc and acetic acid.

$$RNH_2 \xrightarrow[Py]{C_6H_5COCH_2SO_2Cl} RNHSO_2CH_2COC_6H_5 \xrightarrow{R'X} R-\underset{\displaystyle R'}{N}SO_2CH_2COC_6H_5 \xrightarrow{Zn,\ HOAc,\ H^+} R\underset{\displaystyle R'}{N}H$$

[1]W. E. Truce and C. W. Vriesen, *Am. Soc.*, **75**, 2525 (1953)
[2]W. E. Truce and C. C. Alfieri, *ibid.*, **72**, 2740 (1950)
[3]J. B. Hendrickson and R. Bergeron, *Tetrahedron Letters*, 345 (1970)

Phenylboronic acid, 1, 833–834; **2,** 317.

Phenylboronic acid reacts with nucleosides in pyridine at reflux temperature

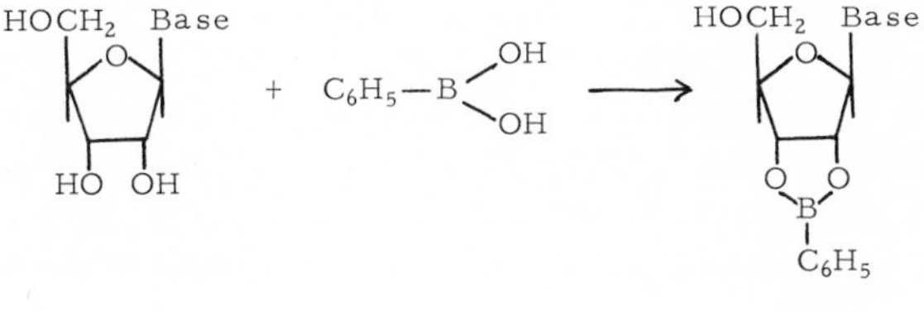

(2 hours) to give 2′,3′-O-phenylboronates. This reaction is useful for protection of the 2′- and 3′-hydroxyl groups in the synthesis of 5′-O-derivatives of nucleosides. The protecting group can be removed under mild conditions with propane-1,3-diol in an anhydrous medium.[1]

[1]A. M. Yurkevich, I. I. Kolodkina, L. S. Varshavskaya, V. I. Borodulina-Shvetz, I. P. Rudakova, and N. A. Preobrazhenski, *Tetrahedron*, **25**, 477 (1969)

Phenyl(dibromochloromethyl)mercury, $C_6H_5HgCBr_2Cl$. Mol. wt. 485.01, m.p. (dec.), 107–109°.

Preparation. The reagent is prepared in 75% yield by the reaction of phenylmercuric chloride, dibromochloromethane, and the *t*-butanol monosolvate of commercial potassium *t*-butoxide.[1]

Bromochlorocarbene.[2] The reaction of phenyl(dibromochloromethyl)mercury in refluxing benzene with olefins gives *gem*-bromochlorocyclopropanes in generally good yield (50–98%). Note that bromochlorocyclopropanes are reduced in

$$R^1-CH{=}CH-R^2 \ + \ C_6H_5HgCBr_2Cl \ \longrightarrow \ \overset{\displaystyle R^1-CH-\!\!-\!\!-CH-R^2}{\underset{\displaystyle Br \diagup \ \diagdown Cl}{\diagdown \underset{C}{} \diagup}} \ + \ C_6H_5HgBr$$

high yield by tri-*n*-butyltin hydride to the corresponding chlorocyclopropanes (**1**, 1193),[3] which are not readily available directly.

[1]D. Seyferth and R. L. Lambert, Jr., *J. Organometal. Chem.*, **16**, 21 (1969)
[2]D. Seyferth, S. P. Hopper, and T. F. Julia, *ibid.*, **17**, 193 (1969)
[3]D. Seyferth, H. Yamazaki, and D. L. Alleston, *J. Org.*, **28**, 703 (1963)

Phenyl(dihalocarbomethoxymethyl)mercury. $C_6H_5HgCCl_2COOCH_3$: Mol. wt. 419.67; m.p. 140–144°. $C_6H_5HgCBr_2COOCH_3$: mol. wt. 508.59; m.p. 154–156° dec.

Preparation.[1] These organomercury compounds are prepared by the procedure used for synthesis of phenyl(trihalomethyl)mercury (This volume).

$$C_6H_5HgCl \ + \ (CH_3)_3COK \ + \ HCX_2COOCH_3 \ \xrightarrow[-60\,-\,-50^0]{THF} (CH_3)_3COH + KCl \ + \ C_6H_5HgCX_2COOCH_3$$

X = Cl, 75% yield; X = Br, 59% yield

Precursor of halocarbomethoxycarbenes. Chlorocarboethoxycarbene, $ClCCOO\text{-}C_2H_5$, has been generated in low yield by photolysis of ethyl chlorodiazoacetate, $ClC({=}N_2)COOC_2H_5$.[2]

Seyferth[1] reports that phenyl(dihalocarbomethoxymethyl)mercury compounds, although more stable than the corresponding trihalomethyl analogs, are good sources of halocarbomethoxycarbenes. Thus, when phenyl(dibromocarbomethoxymethyl)mercury is refluxed with cyclooctene (**1**) in chlorobenzene under nitrogen for 43 hours, phenylmercuric bromide is formed (87% yield) and the two isomeric 9-bromo-9-carbomethoxybicyclo[6.1.0]nonanes (2) are obtained in $1:2.3$ ratio in 50% yield.

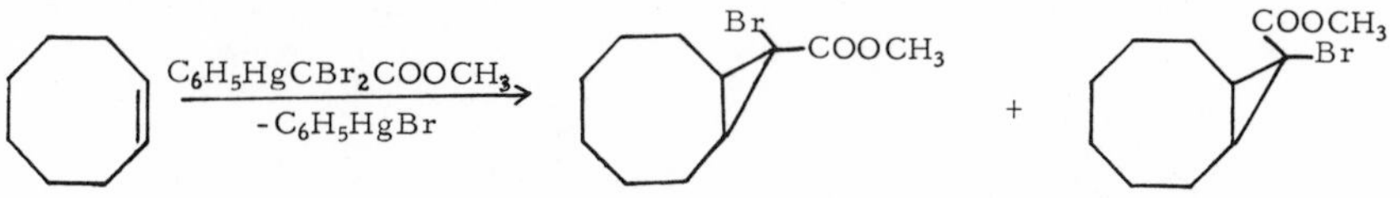

(1)   (2)

The mercurial $C_6H_5HgCClBrCF_3$ has also been prepared from $CF_3CClBrH$ ("Fluothane", Ayerst Laboratories) and used as a precursor of chlorotrifluoro-methylcarbene, $ClCCF_3$.[1]

[1]D. Seyferth, D. C. Mueller, and R. L. Lambert, Jr., *Am. Soc.*, **91**, 1562 (1969)
[2]W. Schöllkopf, F. Gerhart, M. Reetz, H. Frasnelli, and H. Schumacher, *Ann.*, **716**, 204 (1968)

o-**Phenylene phosphorochloridate**, **1**, 837–838; **2**, 321. M.p. 59–61°. The reagent is sensitive to moisture, but can be kept indefinitely in a sealed vessel. Additional supplier: Aldrich.

   Phosphorylation. The definitive paper on the conversion of alcohols into the monophosphate esters has been published.[1]

[1]T. A. Khwaja, C. B. Reese, and J. C. M. Stewart, *J. Chem. Soc.*, (C), 2092 (1970)

4-Phenyl-1,2,4-triazoline-3,5-dione, **1**, 849–850; **2**, 324–326.

   Preparation. In a newer preparative procedure, 4-phenylurazole (3) is obtained by the method here formulated, and oxidized with *t*-butyl hypochlorite in ethyl acetate.[1]

$$H_2NNH_2 + CO(OC_2H_5)_2 \xrightarrow[77.5-82\%]{} H_2NNHCO_2C_2H_5 \xrightarrow[96-99\%]{C_6H_5NC=O} C_6H_5NHCONHNHCO_2C_2H$$

(1)   (2)

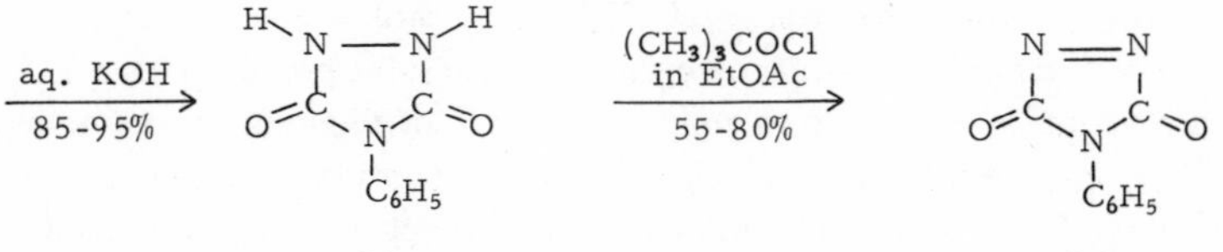

(3)   (4)

   Protection of ring B diene system of ergosterol. This dienophile reacts instantly with ergosterol (1) in acetone–benzene solution at 0° to give the adduct (2) in 85% yield.[2] Barton et al.[3] report that this adduct (or the corresponding adduct of ergosteryl acetate) is reconverted into ergosterol in 99% yield by lithium aluminum hydride reduction; a simple selective and reversible protection of the $\Delta^{5,7}$-diene system of ergosterol can be achieved in this way.

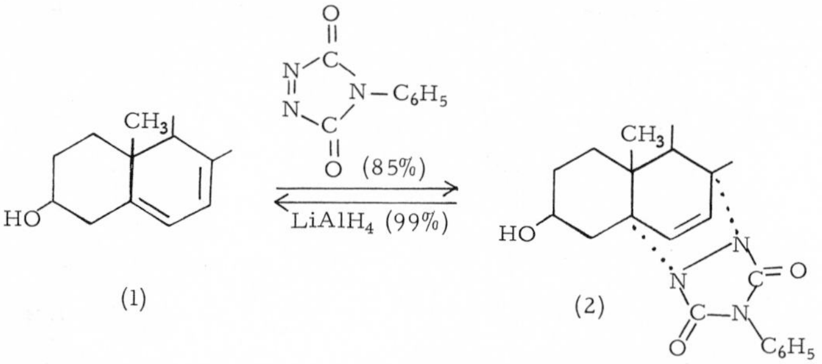

(1)   (2)

Barton *et al.* used this method of protection in a synthesis of a labeled form of $\Delta^{5,7,22,24(28)}$-ergostatetraene-3β-ol (3) and showed that this tetraene is an efficient

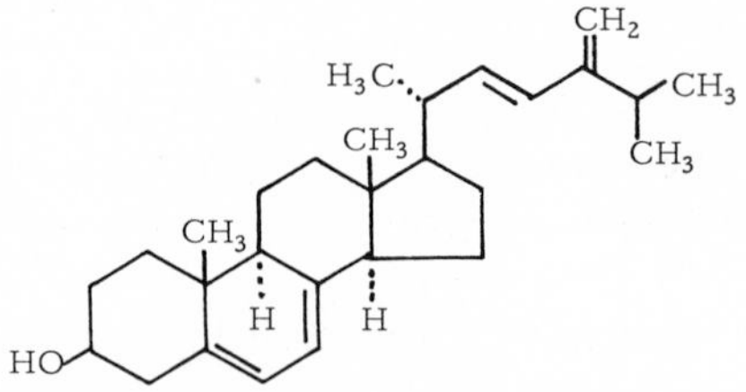

(3)

precursor of ergosterol in yeast. The precursor was synthesized from the Diels-Alder adduct (2) by ozonization to the hexanoraldehyde followed by a Wittig reaction of this aldehyde with the phosphorane derived from a radioactive form of 2-allylisopropyl bromide.

[1]R. C. Cookson, S. S. Gupte, I. D. R. Stevens, and C. T. Watts, procedure submitted to *Org. Syn.* 1969
[2]S. S. H. Giliani and D. J. Triggle, *J. Org.*, **31**, 2397 (1966)
[3]D. H. R. Barton, T. Shioiri, and D. A. Widdowson, *Chem. Commun.*, 939 (1970)

Phenyl(trifluoromethyl)mercury, $C_6H_5HgCF_3$. Mol. wt. 346.72, m.p. 140–143°.

Preparation.[1] This organomercury compound is prepared in 70–75% yield by the reaction at −65° of one molar equivalent of phenyltribromomethylmercury[2] with 3 molar equivalents of phenylmercuric fluoride[3] in the presence of a small quantity of 48% aqueous hydrogen fluoride, which serves as a reaction moderator.

$$C_6H_5HgCBr_3 + 3\ C_6H_5HgF \longrightarrow C_6H_5HgCF_3 + 3\ C_6H_5HgBr$$

The reaction mixture is allowed to warm to room temperature and stirred for an additional hour.

Difluorocarbene.[1] The reagent is very stable thermally but, in the presence of sodium iodide,[4] difluorocarbene is generated. The reaction is carried out in boiling benzene or excess olefin. No solvent is required for the sodium iodide. The reagent

$$C_6H_5HgCF_3 \ + \ NaI \ + \ >C=C< \ \xrightarrow{\text{Benzene, reflux}} \ \begin{array}{c} \backslash \diagup \\ C \\ | \quad \diagdown CF_2 \\ C \diagup \\ \diagup \backslash \end{array} + \ C_6H_5HgI \ + \ NaF$$

converts cyclohexene into 7,7-difluoronorcarane in 83% yield (gas–liquid partition chromatography). 1,1-Difluoro-2-n-amylcyclopropane was formed from 1-heptene in 70% yield.

[1]D. Seyferth, S. P. Hopper, and K. V. Darragh, *Am. Soc.*, **91**, 6536 (1969)
[2]D. Seyferth and R. L. Lambert, Jr., *J. Organometal. Chem.*, **16**, 21 (1969)
[3]G. F. Wright, *Am. Soc.*, **58**, 2653 (1936)
[4]*See* Trimethyl(trifluoromethyl)tin, **1**, 1236–1237; **2**, 442.

Phenyl(trihalomethyl)mercury, **1**, 851–854; **2**, 326–328.

Improved synthesis.[1] In Seyferth's original synthesis, benzene was used as the

$$C_6H_5HgCl \ + \ CHX_3 \ + \ \underline{t}\text{-BuOK} \ \longrightarrow \ C_6H_5HgCX_3 \ + \ KCl \ + \ \underline{t}\text{-BuOH}$$

solvent. However, phenylmercuric chloride and potassium t-butoxide have limited solubility in this solvent. Consequently a t-butanol solvate of the base had to be prepared; a high-speed stirrer was required. In a new procedure the solvent is THF, in which both phenylmercuric chloride and commercial potassium t-butoxide are appreciably soluble. A large excess of the haloform is not necessary nor is a high-speed stirrer. Yields are 72–75%.

[1]D. Seyferth and R. L. Lambert, Jr., *J. Organometal. Chem.*, **16**, 21 (1969)

Phosgene, **1**, 856–859; **2**, 328–329.

Anhydrides. The preparation of nicotinic anhydride by reaction of nicotinic acid with phosgene has been published.[1]

[1]H. Rinderknecht and M. Gutenstein, *Org. Syn.*, **47**, 89 (1967)

Phosphatolead(IV) acids.

Preparation.[1] Lead oxide, Pb_3O_4, 30 g., is added with cooling to 85% phosphoric acid, 300 ml. The mixture is then warmed to 70° for three hours. The product is obtained by centrifugation. It consists of a mixture of 2 parts of $Pb(H_2PO_4)_2$ and 1 part of $H_2[Pb(H_2PO_4)_2(HPO_4)_2]$ ("mixture A").

Oxidation.[1] This mixture is useful for various oxidations in aqueous or methanolic solution. Since the reagent is not soluble in this solvent, the reaction is heterogeneous. Practically no oxidation occurs in aromatic hydrocarbons. The substrate and oxidant are mixed with cooling and then the mixture is warmed to

$$HOCH_2CH_2OH \ \xrightarrow[95\%]{50°} \ 2\,HCHO$$

$$(H_3C)_2C=C(CH_3)_2 \ \longrightarrow \ 2\,(CH_3)_2C=O$$

50–80°. 1,2-Diols are cleaved in high yield; unsaturated compounds are also cleaved, probably by a two-step reaction involving hydroxylation followed by glycol fission. Tartaric acid is oxidized to glyoxylic acid in 87% yield:

$$(CHOH)_2(COOH)_2 \xrightarrow[87\%]{20-40°} 2\ OCHCOOH$$

Hydroquinone is rapidly oxidized to p-benzoquinone even at 0° (yield 93%).

[1]F. Huber and M. S. A. El-Meligy, *Ber.*, **102**, 872 (1969)

Phosphorus pentasulfide, 1, 870–871. Additional supplier: British Drug Houses.

Aromatization by dehydration of Diels-Alder furane adducts. In a study of the Diels-Alder adduct of 1,3-diphenylisobenzofurane (1) with norbornene (2), Cava[1] obtained the adduct (3) in good yield. Conversion into the aromatic system (4) was effected by treatment with concd. hydrochloric acid in acetic acid (steam bath, 12 hrs.) and, also, more surprisingly, by treatment with phosphorus pentasulfide in carbon disulfide at room temperature, 48 hrs., 84% yield. The latter method was found to be superior to acid treatment for dehydration of the bis adduct (6) of (1) with norbornadiene (5).

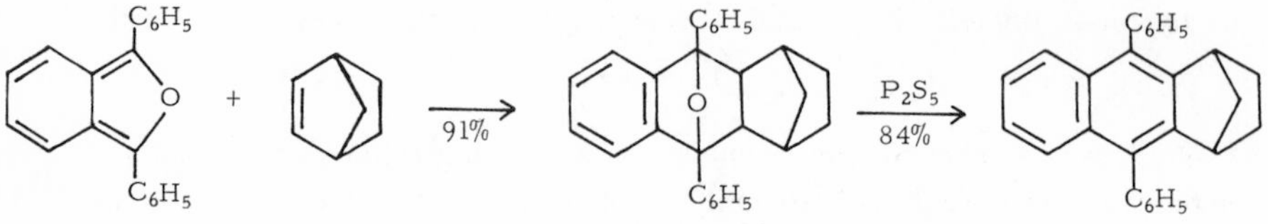

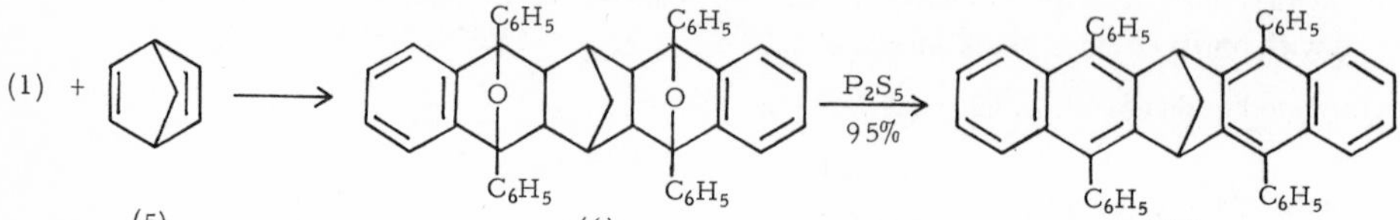

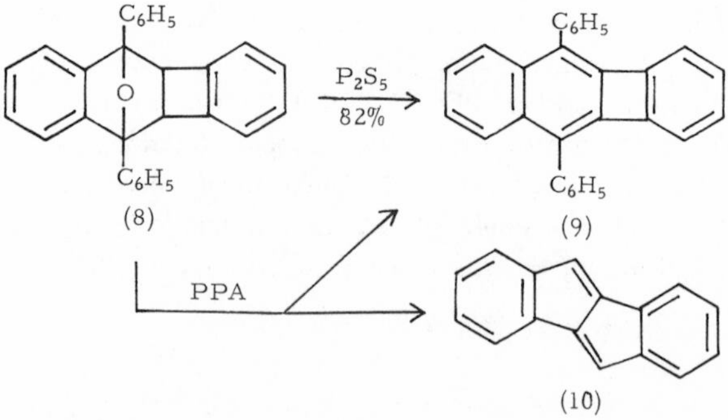

The phosphorus pentasulfide method was also found useful for dehydration of the adduct (8) of 1,3-diphenylbenzofurane with benzocyclobutadiene to give 5,10-diphenylbenzo[*b*]biphenylene (9).[2] In this case, use of polyphosphoric acid gave an inseparable mixture of (9) and the dibenzopentalene (10). Similar results were obtained in dehydration of the related adduct (11), obtained from 1,3-diphenylnaphtho[2.3-*c*]furane.

C_6H_5

O

C_6H_5

(11)

Preparation of thioketones. The reagent is used in the first step of the conversion of adamantanone into 2-adamantanethiol.[3] A 1-l. three-necked round-

$$\xrightarrow[82-91\%]{P_4S_{10}-Py}$$

$$\xrightarrow[74-85\%]{NaBH_4}$$

bottomed flask with a mechanical glass stirrer equipped with a Teflon paddle and a reflux condenser with drying tube is charged with 48.0 g. (0.32 mole) of adamantanone[4] and 300 ml. of pyridine. The flask is heated in an oil bath kept at 90° and, when the stirred solution has reached that temperature, 17.8 g. (0.04 mole) of phosphorus pentasulfide (P_4S_{10}) is added in portions in 10 minutes. Most of it dissolves, and the yellow solution is stirred at 90° for 11 hours, when it becomes dark red. The reaction mixture is cooled and poured into 1.5 l. of petroleum ether and the initially milky orange mixture is washed with four 500-ml. portions of water, followed by two 300-ml. portions of 2 N hydrochloric acid and again two 500-ml. portions of water. The clear orange solution is dried with anhydrous magnesium sulfate, filtered, and evaporated, eventually at reduced pressure, giving crude adamantanethione (22.5 g.) suitable for reduction in 1,2-dimethoxyethane (170 ml.) with sodium borohydride (3.0 g.), added with vigorous stirring and cooling below 45° in 3 minutes. Excess borohydride is decomposed by the dropwise addition of 10 ml. of water, and then concentrated hydrochloric acid is added very carefully until a pH of 2 is reached. The mixture is extracted with four 200-ml. portions of carbon tetrachloride and the combined extract is washed, dried, and evaporated at 20–40° to a thick paste of solid. A solution of this in 150 ml. of 1,2-dimethoxyethane and 100 ml. of 95% ethanol is poured slowly into a stirred solution of 25.8 g. of lead acetate trihydrate in 250 ml. of 70% ethanol. The bright yellow lead thiolate which precipitates is collected after 3 hours and washed intimately with two 300-ml. portions of water and then with two 300-ml. portions of acetone. The air-dried, lump-free lead salt is suspended in 150 ml. of water in a 1-l. Erlenmeyer flask and 250 ml. of petroleum ether (b.p. 30–60°) is added. A gentle stream of hydrogen sulfide is bubbled through the aqueous layer for 1 hour, and the precipitated lead sulfide is collected on a Büchner funnel and washed with two 50-ml. portions of petroleum ether. The solution of thiol in petroleum ether is separated, washed with three 50-ml. portions of water, and dried over anhydrous magnesium sulfate. Removal of the solvent with a rotary evaporator affords 14.2-19.2 g. (74–85%) of 2-adamantanethiol, m.p. 151-158° (sealed capillary).

[1]M. P. Cava and F. M. Scheel, *J. Org.*, **32**, 1304 (1967)
[2]M. P. Cava and J. P. VanMeter, *ibid.*, **34**, 538 (1969)
[3]J. W. Greidanus (Univ. Calgary, Alberta, Canada), procedure submitted to *Org. Syn.* 1970
[4]H. W. Geluk and J. L. M. A. Schlatmann, *Tetrahedron*, **24**, 5361 (1968)

Phosphoryl chloride, 1, 876–882; 2, 330–331.

5'-Phosphorylation of ribofuranosyl nucleosides.[1] Japanese chemists found that phosphoryl chloride is an excellent catalyst for the conversion of ribonucleosides into the 2',3'-O-isopropylidene derivatives. These, on addition of pyridine, are esterified at the primary 5'-position. Aqueous hydrolysis then gives the nucleotide. The method can be used for a single-step conversion of ribonucleosides into ribonucleotides by using a combination of POCl$_3$, acetone, and pyridine without isolation of the isopropylidene derivative.

[1]J. Fujimoto and M. Naruse, *J. Pharm. Soc. Japan*, **87**, 270 (1967)

Phosphoryl chloride–N,N-Dimethylformamide.

Dehydration of amides to nitriles. Lawton and McRitchie (**1**, 287–288) introduced the use of thionyl chloride and DMF for dehydration of amides. Merck chemists[1] obtained improved yields by use of phosphoryl chloride in DMF. For

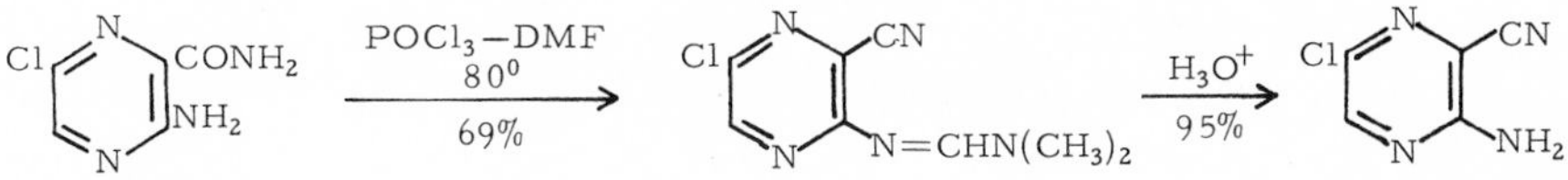

(1)   (2)   (3)

example, the carboxamide (1) was treated with the reagent (80°); dehydration occurs together with conversion of the amino group into a formamidine group (2). Hydrolysis gives the desired nitrile (3) in high overall yield.

[1]J. H. Jones and E. J. Cragoe, Jr., *J. Medicin. Chem.*, **11**, 322 (1968); *see also* A. Albert and K. Ohta, *Chem. Commun.*, 1168 (1969)

Phosphoryl chloride–Trimethyl phosphate. $POCl_3$–$(CH_3O)_3PO$.

Phosphorylation of nucleosides. Direct phosphorylation of unprotected nucleosides generally leads to a mixture of the three possible phosphates. However, treatment of 2′,3′-O-isopropylidene nucleosides with phosphoryl chloride gives 5′-phosphorodichloridates in fair yield. The reaction is greatly improved by use of trimethyl phosphate, $(CH_3O)_3PO$, or triethyl phosphate, in which the substrates are moderately soluble. After hydrolysis, 5′-nucleotides are obtained in nearly quantitative yield.[1] Adenine nucleosides, such as (1), are readily converted into the 3′,5′-cyclic monophosphate, such as (2), by this procedure.[2]

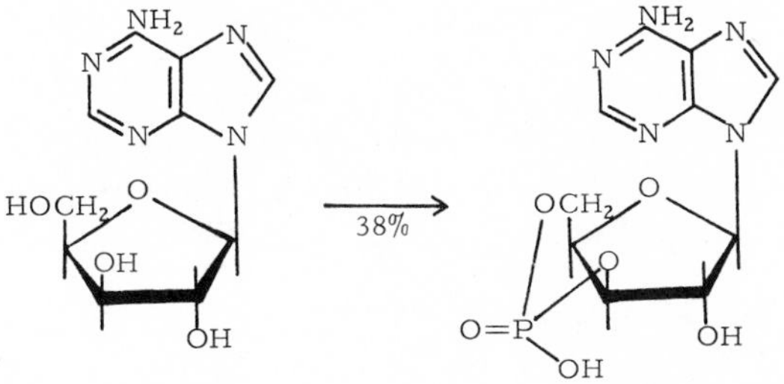

(1)   (2)

[1]M. Yoshikawa, T. Kato, and T. Takenishi, *Tetrahedron Letters*, 5065 (1967)
[2]M. Hubert-Habart and L. Goodman, *Chem. Commun.*, 740 (1969)

Polyphosphate ester (PPE), $C_8H_{20}O_{12}P_4$, **1**, 892–894; **2**, 333–334.

Preparation. Cava[1] recommends the following procedure. Phosphorus pentoxide (150 g.) is added to a solution of 300 ml. of anhydrous ether and 150 ml. of alcohol-free chloroform. The mixture is refluxed under nitrogen for 4 days; then the resulting solution is decanted from a small residue and concentrated to a sirup in a rotary evaporator. Final traces of solvent are removed by heating for 36 hours at 40° *in vacuo*.

Japanese investigators[2] have prepared a series of polyphosphate esters by the reaction of various alcohols with phosphorus pentoxide and have used them for alkylation of 2,3-disubstituted indoles to form 3-alkyl-3*H*-indoles.

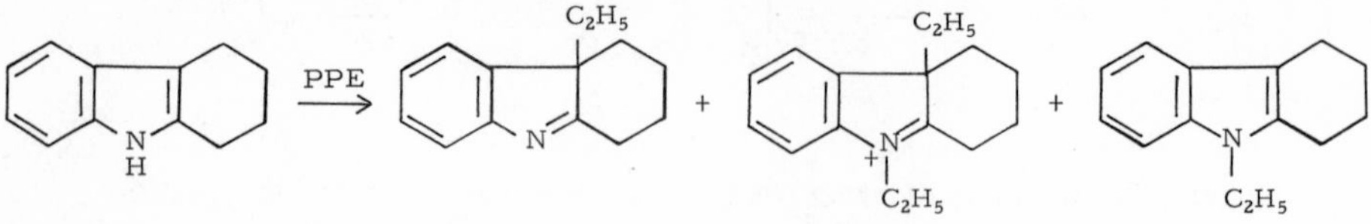

43%   4%   11%

Cyclodehydration. Cava[1] used PPE for cyclization of the amide (1) to the dihydroisoquinoline (2) in 79% yield. This reaction had been conducted previously in unspecified yield with phosphorus pentachloride.[3]

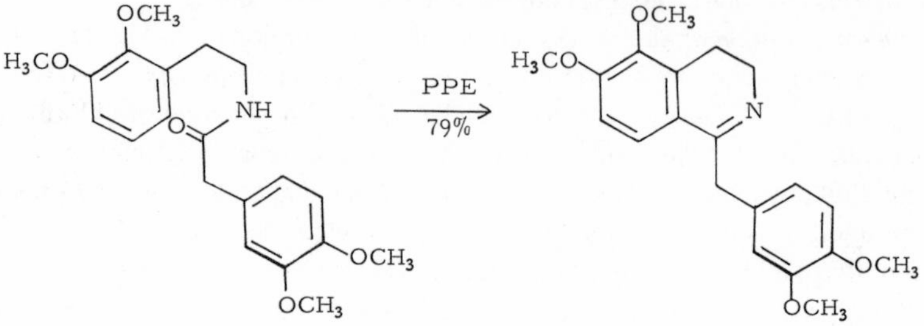

(1)   (2)

Beckmann rearrangement.[4] Beckmann rearrangement of the oxime of adamantanone (1) to the lactam (2) by conventional reagents led to yields of less than 25%; usually a considerable amount of (3) was formed as a result of a second-order Beckmann rearrangement. However, use of PPE in refluxing chloroform

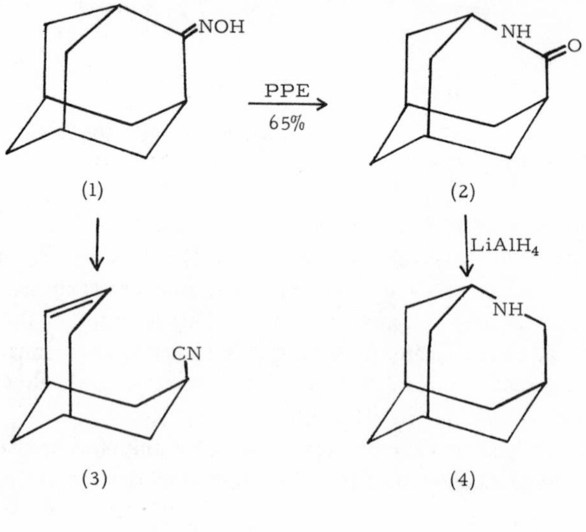

(1)   (2)

(3)   (4)

(5–6 minutes) gave the desired lactam in 65% yield (pure). The product was reduced by lithium aluminum hydride to 4-azahomoadamantane (4).

Fischer indole synthesis. PPE is useful for Fischer synthesis of indoles from phenylhydrazones; gentle refluxing for 5 minutes on a water bath generally completes the reaction.[5] Thus cyclohexanone phenylhydrazone (1) under these conditions gives 1,2,3,4-tetrahydrocarbazole (2) in 86% yield. When the reaction mixture was heated to 160°, ethylation occured as well to give 3-ethyl-2,3-tetramethyleneindolenine (3) in "fairly good yield." This alkylation reaction has

(1)   (2)   (3)

been applied to other 2,3-disubstituted indoles. Since PPE is known to decompose at 150–160° with formation of ethylene, this alkene presumably is involved in the alkylation reaction.[6]

Dehydration of amides. Nitriles are conveniently prepared by dehydration of amides with PPE.[7] The amide (1 part) and PPE (5 parts) are refluxed in chloroform for about two hours. Yields are in the range 35–90%.

Benzoxazoles. Benzoxazoles can be prepared in 45–75% yield by heating an *o*-aminophenol and a carboxylic acid with the reagent for 30 minutes at 100°.[8] An example is the synthesis of 2-phenylbenzoxazole (1). Use of *o*-aminothiophenols gives benzthiazoles.

(1)

[1]M. P. Cava, M. V. Lakshmikantham, and M. J. Mitchell, *J. Org.*, **34**, 2665 (1969)
[2]Y. Kanaoka, K. Miyashita, and O. Yonemitsu, *Tetrahedron*, **25**, 2757 (1969)
[3]A. Lindenmann, *Helv.*, **32**, 69 (1949)
[4]V. L. Narayanan and L. Setescak, *J. Heterocyclic Chem.*, **6**, 445 (1969)
[5]Y. Kanaoka, Y. Ban, K. Miyashita, K. Irie, and O. Yonemitsu, *Chem. Pharm. Bull. Japan*, **14**, 934 (1966)
[6]O. Yonemitsu, K. Miyashita, Y. Ban, and Y. Kanaoka, *Tetrahedron*, **25**, 95 (1969)
[7]Y. Kanaoka, T. Kuga, and K. Tanizawa, *Chem. Pharm. Bull. Japan*, **18**, 397 (1970)
[8]Y. Kanaoka, T. Hamada, and O. Yonemitsu, *ibid.*, **18**, 587 (1970)

Polyphosphoric acid, 1, 894–905; 2, 334–336.

Cyclization of acid chlorides. Aryl alkanecarbonyl chlorides, such as 4-phenyl-butyryl chloride (1) are cyclized by PPA to cyclic ketones.[1] The cyclization to

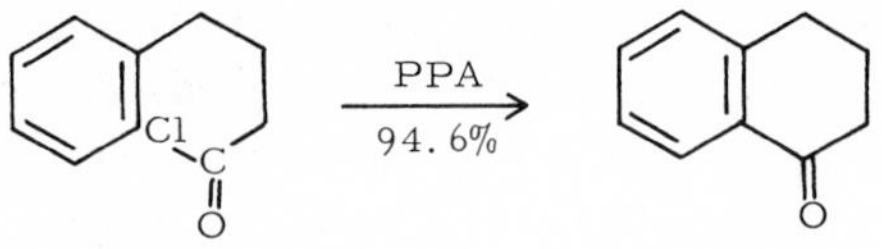

(1)   (2)

1-tetralone proceeds in 94.6% yield; but under the same conditions 1-indanone is obtained from 3-phenylpropionyl chloride in only 57.3% yield.

[1]A. Bhati and N. Kale, *Angew. Chem., internat. Ed.*, **6**, 1086 (1967)

Potassium amide, 1, 907–909; 2, 936.

Carboethoxylation.[1] Potassium amide is prepared as previously described[2] from 8 g. of potassium and 400 ml. of liquid ammonia in a 1-l. three-necked flask fitted with a dry ice–acetone condenser, a glass stirrer,[3] and glass stopper

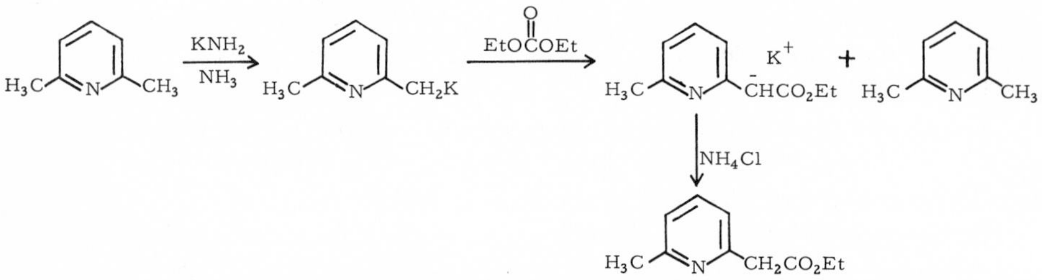

(*caution*[4]). The glass stopper is replaced by an addition funnel containing a solution of 21.4 g. (0.2 mole) of 2,6-lutidine in 20 ml. of ether. The lutidine is run into the amide solution and the funnel rinsed with a little ether. The resulting solution of potassiolutidine is stirred for 30 min., and 11.8 g. (0.1 mole) of freshly distilled diethyl carbonate (**1**, 247) is added rapidly; the color changes to green. After 5 min. the reaction mixture is neutralized by the addition of 10.7 g. (0.2 mole) of ammonium chloride. The green color is discharged and the final color is gray. The condenser is removed and the ammonia is allowed to evaporate; a steam bath or hot air gun can be used with care to speed up the evaporation. The residue is stirred with 500 ml. of ether and filtered, and the solid is stirred with a further 100 ml. of ether and filtered. The extracts are combined and evaporated on a rotary evaporator and the residue is distilled from a modified Claisen flask. Lutidine (10 g., 47%) is collected at 34–42° at 1.4 mm. and the ester (8.9 g., 50%) is collected at 87° at 0.7 mm. The ester is bright yellow liquid, n^{25}D 1.4995, d_4^{20} 1.0608.

[1]W. G. Kofron and L. M. Baclawski, *Org. Syn.* submitted 1970
[2]Vol. **1**, 907.
[3]A Teflon stirrer should not be used since Teflon is attacked by alkali metals, amides, and carbanions.

[4]Potassium is a silver-gray metal with a blue-violet cast, but if it shows an orange or red color or acquires an appreciable oxide coating it should be considered extremely hazardous. See J. F. Short, *Chem. Ind.*, 2132 (1964); D. P. Mellor, *ibid.*, 723 (1965); M. S. Bil, *ibid.*, 812 (1965).

Potassium *t*-butoxide, 1, 911–927; 2, 336–344.

Oxepin formation.[1] A 1-l. three-necked flask fitted with a stirrer and thermometer and connected by means of a section of Gooch rubber tubing to a 125-ml. Erlenmeyer flask containing 33.2 g. of potassium *t*-butoxide (MSA Research Corp.) is charged with 42.1 g. of 4,5-dibromo-1,2-dimethyl-1,2-epoxycyclohexane (1) in 500 ml. of ether and cooled to 0°. The potassium *t*-butoxide is added

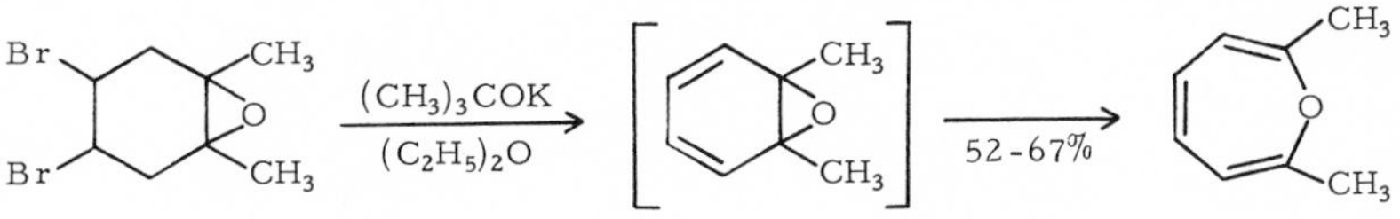

portionwise through the Gooch tubing over a period of 1 hour while maintaining the temperature below 5°. The resulting mixture is stirred for 30 minutes and filtered. The ether is removed under reduced pressure, and the residual liquid is distilled to give 2,7-dimethyloxepin as an orange oil, b.p. 49–50°.

Ring closure. Di-*t*-butyldiaziridinone (3), a new three-membered ring heterocycle, has been prepared from 1,3-di-*t*-butylurea (1) by chlorination (2, *t*-butyl hypochlorite) followed by ring closure with potassium *t*-butoxide in *t*-butanol.[2] The overall yield is 90%. The intermediate 1-chlorourea can be isolated, but improved yields are obtained without isolation. Use of potassium metal in pentane for the cyclization step lowered the yields to 48%. Apparently the two alkyl

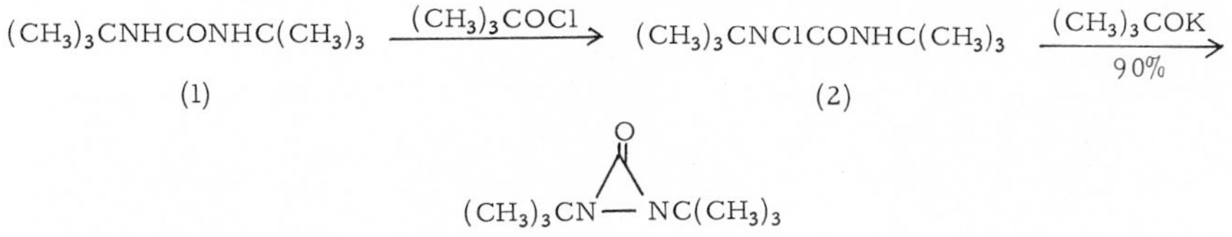

groups attached to N must be tertiary for the synthesis of be successful.

Isomerization of unsaturated compounds (1, 913–914; 2, 336). 1-Methylcyclopropene is isomerized to methylenecyclopropane in quantitative yield by treatment with a catalytic amount of potassium *t*-butoxide in DMSO for 2 hours at room temperature.[3]

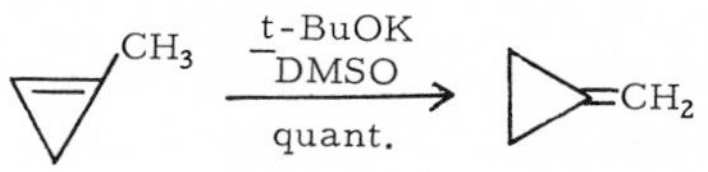

Dehydrotosylation. Dehydration of the terpene alcohol nopol (1a) was originally carried out with potassium hydroxide with a high-boiling solvent and found to yield a mixture of isomers.[4] The difficulty is overcome by converting the

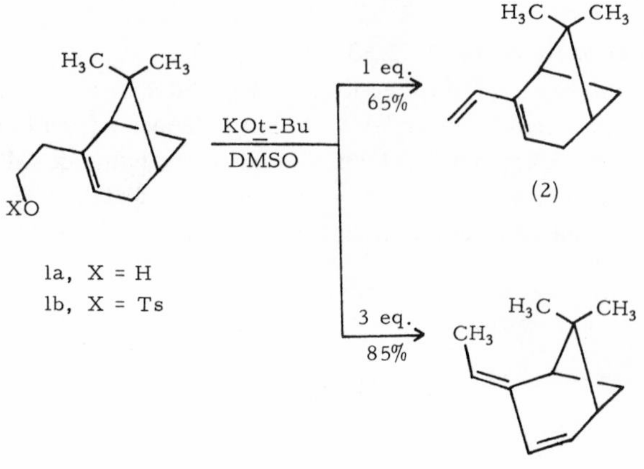

la, X = H
lb, X = Ts

alcohol into the tosylate (1b) and using potassium *t*-butoxide as base in DMSO as solvent. Use of 1 equivalent of base and a temperature of 75° gives nopodiene (2) in 65% yield. The more stable isomeric diene (3) is obtained if 3 equivalents of base are used (50°).

Olefin addition: Addition of dibromocarbene to *cis*-cyclooctene:[5]

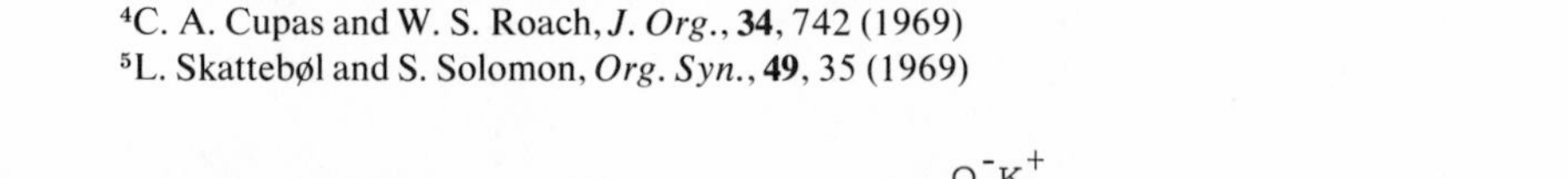

[1]L. A. Paquette and J. H. Barrett, *Org. Syn.*, **49**, 62 (1969)
[2]F. D. Greene, J. C. Stowell, and W. R. Bergmark, *J. Org.*, **34**, 2254 (1969)
[3]I. S. Krull and D. R. Arnold, *Org. Prep. Proc.*, **1**, 283 (1969)
[4]C. A. Cupas and W. S. Roach, *J. Org.*, **34**, 742 (1969)
[5]L. Skattebøl and S. Solomon, *Org. Syn.*, **49**, 35 (1969)

Potassium 2,6-di-*t*-butylphenoxide, $(CH_3)_3C$ $C(CH_3)_3$ Mol. wt. 244.42.

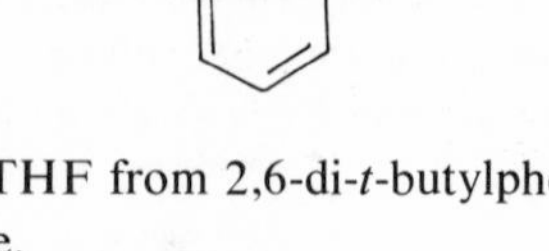

The base is prepared *in situ* in THF from 2,6-di-*t*-butylphenol (Aldrich, Eastman, etc.) and potassium *t*-butoxide.

Reactions of organoboranes. Brown *et al.* have found that trialkylboranes (**2**, 343)[1] and B-alkyl-9-borabicyclo[3.3.1]nonanes (**3**, 25)[2,3] react with α-halo ketones to give α-monoalkylated ketones in good yield. In the original work, potassium *t*-butoxide was used as base and THF as solvent.

$$\text{(cyclohexanone-}\alpha\text{-Br)} \xrightarrow[\text{KO-}\underline{t}\text{-Bu, THF}]{(C_2H_5)_3B} \text{(2-ethylcyclohexanone)} + \underline{t}\text{-BuOB}(C_2H_5)_2 + KBr$$

68%

$$(CH_3)_3CCOCH_2Br + \text{(B-cyclopentyl)} \xrightarrow{KO\,t\text{-Bu}} (CH_3)_3CCOCH_2\text{-(cyclopentyl)} + \underline{t}\text{-BuOB} \text{(bicyclic)} + KBr$$

Unfortunately the condensation failed with certain α-halo ketones, for example α-bromoacetone. Successful alkylation of this ketone would provide a useful route to methyl ketones. Brown et al.[4] then explored use of a number of other bases; the most satisfactory proved to be 2,6-di-*t*-butyl phenoxide. This is a weak base and α-haloketones are stable in its presence for a considerable length of time. The presence of the bulky alkyl substituents also appears to be highly favorable. Thus the following yields of *n*-butyrophenone were obtained from the reaction of phenacyl bromide and triethylborane with the bases indicated: potassium phenoxide (2%), potassium 2-methylphenoxide (29%), potassium 2,6-dimethylphenoxide (75%), and potassium 2,6-di-*t*-butylphenoxide (98%).

$$(C_2H_5)_3B + C_6H_5COCH_2Br + \text{(2,6-di-}t\text{-butylphenoxide K}^+) \xrightarrow{0°,\ THF}$$

$$C_6H_5COCH_2CH_2CH_3 + (C_2H_5)_2BOR + KBr$$

98%

The base was used successfully for alkylation of α-bromoacetone as indicated.

$$\underline{n}\text{-Bu}_3B + CH_2BrCOCH_3 \xrightarrow{84\%} \underline{n}\text{-BuCH}_2COCH_3$$

$$\text{(B-cyclopentyl)} + CH_2BrCOCH_3 \xrightarrow{73\%} \text{(cyclopentyl)}\text{-CH}_2COCH_3$$

Brown[5] then examined the use of the new base for the related alkylation of chloroacetonitrile, which had been unsuccessful with potassium *t*-butoxide. The reactions proceed in good yield (60–95%) and thus provide a convenient route to nitriles (with addition of two carbon atoms). The method is applicable both to trialkylboranes and to B-alkyl-9-borabicyclo[3.3.1]nonanes.

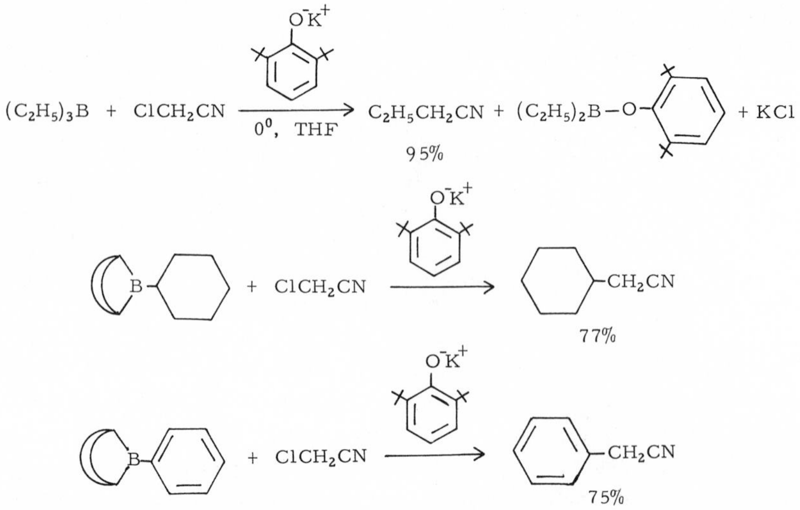

The new base is also superior to potassium *t*-butoxide for the reaction of organoboranes with ethyl bromoacetate (**2**, 192–193) and with ethyl dibromo-acetate (**2**, 195).[6] Yields are higher and it is not necessary to avoid an excess of the base. In the case of trialkylboranes the usual reaction conditions were used. For an unknown reason the reaction with B-alkyl-9-borabicyclo[3.3.1]nonanes in THF was unsuccessful. However, if the reaction is carried out in a solution roughly 50:50 in THF and *t*-butanol, satisfactory yields are obtained.

Use of the new base makes it possible to extend the reaction of organoboranes to ethyl 4-bromocrotonate and thus effect a four-carbon-atom homologation. As an example, the reaction of triethylborane with ethyl 4-bromocrotonate in the presence of one equivalent of the base affords ethyl 3-hexenoate (79% *trans*):

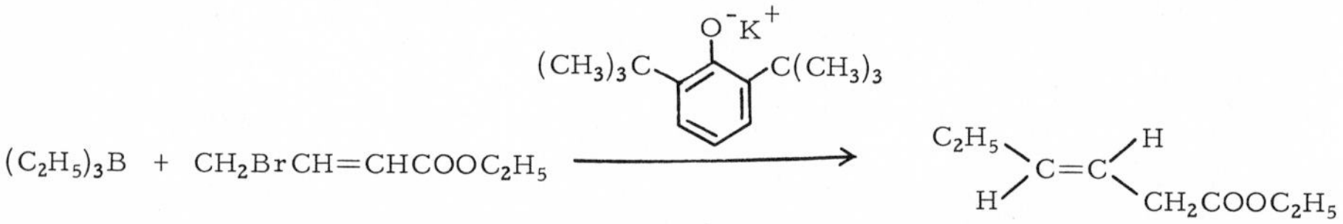

Unexpectedly the reaction is accompanied by migration of the double bond out of conjugation with the ester group, but this is actually a desirable feature since nonconjugated unsaturated esters are less readily available than conjugated ones. The apparent stereoselectivity of the migration is also noteworthy.[7]

This base has also been used successfully for mono- and dialkylation of dichloroacetonitrile.[8] Monoalkylation of dichloroacetonitrile with triethylborane is achieved in 89% yield by addition of potassium *t*-butoxide in THF to an equi-molar mixture of dichloroacetonitrile, triethylborane, and 2,6-di-*t*-butylphenol in the same solvent:

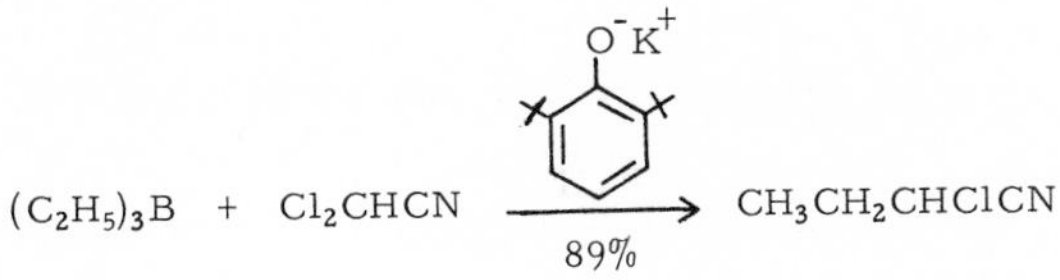

$$(C_2H_5)_3B \;+\; Cl_2CHCN \;\xrightarrow[89\%]{} \; CH_3CH_2CHClCN$$

Dialkylation is carried out by using 2 moles of base and 2 moles of the organoborane:

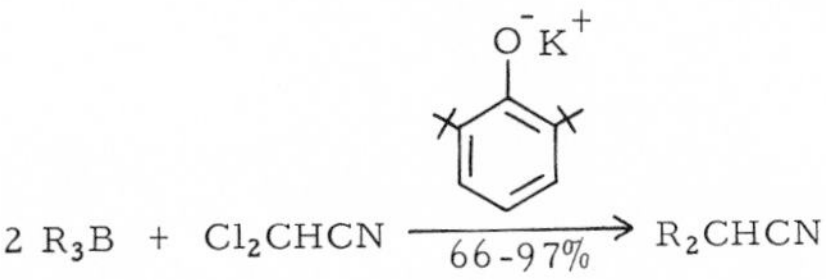

$$2\,R_3B \;+\; Cl_2CHCN \;\xrightarrow[66-97\%]{} \; R_2CHCN$$

Alternatively a dialkylacetonitrile with two different alkyl groups is readily obtained by two successive monoalkylations. The usual mild conditions are employed to introduce two primary alkyl groups; introduction of more hindered alkyl groups requires reflux in THF for extended periods of time. In practice it is not necessary to isolate the intermediate monoalkyl derivative. Since dialkylacetonitriles are readily hydrolyzed to carboxylic acids, this new procedure provides an alternative to the classical malonic ester synthesis.[9]

Nambu and Brown[10] were unable to effect the reaction of diethyl chloromalonate $(C_2H_5O_2CCHClCO_2C_2H_5)$ with triethylborane under the influence of this base; however, ethyl bromocyanoacetate and bromomalononitrile were successfully alkylated in high yield.

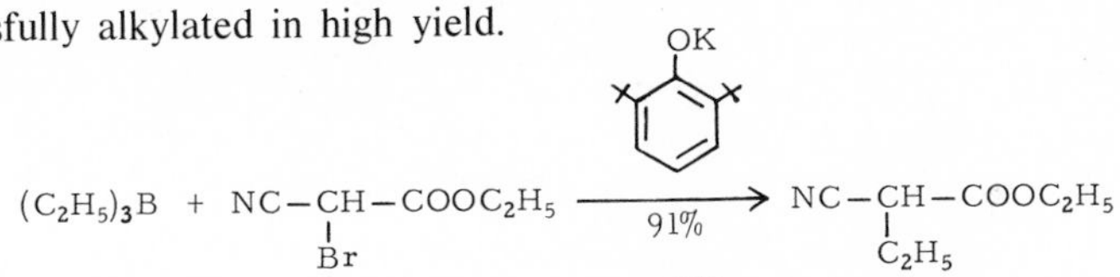

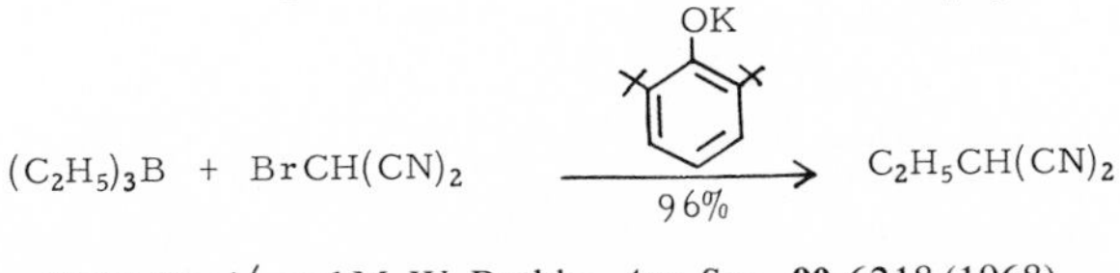

$$(C_2H_5)_3B \;+\; \underset{\underset{Br}{|}}{NC-CH}-COOC_2H_5 \;\xrightarrow[91\%]{} \; \underset{\underset{C_2H_5}{|}}{NC-CH}-COOC_2H_5$$

$$(C_2H_5)_3B \;+\; BrCH(CN)_2 \;\xrightarrow[96\%]{} \; C_2H_5CH(CN)_2$$

[1] H. C. Brown, M. M. Rogić, and M. W. Rathke, *Am. Soc.*, **90**, 6218 (1968)
[2] H. C. Brown, M. M. Rogić, H. Nambu, and M. W. Rathke, *ibid.*, **91**, 2147 (1969)

[3] 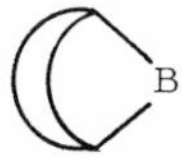 BH is used by Brown as a symbol for 9-borabicyclo[3.3.1]nonane.

[4] H. C. Brown, H. Nambu, and M. M. Rogic, *ibid.*, **91**, 6852 (1969)
[5] *Idem, ibid.*, **91**, 6854 (1969)
[6] *Idem, ibid.*, **91**, 6855 (1969)
[7] H. C. Brown and H. Nambu, *ibid.*, **92**, 1761 (1970)
[8] H. Nambu and H. C. Brown, *ibid.*, **92**, 5790 (1970)
[9] A. C. Cope, H. L. Holmes, and H. O. House, *Org. Reactions*, **9**, 107 (1957)
[10] H. Nambu and H. C. Brown, *Organometall. Chem. Syn.*, **1**, 95 (1970)

Potassium hydroxide, 1, 935–937; **2**, 346–347.

Reduction. Ketones are reduced to secondary alcohols when refluxed in a mixture of potassium hydroxide and ethylene glycol for 24 hours.[1] Thus benzophenone was reduced to benzhydrol in 92.7% yield. This unexpected reduction was discovered in an attempt to isomerize 3-*endo*-phenyl-2-norbornanone (1) to the 3-*exo*-isomer. Instead the three alcohols formulated were obtained. Actually this method is far superior to earlier methods for preparation of (3) and (4).

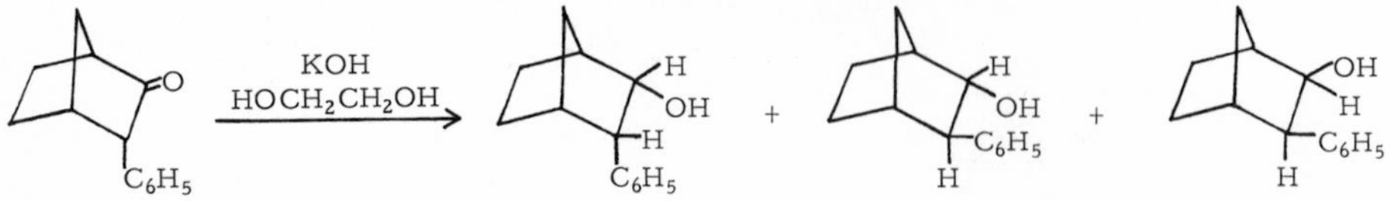

(1)

$$\xrightarrow[\text{HOCH}_2\text{CH}_2\text{OH}]{\text{KOH}}$$

(2, 21.9%)   (3, 43.2%)   (4, 4.6%)

[1]D. C. Kleinfelter, *J. Org.*, **32**, 840 (1967)

Potassium nitrosodisulfonate, 1, 940–942; **2**, 347–348.

Dehydrogenation of enediones. The adduct (1) of 2,3-dimethoxyquinone and 1,3-dimethoxycyclohexa-1,3-diene is converted into the quinone (2) by enolization with potassium hydrogen carbonate and oxidation with potassium nitrosodisulfonate. The ethano bridge is lost on pyrolysis at 150° to give the tetramethyl ether of spirochrome-B (3).[1]

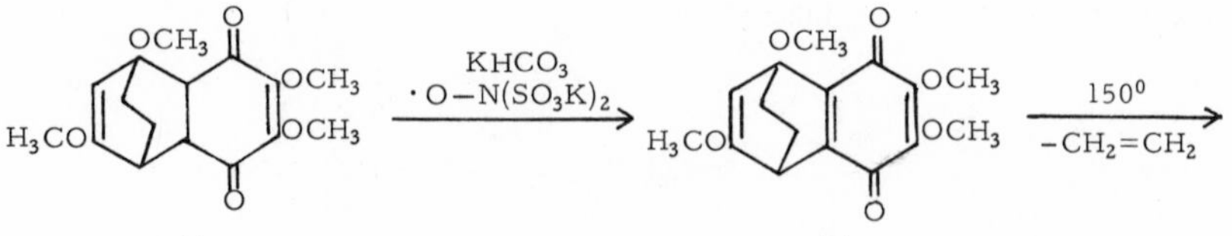

(1)   (2)

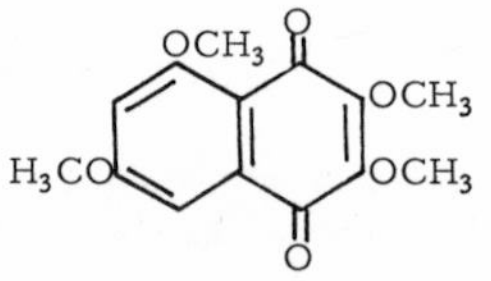

(3)

The oxidation is general and proceeds in yields of 80–95%.[2]

[1]A. J. Birch and V. H. Powell, *Tetrahedron Letters*, 3467 (1970)
[2]V. H. Powell, *ibid.*, 3463 (1970)

Potassium persulfate, ammonium persulfate, 1, 952–954; **2**, 348.

Oxidative decarboxylation. Fichter *et al.*[1] some time ago studied the decarboxylation of salts of carboxylic acids by persulfate ion. Kochi[2] now finds that the reaction is markedly catalyzed by silver(I) ion (silver trifluoroacetate, silver perchlorate). The major products are the alkane and carbon dioxide. However, addition of traces of copper sulfate leads to carbon dioxide and the alkene as major products. Kinetic studies indicate that a radical mechanism is involved.

[1]F. Fichter *et al.*, *Helv.*, **12**, 993 (1929); **15**, 996 (1932); **16**, 338 (1933)
[2]J. M. Anderson and J. K. Kochi, *Am. Soc.*, **92**, 1651 (1970)

Pyridine dichromate, $(C_5H_5NH)_2Cr_2O_7$. Mol. wt. 376.24, m.p. 145–148°.

Preparation.[1] The reagent is prepared *in situ* from sodium dichromate, $Na_2Cr_2O_7\cdot2\ H_2O$, aqueous hydrochloric acid, and an excess of pyridine. It can also be prepared from sodium dichromate, pyridine hydrochloride, and an excess of pyridine. The reagent can be isolated and stored if desired.

Oxidizing agent.[1] Pyridine dichromate is similar to the Sarett reagent (**1**, 145–146) but has the advantage that the preparation is relatively safe. It is used in the same way and yields are comparable.

[1]W. M. Coates and J. R. Corrigan, *Chem. Ind.*, 1594 (1969)

Pyridine hydrochloride, **1**, 964–966; **2**, 352–353.

Dehydration of epoxides. When the triterpenoid epoxide (1), or the isomeric α-oxide, is refluxed with pyridine hydrochloride in pyridine it is converted in high yield into the diene (2). Use of hydrochloric acid in ethanol is attended with re-

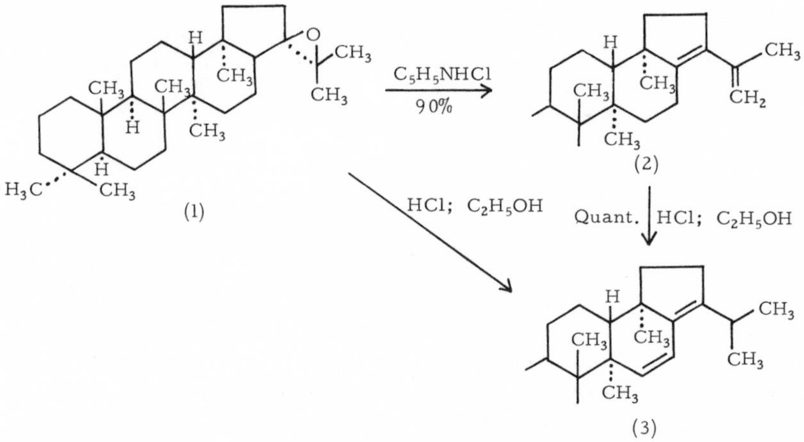

arrangement to give the rearranged diene (3). This diene is also obtained by the same treatment from the diene (2). Even more extensive rearrangements take place when boron trifluoride is used.[1]

N-Dealkylation. When the N-methyl- or N-ethylphenothiazine (1) is heated to melting with pyridine hydrochloride or hydrobromide, the N-alkyl group is cleaved to give (2) in high yield. The method does not appear to be general since N-alkylcarbazoles (3) are not affected by the reagent.[2]

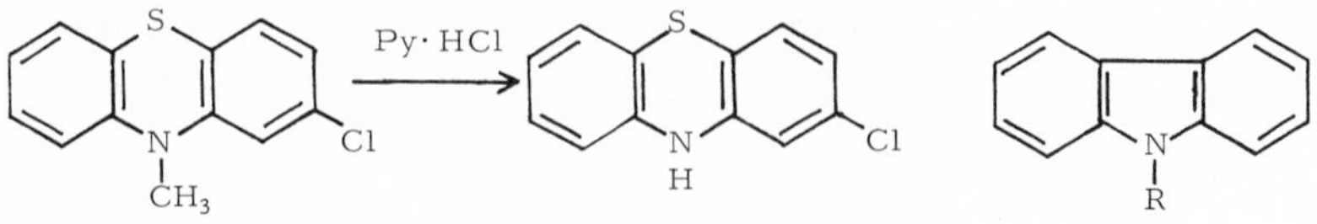

(1)   (2)   (3)

[1] I. Morelli and A. Marsili, *J. Org.*, **35**, 567 (1970)
[2] N. P. Buu-Hoi, G. Saint-Ruf, and B. Lobert, *Bull. soc.*, 1769 (1969)

Pyrrolidine, 1, 972–974; 2, 354–355.

Enamines. In a synthesis of 19β-norsteroids, Habermehl and Haaf[1] found that the 19β-hydroxymethyl group of (1) is eliminated as formaldehyde on formation of the enamine (2). Thus, when (1) is heated to reflux with pyrrolidine in methanol, the enamine (2) separates within minutes. The enamine is hydrolyzed to the Δ^4-3-ketone (3) by treatment with sodium acetate in acetic acid–methanol. Such an elimination had been observed previously by retroaldol

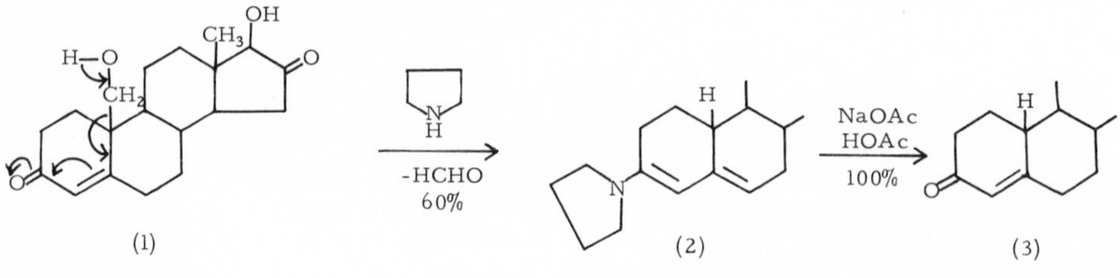

(1)   (2)   (3)

condensation of a steroid such as (1) with potassium hydroxide in methanol, but the yield by this procedure was very low.

[1] G. Habermehl and A. Haaf, *Ber.*, **102**, 186 (1969)

Pyrrolidone-2 hydrotribromide (PHT), $(C_4H_7NO)_3 \cdot HBr \cdot Br_2$. Mol. wt. 496.07, red crystals, m.p. 89–91°. Supplier: Aldrich.

Preparation. This stable complex is prepared by the reaction of pyrrolidone-2 with bromine in chloroform at 60°, or by the reaction of pyrrolidone-2, hydrogen bromide, and bromine in the molar ratios 3 : 1 : 1 in methanol.[1]

Bromination of ketones.[2] The complex is superior to phenyltrimethylammonium perbromide (**1**, 855–856; **2**, 328) for selective bromination of a ketone in the presence of a double bond. Thus benzalacetone (1) was converted smoothly by the reagent in THF into bromomethylstyryl ketone (2). Selectivity is enhanced upon dilution and by acid catalysis (H_2SO_4).

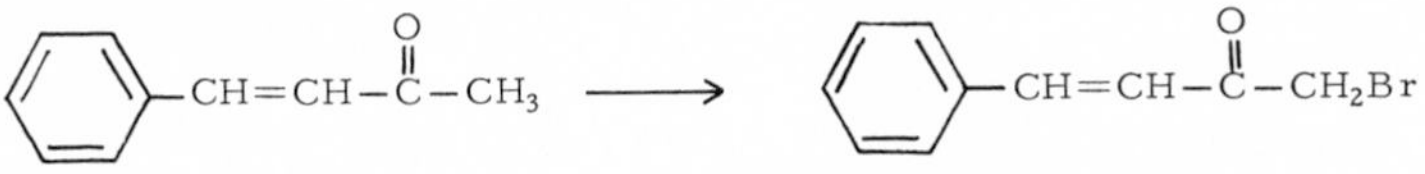

(1)   (2)

[1]W. E. Daniels, M. E. Chiddix, and S. A. Glickman, *J. Org.*, **28**, 573 (1963)
[2]D. V. C. Awang and S. Wolfe, *Canad. J. Chem.*, **47**, 706 (1969)

R

Rhodium trichloride hydrate, $RhCl_3 \cdot 3\, H_2O$, **2**, 357.

Olefin isomerization. A French patent[1] noted that myrcene (1) can be iso-merized to ocimene (2) by certain group VII metal salts including rhodium trichloride. The observation prompted Swiss chemists[2] to examine the isomeriza-

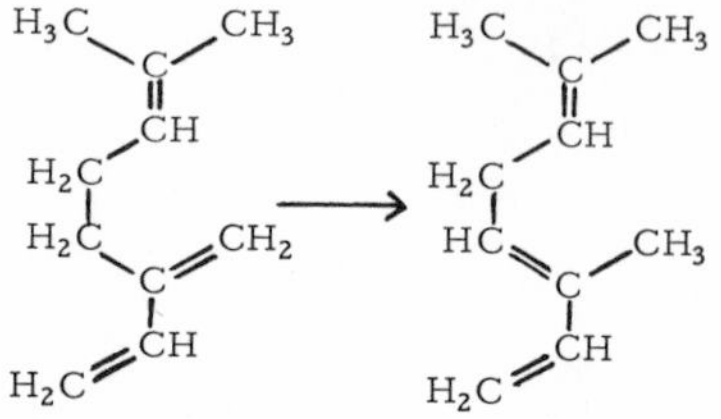

(1)   (2)

tion of the all-*trans*-tetraene ester (3) as a possible route to α-sinensal (5). Treat-ment of (3) with a catalytic quantity of rhodium trichloride hydrate in ethanol at 70° for 70 min. gave a product (62% yield) consisting of two components in the ratio 7:3. The nmr indicated that the major component was the desired all-*trans*-tetraene ester (4) and that the minor component was the Δ^9-*cis*-tetraene isomer. They then took advantage of the finding of Pettit[3] that reaction of *cis*-1,3-penta-diene with iron pentacarbonyl leads to *trans*-1,3-pentadiene–iron tricarbonyl.

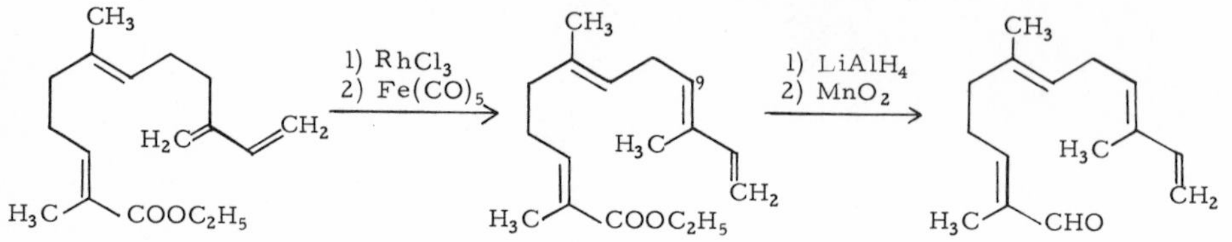

(3)   (4)   (5)

Thus they treated the *cis/trans* mixture obtained by isomerization with iron penta-carbonyl and after decomposition of the complex with ferric chloride obtained the desired all-*trans*-tetraene ester (4) in 51% yield. The remaining steps in the synthesis involved lithium aluminum hydride reduction (80% yield) and MnO_2 oxidation (52% yield).

This isomerization reaction was also found to provide the most useful route, short of total synthesis, to the relatively rare *trans,trans*-α-farnesene (7).[4] Rhod-ium trichloride catalyzed isomerization of the relatively abundant *trans*-β-farnesene (6) gave the desired α-farnesene (7) in 57% yield. Products having a more extensive system of conjugation were also formed to a minor extent.

(6)   (7)

[1]M. S. Lemberg, French Pat. 1,456,900 (19.9.1966)
[2]E. Bertele and P. Schudel, *Helv.*, **50**, 2445 (1967)
[3]G. F. Emerson, J. E. Mahler, R. Kochhar, and R. Pettit, *J. Org.*, **29**, 3620 (1964)
[4]G. Brieger, T. J. Nestrick, and C. McKenna, *ibid.*, **34**, 3789 (1969)

Ruthenium tetroxide, 1, 986–989; **2,** 357–359.

Preparation. The reagent is conveniently prepared by using a catalytic amount of RuO_2 in combination with sodium metaperiodate (**1,** 986). It has also been prepared in the same way from ruthenium trichloride (Alfa Inorganics, Research Organic/Inorganic Chemical Corp).[1] In a recent procedure[2] ruthenium tetroxide is generated by oxidation of ruthenium trichloride with ordinary household bleach (5.25% aqueous solution of sodium hypochlorite). In a typical procedure a solution of cyclohexanol (10 mmole) in water containing 0.5 ml. of 2% aqueous $RuCl_3$ (black) is titrated at 0° with 1.51 N sodium hypochlorite until a yellow end point is reached. Cyclohexanone was isolated as the 2,4-dinitrophenylhydrazone in 90–95% yield.

Oxidation of carbohydrates. Canadian chemists[3] have published an improved procedure for oxidation of carbohydrate derivatives with ruthenium tetroxide. The reagent is generated *in situ* from a catalytic amount of ruthenium dioxide by potassium periodate (this is less soluble than sodium periodate, and hence overoxidation to lactones is minimized) in aqueous chloroform with control of the pH by addition of potassium carbonate. Yields of 83–95% were reported for five oxidations. The method was used successfully for oxidation of (1) to (2) in 84% yield.[4]

(1)   (2)

Oxidation of secondary alcohols in a two-phase system. Moriarty *et al.*[5] report that the oxidation of the hydroxylactone (1) to the ketone (2) is exceedingly difficult; in fact, fifteen standard procedures failed, possibly owing to the acid

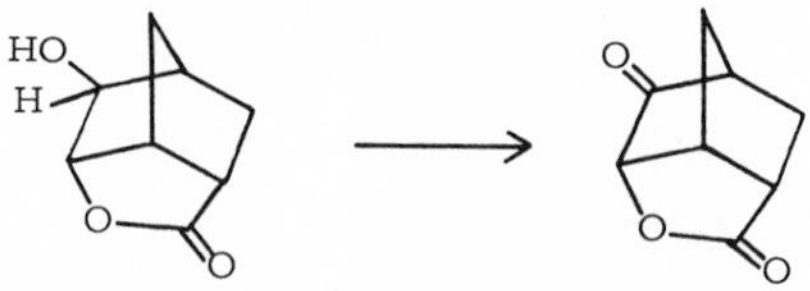

(1)   (2)

sensitivity of (2). They achieved the desired oxidation in 80% yield by using RuO_2–$NaIO_4$ in a neutral two-phase system, chloroform–water. The hydroxy-lactone is dissolved in water, and a suspension of ruthenium dioxide in chloro-form is added. Then an aqueous solution of sodium periodate is added dropwise with vigorous stirring until the yellow color of ruthenium tetroxide persists. The excess oxidant is destroyed, and the ketolactone (2) isolated from the chloroform layer. The reaction was applied successfully to related hydroxylactones.

One limitation is the concurrent oxidation of α-methylene groups as shown in the following example:

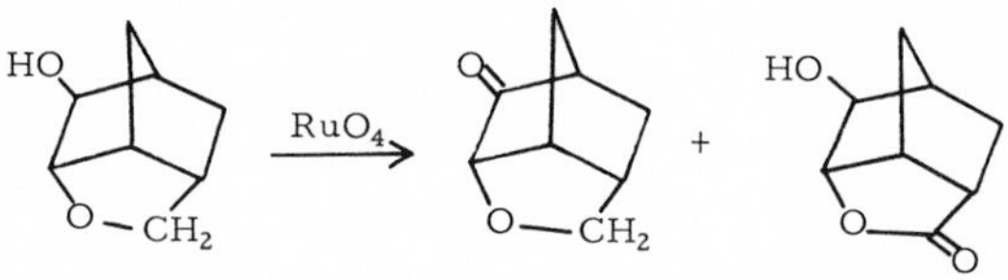

(3)   (4, 30%)   (5, 40%)

The same paper also reports successful oxidation of lactones to keto acids, for example (6) → (7). In this case the lactone is first hydrolyzed with one molar

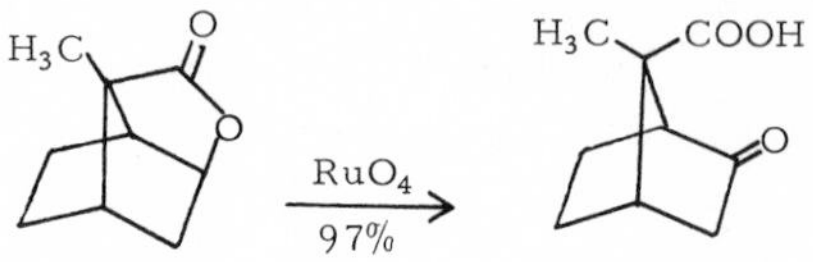

(6)   (7)

equivalent of aqueous base. Then a 0.02 molar equivalent of ruthenium dioxide is added, followed, with stirring, by an aqueous solution of one molar equivalent of sodium periodate.

[1]J. A. Caputo and R. Fuchs, *Tetrahedron Letters*, 4729 (1967)
[2]S. Wolfe, S. K. Hasan, and J. R. Campbell, *Chem. Commun.*, 1420 (1970)
[3]B. T. Lawton, W. A. Szarek, and J. K. N. Jones, *Carbohydrate Res.*, **10**, 456 (1969)
[4]E. H. Williams, W. A. Szarek, and J. K. N. Jones, *Canad. J. Chem.*, **47**, 4467 (1969)
[5]R. M. Moriarty, H. Gopal, and T. Adams, *Tetrahedron Letters*, 4003 (1970)

S

Salcomine, 2, 360.

Autoxidation of 2,6-disubstituted phenols. General Electric chemists[1] found that salcomine monopyridinate is superior to salcomine itself for autoxidation of 2,6-disubstituted phenols to quinones. This complex, red crystals, $C_{21}H_{18}CoN_2O_2$, mol. wt. 404.33, has been prepared in 90% yield by the reaction of 3-ethoxy-salicylaldehyde, ethylenediamine, pyridine, and cobaltous acetate.[2] Dipheno-quinones are formed to a small extent. These become more important products if the O_2-bridged salcomine dimer is used in low concentration and at high temperature.

It is noteworthy that related complexes containing a metal other than cobalt have not been found to be reversible oxygen carriers.

Japanese chemists[3] have used the related cobalt catalyst (di-(3-salicylidene-aminopropyl)amine cobalt(II) (1) for catalytic autoxidation of 4-alkyl-2,6-di-*t*-butylphenols (2) to the corresponding *p*-quinols (4) in good yield. Hydroperoxides (3) are formed as intermediates but during workup are reduced to quinols (4).

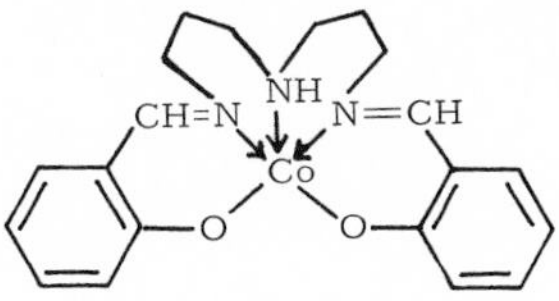

(1)

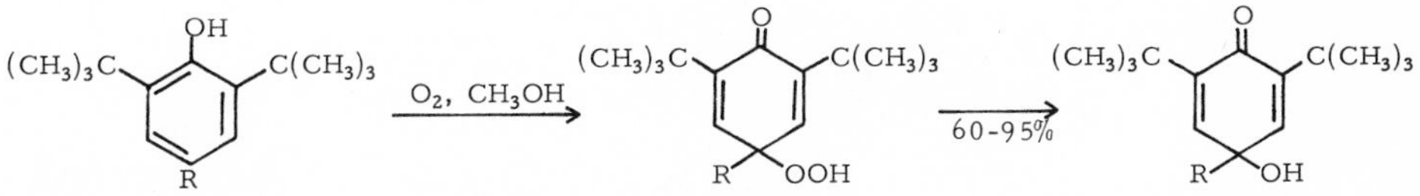

(2)   (3)   (4)

Cobalt catalyst (1) differs from salcomine in that it forms a 1:1 complex with oxygen.

[1]L. H. Vogt, Jr., J. G. Wirth, and H. L. Finkbeiner, *J. Org.,* **34,** 273 (1969)
[2]R. H. Bailes and M. Calvin, *Am. Soc.,* **69,** 1886 (1947)
[3]T. Matsuura, K. Watanabe, and A. Nishinaga, *Chem. Commun.,* 163 (1970)

Selenium dioxide, SeO_2, **1,** 992–1000; **2,** 360–362.

Acyloin oxidation. The final step in the synthesis of 1,3-dioxapane-5,6-dione (2) involved oxidation of the acyloin (1). This proved to be difficult since protic

(1)   (2)

solvents could not be used because they would add to the carbonyl groups of (2). Cupric acetate, bismuth oxide, manganese dioxide, lead dioxide, lead tetraacetate, chromic acid, dinitrogen tetroxide, and DMSO–Ac$_2$O all failed to give an isolable product. However, selenium dioxide in refluxing toluene containing a little acetic acid gave (2) in 41% yield.[1] There is little oxidation in the absence of acetic acid.

Allylic hydroxylation. Contrary to the rules of Guillemonat (**1**, 994, ref. 11) the major product of oxidation (70% yield) of carvone (1) is the tertiary alcohol (2); the expected product (3) is formed in only 10% yield.[2]

(1)   (2, 70%)   (3, 10%)

Acetoxylation of olefins. The oxidation of olefins with selenium dioxide in acetic acid results in allylic oxidation (**1**, 994–996; **2**, 360–361); however, if the reaction is catalyzed by sulfuric acid, the main product results from acetoxylation. Thus oxidation of cyclohexene under these conditions (110°, autoclave) gives 1,2-cyclohexanediol diacetate as a mixture of *cis-* (55%) and *trans-* (45%) isomers in 32% yield. Similarly, oxidation of 1-hexene gives 1,2-hexanediol

cis:trans = 55%:45%

diacetate (35% yield), 3-acetoxy-1-hexene (12% yield), and 1-acetoxy-3-hexene (5% yield).[3]

Oxidative cleavage of allyl and propargyl ethers. Hydroxyl and phenolic groups have been protected as the allyl ethers. Hydrolytic cleavage of protective group involved isomerization to the propenyl ether by potassium *t*-butoxide in DMSO followed by hydrolysis (**1**, 300; **2**, 158). Japanese chemists[4] now find that allyl ethers can be cleaved in one step by oxidation with a slight excess of selenium dioxide in acetic acid–dioxane (reflux, 1 hr.). Presumably a hemiacetal of acrolein

$$ROCH_2-CH=CH_2 \xrightarrow{SeO_2} [ROCHCH=CH_2] \longrightarrow ROH + CH_2=CHCHO$$
$$\qquad\qquad\qquad\qquad\qquad\quad \underset{OH}{|}$$

is an intermediate. Yields are about 50%. Aryl propargyl ethers are cleaved similarly in somewhat higher yield.

$$ArOCH_2C\equiv CH \xrightarrow{\text{SeO}_2} ArOH + CH\equiv CH-CHO$$

[1]M. W. Miller, *Tetrahedron Letters*, 2545 (1969)
[2]G. Büchi and H. Wüest, *J. Org.*, **34**, 857 (1969)
[3]K. A. Javaid, N. Sonoda, and S. Tsutsumi, *Tetrahedron Letters*, 4439 (1969); *Bull. Chem. Soc. Japan*, **42**, 2056 (1969)
[4]K. Kariyone and H. Yazawa, *Tetrahedron Letters*, 2885 (1970)

Silver carbonate–Celite, 2, 363.

Oxidation of diols. Primary 1,4-, 1,5-, and 1,6-diols are oxidized to the corresponding lactones in high yield.[1]

Examples:

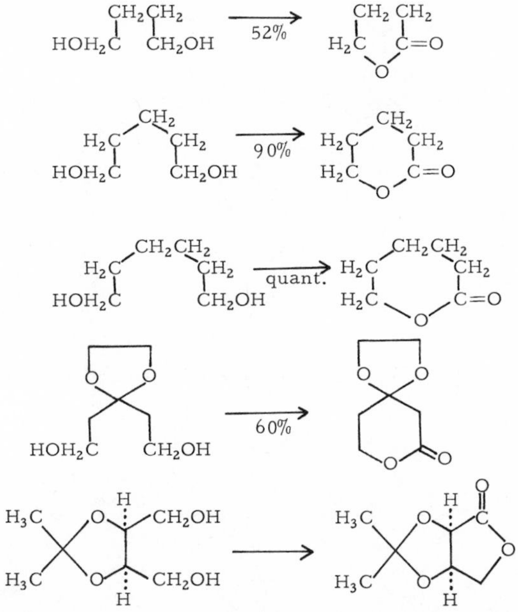

The last two examples show that acid-sensitive functional groups can be present. The reaction was used in a convenient synthesis of (±)-mevalonolactone (3). The Grignard reaction of ethyl acetate and allylmagnesium bromide gives the tertiary alcohol (1). This was ozonized and the crude ozonide reduced with

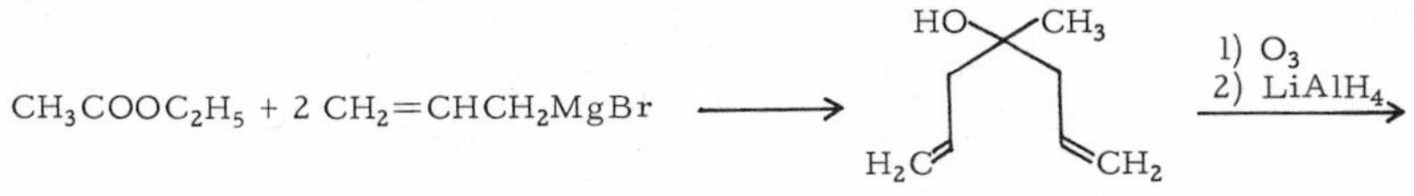

(1)

lithium aluminum hydride in THF to give the triol (2). Oxidation with silver carbonate–Celite gave (±)-mevalonolactone (3).

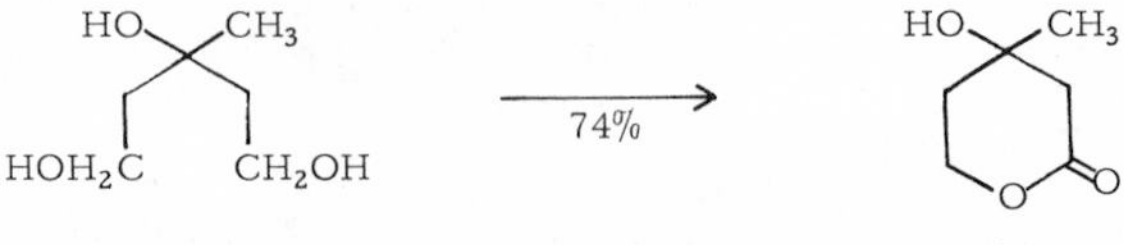

(2)   (3)

The oxidation of α-, β-, and γ-disecondary diols leads to hydroxy ketones, unless forcing conditions are used (large excess silver carbonate, many hours reflux); 1,7- and 1,8-diols also give hydroxy ketones.[2]

Examples:

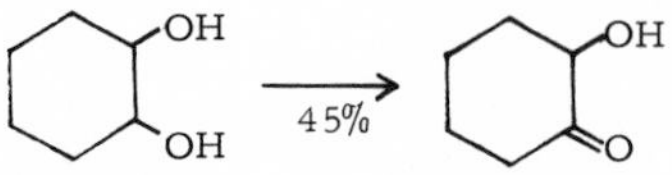

$CH_3CH(OH)CH_2CH_2OH \longrightarrow CH_3COCH_2CH_2OH$
                        80%

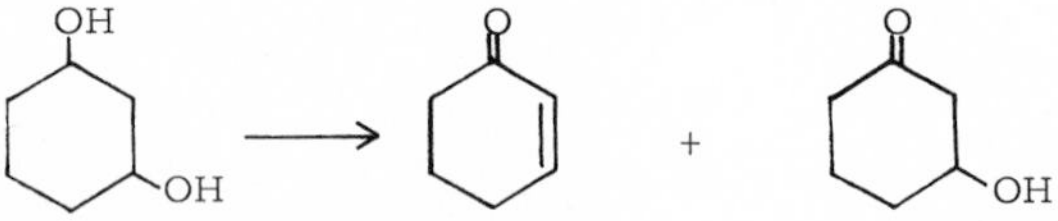

       24%   58%

$HOCH_2CH_2CHOHCH_2CH_2OH \longrightarrow HOCH_2CH_2COCH_2CH_2OH$
                            43%

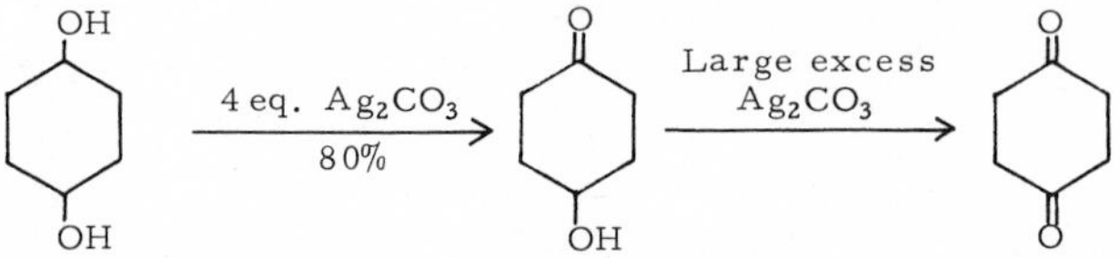

Selective oxidation of a secondary allylic hydroxyl group. D-Glucol (1) is selectively oxidized to the enone (2) in 60–80% yield by the reagent.[3] This oxidation had been effected previously in 1.5% yield by catalytic air oxidation.[4]

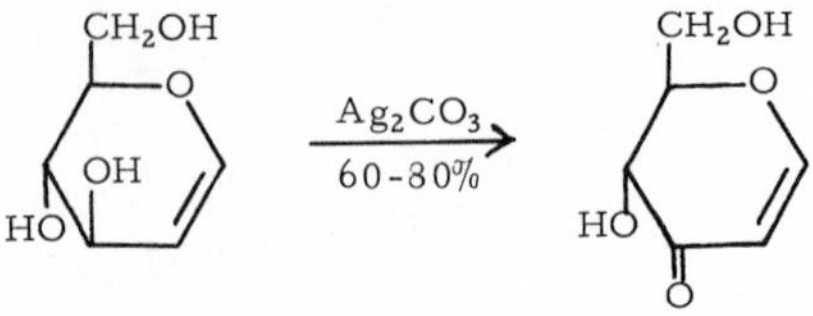

(1)   (2)

Oxidative coupling of phenols.[5] Silver carbonate precipitated on Celite is an excellent reagent for oxidative coupling of phenols; thus 2,6-dimethylphenol (1) gives the *p*-diphenoquinone (2) in 98% yield.

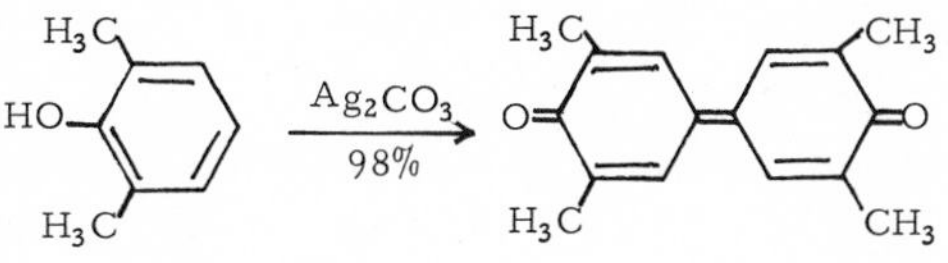

(1)   (2)

2,4,6-Trimethylphenol (3) is oxidized to the stilbenequinone (4) in 93% yield.

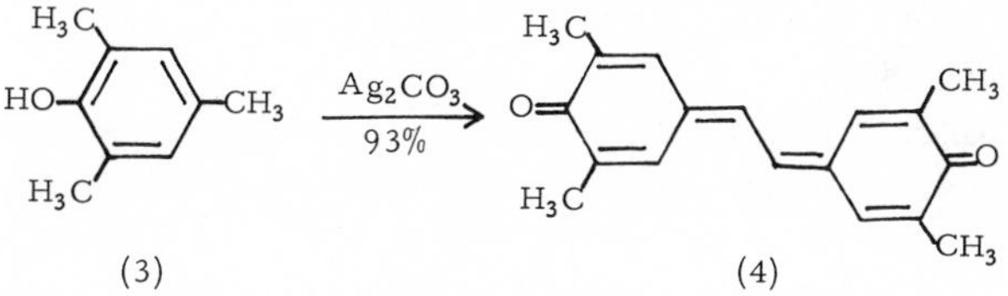

(3)   (4)

2,4,6-*t*-Tributylphenol (5) is oxidized to the deep blue radical (6), which affords the peroxide (7) in the presence of oxygen in 90% yield.

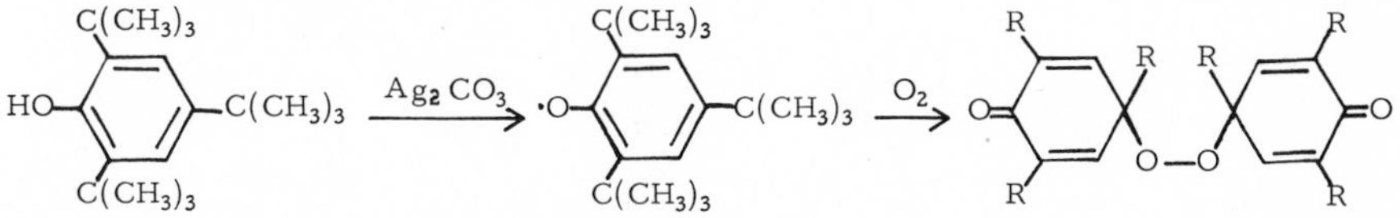

(5)   (6)   (7) R = C(CH₃)₃

[1]M. Fétizon, M. Golfier, and J.-M. Louis, *Chem. Commun.*, 1118 (1969)
[2]*Idem, ibid.*, 1102 (1969)
[3]J. M. J. Tronchet, J. Tronchet, and A. Birkhäuser, *Helv.*, **53**, 1489 (1970)
[4]K. Heyns and H. Gottschalck, *Ber.*, **99**, 3718 (1966)
[5]V. Balogh, M. Fétizon, and M. Golfier, *Angew. Chem., internat. Ed.*, **8**, 444 (1969)

Silver difluorochloroacetate, $ClF_2CCOOAg$. Mol. wt. 237.36. Supplier: K and K Laboratories.

—*Br* → —*Cl*. Chloro compounds are readily converted into bromo compounds, but the reverse reaction is difficult. Vida[1] has observed transhalogenation of bromo compounds with silver difluorochloroacetate. Thus the bisbromomethyl compound (1) is converted into (2) in 70% yield by reaction with 2

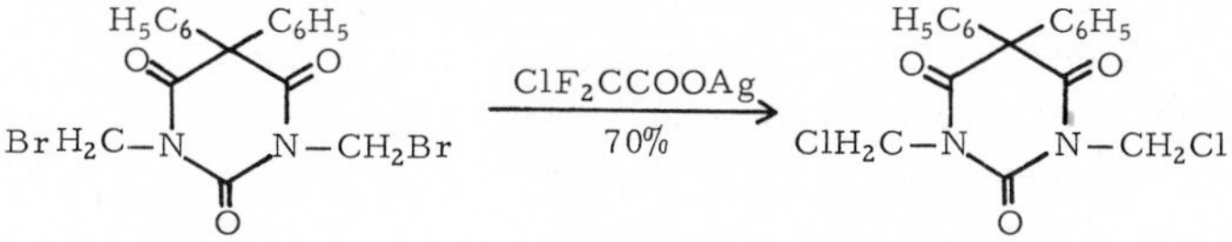

(1)   (2)

equivalents of silver difluorochloroacetate in refluxing acetonitrile. The reaction is general and yields are in the range 70–95%.

[1]J. A. Vida, *Tetrahedron Letters*, 3447 (1970)

Silver fluoroborate (silver tetrafluoroborate), 1,1015–1018; **2,** 365–366. Additional suppliers: Alfa Inorganics, Research Organic/Inorganic Chem. Corp.

Preparation. Meerwein[1] prepared the reagent by the reaction of silver oxide and boron trifluoride etherate in nitromethane and removal of a residue containing silver metaborate and unchanged silver oxide. Lemal and Fry[2] report that on one occasion this residue detonated with terrific brisance. They strongly recommend preparation by the method of Olah[3] in which argentous fluoride is employed instead of the oxide.

Rearrangements of strained σ bonds. Silver(I) ion appears to be a unique catalyst for skeletal rearrangements of highly strained σ bonds by paths which are ordinarily forbidden by orbital symmetry restrictions.[4] Thus treatment of homo-cubane(1) in dilute $CDCl_3$ or acetone-d_6 solution with catalytic amounts of silver fluoroborate results in quantitative conversion into (2), pentacyclo[4.3.0.0^{2,4}.0^{3,8}.0^{5,7}]nonane, within 1 day at 25°. In the absence of silver ion, (1) is stable to 240°.[5]

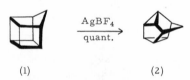

(1)   (2)

Similarly, the pentacyclic diester (3) is transformed into (4). This rearrangement has also been achieved independently by Dauben *et al.*[6] by elution of (3) through

(3)   (4)

a silica gel column impregnated with silver nitrate (see **Silver nitrate,** This volume).

The rearrangement has been extended to a convenient synthesis of the rather elusive semibullvalene (8).[7] Thus silver ion catalyzed rearrangement of (5)[8] gives (6) in 80% yield. The remaining steps involve hydrolysis with excess potassium hydroxide in aqueous ethylene glycol to give the azo compound (7), which on mild air oxidation loses nitrogen to give semibullvalene (8) as the only volatile product.

A related rearrangement is the isomerization of cubane (9) to cuneane (10),[9] promoted by silver perchlorate. The name cuneane, from the Latin *cuneus,* a wedge, is suggested for simplicity. The systematic name is pentacyclo[3.3.0.0^{2,4}.

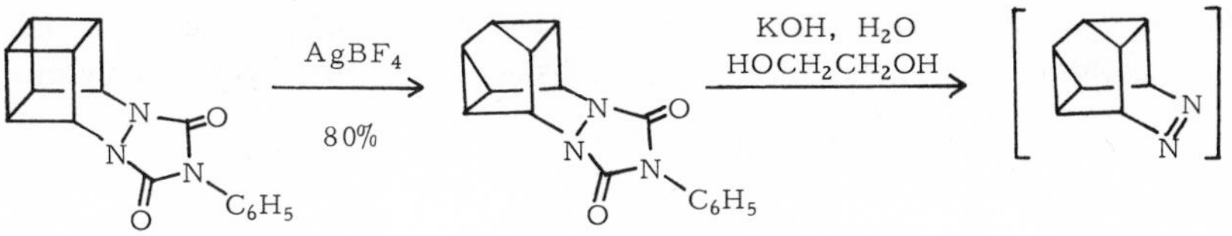

(5)   (6)   (7)

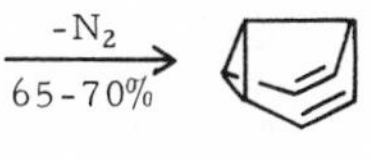

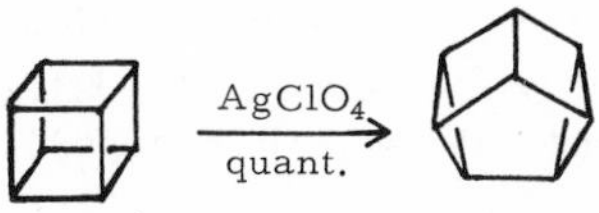

(8)   (9)   (10)

$0^{3,7}.0^{6,8}$]octane. Palladium(II) also catalyzes the same rearrangement. On the other hand, rhodium(I) complexes effect isomerization of cubane (9) to *syn*-tricyclooctadiene (11)[10] and of cuneane (10) to semibullvalene (8).

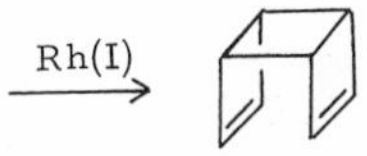

(11)

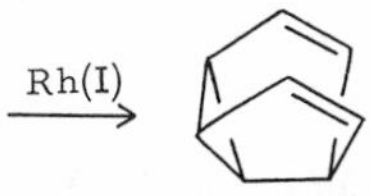

(8)

Glycosyl fluorides. Glycosyl chlorides react with the reagent in anhydrous ether at 0° to give anomeric mixtures of glycosyl fluorides in 95–100% yields.[11]

[1]H. Meerwein, V. Hederich, and K. Wunderlich, *Arch. Pharm.*, **291**, 541 (1958)
[2]D. M. Lemal and A. J. Fry, *Tetrahedron Letters*, 775 (1961)
[3]G. A. Olah and H. W. Quinn, *J. Inorg. Nucl. Chem.*, **14**, 295 (1960)
[4]R. B. Woodward and R. Hoffmann, *Angew. Chem., internat. Ed.*, **8**, 781 (1969)
[5]L. A. Paquette and J. C. Stowell, *Am. Soc.*, **92**, 2584 (1970)
[6]W. G. Dauben, M. G. Buzzolini, C. H. Schallhorn, D. L. Whalen, and K. J. Palmer, *Tetrahedron Letters*, 787 (1970)
[7]L. A. Paquette, *Am. Soc.*, **92**, 5765 (1970)
[8]Available from cyclooctatetraene.
[9]L. Cassar, P. E. Eaton, and J. Halpern, *Am. Soc.*, **92**, 6366 (1970)
[10]Idem, *ibid.*, **92**, 3515 (1970)
[11]K. Igarashi, T. Honma, and J. Irisawa, *Carbohydrate Res.*, **13**, 49 (1970)

Silver iododibenzoate, 1, 1007–1008.

Cleavage of α-glycols. Raman[1] noted that the combination of silver benzoate and iodine cleaves α-glycols to give carbonyl compounds. Lüning and Paulsson[2] used this combination for cleavage of DL-pentadecane-1,2,15-triol (1) to tetra-decanal-14-ol (2). The triol (0.026 mole) was added under anhydrous conditions

$$HO(CH_2)_{13}CHOHCH_2OH \xrightarrow[60\%]{} HO(CH_2)_{13}CHO$$

$$(1) \qquad\qquad\qquad (2)$$

to silver benzoate (0.053 mole) and iodine (0.12 mole) and benzene, and the reaction mixture was refluxed for 4 hours. In this case periodate, periodic acid, and lead tetraacetate gave only polymeric gums.

[1]P. S. Raman, *Proc. Indian Acad. Sci.*, **44A**, 321 (1956) [C.A., **51**, 8010f (1957)]
[2]B. Lüning and L. Paulsson, *Chem. Scand.*, **21**, 829 (1967)

Silver nitrate, 1, 1008–1011; **2**, 366–368.

Isomerization. The 1,1'-bishomocubane derivative (1) is rearranged to (2), dimethyl *cis*-pentacyclo[4.4.0.0^{2,10}.0^{3,5}.0^{4,9}]decane-7,8-dicarboxylate, in quantitative yield by stirring with an aqueous methanolic solution of silver nitrate.[1] This transformation had been carried out unknowingly by French chemists[2] by

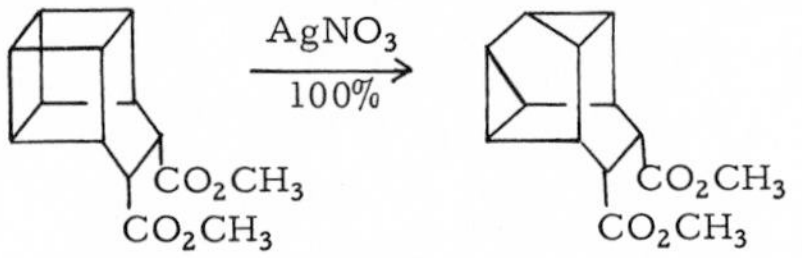

$$(1) \qquad\qquad\qquad (2)$$

silver nitrate silica gel chromatography of (1). The transformation of (1) into (2) can also be carried out by thermolysis at 235° under one atmosphere of nitrogen; yield 70–100%. The structure of (2) was deduced mainly from nmr data. Silver(I) ion is apparently specific for this isomerization.

[1]W. G. Dauben, M. G. Buzzolini, C. H. Schallhorn, D. L. Whalen, and K. J. Palmer, *Tetrahedron Letters*, 787 (1970)
[2]R. Furstoss and J.-M. Lehn, *Bull. soc.*, 2497 (1966)

Silver oxide, 1, 1011–1015; **2**, 368.

3,3,6,6-Tetramethyl-1-thiacycloheptyne (2).[1] Cycloheptyne itself has never been prepared, but this seven-membered ring thiaalkyne, has been obtained in low yield by oxidation of the bishydrazone (1) with silver oxide in THF. It is

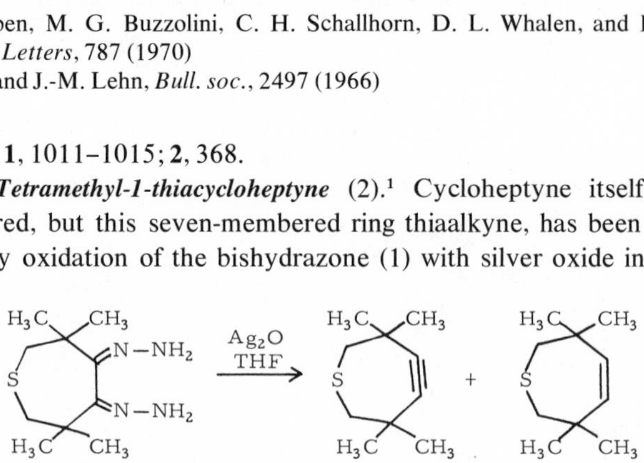

$$(1) \qquad\qquad (2,\ 5.5\%) \qquad (3,\ 6.9\%)$$

accompanied by the *cis*-olefin (3); these two products are separable by gas chromatography.

Tetrahydrofuranes. Primary and secondary aliphatic alcohols (1) are converted into tetrahydrofuranes (2) in 60–73% yields by silver oxide and bromine or by mercuric oxide and iodine.[2] The only other major product is the carbonyl compound (3). Tetrahydropyranes (4) are formed in only small amounts.

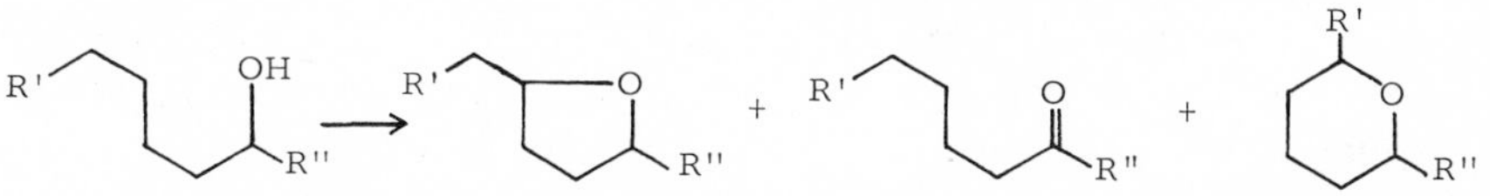

(1)   (2)   (3)   (4)

R' = H, Me, Et, n-Pr   60-73%   5-23%
R'' = H, Me

Synthesis of a spiroheptadiene.[3] Treatment of 2,6-bisdimethylaminospiro[3.3]-heptane dimethiodide (1) with an aqueous suspension of silver oxide followed by pyrolysis gives spiro[3.3]hepta-1,5-diene (2) in 27% yield (crude). The diene (2) is a rigid dissymmetric system and should therefore be capable of existing in two optically active forms.

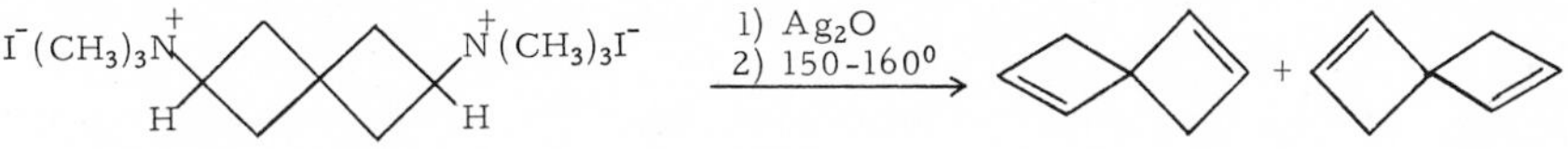

(1)   (2)

α-Diazoketones and α-diazoesters.[4] α-Diazoketones and esters which have a hydrogen atom at the carbon atom bearing the diazo group react with silver oxide (in ether) to give a suspension of silver derivatives (explosive if isolated as a dry powder). These can be C-alkylated on treatment with allyl halides or benzyl bromide in fair to good yield.

Examples:

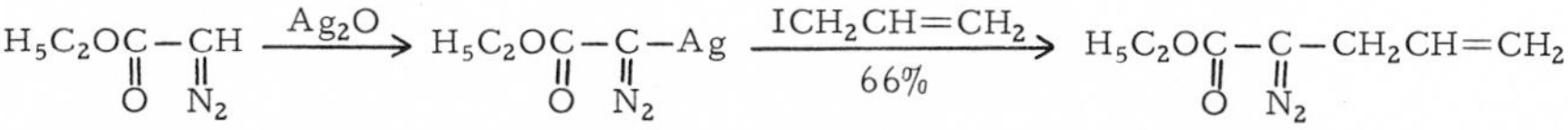

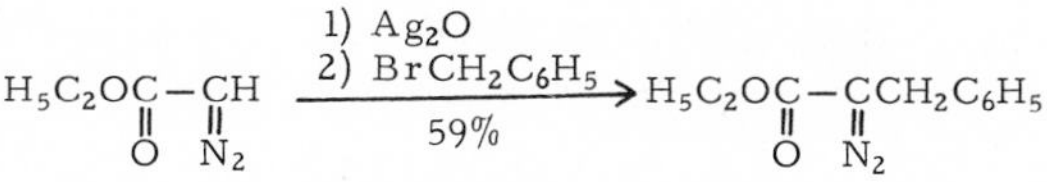

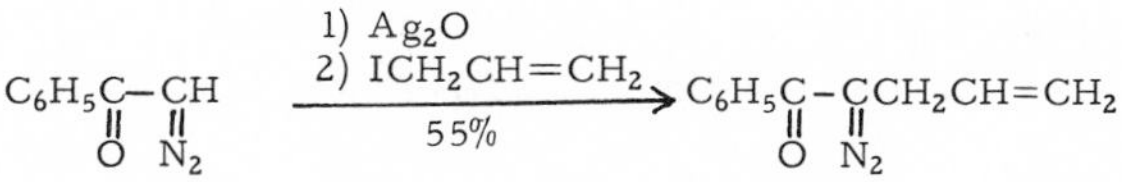

[1]A. Krebs and H. Kimling, *Tetrahedron Letters*, 761 (1970)
[2]M. Lj. Mihailović, Ž. Čeković, and J. Stanković, *Chem. Commun.*, 981 (1969)
[3]J. P. M. Houbiers, L. A. Hulshof, and H. Wynberg, *ibid.*, 91 (1969)
[4]U. Schöllkopf and N. Rieber, *Ber.*, **102**, 488 (1969)

Silver sulfate, 1, 1015.

Reaction with dibromocyclopropanes. Dibromocarbene adducts of cyclic olefins react with silver sulfate in 98% sulfuric acid to give isomeric mixtures of cyclic unsaturated ketones with ring enlargement. Thus dibromonorcarane (1) gives a mixture of two cycloheptenones in a total yield of 29% in which Δ^2-cycloheptenone (2) predominates. It is suggested that the reaction involves loss of bromide ion to give the allylic carbonium ion, then loss of a proton to give a diene, followed by hydrolysis of the bromovinyl group. The minor product, presumed to be Δ^4-

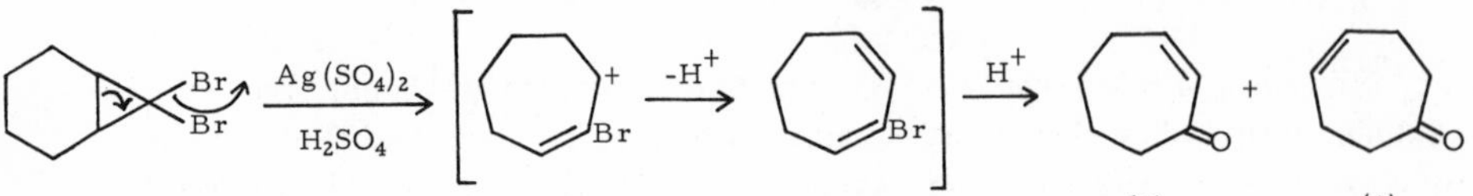

(1)   (2)   (3)

cycloheptenone (3), could arise by transannular hydride shift in the allylic carbonium ion. The same unsaturated ketones were obtained from dichloronorcarane.[1]

The reaction, as applied to the more complicated system (4), does not lead to the expected tropone derivative (5), but to (6), probably formed by decarbonylation of (5).[2] The reaction has been used in the synthesis of an optically active tribenzooxepin.

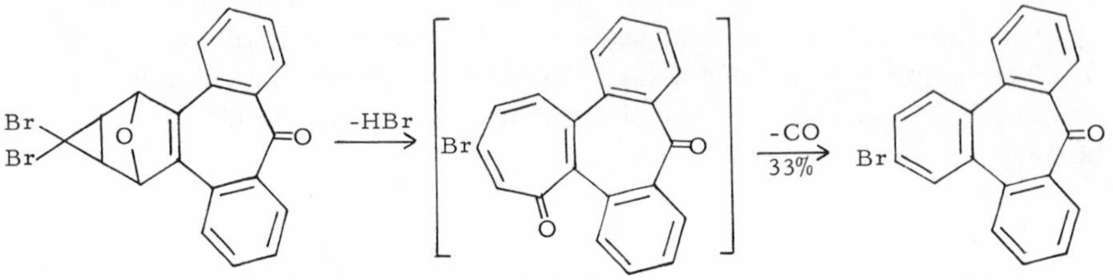

(4)   (5)   (6)

[1]A. J. Birch, G. M. Iskander, B. I. Magboul, and F. Stansfield, *J. Chem. Soc.*, (C), 358 (1967)
[2]W. Tochtermann, C. Franke, and D. Schäfer, *Ber.*, **101**, 3122 (1968); *see also* W. Tochtermann and C. Franke, *Angew. Chem., internat. Ed.*, **8**, 68 (1969)

Silver tosylate, 1, 1018; 2, 370–371.

Preparation. Correction: 1, p. 1018, line 5. Read: After one-half hour silver tosylate was isolated by filtration and evaporation of solvent, and dried in vacuum at 65°.

Simmons-Smith reagent, 1, 1019–1022; **2,** 371–372.

Preparation. Rawson and Harrison[1] find that it is not necessary to use a zinc–copper couple, but that a mixture of zinc dust and a cuprous halide (copper powder is less effective) is even more effective. This modification reduces the experimental difficulties to a level of those encountered in the Grignard reaction. In this way cyclohexene was converted into norcarane in 92% yield.

Reaction with acetylenes (**1,** 1021–1022). Gensler *et al.*[2] as well as several other groups report that they were unable to repeat the synthesis of sterculic acid by addition of methylene across the triple bond of stearolic acid. Gensler has effected the synthesis in about 30% overall yield by the following six-step route:

$$CH_3(CH_2)_7C\equiv C(CH_2)_7COOCH_3 \xrightarrow[\text{60–70\%}]{\substack{1)\ N_2CHCOOC_2H_5 \\ 2)\ KOH}} CH_3(CH_2)_7C\overset{\overset{\displaystyle COOH}{\overset{\displaystyle |}{CH}}}{=\!\!=\!\!=}C(CH_2)_7COOH \xrightarrow{(COCl)_2}$$

$$(1) \qquad\qquad\qquad\qquad\qquad\qquad\qquad\qquad\qquad (2)$$

$$CH_3(CH_2)_7C\overset{\overset{\displaystyle COCl}{\overset{\displaystyle |}{CH}}}{=\!\!=\!\!=}C(CH_2)_7COCl \xrightarrow{ZnCl_2} \left[CH_3(CH_2)_7C\overset{\overset{\displaystyle +}{\overset{\displaystyle CH}{}}}{=\!\!=\!\!=}C(CH_2)_7COCl \right.\!\!\!\!\!\! \xrightarrow{CH_3OH}$$

$$(3) \qquad\qquad\qquad\qquad\qquad\qquad\qquad\qquad (4)$$

$$CH_3(CH_2)_7C\overset{\overset{\displaystyle +}{\overset{\displaystyle CH}{}}}{=\!\!=\!\!=}C(CH_2)_7COOCH_3 \left.\right] \xrightarrow[\substack{40\% \\ \text{from (2)}}]{NaBH_4} CH_3(CH_2)_7C\overset{\overset{\displaystyle CH_2}{}}{=\!\!=\!\!=}C(CH_2)_7COOCH_3$$

$$(5) \qquad\qquad\qquad\qquad\qquad\qquad\qquad\qquad (6)$$

The reaction of methyl stearolate (1) with diazoacetic ester in the presence of copper bronze produces a diester, which is hydrolyzed to the diacid (2). This is converted into the diacid chloride (3). Treatment with zinc chloride or other Lewis acid effects selective decarbonylation to give the cyclopropenium ion (4). Methanol is then added to convert the acid chloride grouping into the methyl ester (5). Finally reduction with sodium borohydride gives methyl sterculate (6).

A similar sequence was used by Gensler[3] for the synthesis of the methyl ester of malvalic acid (8), a component of cottonseed oil, from 1-chloro-7-hexadecyne (7).

$$C_8H_{17}C\equiv C(CH_2)_6Cl \xrightarrow[23\%]{} C_8H_{17}C\overset{\overset{\displaystyle CH_2}{}}{=\!\!=\!\!=}C(CH_2)_6COOCH_3$$

$$(7) \qquad\qquad\qquad\qquad (8)$$

Reaction with α,β-unsaturated ketones. Early work on the reaction of the Simmons-Smith reagent with α,β-unsaturated ketones suggested that the reaction fails or gives very poor yields of cyclopropylketones.[4] Actually, some α,β-unsaturated ketones and cross-conjugated ketones give good to excellent yields of products, as shown by the examples:

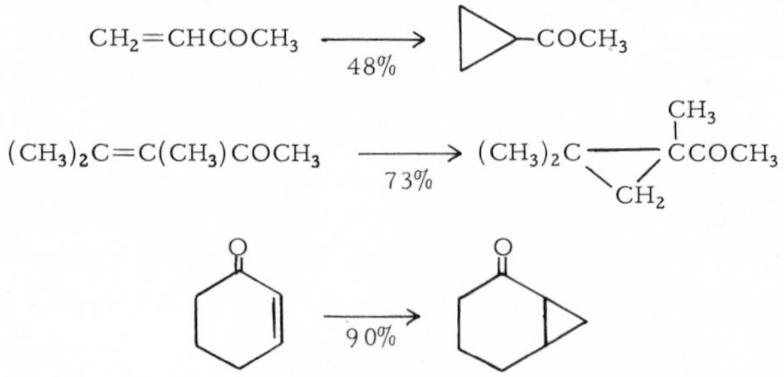

$$CH_2=CHCOCH_3 \xrightarrow[48\%]{} \triangleright\!\!-COCH_3$$

$$(CH_3)_2C=C(CH_3)COCH_3 \xrightarrow[73\%]{} (CH_3)_2C\!\!-\!\!\overset{\overset{\displaystyle CH_3}{|}}{\underset{\underset{\displaystyle CH_2}{\diagdown}}{C}}COCH_3$$

On the other hand, some α,β-unsaturated ketones give only resins or do not react.

$$(CH_3)_2C=CHCOCH_3$$

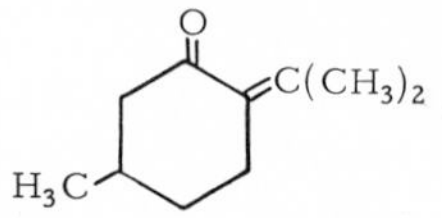

The differing behavior appears to be due to ease of enolization of the starting ketone. α,β-Unsaturated ketones that do not enolize readily react normally; those that enolize readily form an enolate complex with the reagent which prevents methylene transfer.[5] (See also **2**, 371–372.)

Conversion of aromatic aldehydes to styrenes. Japanese investigators[6,7] report that aromatic aldehydes react with the Simmons-Smith reagent in THF to give styrenes in about 50% yield (isolated). Somewhat higher yields are obtained if

$$ArCHO + CH_2X_2 \xrightarrow{Zn/Cu} ArCH=CH_2$$

methylene iodide is replaced by chloroiodomethane, CH_2ClI. The reaction was also applied successfully to *n*-heptaldehyde to give 1-octene in 47% yield (isolated). The α,β-unsaturated aldehyde cinnamaldehyde gave *trans*-phenylbutadiene in 18% yield (isolated); no cyclopropane derivative was detected.

Angular methylation. Whitlock[8] has described a new procedure for angular α-methylation of ketones. For example, the ketone (1) is converted into the enol benzoate (2) by reaction with benzoic anhydride catalyzed by perchloric acid. This is then converted into the lithium enolate by reaction with methyllithium in

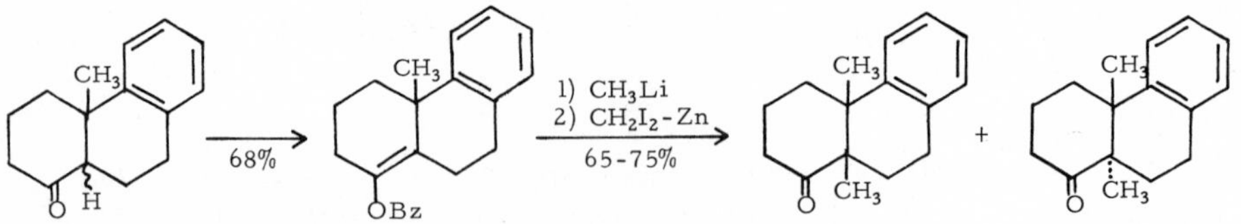

(1)   (2)   (3)   3:1   (4)

dimethoxyethane containing a few crystals of triphenylmethane.[9] A solution of the Simmons-Smith reagent in ether is then added. The reaction gives a 3:1 mixture of (3) and (4). Methylation with methyl iodide gives an inseparable mixture of un-, mono-, di-, and trimethylated ketones.

Wenkert and Berges[10] achieved a synthesis of the interesting sesquiterpene *l*-valeranone (8), which has an anomalous structure. The unnatural 10-angular methyl group was introduced stereospecifically by way of the Simmons-Smith reagent. Thus Simmons-Smith reaction of the allylic alcohol (5) gives the methoxy-cyclopropane (6). Jones oxidation followed by Wolff-Kishner reduction gives the tricyclic ether (7). Protonolysis (aqueous, methanolic hydrochloric acid) converts

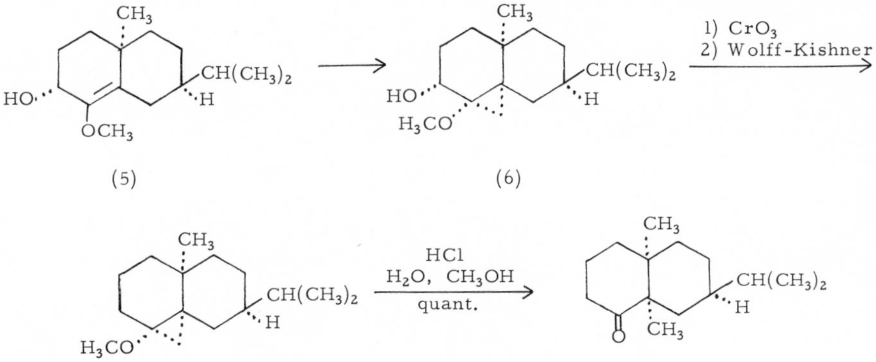

this intermediate into *l*-valeranone in quantitative yield.

The method has now been used to synthesize 9,10-dimethyl-*trans*-1-decalones,[11] which have the characteristic C/D ring system of pentacyclic triterpenes. Thus reaction of the hydroxy enol ether (9) with the Simmons-Smith reagent gives the cyclopropyl ether (10). Cleavage with 7% methanolic hydrochloric acid leads to the hydroxy ketone (11), convertible by Wolff-Kishner reduction followed by oxidation into (12).

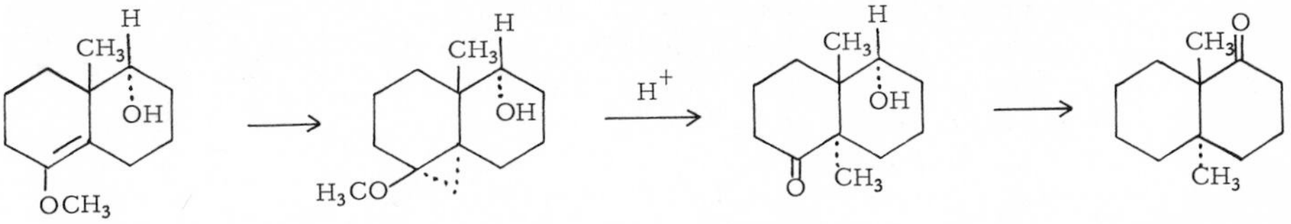

Related reagent, ethyliodomethylzinc, $C_2H_5ZnCH_2I$.[12] Mol. wt. 235.38. The reagent is prepared by the reaction of ethyl iodide with zinc–copper couple in absolute ether to give ethylzinc iodide (1), probably in equilibrium with diethyl-zinc and zinc iodide. This stock solution can be stored at room temperature for a week. Methylene iodide is added at 30–35° (1 hour) to form the organozinc reagent (3). Subsequent addition of the olefin substrate affords, after a reaction period of 0.5–5 hours, the cyclopropane derivative (4). This modification usually gives

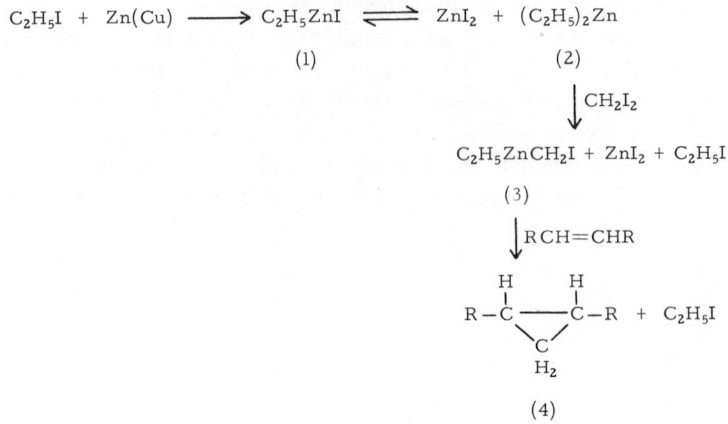

$$C_2H_5I + Zn(Cu) \longrightarrow C_2H_5ZnI \rightleftharpoons ZnI_2 + (C_2H_5)_2Zn$$

$$(1) \hspace{5cm} (2)$$

$$\downarrow CH_2I_2$$

$$C_2H_5ZnCH_2I + ZnI_2 + C_2H_5I$$

$$(3)$$

$$\downarrow RCH=CHR$$

$$R-\overset{H}{\underset{}{C}} \underset{\underset{H_2}{C}}{\diagup} \overset{H}{\underset{}{C}}-R + C_2H_5I$$

$$(4)$$

higher yields than those obtained by the original procedure of Simmons and Smith.

Examples:

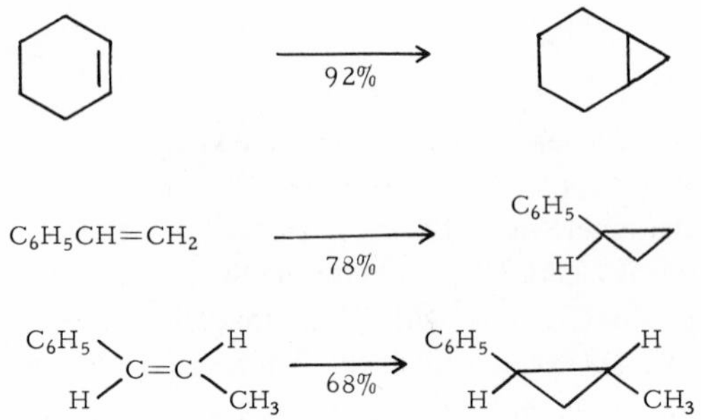

92%

$C_6H_5CH=CH_2$ → 78%

68%

[1] R. J. Rawson and I. T. Harrison, *J. Org.*, **35**, 2057 (1970)
[2] W. J. Gensler, M. B. Floyd, R. Yanase, and K. W. Pober, *Am. Soc.*, **92**, 2472 (1970)
[3] W. J. Gensler, K. W. Pober, D. M. Solomon, and M. B. Floyd, *J. Org.*, **35**, 2301 (1970)
[4] G. Wittig and F. Wingler, *Ber.*, **97**, 2139, 2146 (1964); Y. Armand, R. Perraud, J.-L. Pierre, and P. Arnaud, *Bull. soc.*, 1893 (1965)
[5] J.-C. Limasset, P. Amice, and J.-M. Conia, *ibid.*, 3981 (1969)
[6] H. Hashimoto, M. Hida, and S. Miyano, *J. Organometal. Chem.*, **10**, 518 (1967)
[7] S. Miyano, M. Hida, and H. Hashimoto, *ibid.*, **12**, 263 (1968)
[8] H. W. Whitlock, Jr., and L. E. Overman, *J. Org.*, **34**, 1962 (1969)
[9] Procedure of H. O. House and B. M. Trost, *ibid.*, **30**, 2502 (1965)
[10] E. Wenkert and D. A. Berges, *Am. Soc.*, **89**, 2507 (1967)
[11] R. E. Ireland, D. R. Marshall, and J. W. Tilley, *ibid.*, **92**, 4756 (1970)
[12] S. Sawada and Y. Inouye, *Bull. Chem. Soc. Japan*, **42**, 2669 (1969)

Sodium aluminum hydride, 1, 1030. Supplier: Alfa Inorganics.

Sodium aluminum hydride has been found to be superior to lithium aluminum hydride for reduction of dimethylamides of carboxylic acids to aldehydes.

$$4\ RCON(CH_3)_2 \xrightarrow{NaAlH_4} \left(\begin{array}{c} H \\ | \\ R\,C-O \\ | \\ N(CH_3)_2 \end{array} \right)_4 AlNa \xrightarrow{H_2O} 4\ RCHO$$

Optimum yields (70–90%) are obtained by addition of a slight excess of the reducing agent to the amide at 0–20° in THF.[1]

[1]L. I. Zakharkin, D. N. Maslin, and V. V. Gavrilenko, *Tetrahedron*, **25**, 5555 (1969)

Sodium amalgam, 1, 1030–1033; **2**, 373.

Reduction of phthalic acid (**2**, 373, ref. 14). The reduction of phthalic acid to *trans*-1,2-dihydrophthalic acid with sodium amalgam has now been published. The procedure also gives a detailed description of the preparation of 3% sodium amalgam.[1]

[1]R. N. McDonald and C. E. Reineke, *Org. Syn.*, **50**, 50 (1970)

Sodium–Ammonia, 1, 1041; **2**, 374–376.

Birch reduction.[1] A flask is charged with 2.5 l. of liquid ammonia and fitted with a dry ice condenser and stirrer; 450 g. of precooled anhydrous diethyl ether, 460 g. (10 moles) of precooled absolute ethanol, and 318.5 g. (3.0 moles) of *o*-xylene are added slowly in that order. Then 207 g. of sodium is added in small pieces over a 5-hour period. The ammonia is allowed to evaporate overnight.

$$\xrightarrow[\substack{NH_3/C_2H_5OH/(C_2H_5)_2O \\ \text{yield: } 77-92\%}]{Na}$$

The flask is then equipped with a reflux condenser and 800 ml. of ice water is run in slowly with stirring to dissolve the salts (exothermic reaction). The organic layer is washed well with water, dried, and the 1,2-dimethyl-1,4-cyclohexadiene is distilled.

[1]L. A. Paquette and J. H. Barrett, *Org. Syn.*, **49**, 62 (1969)

Sodium azide, 1, 1041–1044; **2**, 376.

Reaction with adamantanone. Japanese chemists[1] treated adamantanone (1) with sodium azide in methanesulfonic acid with the expectation of achieving the Schmidt reaction. However, the unexpected 4-methylsulfonoxyadamantanone was obtained in 90% yield. Alkaline hydrolysis cleaves (2) to Δ²-bicyclo[3.3.1]-nonene-7-carboxylic acid (3) in 85% yield by a quasi-Favorsky reaction. Adamantane itself does not undergo this unusual substitution reaction.

(1)   (2)   (3)

[1]T. Sasaki, S. Eguchi, and T. Toru, *Am. Soc.*, **91**, 3390 (1969)

Sodium bicarbonate, $NaHCO_3$. Mol. wt. 84.02. Suppliers: B, F, MCB.

Anhydrides of aromatic acids. Treatment of the acid chloride of an aromatic acid with aqueous sodium bicarbonate in the presence of a weak tertiary base such as pyridine (20°, 30 minutes) leads to the anhydride in very good yield. Considerably lower yields are obtained if an alkali hydroxide is used.[1]

[1]P. Rambacher and S. Mäke, *Angew. Chem., internat. Ed.*, **7**, 465 (1968)

Sodium bis-(2-methoxyethoxy)aluminum hydride, $NaAlH_2(OC_2H_4OCH_3)_2$. Mol. wt. 202.16, dec. 205°. Suppliers: Aldrich, Eastman (70% benzene solution).

Review. A comprehensive review of this reagent by J. Vit is available in Eastman Organic Chemical Bulletin, **42**, no. 3, 1970.

Preparation.[1] The most convenient preparation is a direct synthesis in benzene solution at temperatures above 100° under hydrogen pressure:

$$Na + Al + 2\ CH_3OCH_2CH_2OH \longrightarrow NaAlH_2(OCH_2CH_2OCH_3)_2$$

Reduction. Sodium bis-(2-methoxyethoxy)aluminum hydride has been introduced recently as a substitute for lithium aluminum hydride. It has certain advantages. It does not ignite when exposed to moist air or oxygen; it is stable to dry air. It is highly soluble in aromatic solvents and in ethers. Reactions can be carried out at temperatures as high as 200°. It dries aromatic hydrocarbons and ethers completely and quickly. In reducing activity the new reagent is fully comparable to lithium aluminum hydride. Thus it reduces aldehydes and ketones to the corresponding alcohols in high yield, a 5–10% excess being used to ensure full reduction.[2] It reduces saturated and α,β-unsaturated acids, esters, chlorides, and anhydrides. Isolated double bonds are not reduced. Lactones are reduced to diols.[3] Oximes are reduced satisfactorily to primary amines. A nitrile group attached directly to an aromatic ring is reduced, but arylaliphatic nitriles are reduced only in low yield, while aliphatic nitriles are not reduced to any extent.[4]

Aromatic aldehydes, acids and esters having a hydroxyl group in the *ortho*- or *para*-position undergo hydrogenolysis to *ortho*- or *para*-cresols. If the reduction is carried out for a short time and at temperatures below 80°, hydroxy-substituted benzyl alcohols can be obtained and are evidently intermediates in the further reduction.[5] Note that Conover and Tarbell[6] had observed similar hydrogenolysis by LiAlH₄ of aromatic acids, esters, and aldehydes bearing a strong electron-donating group in the *ortho*- or *para*-position.

Esters (0.1 mole) can be reduced to aldehydes in good yield by the reagent (0.05 mole) at temperatures in the range −50 to −70°. At these temperatures the result-

$$2\ RCOOR' + NaAlH_2(OCH_2CH_2OCH_3)_2 \longrightarrow 2\ RCHO + NaAl(OR')_2(OCH_2CH_2OCH_3)_2$$

ing aldehyde is practically unreactive, whereas the starting esters are reduced fairly readily. Toluene, ether, or THF is used as solvent.[7]

Organosilicon halides are readily reduced to silanes. Almost all aliphatic halides and some aryl halides can be reduced with the reagent.[8] It appears to be slightly more reactive than LiAlH₄ for this purpose. Yields can be increased by raising the boiling point of the solvent. Also older, partly hydrolyzed solutions of the reagent give higher yields.[9]

[1]J. Vit. B. Čásenský, and J. Macháček, French Pat. 1,515,582
[2]M. Čapka, V. Chvalovský. K. Kochloefl, and M. Kraus, *Czech. Commun.*, **34**, 118 (1969)
[3]M. Černý, J. Málek, M. Čapka, and V. Chvalovský, *ibid.*, **34**, 1025 (1969)
[4]*Idem, ibid.*, **34**, 1033 (1969)
[5]M. Černý and J. Málek, *Tetrahedron Letters*, 1739 (1969)
[6]L. H. Conover and D. S. Tarbell, *Am. Soc.*, **72**, 3586 (1950)
[7]J. Vit. in press.
[8]M. Čapka and V. Chvalovský, *Czech. Commun.*, **34**, 2782 (1969)
[9]*Idem, ibid.*, **34**, 3110 (1969)

Sodium bistrimethylsilylamide [(CH₃)₃Si]₂N·Na, **1**, 1046–1047.

Enolates of prednisone–BMD. Two groups[1,2] have independently studied the preparation of enolates of prednisone–BMD (1), (17,20;20,21-bismethylene-dioxyprednisone). Treatment of (1) with the more usual bases (triphenylmethyl-

(1)

(2a) R = Na
(2b) R = Bz

(3a) R = Na
(3b) R = Bz

sodium or ethynylsodium) followed by benzoic anhydride gives mixtures of unchanged (1) and variable amounts of the 9(11)-enol benzoate (2b) and 3-enol benzoate (3b). Sodium bistrimethylsilylamide is a much more useful base. Treatment of (1) with a modest excess of this base affords largely the 9(11)-enolate (2). It was then found that the 3-enolate (3a) rearranges to (2a) in the presence of (1). Methods were then developed for preparation of either enolate as desired. Slow addition of (1) to a solution containing more than one equivalent of the base gives mainly the 3-enolate (3a isolated as 3b in >80% yield). Slow addition of slightly less than 1 equivalent of the base gives the 9(11)-enolate (2a, isolated as 2b in 80% yield). Thus kinetically controlled enolization occurs at C_6 followed by rearrangement catalyzed by the starting ketone to the C_9-anion. Fluorination of (2b) and (3b) with fluoroxytrifluoromethane (**2**, 200) affords 9-fluoro- and 6-fluoro-corticosteroids, respectively.

Reaction of (1) with lithium bismethylsilylamide[3] followed by quenching with benzoic anhydride affords only the 3-enol benzoate (3b). This observation is in accord with the known reluctance of lithium enolates to isomerize.

[1]D. H. R. Barton, R. H. Hesse, G. Tarzia, and M. M. Pechet, *Chem. Commun.*, 1497 (1969)
[2]M. Tanabe and D. F. Crowe, *ibid.*, 1498 (1969)
[3]U. Wannagat and H. Niederprüm, *Ber.*, **94**, 1540 (1961)

Sodium borohydride, **1**, 1049–1055; **2**, 377–378.

Reduction of halides. Primary, secondary, and certain tertiary halides and tosylates can be selectively reduced by sodium borohydride in DMSO or sulfolane.[1] 1,2-Dibromides are reduced to hydrocarbons in moderately good yields. The method complements that of Bell and Brown,[2] which uses sodium borohydride in 65% aqueous diglyme for reduction of secondary and tertiary alkyl halides which are capable of forming relatively stable carbonium ions (**1**, 1054). Lithium aluminum hydride reduction affords only olefins. Reduction of optically active tertiary alkyl halides proceeds with racemization, presumably by way of an elimination–addition mechanism.[3]

Primary and secondary alkyl iodides, primary bromides, and benzyl and allyl chlorides are smoothly reduced by sodium borohydride in DMF at room temperature:

$$RX + NaBH_4 \longrightarrow RH + NaX + BH_3$$

Kinetic evidence indicates an S_N2 mechanism for the reduction.[4]

Reduction of phenolic ketones. The carbonyl group of hydroxyacetophenones can be reduced with boiling aqueous alkaline sodium borohydride. When the carbonyl group is *ortho* or *para* to a phenolic hydroxyl reduction proceeds to the methylene derivative (1 → 2), but stops at the carbinol stage when the phenolic hydroxyl and carbonyl group are *meta*-oriented (3 → 4).[5]

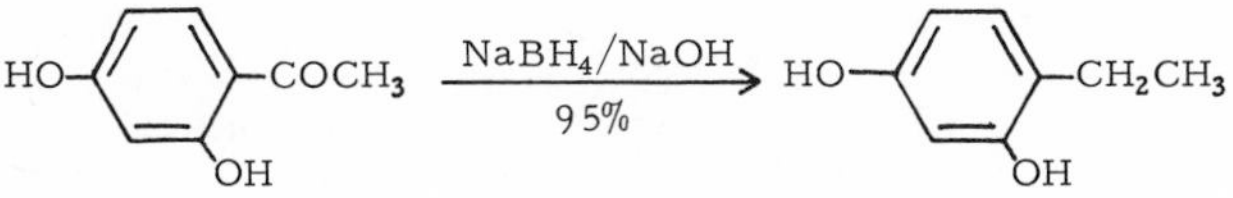

(1)   (2)

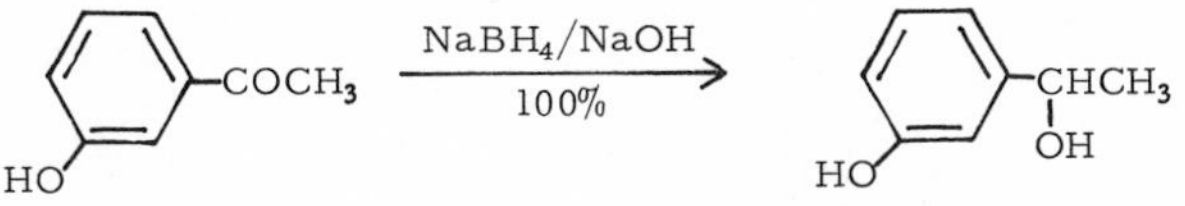

(3)   (4)

Other examples:

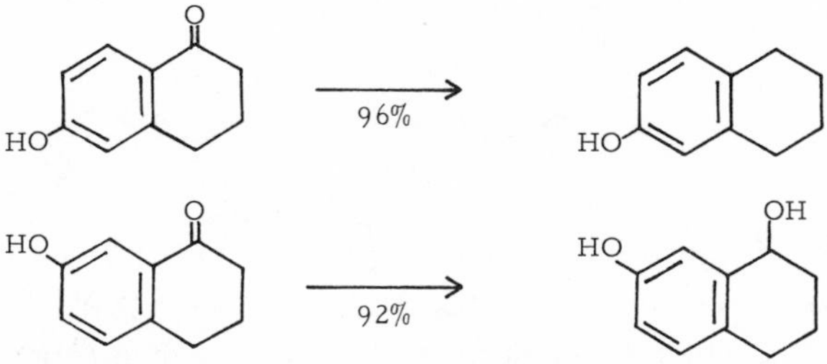

96%

92%

Amines and nitriles from amides.[6] Tertiary aliphatic and aromatic amides are reduced in moderate yield to amines by sodium borohydride in refluxing pyridine:

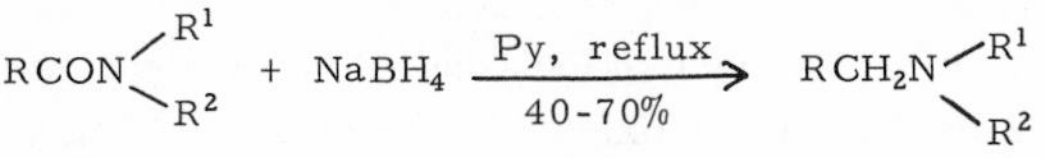

$$RCON\begin{smallmatrix}R^1\\R^2\end{smallmatrix} + NaBH_4 \xrightarrow[40-70\%]{Py, reflux} RCH_2N\begin{smallmatrix}R^1\\R^2\end{smallmatrix}$$

No reaction occurs with secondary amides. Primary amides, on the other hand, are dehydrated to nitriles:

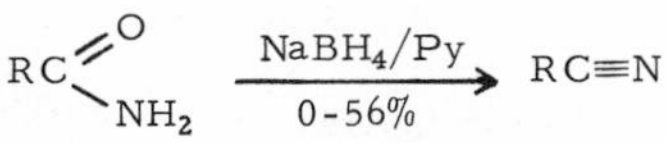

$$RC\begin{smallmatrix}O\\NH_2\end{smallmatrix} \xrightarrow[0-56\%]{NaBH_4/Py} RC{\equiv}N$$

In the case of nicotinamide, partial hydrogenation also occurs to give 3-cyano-1,4,5,6-tetrahydropyridine.

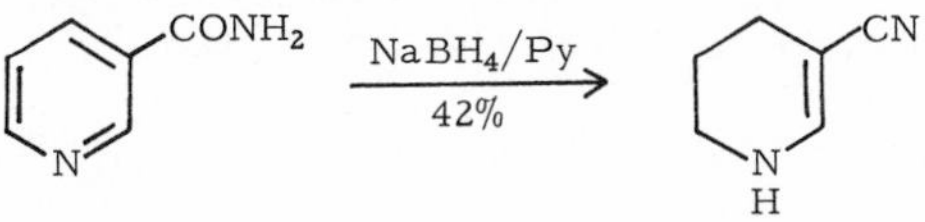

$$\xrightarrow[42\%]{NaBH_4/Py}$$

Reduction of cyclic anhydrides. Although acid anhydrides are reduced to only a slight extent by sodium borohydride, cyclic anhydrides are reduced to δ- and γ-lactones in 51–97% yield. Hydride attack occurs principally at the carbonyl group adjacent to the more highly substituted carbon atom.[7]

Examples:

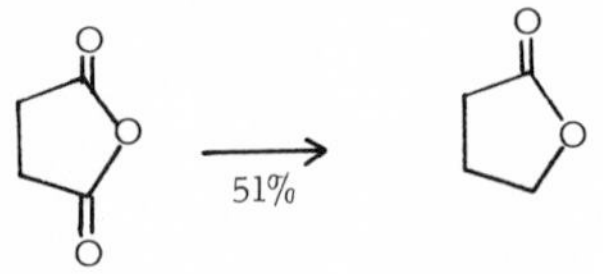

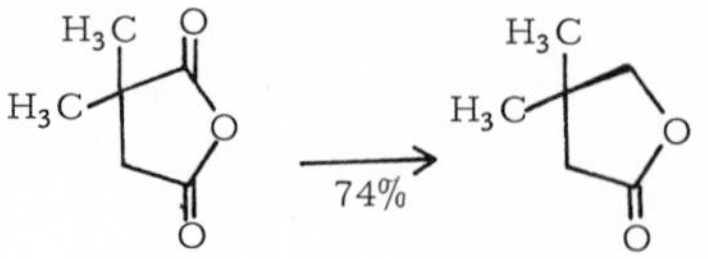

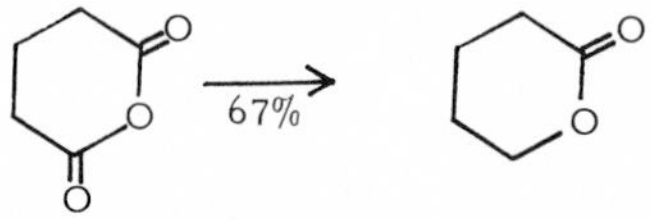

Sulfurated sodium borohydride, $NaBH_2S_3$. Mol. wt. 132.03. The reagent is prepared by the reaction of sodium borohydride with sulfur in THF (25°, cooling). Soluble in THF and HMPT, but insoluble or sparingly soluble in common organic solvents.[8]

The reagent reduces aldehydes[9] and ketones[10] to the corresponding alcohols at room temperatures or below in good yield. It is somewhat more reactive than sodium borohydride. It reduces oximes to the corresponding amines[11] but does not reduce nitro, nitrile, halogen, or ester groups.

[1]R. O. Hutchins, D. Hoke, J. Keogh, and D. Koharski, *Tetrahedron Letters*, 3495 (1969); H. M. Bell, C. W. Vanderslice, and H. Spehar, *J. Org.*, **34**, 3923 (1969)

[2]H. M. Bell and H. C. Brown, *Am. Soc.*, **88**, 1473 (1966)

[3]J. Jacobus, *Chem. Commun.*, 338 (1970)

[4]M. Vol'pin, M. Dvolaitzky, and I. Levitin, *Bull. soc.*, 1526 (1970)

[5]K. H. Bell, *Australian J. Chem.*, **22**, 601 (1969)

[6]Y. Kikugawa, S. Ikegami, and S. Yamada, *Chem. Pharm. Bull. Japan*, **17**, 98 (1969)

[7]D. M. Bailey and R. E. Johnson, *J. Org.*, **35**, 3574 (1970)

[8]J. M. Lalancette, A. Frêche, and R. Monteux, *Canad. J. Chem.*, **46**, 2754 (1968)

[9]J. M. Lalancette and A. Frêche, *ibid.*, **47**, 739 (1969)

[10]*Idem, ibid.*, **48**, 2366 (1970)

[11]J. M. Lalancette and J. R. Brindle, *ibid.*, **48**, 735 (1970)

Sodium borohydride–Cobaltous chloride ($CoCl_2$).

Reduction. Nitriles and amides are not reduced by sodium borohydride alone but are reduced to primary amines in satisfactory yields by the combination of sodium borohydride–cobaltous chloride in both hydroxylic and nonhydroxylic solvents.[1] Aromatic nitro compounds are reduced by this system to azoxy compounds.[2] Amines are obtained if cupric chloride is used in place of cobaltous chloride.

[1]T. Satoh, S. Suzuki, Y. Suzuki, Y. Miyaji, and Z. Imai, *Tetrahedron Letters*, 4555 (1969)
[2]T. Satoh, S. Suzuki, T. Kikuchi, and T. Okada, *Chem. Ind.*, 1626 (1970)

Sodium chlorodifluoroacetate, 1, 1058; 2, 379–381.

Generation of difluorocarbene. In an early study[1] of the reaction of difluorocarbene with steroidal double bonds, the methylenation was conducted by portionwise addition of an excess of dry sodium chlorodifluoroacetate to a diglyme solution of the substrate at temperatures of 120–150°. The same laboratory (Syntex)[2] now reports that the reaction is improved by addition of a solution of the salt in diglyme to a refluxing diglyme solution of the substrate. The reaction with $\Delta^{4,6}$- and $\Delta^{1,4,6}$-3-ketosteroids results mainly in addition to the Δ^6-double bond to give $6\alpha,7\alpha$-difluoromethylene-Δ^4- and $\Delta^{1,4}$-3-ketones, respectively.

[1]L. H. Knox, E. Velarde, S. Berger, D. Cuadriello, P. W. Landis, and A. D. Cross, *Am. Soc.*, **85**, 1851 (1963)
[2]C. Beard, B. Berkoz, N. H. Dyson, I. T. Harrison, P. Hodge, L. H. Kirkham, G. S. Lewis, D. Giannini, B. Lewis, J. A. Edwards, and J. H. Fried, *Tetrahedron*, **25**, 1219 (1969)

Sodium dichromate dihydrate, 1, 1059–1064.

Oxidation of primary alcohols to aldehydes. The oxidation of primary alcohols to aldehydes is complicated by the fact that aldehydes are easily oxidized further to carboxylic acids. Lee and Spitzer[1] now find that the oxidation can be effected in fair to good yield by neutral aqueous dichromate at reflux temperature (3 hours). Usually the alcohol is not completely soluble, and with some alcohols it is advisable to add a small amount of acetone to prevent sublimation. In the majority of cases a 1 : 1 mole ratio of $Na_2Cr_2O_7$ to alcohol was employed; 100 ml. of water was usually used for each 0.10 mole of components.

[1]D. G. Lee and U. A. Spitzer, *J. Org.*, **35**, 3589 (1970)

Sodium dicyanocuprate, $NaCu(CN)_2$. Mol. wt. 138.58.

The reagent is prepared[1] *in situ* from sodium cyanide and copper(I) cyanide in DMF. A 1 M solution of the salt in DMF can be prepared. The potassium and lithium salts are less soluble.

The reagent can be used for conversion of aryl and vinyl halides into the corresponding cyano compounds.[1] Actually sodium dicyanocuprate is somewhat less reactive than copper(I) cyanide but does have the advantage that the reaction is homogeneous.

[1]H. O. House and W. F. Fischer, Jr., *J. Org.*, **34**, 3626 (1969)

Sodium ethoxide, 1, 1065–1073.

Basification. Use of the reagent as an alcohol-soluble base is illustrated by the synthesis of 2-methylthiazoline[1] from cysteamine hydrochloride.[2] A 2-l. round-bottomed flask fitted with a reflux condenser is mounted in an oil bath or heating mantle in the hood, charged with 720 ml. of absolute ethanol and 72 g. (1.06 moles) of sodium ethoxide (MCB), and heated at reflux for 30 minutes to

$$HSCH_2CH_2NH_2 \cdot HCl + NaOC_2H_5 \longrightarrow HSCH_2CH_2NH_2 + C_2H_5OH + NaCl$$

$$HSCH_2CH_2NH_2 + CH_3CN \longrightarrow \begin{array}{c} H_2C-S \\ | \qquad\qquad C-CH_3 \\ H_2C-N \end{array} + NH_3$$

bring most of the sodium ethoxide into solution. The solution is allowed to cool slightly and a solution of 120 g. of cysteamine hydrochloride in 240 ml. of absolute ethanol is added cautiously with swirling. The mixture is heated at reflux for 45 minutes and then is cooled to room temperature or below. The cooled mixture is filtered with suction to remove the precipitated sodium chloride. The solid is washed with three 30-ml. portions of absolute ethanol and the washings are combined with the filtrate. The residual sodium chloride, which still smells of cysteamine and contains about 2% of this base by titration, is discarded.

The filtrate containing the free base is placed in a clean 2-l. round-bottomed flask, 130 g. (170 ml., 3.2 moles) of acetonitrile[3] is added and the mixture is refluxed for 6 hours.[4] Heating is interrupted, the mixture is allowed to cool slightly, the condenser is replaced by a distilling head bearing a thermometer, and the condenser is arranged for downward distillation. Heating is resumed, and ethanol and excess acetonitrile are distilled until a vapor temperature of 95° is attained. The residue is cooled to room temperature, filtered if necessary to remove precipitated solid, and transferred to a 250-ml. distilling flask. Distillation is resumed and a fraction boiling at 135–150° is collected. This material (about 80 g. of yellow oil) is redistilled from a 125-ml. flask. The 2-methylthiazoline is collected as a nearly colorless liquid boiling at 142–146°. Yield 72–80 g. (67–73%).[5]

[1]F. E. Condon and J. P. Trivedi, procedure submitted to *Org. Syn.* 1970
[2]Available from Evans Chemetics, Inc., Waterloo, New York 13165
[3]With only 1 mole of acetonitrile the yield of thiazoline was only 45–48%.
[4]Refluxing beyond this point results in considerable decomposition and lowered yield.
[5]When the base used was sodium methoxide, 25% in methanol (Technical grade, MCB), the yield of thiazoline was only 53%.

Sodium hydrosulfide, NaSH. Mol. wt. 56.07. Supplier: Matheson, Coleman and Bell.

Preparation.[1] The reagent can be prepared from sodium sulfide and sodium bicarbonate.

Selective reduction of polynitro compounds. Hodgson and Ward[2] reported some time ago that methanolic sodium hydrosulfide is useful for monoreduction of polynitronaphthalenes, but the reagent has since attracted little use. Idoux[3] now finds that the reagent is superior to ethanolic sodium polysulfide or ethanolic ammonium sulfide for selective reduction of x,y'-dinitrodiphenyls. For example, 4,4'-dinitrodiphenyl is reduced to 4-amino-4'-nitrodiphenyl in 79% yield. Elemental sulfur, which is difficult to remove, is not formed; moreover the reagent is easy to prepare in known concentration, which is important in controlling the extent of the reduction. The reductions are run in methanol–acetone if the solubility is sufficient, otherwise in methanol–toluene.

[1]H. H. Hodgson and E. R. Ward, *J. Chem. Soc.*, 242 (1948)
[2]*Idem, ibid.*, 794 (1945)
[3]J. P. Idoux, *ibid.*, (C), 435 (1970)

Sodium iodide, 1, 1087–1090; 2, 384.

Pschorr synthesis. The Pschorr synthesis of phenanthrene derivatives is usually carried out with copper powder. Australian chemists[1] report that a diazotized 2-phenyl-3-(2-aminophenyl)acrylic acid of type (1) undergoes practically instantaneous cyclization in the presence of sodium iodide (acetone is used as solvent). When copper powder is added the Pschorr method requires more than ten hours and the yield is 55%.

H_3CO H $COOH$ $C=C$ $\xrightarrow[72\%]{NaI}$ H_3CO $COOH$ H_3CO N_2^+ Cl^- H_3CO

(1)   (2)

[1]B. Chauncy and E. Gellert, *Australian J. Chem.*, **22**, 993 (1969)

Sodium persulfate, 1, 1102.

Oxidation of primary amines. A primary aliphatic amine of the type RCH_2NH_2 is oxidized by aqueous sodium persulfate in an alkaline medium and in the presence of silver nitrate as catalyst to an aldimine, $RCH{=}NCH_2R$, which on hydrolysis gives the aldehyde, $RCHO$, and the amine, RCH_2NH_2. In theory 0.5 mole of aldehyde is formed from 1 mole of amine. Yields of aldehydes range from 15 to 95% among C_3–C_9 amines.[1] In the same way a primary amine of the type $R'R''CHNH_2$ is converted into a ketone, $R'R''CO$. Secondary amines give inferior yields. Argentic picolinate has also been used for the same purpose, but yields are lower with this oxidant.[2]

[1]R. G. R. Bacon and D. Stewart, *J. Chem. Soc.*, (C), 1384, 1388 (1966)
[2]R. G. R. Bacon and W. J. W. Hanna, *ibid.*, 4962 (1965)

Sodium tetracarbonylferrate(-II), Na₂Fe(CO)₄. Mol. wt. 213.88.

Preparation.[1] The reagent (1) is prepared by the reaction of iron pentacarbonyl[2] with 1% sodium amalgam in dry THF (nitrogen); carbon monoxide is evolved and the colorless salt (1) separates partially.

$$Fe(CO)_5 \xrightarrow[\text{THF}]{\text{Na-Hg}} Na_2Fe(CO)_4$$

$$(1)$$

RBr → RCHO. The reagent reacts with primary bromides in the presence of triphenylphosphine (25°) to give the corresponding aldehyde in high yield (75–85%, isolated) after protonation with acetic acid. The reaction is considered to involve oxidative addition, migratory insertion, and reductive elimination.

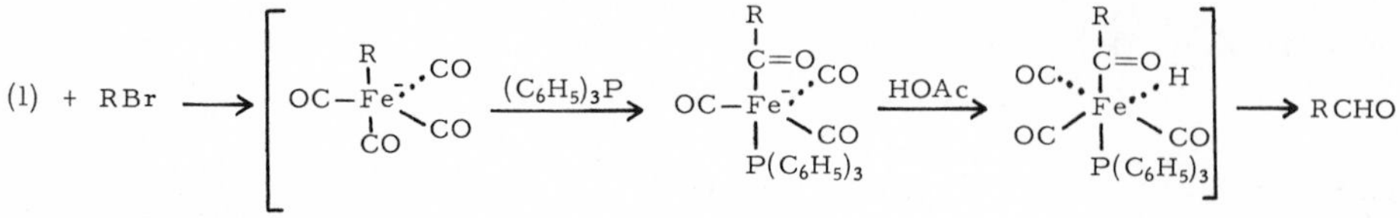

Yields of aldehydes from secondary and benzylic bromides are low.

[1] M. P. Cooke, Jr., *Am. Soc.*, **92**, 6080 (1970)
[2] Pressure Chemical Co., Pittsburgh, Pa.; toxic.

Sodium thiophenoxide, 1, 1106–1107.

Debenzylation of quaternary ammonium salts. The selective demethylation of quaternary ammonium salts noted by Shamma (**1**, 1107, ref. 3) has been found by Japanese chemists[1] to be applicable to selective debenzylation of such salts. Thus N,N-dibenzyl-N,N-dimethylammonium chloride is converted in 85% yield into N,N-dimethylbenzylamine. The reagent also effects deallylation of quaternary ammonium salts.

[1] T. Kametani, K. Kigasawa, M. Hiiragi, N. Wagatsuma, and K. Wakisaka, *Tetrahedron Letters*, 635 (1969)

Sodium trimethoxyborohydride, 1, 1108–1109. Additional supplier: Alfa Inorganics.

Reduction of esters. The secondary ester group of the diester-acid (1) can be selectively reduced by use of a three mole-equivalent excess of this reducing agent in refluxing dimethoxyethane. In model experiments partial selectivity

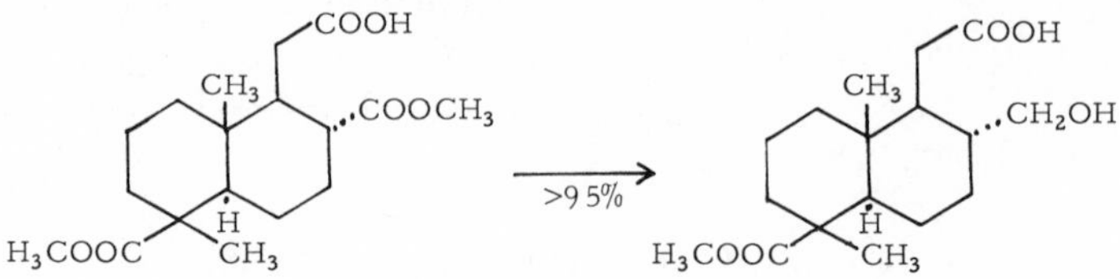

(1)   (2)

between primary and secondary ester groups was noted, primary esters being reduced somewhat more rapidly.[1]

R. A. Bell and M. B. Gravestock, *Canad. J. Chem.*, **47**, 2099 (1969)

Stannic chloride, 1, 1111–1113.

Cyclization of cyclopropyl ketones. Stork *et al.*[1] have developed an interesting new synthetic route to steroids–terpenoids involving acid-catalyzed concerted cyclization of cyclopropyl ketones. For example, the diazoketone (1) undergoes internal diazoketone insertion (see **2**, 82–84) to give the cyclopropyl ketone (2) in good yield. This is then treated overnight at room temperature with benzene containing stannic chloride and a trace of water. The products, obtained in about 80% yield, are a mixture of 3 and 4 (5:1 ratio). The high stereospecificity in the last step argues that the cyclization accompanying cyclopropyl ring opening is a concerted reaction.

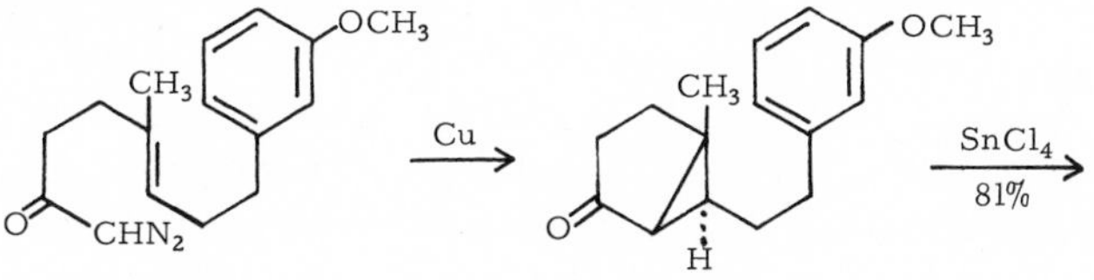

(1)   (2)

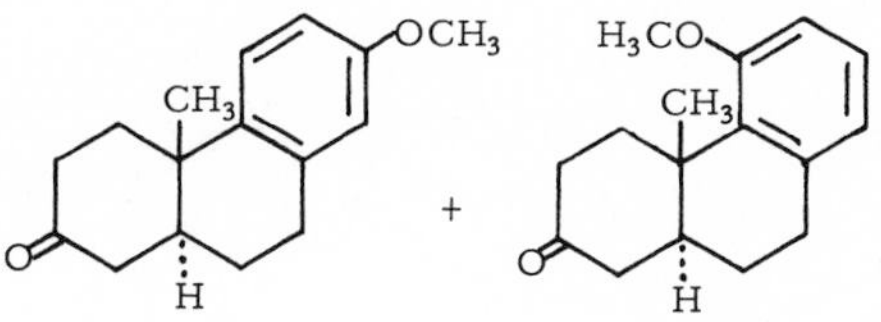

(3)   5:1   (4)

Another example:

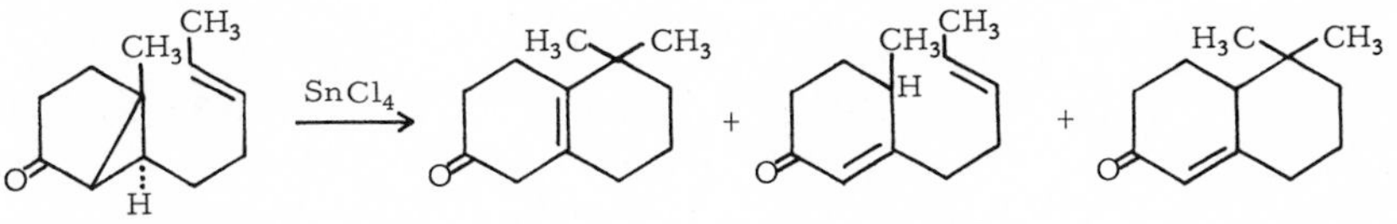

1 : 5 : 3

[1]G. Stork and M. Marx, *Am. Soc.*, **91**, 2371 (1969); G. Stork and M. Gregson, *ibid.*, **91**, 2373 (1969); G. Stork and P. A. Grieco, *ibid.*, **91**, 2407 (1969)

Sulfides

(a) Sodium sulfide nonahydrate, crystalline, $Na_2S \cdot 9\ H_2O$, mol. wt. 240.18, supplied by Fisher.

(b) Technical fused chip sodium sulfide, 60% Na_2S (mol. wt. 78.06), supplied by Baker and Adamson.

(c) Sodium polysulfide. Sulfur (at. wt. 32.07) is dissolved by heating and stirring in aqueous or alcoholic sodium sulfide.

(d) Ammonium sulfide. Aqueous or alcoholic ammonia is saturated with hydrogen sulfide.

Reduction of nitro compounds. A solution of *p*-nitrophenylacetic acid in 6 *N* aqueous ammonia is cooled in ice and saturated with hydrogen sulfide.[1] The solution is gently boiled until nearly all the excess hydrogen sulfide and ammonia

$$O_2N\text{-}C_6H_4\text{-}CH_2CO_2H + 3(NH_4)_2S + 3H_2O \xrightarrow[83-84\%]{} H_2N\text{-}C_6H_4\text{-}CH_2CO_2NH_4 + 3S + 5NH_4OH$$

$$\downarrow CH_3CO_2H$$

$$H_2N\text{-}C_6H_4\text{-}CH_2CO_2H + CH_3CO_2NH_4$$

have escaped and the solution has changed from dark orange-red to pale yellow. The deposited sulfur is removed by filtration and 40 ml. of acetic acid is added to the filtrate. The *p*-aminophenylacetic acid which separates is freed from a little sulfur by recrystallization from water.

Partial reduction of polynitro compounds. In one example[2] a 1-l. round-bottomed flask equipped with a heating mantle and a condenser is charged with 42 g. of 2,2'-dinitrodiphenyl (1)[3] and 600 ml. of 95% ethanol. The solution is heated to reflux and a solution prepared by heating 48 g. of sodium sulfide nona-

$$(1) \xrightarrow[40-50\%]{Na_2S_X} (2)$$

(1)   (2)

hydrate, and 12 g. of sulfur in 150 ml. of water is added slowly to the alcoholic solution. After refluxing for 3 hours the solution is left to cool and stand overnight. About 400 ml. of solvent is removed by distillation and the residue is poured into 1 l. of ice water and the mixture let stand overnight. A solid separates and is collected by filtration. Extraction with ether and further processing gives 2'-nitro-2-aminodiphenyl (2) as a fluffy yellow solid, m.p. 63–65°, in moderate yield.

Another example is the reduction of 2,4-dinitrophenol to 2-amino-4-nitrophenol with aqueous sodium sulfide and ammonium chloride.[4] A suspension of the dinitrophenol in aqueous ammonia and ammonium chloride is stirred at 70° and treated with 60% fused sodium sulfide in 2–3 portions and the temperature is raised to 85°. The mixture is filtered through a hot Büchner funnel and the product which separates on cooling is recrystallized from water after acidification with acetic acid. Here, the nitro group selectively reduced is that *ortho* to the phenolic hydroxyl group. The same relationship applies in the selective reduction of picric acid to picramic acid[5] by treating the substrate in alcoholic alkali at 65° with the calculated amount of sodium sulfide.

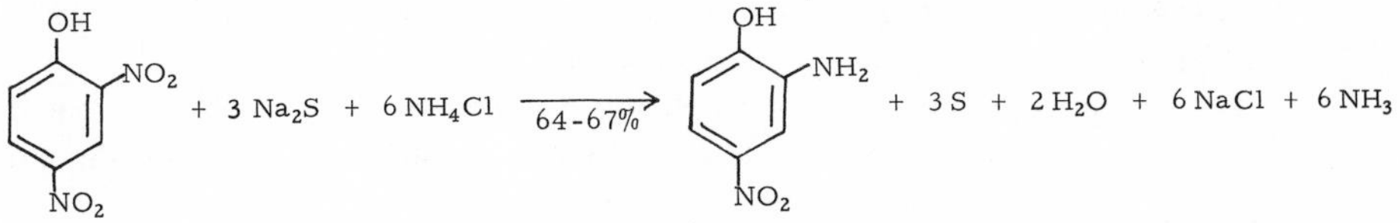

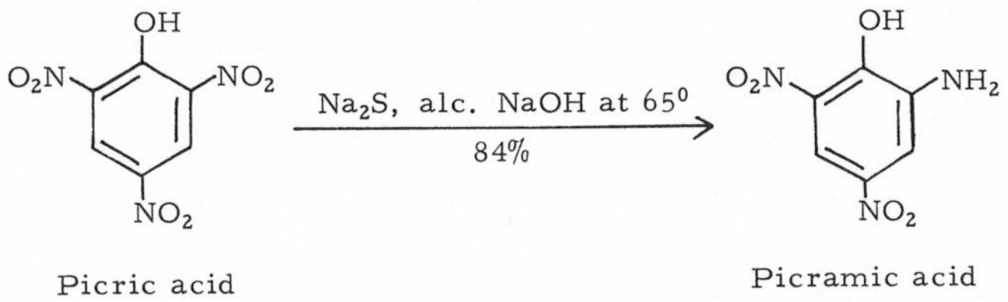

Picric acid   Picramic acid

Note that in the partial reduction of 2,4-dinitro-1-naphthol with alcoholic stannous chloride to 2-nitro-4-amino-1-naphthol[6] the nitro group selectively reduced is not that *ortho* to the hydroxyl group but that in the reactive α-position of the naphthalene nucleus.

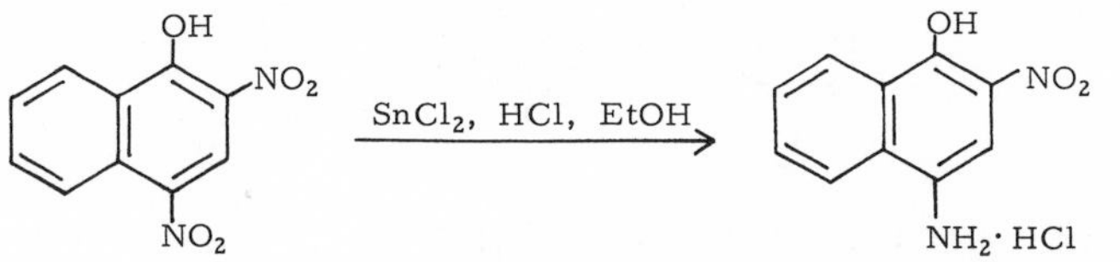

The selective reduction of one of the two equivalent nitro groups of 1,3-dinitro-4,6-diaminobenzene was accomplished[7] by refluxing 22.5 g. of the compound with

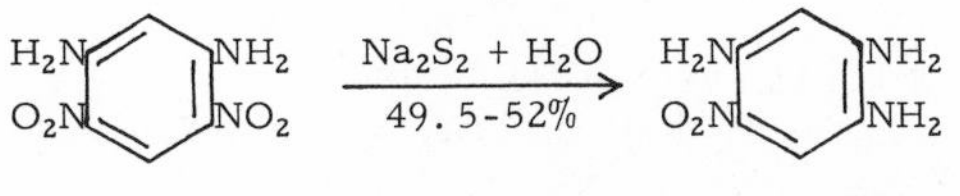

a clear orange-red solution of sodium polysulfide prepared by heating 30 g. of sodium sulfide nonahydrate and 7.25 g. of sulfur with 125 ml. of water.

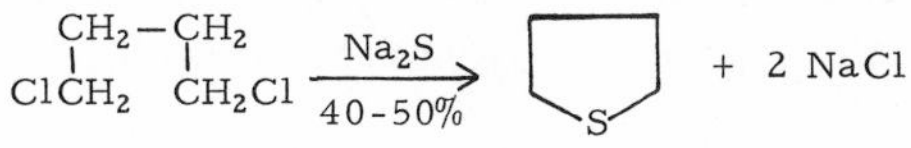

Displacements. 1,4-Dichlorobutane reacts in dimethylformamide with 60% sodium sulfide to give tetrahydrothiophene in 40–50% yield.[8] Somewhat lower yields were obtained when an appropriate amount of the crystalline $Na_2S \cdot 9\ H_2O$ was employed instead of the technical material.

For the preparation of di-*o*-nitrophenyl disulfide, Bogert and Stull[9] treated *o*-nitrochlorobenzene with an alcoholic solution of sulfur in sodium sulfide and

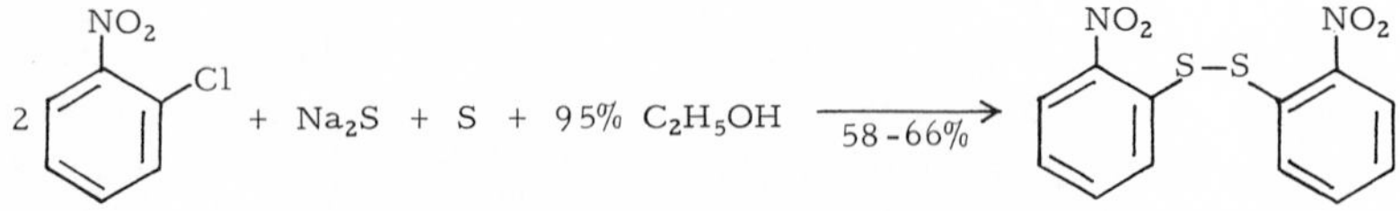

heated the mixture for 2 hours. Similar displacements are involved in the reaction of sodium sulfide with diazotized anthranilic acid,[10] and with ethylene chloro-hydrin:[11]

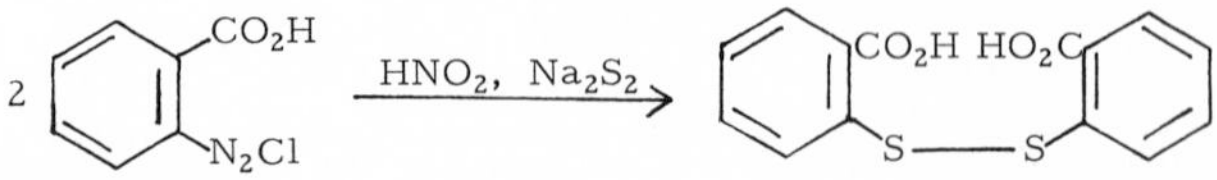

$$HOCH_2CH_2Cl\ +\ Na_2S \cdot 9\ H_2O\ \xrightarrow[79-86\%]{}\ HOCH_2CH_2SCH_2CH_2OH$$

Disproportionation. *p*-Nitrotoluene, heated with a solution of 0.125 mole of sodium sulfite nonahydrate (freshly opened), 0.47 gram atom of flowers of sulfur, and 0.67 mole of sodium hydroxide pellets in 600 ml. distilled water, undergoes disproportionation to give golden yellow needles of *p*-aminobenzaldehyde.[12]

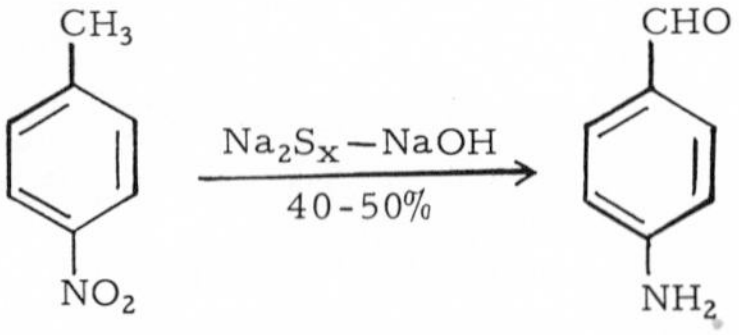

[1]G. Ross Robertson, *Org. Syn., Coll. Vol.*, **1**, 52 (1941)
[2]S. D. Lewis, procedure submitted to *Org. Syn.* 1969
[3]Also known as 2,2'-dinitrobiphenyl, *Org. Syn., Coll. Vol.*, **3**, 339 (1955); supplier: Aldrich.
[4]W. W. Hartman and H. L. Silloway, *ibid., Coll. Vol.*, **3**, 82 (1955)
[5]K. Brand, *J. prakt. Chem.*, **74**, 471 (1906)
[6]*Reagents*, Vol. **1**, 1113
[7]J. H. Boyer and R. S. Buriks, *Org. Syn.*, **40**, 96 (1960)
[8]J. K. Lawson, W. K. Easley, W. S. Wagner, *Org. Syn., Coll. Vol.*, **4**, 892 (1963)
[9]M. T. Bogert and A. Stull, *ibid., Coll. Vol.*, **1**, 220 (1941)
[10]C. F. H. Allen and D. D. MacKay, *ibid., Coll. Vol.*, **2**, 580 (1949)
[11]E. M. Faber and G. E. Miller, *ibid., Coll. Vol.*, **2**, 576 (1943)
[12]E. Campaigne, W. M. Budde, and G. F. Schaeffer, *ibid., Coll. Vol.*, **4**, 31 (1963)

3-Sulfolene, 2, 389–390.

Source of butadiene (2, 390, ref. 2). The preparation of diethyl *trans*-Δ^4-tetra-hydrophthalate by the reaction of 3-sulfolene with diethyl fumarate has been published.[1]

Diels-Alder dienophile. 3-Sulfolene has been used for the first time by Cava[2] as a dienophile, but highly reactive dienes are probably necessary. Thus 3-sulfolene (1) reacts with 1,3-diphenylisobenzofurane (2) in refluxing benzene (64 hours) to give the Diels-Alder adduct (3). Yields as high as 97% can be ob-

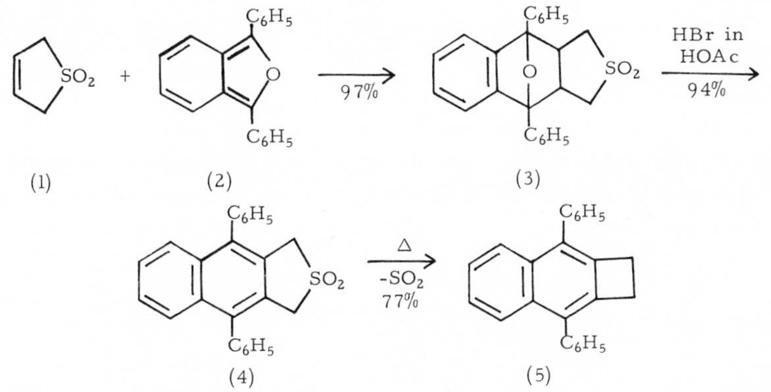

(1)   (2)   (3)

(4)   (5)

tained by use of excess (1). Aromatization of (3) to (4) can be effected in high yield by 48% hydrobromic acid in acetic acid (steam bath). When (4) is refluxed in diethyl phthalate for more than 5 hours, sulfur dioxide is lost with formation of (5), 3,8-diphenylnaphtho[*b*]cyclobutene. The yield is 77% on the basis of changed (4); 26% of (4) is recovered. This synthesis of (5) is more practical than the only previous one employing cyclobutene as starting material.

The same sequence of reactions has been applied to the benz analog of (2), namely, 1,3-diphenylnaphtho[2.3-*c*]furane.

[1] T. E. Sample, Jr., and L. F. Hatch, *Org. Syn.*, **50**, 43 (1970)
[2] M. P. Cava and J. P. VanMeter, *J. Org.*, **34**, 538 (1969)

Sulfur, S. At. wt. 32.07, **1**, 1118–1121.

Dehydrogenation. In an experiment for students,[1] commercial guaiene is dehydrogenated by refluxing with sulfur in triethylene glycol dimethyl ether ("triglyme," b.p. 222°, miscible with water), a high-boiling solvent in which sulfur is moderately soluble. The brilliantly blue guaiazulene is isolated by steam distillation and conversion into the crystalline picric acid salt, or picrate.

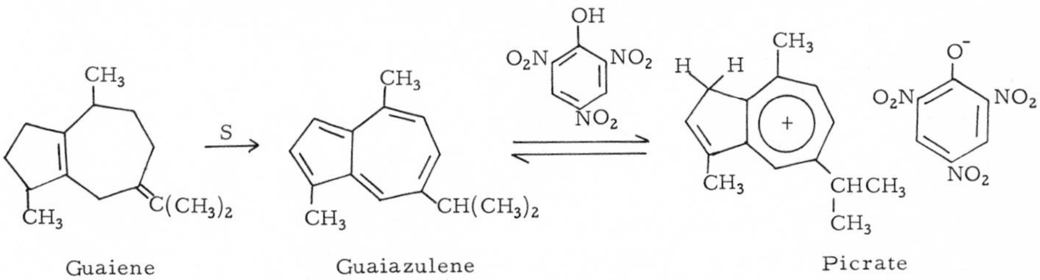

Guaiene   Guaiazulene   Picrate

Wynberg[2] showed that 3-phenyl-2,5-dihydrofurane can be dehydrogenated to 3-phenylfurane in high yield by refluxing a homogeneous solution in dimethyl-formamide containing 2–3% sulfur for 30 min. and steam-distilling the product.

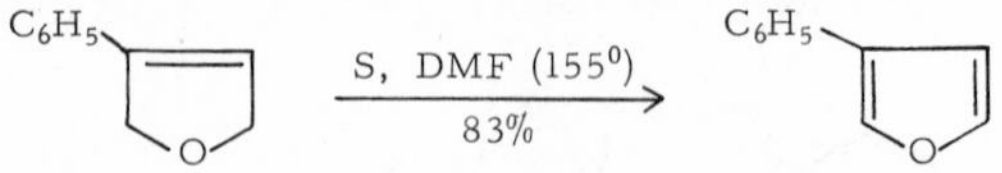

Using only sulfur, a 74% yield of phenylfurane could be obtained, while chloranil in boiling ethylene glycol furnished the aryl furane in 10% yield.

Pettit et al.[3] and Engel et al.[4] have recently developed successful syntheses of a bufadienolide. In both cases the final step involved dehydrogenation of a bufa-20(21)-enolide (1) to a bufa-20,22-dienolide (2). Both laboratories first tried

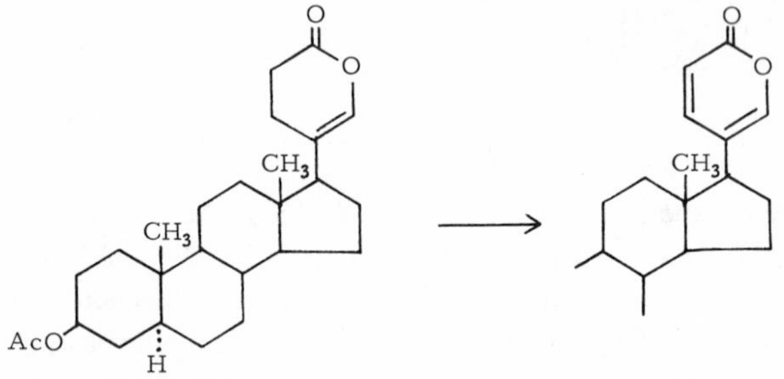

dehydrogenation with 10% palladium-on-charcoal in p-cymene at reflux temperature, a method used successfully by Fried and co-workers in the degradation of eburicoic acid to 14-methylpregnane (**1**, 779–780). In the present case both groups obtained low and variable yields. Pettit et al. then found that the desired dehydrogenation could be effected in 60–70% yield by heating (1) with sulfur at 221–227° for 30 minutes under nitrogen. Engel's group accomplished the dehydrogenation with selenium dioxide in t-butanol (**1**, 997) in 34% yield or by free-radical bromination with NBS followed by dehydrobromination with DMF and lithium chloride (**1**, 609) in 31% yield.

Willgerodt-Kindler reaction, (**1**, 1120–1121). A variation of the Willgerodt-Kindler reaction gives thioamides in good yield *at room temperature and atmospheric pressure* when run in DMF.[5]

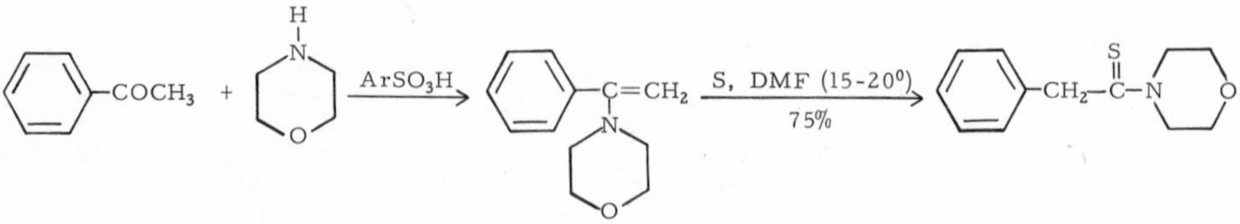

[1]L. F. Fieser, *Org. Expts.*, 2nd Ed., 290–294 (1968)

[2] H. Wynberg, *Am. Soc.*, **80**, 364 (1958)
[3] G. R. Pettit, D. C. Fessler, K. D. Paull, P. Hofer, and J. C. Knight, *J. Org.*, **35**, 1398 (1970)
[4] C. R. Engel, R. Bouchard, A. F. deKrassny, L. Ruest, and J. Lessard, *Steroids*, **14**, 637 (1969)
[5] R. Mayer and J. Wehl, *Angew. Chem., internat. Ed.*, **3**, 705 (1964)

Sulfur monochloride, 1, 1122–1123. Additional supplier: Hooker Chem. Co.

Preparation.[1] Pure sulfur (American Smelting and Refining Co.) is heated to 165°, and pure chlorine is passed through the melt until all sulfur remains dissolved when the solution is cooled to room temperature. The product is then fractionated; S_2Cl_2, b.p. 40°/22 mm.

Episulfides.[1] Sulfur monochloride adds to alkenes to form mixtures of β-chloroalkyl mono-, di-, and trisulfides; the adducts are then converted into the episulfide by reduction and dehydrohalogenation with sodium sulfide (**1,** 1104–1105) or aluminum amalgam.

n = 1, 2, 3

[1] F. Lautenschlaeger and N. Y. Schwartz, *J. Org.*, **34**, 3991 (1969)

Sulfur trioxide–Pyridine, 1, 1127–1128; **2,** 393–394. Supplier: Aldrich.

Deoxygenation of allylic and benzylic alcohols.[1] Allylic and benzylic alcohols are conveniently reduced to the corresponding hydrocarbons in two steps. The alcohol is first converted into a pyridinium alkyl sulfate by reaction with sulfur trioxide–pyridine in THF at 0–3° (3–20 hours). Then lithium aluminum hydride (or $LiAlH_4$–$AlCl_3$, 3 : 1) in THF is added and the mixture stirred for 1 hour at 0° and then at 25° for 3–5 hours. Yields are high. In the case of allylic alcohols

$$\text{ROH} \xrightarrow[\text{THF}]{\text{SO}_3\cdot\text{C}_5\text{H}_5\text{N}} \text{ROSO}_3^-\,\text{C}_5\text{H}_5\text{NH}^+ \xrightarrow{\text{LiAlH}_4} \text{RH}$$

(geraniol and farnesol) neither *cis-trans* isomerization nor allylic transposition is observed to a significant extent.

The method was developed for use in the final step of a synthesis of the sesquiterpene sesquicarene (1).[2]

(1)

[1]E. J. Corey and K. Achiwa, *J. Org.*, **34**, 3667 (1969)
[2]*Idem, Tetrahedron Letters*, 1837 (1969)

Sulfuryl chloride, 1, 1128–1131; **2,** 394–395.

α-*Chlorination of sulfoxides.*[1] Sulfoxides are chlorinated rapidly and generally in good yield by sulfuryl chloride in methylene chloride at −78 to 0°. Thus ethyl phenyl sulfoxide is converted into α-chloroethyl sulfoxide in 89% yield. Monochlorinated sulfoxides are less reactive than the precursors and are readily isolated, but dichlorinated sulfoxides can be prepared.

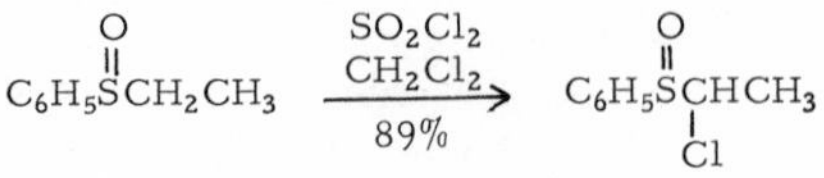

$$
\underset{\displaystyle C_6H_5\overset{\textstyle O}{\underset{\textstyle \|}{S}}CH_2CH_3}{} \quad \xrightarrow[\text{89\%}]{\overset{\text{SO}_2\text{Cl}_2}{\text{CH}_2\text{Cl}_2}} \quad C_6H_5\overset{\textstyle O}{\underset{\textstyle \|}{S}}\underset{\underset{\textstyle Cl}{\textstyle |}}{CH}CH_3
$$

[1]K.-C. Tin and T. Durst, *Tetrahedron Letters*, 4643 (1970)

T

Tetra-*n*-butylammonium acetate, $(C_4H_9)_4\overset{+}{N}(\overset{-}{O}COCH_3)$. Mol. wt. 301.50, m.p. 95–98°.

Prepared by neutralization of tetra-*n*-butylammonium hydroxide with acetic acid.[1,2]

Axial 3-hydroxysterols.[2] S_N2-Displacement of the tosylates of equatorial 3-hydroxysterols gives the axial 3-acetoxysteroid in 80–90% yield together with a small amount of olefin. The method is useful in the case of unsaturated sterols where catalytic reduction of the 3-keto group cannot be used. Tetraethylammonium acetate (**1**, 1136–1137; **2**, 397) has been more commonly used for such epimerizations.

[1]S. Winstein, E. C. Friedrich, R. Baker, and Y. Lin, *Tetrahedron*, **22**, Suppl. 8, 621 (1966)
[2]R. Baker, J. Hudec, and K. L. Rabone, *J. Chem. Soc.*, (C), 1605 (1969)

Tetrachlorosilane, $SiCl_4$. Mol. wt. 169.9, b.p. 57°. Supplier: Pierce Chem. Co.

Amides.[1] Tetrachlorosilane has been used as a coupling reagent for the formation of an amide from a carboxylic acid and an amine. Pyridine is used as solvent. Aromatic amines react at room temperature, but aliphatic amines require a reflux temperature for satisfactory yields. The reported yields are 25–90%. One advantage of this procedure is that the other product formed is silica, which is insoluble in all common solvents.

$$2 \, RCOOH + 2 \, R'NH_2 + SiCl_4 \longrightarrow 2 \, RCONHR' + (SiO_2)_n + 4 \, HCl$$

[1]T. H. Chan and L. T. L. Wong, *J. Org.*, **34**, 2766 (1969)

Tetraethyl dimethylaminomethylenediphosphonate,
$$\begin{array}{c}(C_2H_5O)_2P\overset{O}{\diagup}\\ \diagdown_{CHN(CH_3)_2}\\ (C_2H_5O)_2P\diagup\\ \diagdown_O\end{array}$$
Mol. wt. 299.29, b.p. 114–115°/0.03 mm.

Preparation.[1] The reagent is prepared in 60–62% yield by warming dimethylformamide dimethyl acetal with two equivalents of diethyl phosphite to 60–80° for a short time.

—CHO → —CH₂COOH.[1] The reagent (1) is treated in dioxane with sodium hydride and then with an aldehyde to form a diethyl 1-dimethylaminoalkenylphosphonate (2), which is hydrolyzed in boiling concentrated hydrochloric acid to give a carboxylic acid (3). Ketones do not react with (1).

$$(C_2H_5O)_2P \overset{O}{\diagdown} \diagup CHN(CH_3)_2 \quad \xrightarrow[\substack{60-75\%}]{\substack{1) \ NaH \\ 2) \ RCHO}} \quad R-CH=C \overset{\overset{O}{\underset{\parallel}{P(OC_2H_5)_2}}}{\diagdown N(CH_3)_2} \quad \xrightarrow[75-90\%]{HCl} \quad RCH_2COOH$$

(1)   (2)   (3)

[1]H. Gross and B. Costisella, *Angew. Chem., internat. Ed.*, **7**, 391 (1968)

Tetrahydrofurane, 1, 1140–1141; **2,** 398.

Purification. Seyferth and Spohn[1] recommend that THF be distilled from sodium benzophenone ketyl (*see* **1,** 1034) immediately before use.

[1]D. Seyferth and R. J. Spohn, *Am. Soc.*, **91**, 3037 (1969)

Tetrakis[iodo(tri-*n*-butylphosphine)copper(I)], 2, 400.

Conjugate addition of organometallic reagents. In a study[1] of conjugate addition of methylmagnesium bromide and methyllithium to *trans*-3-pentene-2-one, House found that essentially exclusive conjugate addition took place if this copper(I) iodide–tri-*n*-butylphosphine complex was present. The actual reactant is probably $(n\text{-}C_4H_9)_3P\text{-}CuCH_3$. However, in a later study[2] with 5-methyl-2-cyclohexenone, this procedure gave conjugate addition products in only 34% yield, the major product being a viscous, apparently polymeric, material. In this case the complex of trimethyl phosphite and methylcopper (**2,** 441) was found satisfactory.

Siddall, Biskup, and Fried[3] achieved a simple, direct synthesis of the iso-prenoid (3) by the reaction of the Grignard reagent from 4-methylpent-3-enyl bromide (2) with methyl but-2-ynoate (1) in the presence of 1.5 equivalents of tetrakis[iodo(tri-*n*-butylphosphine)copper(I)]. The reaction must be carried out below −50°. This reaction is an example of conjugate addition of organometallic

$$CH_3C{\equiv}CCOOCH_3 \ + \ (CH_3)_2C{=}CHCH_2CH_2MgBr \ \xrightarrow[57\%]{} \ (CH_3)_2C{=}CHCH_2CH_2\overset{\overset{CH_3}{\underset{|}{}}}{C}{=}CHCOOCH_3$$

(1)   (2)   (3) trans:cis 96:4

reagents to α,β-acetylenic esters (*see* **Dimethylcopperlithium,** This volume).

[1]H. O. House, W. L. Respess, and G. M. Whitesides, *J. Org.*, **31**, 3128 (1966)
[2]H. O. House and W. F. Fischer, Jr., *ibid.*, **33**, 949 (1968)
[3]J. B. Siddall, M. Bikup, and J. H. Fried, *Am. Soc.*, **91**, 1853 (1969)

Tetralin hydroperoxide,   Mol. wt. 164.20, m.p. 54.5°.

Preparation. This hydroperoxide is prepared[1] by passing oxygen through pure tetralin (70°, 24–48 hours.); yield 44–57%.

Epoxidation.[2] The reaction of tetralin hydroperoxide (cumene, cyclohexyl, and *t*-butyl hydroperoxides were also used, but to a lesser extent) with cyclohexyl metaborate (This volume)[3] in the presence of an olefin results in formation of an epoxide from the olefin and generation of tetralol. The new reagent is very similar in behavior to peroxyacids. Transient hydroxonium ions, OH^+, are postu-

$$>C=C< \; + \; ROOH \; \xrightarrow{\Delta} \; >C\overset{O}{\triangle}C< \; + \; ROH$$

lated. Thus oxidation of mesitylene produces mesitol in more than 90% yield. The new system is thus comparable to pertrifluoroacetic acid–boron trifluoride (**1**,

825), which is also believed to generate OH^+.

[1]H. B. Knight and D. Swern, *Org. Syn., Coll. Vol.*, **4**, 895 (1963)
[2]P. F. Wolf and R. K. Barnes, *J. Org.*, **34**, 3441 (1969)
[3]Any alkyl metaborate ester can be used; cyclohexyl metaborate is easily prepared and is soluble in organic solvents.

1,1,3,3-Tetramethylbutylisonitrile (TMBI), (**1**), $(CH_3)_3CCH_2\overset{\overset{\displaystyle CH_3}{|}}{\underset{\underset{\displaystyle CH_3}{|}}{C}}-\overset{..}{N}=C:$ Mol. wt. 139.24, b.p. 55.5–56.5°. Supplier: Columbia Organic   Chemicals.

Preparation. The reagent is prepared[1] in 87% yield from 1,1,3,3-tetramethylbutylamine (Aldrich, Eastman) in two steps: conversion into the formamide by formic acid, and dehydration of the amide with thionyl chloride in DMF.[2]

$$(CH_3)_3CCH_2\overset{\overset{\displaystyle CH_3}{|}}{\underset{\underset{\displaystyle CH_3}{|}}{C}}NH_2 \; \xrightarrow[94\%]{HCOOH} \; (CH_3)_3CCH_2\overset{\overset{\displaystyle CH_3}{|}}{\underset{\underset{\displaystyle CH_3}{|}}{C}}CONH_2 \; \xrightarrow[93\%]{SOCl_2, \; DMF} \; (CH_3)_3CCH_2\overset{\overset{\displaystyle CH_3}{|}}{\underset{\underset{\displaystyle CH_3}{|}}{C}}\overset{..}{N}=C:$$

(1)

Lithium aldimines.[1] Appropriate isonitriles (1,1,3,3-tetramethylbutylisonitrile was used in the initial work) are converted into lithium aldimines (2) by reaction with 1 equivalent of an alkyllithium such as *sec*-butyllithium. These are converted into α-deuterioaldehydes when treated with D_2O followed by acid hydrolysis (oxalic acid). Carbonation of (2) followed by hydrolysis gives an α-keto carboxylic acid.

$$(1) \quad + \quad CH_3CH_2\overset{\underset{\displaystyle CH_3}{|}}{C}HLi \quad \longrightarrow \quad (CH_3)_3CCH_2\overset{\underset{\displaystyle CH_3}{|}}{C}-N=\overset{\displaystyle \cdot Li}{\underset{\displaystyle CHCH_2CH_3}{C}}$$

$$\overset{\underset{\displaystyle CH_3}{|}}{}$$

(2)

$$(2) \quad \xrightarrow[92\%]{D_2O, \ H_3O^+} \quad CH_3CH_2\overset{\underset{\displaystyle CH_3}{|}}{C}HC\overset{\displaystyle O}{\underset{\displaystyle D}{\diagdown}}$$

$$(2) \quad \xrightarrow[80\%]{CO_2, \ H_3O^+} \quad CH_3CH_2\overset{\underset{\displaystyle CH_3}{|}}{C}H-\overset{\displaystyle O}{\overset{\displaystyle \|}{C}}COOH$$

[1]H. M. Walborsky and G. E. Niznik, *Am. Soc.*, **91**, 7778 (1969)
[2]Details of the procedure are available from Dr. H. M. Walborsky upon request.

2,4,4,6-Tetramethyl-5,6-dihydro-1,3-(4H)-oxazine (1). Mol. wt. 141.61, b.p. 58°/25 mm., n^{25}D = 1.4355. Supplier: Columbia Organic Chemicals.

Preparation. The reagent (1) has been prepared[1] in 44% yield by the sulfuric acid-catalyzed condensation of 2-methyl-2,4-pentanediol with acetonitrile. A possible mechanism is outlined:

(1)

Synthesis of aldehydes. Meyers[2] has used this reagent for an interesting new synthesis of aldehydes. The oxazine (1) is converted quantitatively by reaction with an alkyllithium (commercial *n*-butyllithium is satisfactory) in THF at −78° into an anion, which although it is ambident is alkylated exclusively at the C_2-methyl group. Alkyl bromides and iodides are more reactive than alkyl chlorides. The alkylated oxazine (2) can be isolated, but this step is not necessary. Reduction with sodium borohydride in THF–ethanol–water gives a tetrahydro-1,3-oxazine (3) which is hydrolyzed by dilute oxalic acid to an aldehyde (4). The overall transformation is thus RX → RCH$_2$CHO. The methyl group at C_2 in the oxazine (1) can carry a substituent. Thus the sequence has been carried out on the 2-benzyl and 2-carboethoxymethyloxazine derivatives. Furthermore the reduction step can be carried out with sodium borodeuteride; in this way C_1-labeled aldehydes can be synthesized. The overall yields of aldehydes are in the range of 50–70%.

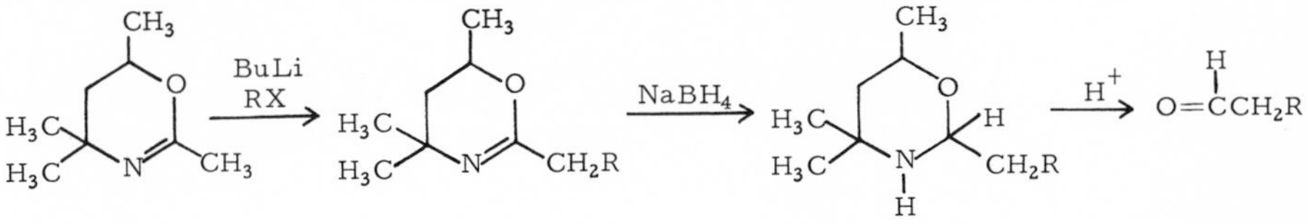

The scheme has been extended to synthesis of α,β-unsaturated aldehydes.[3] Thus reaction of the anion of (1) with carbonyl compounds (aldehydes or ketones) gives an oxazinyl alcohol (5), which on reduction (6) and hydrolysis as before, gives an α,β-unsaturated aldehyde (7) in 48–69% overall yield. The reaction is

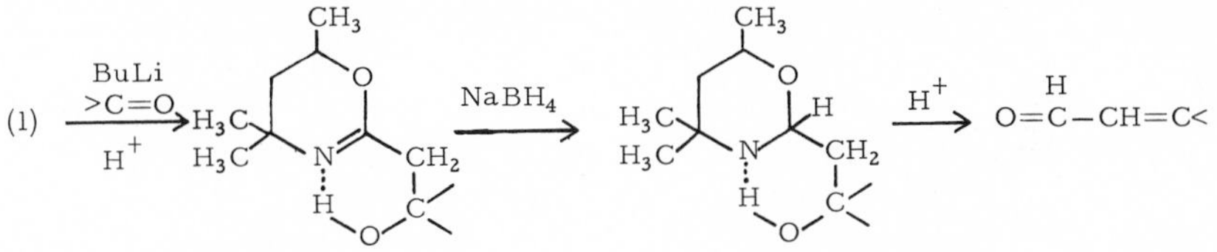

stereoselective; thus β-ionone (8) is converted into *trans*-β-ionylideneacetaldehyde (9) in 54% yield.

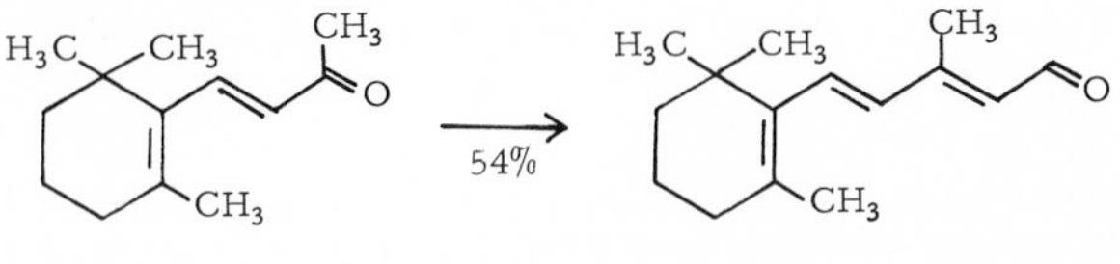

The method has also been used for synthesis of cycloalkanecarboxaldehydes.[4] Reaction of the anion derived from (1) with 1 equiv. of an α,ω-dibromoalkane gives the bromoalkyloxazine (10), which on addition of a second equivalent of base undergoes cyclization to a 2-cycloalkyloxazine (11). Reduction and hydrolysis as before gives the cycloalkanecarboxaldehyde (12). The lithio salt of (1) also reacts at −78° in THF with typical epoxides, for example ethylene oxide, to afford

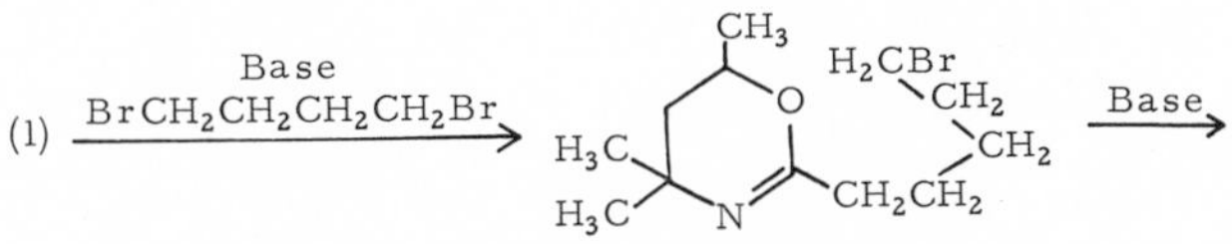

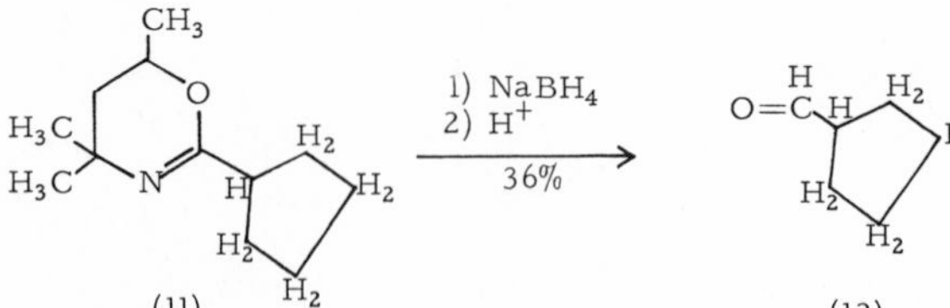

(11)   (12)

an oxazinylcarbinol (13). Reduction with sodium borohydride to a tetrahydro-1,3-oxazine (14), as before, and cleavage with aqueous oxalic acid gives the γ-hydroxyaldehyde (15), which exists primarily as the cyclic hemiacetal (16).[5]

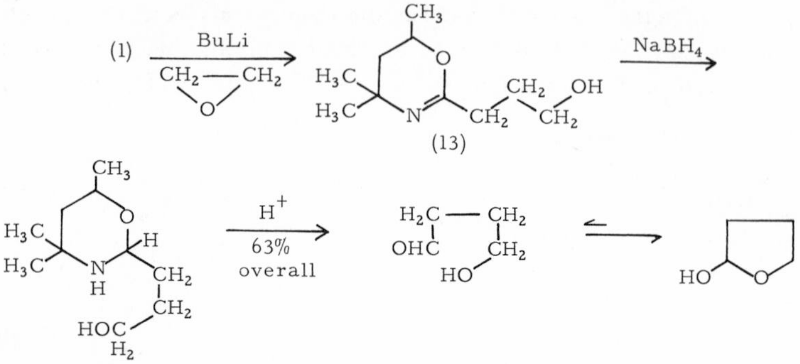

(13)

(14)   (15)   (16)

Synthesis of carboxylic acids. Dihydro-1,3-oxazines are hydrolyzed by aqueous hydrobromic acid to carboxylic acids.[6] This fact, coupled with the fact that this ring system is completely inert to Grignard derivatives, forms the basis for a new synthesis of carboxylic acids.[7] Thus 2,4,4,6-tetramethyl-5,6-dihydro-1,3-(4H)-oxazine (1) is alkylated with 1,5-dibromopentane to produce (2). This is converted into the nitrile (3), then the phenyl ketone (4); finally acid hydrolysis gives 7-benzoylheptanoic acid (5).

Aromatic carboxylic acids were obtained by the same sequence but starting with the 2-phenyl-5,6-dihydro-1,3-oxazine (6).[8]

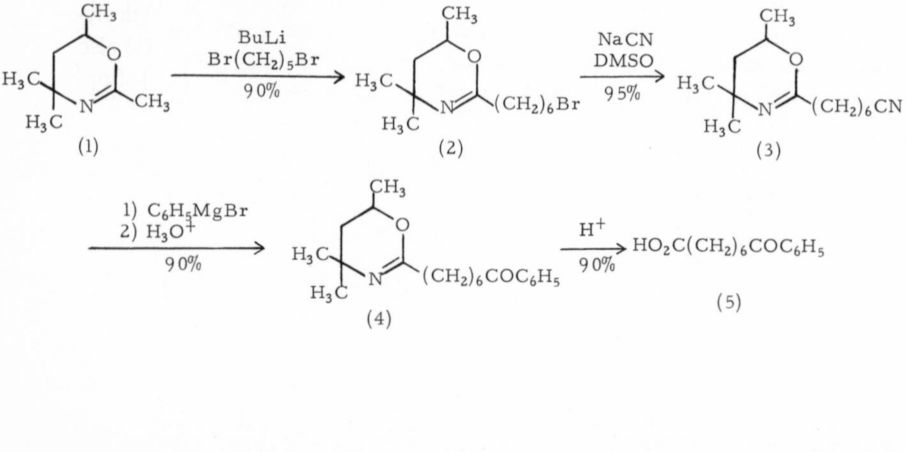

(1)   (2)   (3)

(4)   (5)

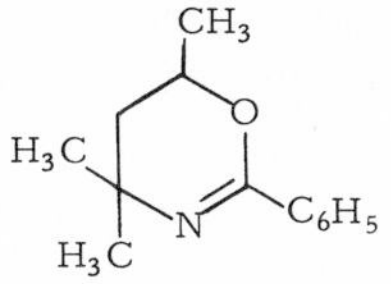

(6)

Synthesis of ketones. 5,6-Dihydro-1,3-oxazines are also useful for synthesis of ketones.[9] Thus the 2-isopropenyl oxazine (1)[10] reacts in THF with Grignard or organolithium reagents to give a 2,2-dialkyltetrahydro-1,3-oxazine (2), which on hydrolysis with oxalic acid gives an α-methyl ketone (3). An important feature of

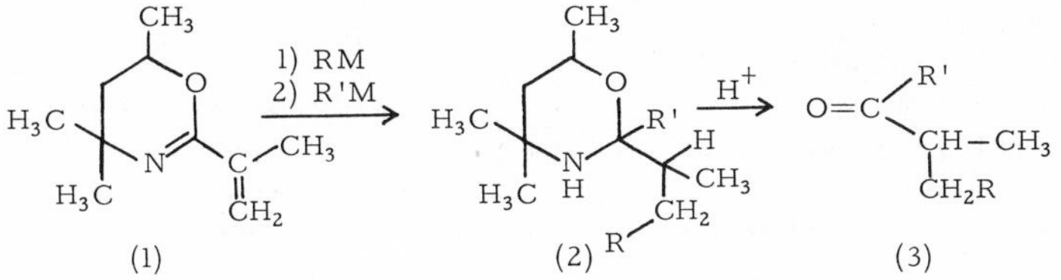

(1)   (2)   (3)

the new synthesis is that two different organometallic reagents can be used. If R = R', (1) is treated with 2.0–2.5 equiv. of the reagent. If R ≠ R', then one reagent is added first and the second after approximately 1 hour. However, R' (but not R) cannot be a bulky group such as isopropyl, *sec*-butyl, cyclohexyl, or *t*-butyl. When only monoalkylation takes place, the resulting oxazine (4) can be reduced and cleaved as usual to an α-methyl aldehyde (5). The formation of (2)

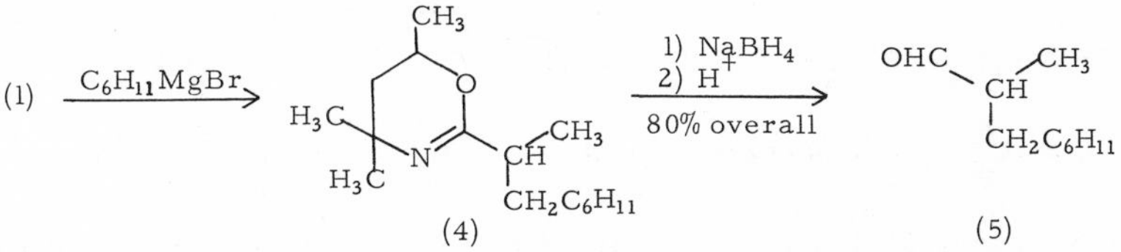

(4)   (5)

from (1) is actually not a simple process as shown by isolation of a ketenimine intermediate (b); the following sequence of intermediates is suggested:

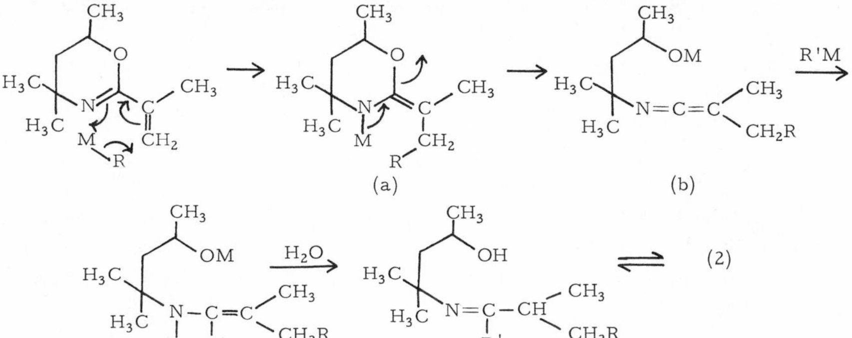

(a)   (b)

(2)

The α-styryldihydro-5,6-dihydro-1,3-oxazine (6)[11] has been used in the same way for the synthesis of α-phenyl aldehydes and ketones.[12]

(6)

[1]E.-J. Tillmanns and J. J. Ritter, *J. Org.*, **22**, 839 (1957)

[2]A. I. Meyers, A. Nabeya, H. W. Adickes, I. R. Politzer, *Am. Soc.*, **91**, 763 (1969)

[3]A. I. Meyers, A. Nabeya, H. W. Adickes, J. M. Fitzpatrick, G. R. Malone, and I. R. Politzer, *ibid.*, **91**, 764 (1969)

[4]A. I. Meyers, H. W. Adickes, I. R. Politzer, W. N. Beverung, *ibid.*, **91**, 765 (1969)

[5]H. W. Adickes, I. R. Politzer, and A. I. Meyers, *ibid.*, **91**, 2155 (1969)

[6]Z. Eckstein and T. Urbánski, *Adv. Heterocyclic Chem.*, **2**, 336 (1963)

[7]A. I. Meyers, I. R. Politzer, B. K. Bandlish, and G. R. Malone, *Am. Soc.*, **91**, 5886 (1969)

[8]Prepared from benzonitrile, 2,4-dimethyl-2,4-pentanediol and 96% sulfuric acid [E.-J. Tillmanns and J. J. Ritter, *J. Org.*, **22**, 839 (1957)]

[9]A. I. Meyers and A. C. Kovelesky, *Am. Soc.*, **91**, 5887 (1969)

[10]Prepared by the acid-catalyzed condensation of 2-methyl-2,4-pentanediol with methacrylonitrile; supplier: Columbia Organic Chemicals.

[11]Prepared by condensation of 2-benzyl-4,4,6-trimethyl-5,6-dihydro-1,3-oxazine (Columbia) with paraformaldehyde catalyzed by trifluoroacetic acid.

[12]A. I. Meyers and A. C. Kovelesky, *Tetrahedron Letters*, 4809 (1969)

N,N,N',N'-Tetramethylethylenediamine (TMEDA), $(CH_3)_2NCH_2CH_2N(CH_3)_2$, 2, 403.

Activation of organolithium compounds. n-Butyllithium and phenyllithium react very slowly with diphenylacetylene. However, the 1:1 complex of either lithium compound and TMEDA reacts with diphenylacetylene at room temperature. For example, the reaction of t-butyllithium under these conditions followed by carbonation gives cis-4,4-dimethyl-2,3-diphenyl-2-pentenoic acid (1) and a trace of 2-phenyl-3-t-butylindone (2). Thus addition takes place as well as metallation.[1]

(1)   (2)

Transmetallation activator. The complex of n-butyllithium and this diamine is recommended for transmetallation of 4-bromo-N,N-dimethylaniline to give p-dimethylaminophenyllithium. Only a moderate yield is obtained with n-butyllithium alone.[2]

Aprotic solvent. The use of this solvent for reactions with organometallic compounds has been reviewed by Agami.[3] The diamine is comparable to monoglyme ($CH_3OCH_2CH_2OCH_3$) in its ability to form complexes with organometallics.

[1] J. E. Mulvaney and D. J. Newton, *J. Org.*, **34**, 1936 (1969)
[2] G. Hallas and D. R. Waring, *Chem. Ind.*, 620 (1969)
[3] C. Agami, *Bull. soc.*, 1619 (1970)

Tetramethyl orthocarbonate, $C(OCH_3)_4$. Mol. wt. 136.15, m.p. $-5.5°$, b.p. 114°, $n^{16}D$ 1.3864.

Preparation. The reagent is prepared by the reaction of chloropicrin (irritant and lachrymator) with sodium methoxide.[1] Compare the preparation of tetraethyl orthocarbonate.[2]

$$CCl_3NO_2 \ + \ 4\ CH_3ONa \longrightarrow C(OCH_3)_4 \ + \ 3\ NaCl \ + \ NaNO_2$$

Ribonucleoside 2′,3′-orthocarbonates. Under acid catalysis (*p*-toluenesulfonic acid) ribonucleosides (1) undergo exchange with tetramethyl orthocarbonate (excess) in anhydrous dioxane solution to give 2′,3′-O-dimethoxymethylidene derivatives (2) in 50–80% yield. The derivatives are converted by mild acid treatment (98% formic acid) into 2′,3′-carbonates (3). The sequence may have useful

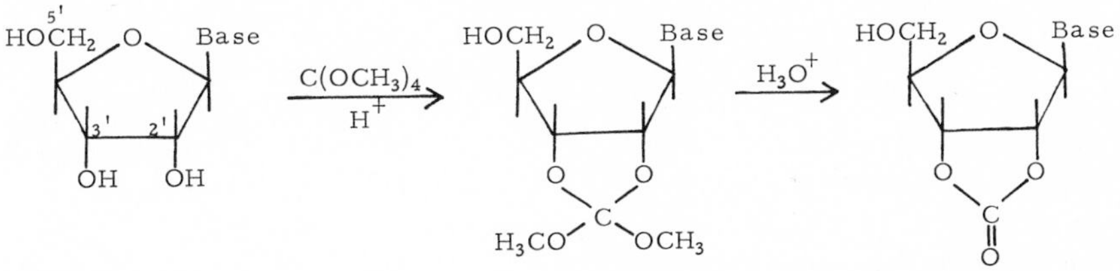

synthetic possibilities in that a base-stable blocking group is transformed into an acid-stable blocking group.[3]

[1] J. D. Roberts and R. E. McMahon, *Org. Syn., Coll. Vol.*, **4**, 457 (1963)
[2] H. Tieckelmann and H. W. Post, *J. Org.*, **13**, 265 (1948)
[3] G. R. Niaz and C. B. Reese, *Chem. Commun.*, 552 (1969)

Thallium, Tl. At. wt. 204.37. Suppliers: Alfa Inorganics, Apache Chemicals.

Azoxy compounds. McKillop, Raphael, and Taylor[1] noted that, in refluxing nitrobenzene, thallium undergoes slow oxidation to thallium(III) oxide with formation of azoxybenzene:

$$2\ C_6H_5NO_2 \ + \ 2\ Tl \longrightarrow C_6H_5\overset{+}{\underset{}{N}}=NC_6H_5 \ + \ Tl_2O_3$$
$$20\text{-}60\%$$

They reasoned that thallium(I) oxide is an intermediate and then carried out the reaction in refluxing ethanol[2] to trap thallium(I) oxide as a stable thallium(I) derivative:

$$2\,ArNO_2 \;+\; 6\,Tl \;+\; 6\,EtOH \;\longrightarrow\; Ar\overset{O^-}{\underset{+}{N}}{=}NAr \;+\; 6\,TlOEt \;+\; 3\,H_2O$$

Yields are in the range 65–93% in the case of nitro compounds with ether or alkyl substituents; however, electron-withdrawing substituents (—CHO, —COR, —COOH, —CN) and phenolic hydroxyl groups totally inhibit the reaction.

[1]A. McKillop, R. A. Raphael, and E. C. Taylor, *J. Org.*, **35**, 1670 (1970)
[2]The role of ethanol is apparently specific. No azoxy compound was isolated when methanol, 2-methyl-2-propanol, or cyclohexanol was used as solvent.

Thallium(I) bromide, **2**, 405–406.

Coupling of aryl and alkyl Grignard reagents. A definitive paper has been published.[1] This paper notes three limitations to the coupling reaction: (a) the reaction is unsuccessful with *ortho*-substituted Grignard reagents; (b) yields are only moderate in the preparation of unsymmetrical biaryls; and (c) aromatic nuclear substituents must be compatible with formation of a Grignard reagent.

[1]A. McKillop, L. F. Elsom, and E. C. Taylor, *Tetrahedron*, **26**, 4041 (1970)

Thallium triacetate, **1**, 1150–1151; **2**, 406.

Aromatic bromides. McKillop, Bromley, and Taylor[1] have described a new, simple, general procedure for bromination of aromatic hydrocarbons using thallium triacetate and bromine (CCl_4). The reaction is remarkable in that a single monobromo product is obtained. Note, however, that the combination of thallium triacetate and chlorine or iodine leads to isomeric mixtures. The method is particularly useful in the case of aromatics that are sensitive to bromine.

Examples:

> Benzene → Bromobenzene (83%)
> Anthracene → 9-Bromoanthracene (91%)
> Diphenyl → 4-Bromodiphenyl (93%)

Oxidative rearrangement of chalcones (ref. 7, **2**, 406). Definitive paper: W. D. Ollis, K. L. Ormand, and I. O. Sutherland, *J. Chem. Soc.* (C), 119 (1970).

[1]A. McKillop, D. Bromley, and E. C. Taylor, *Tetrahedron Letters*, 1623 (1969)

Thallium(III) trifluoroacetate (TTFA), $Tl(OCOCF_3)_3$. Mol. wt. 543.43 (at. wt. Tl 204.37). The reagent is a water-sensitive solid which decomposes above 100° without melting. Supplier: Aldrich.

Preparation[1] by heating under reflux with stirring a suspension of thallium(III) oxide (50 g.) in trifluoroacetic acid (250 ml.) with protection from light.

Thallation of aromatic compounds. A colorless solution of thallium(III) trifluoroacetate (TTFA) in trifluoroacetic acid (TFA) is a powerful reagent for the direct thallation of aromatic compounds:[1]

$$Ar-H + Tl(OCOCF_3)_3 \longrightarrow ArTl(OCOCF_3)_2 + CF_3CO_2H$$

Reaction with substrates which are activated toward electrophilic substitution is generally complete within a few minutes at room temperature. Thallation of mildly deactivated substrates such as the halobenzenes requires longer reaction times at room temperature or about 30 minutes at reflux temperature (73°), and the deactivated substrates benzoic acid and α,α,α-trifluorotoluene are thallated after 21 and 98 hours reflux, respectively. For acid-sensitive substrates (e.g., thiophenes) the conditions of choice are solid TTFA in acetonitrile. The arylthallium ditrifluoroacetates usually crystallize from the reaction mixtures and are isolated by simple filtration, usually in analytically pure form.

Synthesis of aromatic iodides. Treatment of an arylthallium ditrifluoroacetate with aqueous potassium iodide at room temperature results in rapid precipitation of thallium iodide and formation of the corresponding aromatic iodide in high yield:[2]

$$ArTl(OCOCF_3)_2 + 2 KI \longrightarrow ArTlI_2 \xrightarrow{\text{spontaneous}} ArI + TlI + 2 K^+O^-COCF_3$$

Thus a solution of 0.86 g. of mesitylene was added to 10 ml. of TTFA in TFA and the solution allowed to stand at room temperature for one hour; a solution of 8.3 g. of potassium iodide in 25 ml. of water was then added, the mixture was stirred for 15 minutes, *ca.* 1 g. of sodium metabisulfite was added and the mixture stirred until the color changed from blue-black to yellow. The mixture was made basic with sodium hydroxide and extracted with ether. Evaporation of the dried extracts and distillation afforded 1.66 g. (94%) of pure iodomesitylene as a colorless solid, m.p. 29–30°. For acid-sensitive substrates, it is advantageous to use solid TTFA in acetonitrile solution for the thallation step. Addition of aqueous potassium iodide and subsequent workup are carried out as described above. In this manner 2-methylthiophene is converted into 2-methyl-5-iodothiophene in 98% yield.

Synthesis of phenols and aromatic nitriles.[3] In a new phenol synthesis, an aromatic hydrocarbon is first thallated with the reagent in trifluoroacetic acid to give an arylthallium ditrifluoroacetate (which can be isolated) and then oxidized with lead tetraacetate in the presence of 1 equivalent of triphenylphosphine. The resulting aryl trifluoroacetate is then hydrolyzed with dilute base. The triphenyl-

$$ArH \xrightarrow{Tl(OCOCF_3)_3} ArTl(OCOCF_3)_2 \xrightarrow[\text{2) } P(C_6H_5)_3]{\text{1) } Pb(OCOCH_3)_4} ArOCOCF_3 \xrightarrow{\text{dil. NaOH}} ArOH$$

phosphine is added as a specific reducing agent, since TTFA is an efficient reagent for oxidation of phenols. Yields are in the range 40–78%. The reaction is of interest because of control of isomer distribution. The conversion of *p*-cymene into carvacrol shows that steric factors are involved.

Examples:

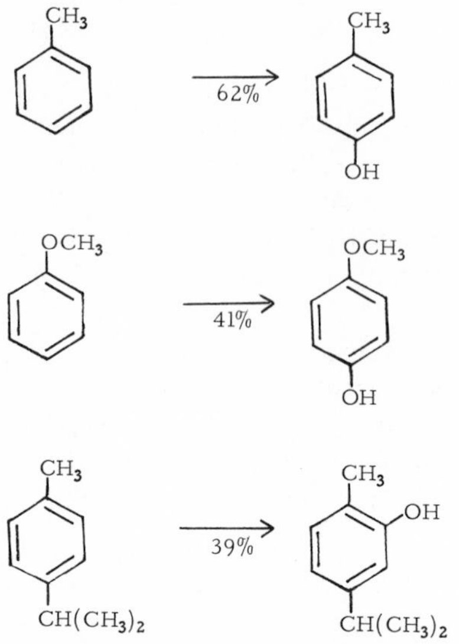

p-Cymene   Carvacrol

Treatment of the arylthallium ditrifluoroacetates with an excess of aqueous potassium cyanide leads to the formation of complex ions $[ArTl(CN)_3]^-K^+$. On photolysis in the presence of excess potassium cyanide these are converted into the nitriles by a radical reaction. Yields are 35–80%.

$$ArTl(OCOCF_3)_2 \xrightarrow[h\nu]{KCN} ArCN$$

An important feature of both synthetic reactions is that the hydroxyl and cyano groups replace the thallium atom.

Unsymmetrical biphenyls.[4] Arylthallium ditrifluoroacetates are phenylated in high yield when irradiated[5] in benzene. The reaction is useful because the replacement of thallium by the phenyl group occurs cleanly. The method is particularly

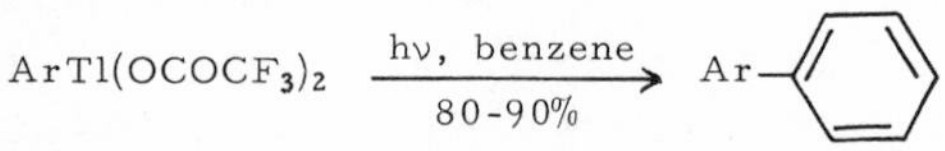

$$ArTl(OCOCF_3)_2 \xrightarrow[80-90\%]{h\nu, \text{ benzene}} Ar\text{—}\bigcirc$$

useful for synthesis of unsymmetrical biphenyls. A radical mechanism operates as in the reaction of arylthallium ditrifluoroacetates with potassium cyanide (see above).

Furthermore photolysis of phenylthallium ditrifluoroacetate in the presence of ammonia gives aniline.

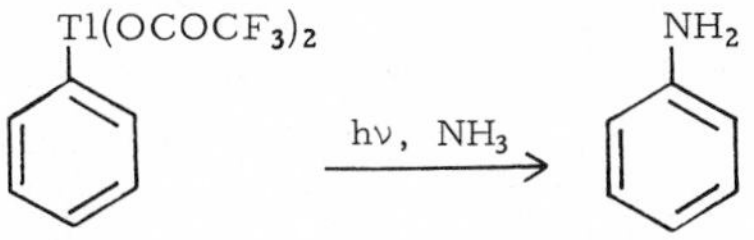

p-**Quinones.** Thallium(III) trifluoroacetate is a very effective oxidation reagent for conversion of 2,6-disubstituted-4-t-butylphenols into the corresponding 2,6-disubstituted p-quinones.[6] The reaction is carried out with a slight excess of two molar equivalents of the reagent in trifluoroacetic acid at room temperature (1–2 hours). Such oxidations are also possible with other groups at C_4 which are capable

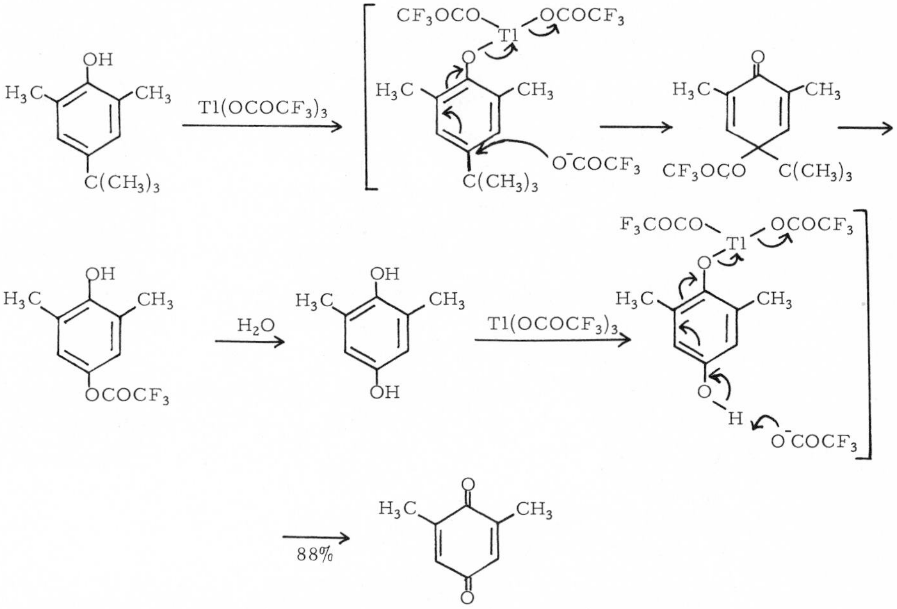

of departing with the bonding electron pair (NH_2, halogen, $OCOCH_3$).

Hydroquinones can also be oxidized by this reaction; in this case, reaction is complete at room temperature within a few minutes. Thus methylhydroquinone is oxidized to methylbenzoquinone (toluquinone) in 77% yield.

[1]A. McKillop, J. S. Fowler, M. J. Zelesko, J. D. Hunt, E. C. Taylor, and G. McGillivray, *Tetrahedron Letters*, 2423 (1969)

[2]*Idem, ibid.*, 2427 (1969)

[3]E. C. Taylor, H. W. Altland, R. H. Danforth, G. McGillivray, and A. McKillop, *Am. Soc.*, **92**, 3520 (1970)

[4]E. C. Taylor, F. Kienzle, and A. McKillop, *ibid.*, **92**, 6088 (1970)

[5]Rayonet photochemical reactor, 3000-Å tubes.

[6]A. McKillop, B. P. Swann, M. J. Zelesko, and E. C. Taylor, *Angew. Chem., internat. Ed.*, **9**, 74 (1970); A. McKillop, B. P. Swann, and E. C. Taylor, *Tetrahedron*, **26**, 4031 (1970)

Thionyl chloride, 1, 1158–1163; **2,** 412.

Purification. Rigby[1] has published a simple method for removing sulfur chlorides and sulfuryl chloride from technical thionyl chloride by distillation of a mixture of thionyl chloride (250 ml.) and dipentene (12 ml. or less). The product is obtained in 80–90% yield and is colorless. It still contains a trace of an impurity which is innocuous for many purposes, but which can be removed by simple distillation. Finally, the purified thionyl chloride can be distilled from 1–2% of linseed oil, and then redistilled through a column; b.p. 76.7°/760 mm., d_{20}^{20} 1.6407. This material remains colorless when stored in the dark for several years.

Synthesis of heterocycles.[2] 2,1-Benzisothiazole is prepared conveniently, if in only very low yield, by adding thionyl chloride (2.2 m.) very slowly from a dropping funnel to a stirred and refluxing solution of *o*-toluidine (1 m.) in a mixture of xylenes, b.p. 141–144°. The reflux condenser is connected to a trap for hydrogen chloride and sulfur dioxide. A yellow solid begins to separate. The mixture is refluxed for 24 hours and the dark brown solution is steam-distilled to give a

mixture of 2,1-benzisothiazole, *o*-toluidine, and water. The aqueous layer is extracted twice with xylene and the combined extract is washed neutral with hydrochloric acid and distilled, b.p. 82–84°/1 mm.

cis-Crotonyl chloride. Published preparations of *cis*-crotonyl chloride give a product contaminated with appreciable amounts of the *trans*-isomer. Even the method of Lee[3] (**1,** 1247; **2,** 445) gives mainly the *trans*-isomer. However, *cis*-chloride of 97% purity can be obtained by the reaction of *cis*-crotonic acid with neat thionyl chloride at 0° (90–150 minutes) followed by distillation. The product tends to isomerize and should be stored at −20°.[4]

[1]W. Rigby, *Chem. Ind.,* 1508 (1969)
[2]M. Davis and E. Homfeld, procedure submitted to *Org. Syn.* 1970
[3]J. B. Lee, *Am. Soc.,* **88,** 3440 (1966)
[4]M. B. Hocking, *Canad. J. Chem.,* **46,** 466 (1968)

Thiourea, 1, 1164–1167; **2,** 412–413.

Inclusion complexes. The explanation which we presented (**2,** 412–413) for the formation of a thiourea inclusion complex from *trans*-4-isopropylcyclohexene-1-carboxylic acid but not from the *cis*-isomer is invalidated by the further finding of the Delft group[1] that the methyl esters of these two acids both form thiourea inclusion complexes.

[1]H. van Bekkum, A. A. B. Kleis, D. Medema, P. E. Verkade, and B. M. Wepster, *Rec. trav.*, **81**, 833 (1962)

Titanium tetrachloride, 1, 1169–1171; **2**, 414–415.

Knoevenagel condensation. The most generally used catalyst for the condensation of aldehydes with malonic ester is pyridine with or without piperidine.[1] Lehnert[2] has reported recently that yields of alkylidenemalonates are improved considerably if a combination of titanium tetrachloride (0.1 mole) and pyridine (0.2 mole) is used for the condensation of the aldehyde (0.05 mole) and malonic

$$\underset{R}{\overset{H}{>}}C{=}O \ + \ CH_2(COOC_2H_5)_2 \ \longrightarrow \ \underset{R}{\overset{H}{>}}C{=}C(COOC_2H_5)_2$$

ester (0.05 mole). The reaction is carried out at 0–22° for 8–70 hours in either THF or dioxane as solvent. Yields by the new procedure range from 75 to 100% and in most cases are considerably higher than those reported in the older literature.

N-Alkyl benzophenone imines. Amines generally react very slowly with aromatic ketones to form imines. However, if titanium tetrachloride is used as catalyst, the reaction of benzophenones proceeds at room temperature in benzene with yields of 75–98%.[3]

$$2 \quad \underset{X'}{\overset{X}{\bigominus}}C{=}O \ + \ 6\,H_2N{-}R \ + \ TiCl_4 \ \longrightarrow 2 \quad \underset{X'}{\overset{X}{\bigominus}}C{=}N{-}R \ + \ TiO_2 \ + \ 4\,[H_3\overset{+}{N}{-}R]Cl^{-}$$

(X = H, Cl, Br)

[1]G. Jones, *Org. Reactions*, **15**, 204 (1967)
[2]W. Lehnert, *Tetrahedron Letters*, 4723 (1970)
[3]I. Moretti and G. Torre, *Synthesis*, 141 (1970)

_p_-Toluenesulfonyl azide (Tosyl azide), 1, 1178–1179; **2**, 415–417.

α-Diazoaldehydes and -ketones; α-diazocarboxylic acid esters. Formylation of a carbonyl compound by methyl formate in ether in the presence of sodium or sodium methoxide gives the formyl derivative (1), which on reaction with tosyl

$$\underset{O}{\overset{}{R\,\underset{\|}{C}CH_2R'}} \ \xrightarrow[\text{Na or NaOCH}_3]{\text{HCOOCH}_3} \ \underset{O \quad CHO}{\overset{}{R\,\underset{\|}{C}{-}\underset{|}{C}HR'}} \ \xrightarrow{\text{Tos-N}_3} \ \underset{O \quad N_2}{\overset{}{R\,\underset{\|}{C}{-}\underset{\|}{C}{-}R'}}$$

$$\qquad\qquad\qquad\qquad (1) \qquad\qquad\qquad (2)$$

azide in methylene chloride (triethylamine) yields an α-diazoaldehyde or -ketone (2). Overall yields are 65–75%. In the case of methyl ketones, the free formyl derivatives are unstable and are isolated as the stable sodium salts, which are then treated with tosyl azide. Yields in this case are in the range 50–75%.[1]

$$R\underset{\underset{O}{\|}}{C}CH_3 \longrightarrow R\underset{\underset{O}{\|}}{C}-\overset{-}{C}H-CHO \ Na^+ \xrightarrow{Tos-N_3} R\underset{\underset{O}{\|}}{C}CH=N_2$$

α-Diazocarboxylic acid esters are obtained in the same way starting with α-monosubstituted acetic acid esters.

$$RCH_2COOC_2H_5 \longrightarrow R\underset{\underset{CHO}{|}}{C}HCOOC_2H_5 \longrightarrow R\underset{\underset{N_2}{\|}}{C}COOC_2H_5$$

A one-step conversion of α-methylene ketones into α-diazoketones is only possible if the methylene group is activated, for example, by an aryl group. However, if the ketone is first formylated the conversion proceeds satisfactorily. Regitz and Rüter[2] have prepared a series of 2-diazocycloalkanones from C_5 to C_{12} in this way (compare **2**, 62). *Caution*: The products explode when warmed.

n = 3 to 10

Azide synthesis (**2**, 415). In the definitive paper[3] describing the conversion of primary amines into azides by a diazo transfer reaction, methyllithium is preferred to methylmagnesium chloride as the base. The reaction has been extended to hydrazones. Thus, when the hydrazones of benzophenone, fluorenone, and acetophenone are treated with methylmagnesium chloride and then with tosyl azide, the corresponding diazoalkanes are obtained in about 20% yield.

[1] M. Regitz and F. Menz, *Ber.*, **101**, 2622 (1968)
[2] M. Regitz and J. Rüter, *ibid.*, **101**, 1263 (1968); the detailed procedure for preparation of 2-diazocyclohexanone has been submitted by M. Regitz, J. Rüter, and A. Liedhegener to *Org. Syn.* 1969
[3] J.-P. Anselme and W. Fischer, *Tetrahedron*, **25**, 855 (1969)

p-Toluenesulfonyl chloride, **1**, 1179–1185.

Tosylates. Tosylates of phenols and acidic alcohols can be prepared conveniently and in high yields by stirring an acetone solution of tosyl chloride and the alcohol or phenol with an excess of aqueous sodium hydroxide (room temp., overnight).[1]

$$\underline{p}\text{-CH}_3C_6H_4SO_2Cl + ROH \xrightarrow[75-98\%]{NaOH, \ acetone} \underline{p}\text{-CH}_3C_6H_4SO_2OR$$

[1] S. E. Wentworth and P. L. Sciaraffa, *Org. Prep. Proc.*, **1**, 225 (1969)

p-Toluenesulfonyl chloride–Lithium chloride.

Allylic chlorides. Stork *et al.*[1] have published a method for conversion of allylic alcohols into allylic chlorides without rearrangement. For example, the sensitive acetal alcohol (1) with geraniol geometry was treated in ether–HMPT with methyllithium in ether and then with *p*-toluenesulfonyl chloride and lithium chloride in ether–HMPT. Workup after standing overnight gave the corresponding chloride (2) in 80% yield with no detectable rearrangement. The same method was equally successful with the *cis*-isomer of (1) with nerol geometry.

$$(CH_3O)_2CHCH_2CH_2 \diagdown C=C \diagup CH_2OH \diagup H \longrightarrow (CH_3O)_2CHCH_2CH_2 \diagdown C=C \diagup CH_2Cl \diagup H$$

(1)   (2)

Allylic chlorides are generally more useful than allylic alcohols for further synthetic operations since they are subject to S_N2 displacements and, particularly, to coupling with Grignard reagents. Thus (2) on coupling with methallyl chloride in THF–HMPT gives the 1,5-diene (3) in high yield. Use of HMPT as solvent is essential for satisfactory yields.

$$\underset{CH_3}{(2) \ + \ CH_2=CCH_2Cl} \longrightarrow \underset{(3)}{(CH_3O)_2CHCH_2 \diagdown C=C \diagup CH_2CH_2\overset{CH_3}{C}=CH_2}$$

[1]G. Stork, P. A. Grieco, and M. Gregson, *Tetrahedron Letters*, 1393 (1969)

p-Toluenesulfonylhydrazine, **1**, 1185–1187; **2**, 417–423.

Acetylene synthesis (**2**, 419–422). Wieland[1] has extended the fragmentation of α,β-epoxyketones to give acetylenes to a new synthesis of acetylenes by the reaction of α-halo or α-sulfonyloxy ketones with *p*-toluenesulfonylhydrazine. Thus reaction of either (1a) or (1b) with *p*-toluenesulfonylhydrazine in the presence of potassium acetate in methylene chloride–acetic acid (4 days) gives the acetylene (2) in 50% yield.

1a: R = O_3SCH_3
1b: R = F

(2)

[1]P. Wieland, *Helv.*, **53**, 171 (1970)

Tri-*n*-butyltin hydride, 1, 1192–1193; **2,** 424.

Preparation. The organotin hydride is prepared conveniently in 88% yield by reduction of bis-[tri-*n*-butyltin] oxide[1] with an equimolar quantity of lithium aluminum hydride.[2]

$$2[(\underline{n}\text{-}C_4H_9)_3Sn]_2O \; + \; LiAlH_4 \longrightarrow 4(\underline{n}\text{-}C_4H_9)_3SnH \; + \; LiAlO_2$$

Generation in situ. Organotin hydrides are generally prepared by reduction of an organotin halide or oxide with $LiAlH_4$. Japanese chemists[3] recently reported the preparation by reaction of organotin oxides with polymethylhydrosiloxane:[4]

$$(R_3Sn)_2O + [CH_3SiHO]_n \rightarrow 2R_3SnH + [CH_3SiO_{1.5}]_n$$

Grady and Kuivila[5] reasoned that it should be possible to carry out a reduction by simply mixing the polysiloxane, an organotin oxide, and substrate. The method was found to be as satisfactory as the conventional method for reduction of alkyl and aryl halides and *gem*-dibromocyclopropanes.

Reduction of α-ketocyclopropanes. Cyclopropyl methyl ketone (1) is reduced by the reagent in methanol under the influence of radical initiators [hν, azobisiso-butyronitrile, **1,** 45] to propyl methyl ketone as the only product of reduction (51% yield). In the absence of an initiator the reaction is slow and results in reduction of the carbonyl group.[6]

Reduction of diazonium fluoroborates. Diazonium fluoroborates are reduced smoothly by the reagent in ether, THF, or acetonitrile. Triethylsilane (**1,** 1218; **2,** 433) is equally satisfactory. The reaction therefore provides a method for replacement by hydrogen of an amino group attached to an aromatic ring.[7]

Review. The reduction of organic compounds by organotin hydrides has been reviewed in depth by Kuivila.[8]

[1]Suppliers: Aldrich; Metal and Thermit Corp., Rahway, N.J. 07065
[2]W. J. Considine and J. J. Ventura, *Chem. Ind.,* 1683 (1962); and H. G. Kuivila, *Synthesis,* 500 (1970)
[3]K. Hayashi, J. Iyoda, and I. Shiihara, *J. Organometal. Chem.,* **10,** 81 (1967)
[4]Union Carbide Co.; General Electric Dri-Film 1040
[5]G. L. Grady and H. G. Kuivila, *J. Org.,* **34,** 2014 (1969)
[6]M. Pereyre and J.-Y. Godet, *Tetrahedron Letters,* 3653 (1970)
[7]J. Nakayama, M. Yoshida, and O. Simamura, *Tetrahedron,* **26,** 4609 (1970)
[8]H. G. Kuivila, *Synthesis,* 499 (1970)

Trichloramine, 1, 1193–1194; **2**, 424–425.

vic-Dichlorides. *vic*-Dichlorides have usually been prepared by the reaction of olefins with sulfuryl chloride.[1] Field and Kovacic[2] report that the reaction of olefins with trichloramine gives *vic*-dichlorides in about 90% yield.

$$3 \ CH_3(CH_2)_3CH{=}CH_2 \ + \ 2 \ NCl_3 \ \xrightarrow[92-94\%]{CH_2Cl_2} \ 3 \ CH_3(CH_2)_3CHClCH_2Cl \ + \ N_2$$

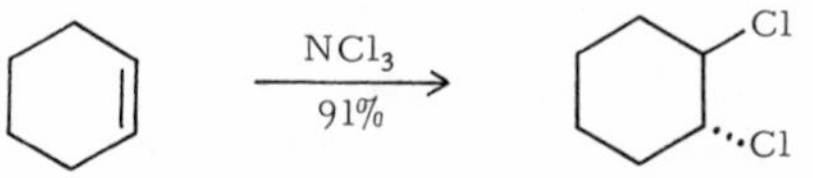

1-Aminoadamantanes. 1-Aminoadamantanes can be prepared in 67–87% yield by the reaction of adamantane or an alkyladamantane with trichloroamine and aluminum chloride in methylene chloride at 10–15°.

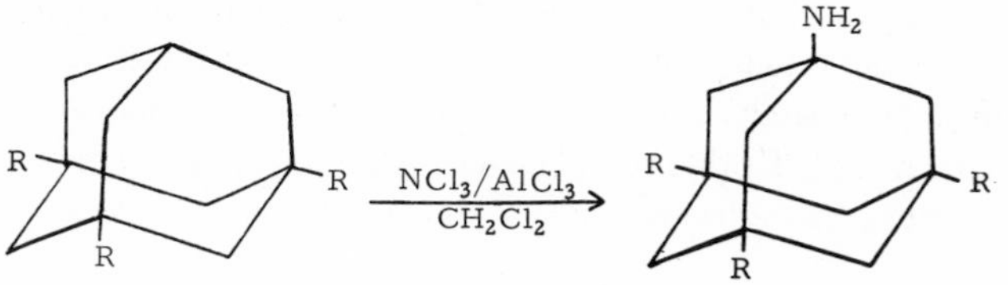

R = H or CH₃

It is also possible to carry out the reaction using tricyclic precursors of adamantanes.[3]

t-Alkylamines. *t*-Alkylamines are obtained easily from the reaction of *t*-alkyl halides with trichloramine and aluminum chloride in methylene chloride at −10°. Skeletal rearrangements occur in the similar reaction with primary halides. Primary and secondary amines, as well as N-alkylaziridines, are generated from *sec*-halides in this reaction.[4]

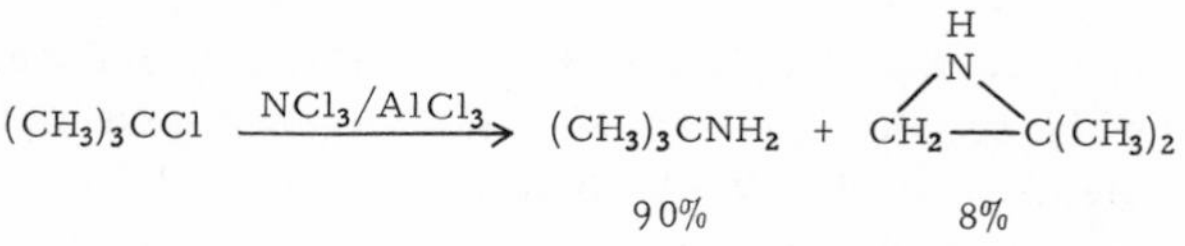

[1]M. S. Kharasch and H. C. Brown, *Am. Soc.*, **61**, 3432 (1939); M. S. Kharasch and A. F. Zavist, *ibid.*, **73**, 964 (1951)
[2]K. W. Field and P. Kovacic, *Org. Syn.* submitted 1969; *Synthesis*, 135 (1969)
[3]P. Kovacic and P. D. Roskos, *Am. Soc.*, **91**, 6457 (1969)
[4]P. Kovacic and M. K. Lowery, *J. Org.*, **34**, 911 (1969)

Trichloroethanol, CCl_3CH_2OH. Mol. wt. 149.40, b.p. 52–54°/10 mm. Supplier: Aldrich.

Trichloroethyl esters. In the total synthesis of cephalosporin C, carboxyl groups were protected as the trichloroethyl esters, prepared by reaction of the acid with trichloroethanol in the presence of DCC (Py). The protective group was

$$-COOH \; \underset{Zn-HOAc}{\overset{\overset{\displaystyle DCC}{\overset{\displaystyle CCl_3CH_2OH}{\longrightarrow}}}{\rightleftarrows}} \; -COOCH_2CCl_3$$

removed by reduction with zinc dust–90% aqueous acetic acid at 0°.[1] These esters have proved useful in other cases, for example, in synthesis of prostaglandins.[2]

[1]R. B. Woodward et al., Am. Soc., **88**, 852 (1966)
[2]J. E. Pike, F. H. Lincoln, and W. P. Schneider, J. Org., **34**, 3552 (1969)

β,β,β-Trichloroethyl chloroformate, 2, 426.

The β,β,β-trichloroethoxycarbonyl group has proved useful in the synthesis of optically active 1,2-diacylglycerols (5).[1] Thus 1,2-isopropylidene–sn-glycerol (1) is treated with β,β,β-trichloroethyl chloroformate and pyridine in dry chloroform. The isopropylidene group of the product (2) is then cleaved with dilute hydrochloric acid to give sn-glycerol-3-β,β,β-trichloroethylcarbonate (3). This is then acylated with an acyl chloride (pyridine–chloroform) (4) and in the final step the protective group is cleaved with activated zinc in acetic acid.[2]

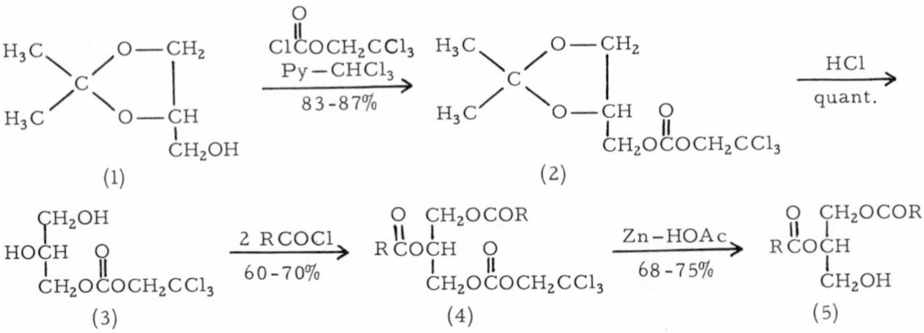

Optically active 2,3-diacyl glycerols (9) can also be prepared from the intermediate (3) in the sequence above.[3] Thus (3) is treated with triphenylmethyl chloride and pyridine in chloroform to give 1-O-triphenylmethyl-sn-glycerol-2,3-carbonate (6). The carbonate group is then hydrolyzed with dilute sodium hydroxide in ethanol to give (7). This is acylated as usual (8); finally the protective group is hydrolyzed with boric acid and trimethyl borate (Matheson, Coleman and Bell).

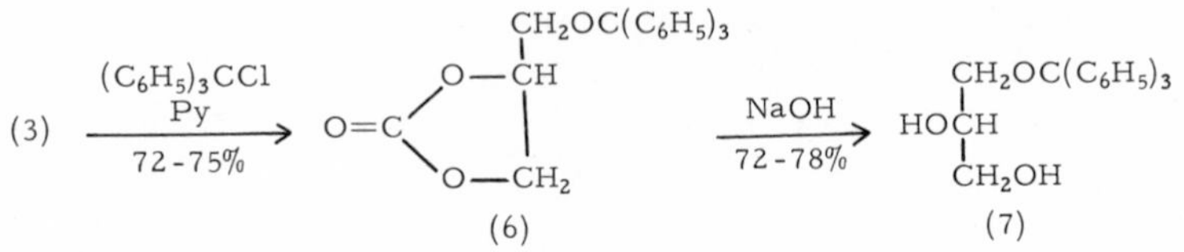

$$\xrightarrow[42-74\%]{2\ RCOCl}\quad \underset{\underset{CH_2OCR}{\overset{\parallel}{O}}}{\overset{CH_2OC(C_6H_5)_3}{\underset{RCOCH}{\overset{\overset{O}{\parallel}}{\underset{}{}}}}O}$$

$$\underset{(8)}{} \qquad \xrightarrow[50-65\%]{\underset{B(OCH_3)_3}{H_3BO_3}} \quad \underset{\underset{CH_2OCR}{\overset{\parallel}{O}}}{\overset{CH_2OH}{\underset{RCOCH}{\overset{\overset{O}{\parallel}}{}}}O}$$

$$(9)$$

[1] The nomenclature of optically active glycerol derivatives is described in *Biochem.*, **6**, 3287 (1967)

[2] F. R. Pfeiffer, C. K. Miac, and J. A. Weisbach, procedure submitted to *Org. Syn.* 1970

[3] F. R. Pfeiffer and J. A. Weisbach, procedure submitted to *Org. Syn.* 1970

Tri(2-chloroethyl) orthoformate, $HC(OCH_2CH_2Cl)_3$. Mol. wt. 216.09, b.p. 154–156°/11 mm.

The reagent is prepared in 40% yield by treating trimethyl orthoformate with 2-chloroethanol at 100° for two hours.

Nucleoside di-2-chloroethyl orthoformates. The reagent has been used for protection of hydroxyl groups of nucleosides. Thus, when 2′,3′-O-dibenzoyluridine is heated with a large excess of reagent at 100° for 2 hours, the corresponding

5′-orthoformate is obtained in 76% yield. The protective group is stable to base but is readily removed by 80% acetic acid at room temperature (1 hour).[1]

[1] T. Hata and J. Azizian, *Tetrahedron Letters*, 4443 (1969)

Trichloroisocyanuric acid, **2**, 426–427.

Chlorination. Ziegler et al.[1] reported one instance of allylic chlorination with the reagent: cyclohexene → 3-chlorocyclohexene (29.2% yield).

The reagent has been used to chlorinate certain saturated cyclic ethers.[2] Thus tetrahydrofurane gives *trans*-2,3-dichlorotetrahydrofurane as the major product (26% yield).

Juenge[3] has now found that the reagent is effective for nuclear or side-chain halogenation of aromatic systems, the former occurring under ionic conditions, the latter under free-radical conditions. Thus benzene and naphthalene are chlorinated under Lewis acid catalysis; 1-chloronaphthalene can be prepared in 58% yield. Toluene under similar conditions gives a mixture of 2- and 4-chlorotoluene in 66% yield. In the presence of benzoyl peroxide, benzyl chloride is obtained in 44% yield.

[1] K. Ziegler et al., *Ann.*, **551**, 80 (1942) [see p. 108]

[2] E. C. Juenge, P. L. Spangler, and W. P. Duncan, *J. Org.*, **31**, 3836 (1966)

[3] E. C. Juenge, D. A. Beal, and W. P. Duncan, *ibid.*, **35**, 719 (1970)

Trichlorosilane, $HSiCl_3$. Mol. wt. 135.5, b.p. 31.5°. Suppliers: Pierce Chem. Co., Pfalz and Bauer.

Reductions. The combination of trichlorosilane and tertiary amines (tri-*n*-propylamine, tri-*n*-butylamine used for the most part) has been used by Benkeser to effect some interesting reductions. Thus polyhalo compounds are reduced cleanly and selectively by this combination:[1]

$$CCl_4 + HSiCl_3 + R_3N \longrightarrow CCl_3H \ (82\%) + SiCl_4 + R_3N$$

$$BrCCl_3 + HSiCl_3 \xrightarrow{R_3N} HCCl_3 \ (quant.) + BrSiCl_3$$

$$CCl_3CCl_3 + HSiCl_3 \xrightarrow{R_3N} Cl_2C{=}CCl_2 \ (86\%) + SiCl_4$$

Carbonyl compounds can be converted into organosilicon derivatives:[2]

$$\underset{C_6H_5}{\overset{C_6H_5}{>}}C{=}O \xrightarrow{HSiCl_3 + R_3N} \underset{C_6H_5}{\overset{C_6H_5}{>}}CHSiCl_3$$

Aromatic carboxylic acids are reduced to benzylic trichlorosilanes, which are cleaved by base to toluenes. It is thus possible to reduce a carboxyl group to a methyl group in two steps.[3]

$$C_6H_5COOH \xrightarrow{HSiCl_3 - (n\text{-}C_3H_7)_3N} C_6H_5CH_2SiCl_3 \ (58\%)$$

$$p\text{-}BrC_6H_4COOH \longrightarrow p\text{-}BrC_6H_4CH_2SiCl_3 \ (58\%)$$

Sulfenyl, sulfinyl, and sulfonyl chlorides are reduced to symmetrical disulfides by the trichlorosilane–tri-*n*-propylamine system.[4]

Examples:

$$C_6H_5SCl \xrightarrow[53\%]{n\text{-}Pr_3N-HSiCl_3} C_6H_5SSC_6H_5$$

$$p\text{-}CH_3C_6H_4SOCl \xrightarrow[91\%]{n\text{-}Pr_3N-HSiCl_3} p\text{-}CH_3C_6H_4SSC_6H_4CH_3\text{-}p$$

$$C_6H_5SO_2Cl \xrightarrow[67\%]{n\text{-}Pr_3N-HSiCl_3} C_6H_5SSC_6H_5$$

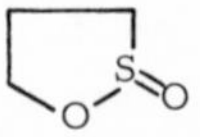

$$\text{(cyclic sulfone)} \xrightarrow[80\%]{n\text{-}Pr_3N-HSiCl_3} HOCH_2CH_2CH_2SSCH_2CH_2CH_2OH$$

Silylation of acid chlorides. The reaction of acid chlorides with trichlorosilane and a tertiary amine in acetonitrile affords bis[trichlorosilyl]alkanes in generally good yields.[5]

$$R-C\overset{\displaystyle O}{\underset{\displaystyle Cl}{\Big\langle}} \;+\; HSiCl_3 \;\xrightarrow[20-72\%]{R'_3N}\; RCH\overset{\displaystyle SiCl_3}{\underset{\displaystyle SiCl_3}{\Big\langle}}$$

[1] R. A. Benkeser and W. E. Smith, *Am. Soc.*, **90**, 5307 (1968)
[2] *Idem, ibid.*, **91**, 1556 (1969)
[3] R. A. Benkeser and J. M. Gaul, *ibid.*, **92**, 720 (1970); R. A. Benkeser, K. M. Foley, J. M. Gaul, and G. S. Li, *ibid.*, **92**, 3232 (1970)
[4] T. H. Chan, J. P. Montillier, W. F. Van Horn, and D. N. Harpp, *ibid.*, **92**, 7224 (1970)
[5] R. A. Benkeser, K. M. Foley, J. M. Gaul, G. S. Li, and W. E. Smith, *ibid.*, **91**, 4578 (1969)

sym-Trichlorotrifluoroacetone, Cl_3CCOCF_3. Mol. wt. 215.40, b.p. 83.5–84.5°.

Preparation.[1] The reagent is prepared by the reaction of chloropentafluoroacetone (Allied Chem. Corp., Peninsular ChemResearch) with aluminum chloride.

$$3\ CF_3COCF_2Cl \;+\; 2\ AlCl_3 \;\longrightarrow\; 3\ CF_3COCCl_3 \;+\; 2\ AlF_3$$

Trifluoroacetyl amino acids. Amino acids and peptides can be N-trifluoroacetylated with this reagent in DMSO.[1]

$$CCl_3COCF_3 \;+\; \underset{\displaystyle NH_2}{RCHCOOH} \;\xrightarrow{DMSO,\ 25^0}\; \underset{\displaystyle \underset{\displaystyle O}{\overset{\displaystyle \parallel}{NHCCF_3}}}{RCHCOOH} \;+\; CHCl_3$$

[1] C. A. Panetta and T. G. Casanova, *J. Org.*, **35**, 4275 (1970)

Triethylaluminum, 1, 1197–1198; 2, 427.

Alkylation of ketones. Tardella[1] reports that addition of one equivalent of triethylaluminum or tributyltin hydride in the alkylation of lithium or potassium enolates favors monoalkylation at the expense of dialkylation. The lower reaction rates necessitate addition of HMPT or dimethoxyethane to achieve reasonably rapid reactions. Triethylaluminum is generally more useful than tributyltin hydride.

[1] P. A. Tardella, *Tetrahedron Letters*, 1117 (1969)

Triethylamine–Borane, $H_3B\cdot N(C_2H_5)_3$. Mol. wt. 112.01, b.p. 76°/4 mm.

The reagent is readily prepared by the reaction of sodium borohydride and triethylamine in liquid sulfur dioxide (85% yield).[1] Pyridine–borane can be prepared in the same way.

The reagent reduces aralkyl halides in liquid sulfur dioxide or in nitromethane. The reduction is limited to tertiary and secondary halides, no reaction taking place with primary halides.[1] The rates and yields decrease in the following order:

$$(C_6H_5)_3CCl > (C_6H_5)_2CHBr > (C_6H_5)_2CHCl \approx (CH_3)_3CI \gg (CH_3)_3CCl \approx C_6H_5CH_2Cl$$

[1] S. Matsumara and N. Tokura, *Tetrahedron Letters*, 4703 (1968)

Triethylammonium formate (TEAF), $5 \ HCOOH \cdot 2 \ N(C_2H_5)_3$. Mol. wt. 432.51, b.p. 95°/15 mm. *See* **Trimethylammonium formate, 1**, 1231; **2**, 435.

Preparation.[1] Triethylamine is introduced into ice-cold 80% formic acid until the solution becomes basic. The solution is concentrated on the water bath under reduced pressure to remove water, and the residue is then distilled.

Reduction. Japanese chemists have used this reagent in place of formic acid as a reducing agent because of its high boiling point and weak acidity. The reagent is effective for selective reduction of α,β-unsaturated ketones of the type (1).[2]

(1)   (2)

[1] K. Ito, *Yakugaku Zasshi*, **86**, 1166 (1966) [C. A., **66**, 75899w (1967)]
[2] M. Sekiya and K. Suzuki, *Chem. Pharm. Bull. Japan*, **18**, 1530 (1970)

Triethyl orthoacetate, $CH_3C(OC_2H_5)_3$. Mol. wt. 162.23, b.p. 142°. Suppliers: Aldrich, Eastman, others.

trans-Trisubstituted olefins. Some time ago Burgstahler[1] noted that there is a marked stereoselectivity in the Claisen rearrangement[2] of vinyl ethers of substituted allyl alcohols. Thus both the *cis*-vinyl ether (1) and the *trans* isomer on Claisen rearrangement give essentially only the *trans* product (2). The vinyl

(1)   (2)

ethers are readily available by mercuric acetate-catalyzed vinyl transetherification of allylic alcohols with ethyl vinyl ether.[3] The stereoselectivity of the reaction has been confirmed by other chemists, particularly by Faulkner and Petersen,[4] who have utilized this rearrangement for synthesis of polyisoprenoids with the characteristic *trans* double bonds of natural products. Thus the allylic alcohol (3) is treated with excess 3-methoxyisoprene (2-methoxy-3-methylbutadiene)[5] and catalytic quantities of oxalic acid and hydroquinone in a sealed tube at 110°. The product is the α,β-unsaturated ketone (4). Apparently none of the *cis* isomer is formed as judged by vapor phase chromatography. Reduction of (4) with sodium borohydride in methanol gives a new allylic alcohol which is amenable to the same reaction sequence, eventually leading to a polyisoprenoid structure.

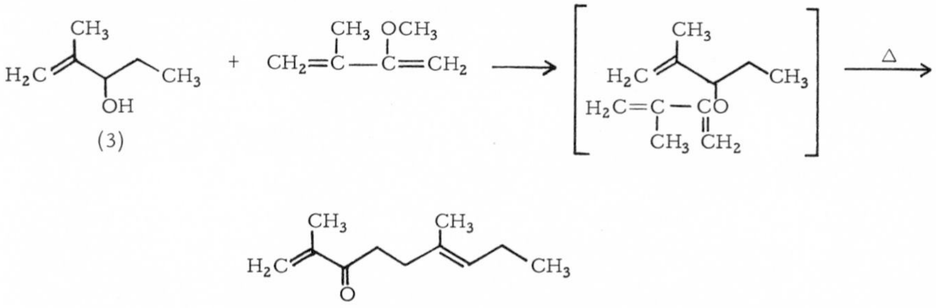

(4)

Johnson, Faulkner, *et al.*[6] have developed another approach to synthesis of polyisoprenoids which also utilizes the Claisen rearrangement to establish *trans*-trisubstituted double bonds. In a model experiment, the allylic alcohol (5) was heated with 7 equivalents of triethyl orthoacetate and 0.06 equivalent of propionic acid at 138° for 1 hour with provision for distillative removal of ethanol; the diene ester (6) was obtained in 92% yield. Analysis by vpc indicated that (6) is the *trans* isomer to the extent of >98% with less than 2% of the *cis* isomer. If the classical

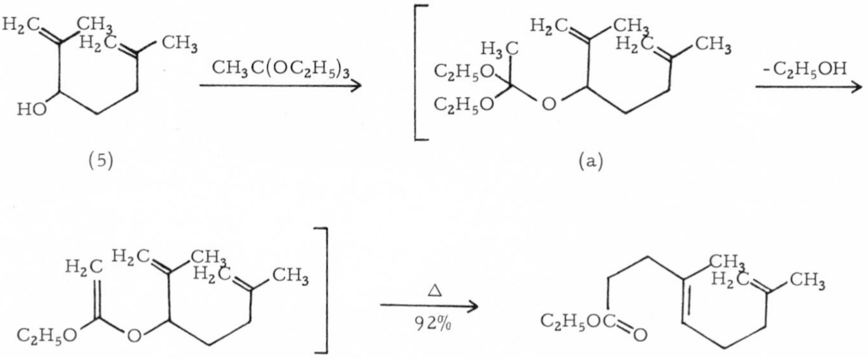

vinyl ether approach is used (reaction with ethyl vinyl ether followed by pyrolysis), the aldehyde corresponding to (6) is obtained in about 60% yield. But in this case the product consists of 86% of the *trans* isomer and 14% of the *cis* isomer. The high stereoselectivity of the orthoester approach is attributed to nonbonded inter-actions between the ethoxy group and the long side chain in the ketene acetal intermediate (b). Another advantage of the new procedure is that only one opera-tion is required. It also lends itself to the synthesis of polyisoprenoids. Thus it has been used for a total synthesis of all-*trans* squalene which is about 98% stereo-selective for each double bond.

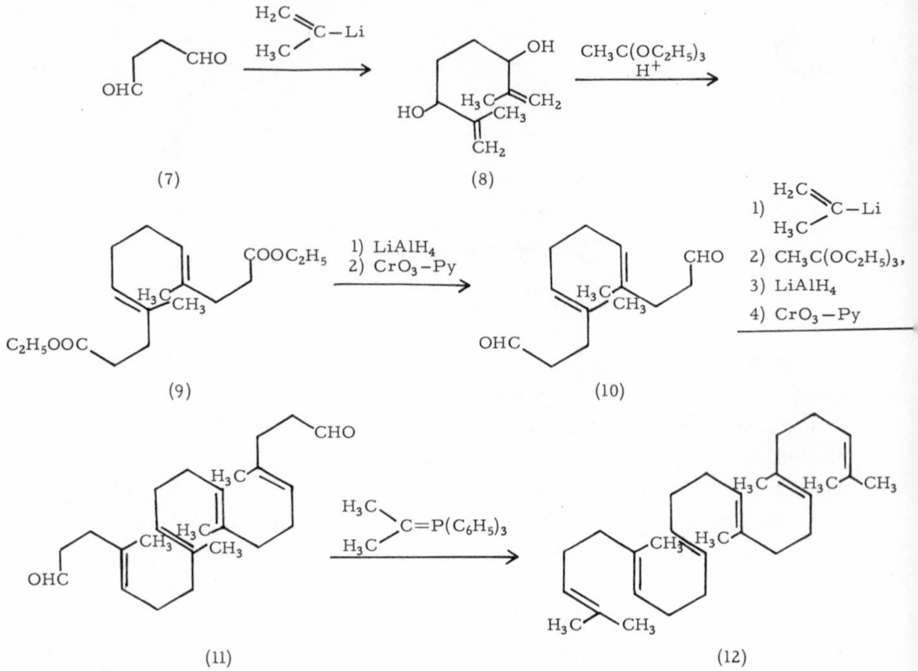

(7)   (8)

(9)   (10)

(11)   (12)

Succinic dialdehyde (7) was used as starting material to supply the central two carbon atoms of the final product. Reaction with 2-propenyllithium gave the dienediol (8). Reaction with triethyl orthoacetate then gave the diene diester (9). Reduction with lithium aluminum hydride followed by oxidation with the modified Sarett reagent (2, 74) gave the diene dialdehyde (10). The first sequence of reactions was then employed to obtain the tetraenedialdehyde (11). The completing step was a Wittig reaction with isopropylidenetriphenylphosphorane. The squalene (12) obtained was 95% pure all-*trans* isomer. The 3-methoxyisoprene approach was also used as an alternative method for conversion of (8) into a tetraene diol in the sequence. Johnson and Faulkner[6] note that the two methods are comparably efficacious, but of course triethyl orthoacetate is more readily available.

[1] A. W. Burgstahler, *Am. Soc.*, **82**, 4681 (1960)

[2] D. S. Tarbell, *Org. Reactions*, **2**, 1 (1944)

[3] W. H. Watanabe and L. E. Conlon, *Am. Soc.*, **79**, 2828 (1957)

[4] D. J. Faulkner and M. R. Petersen, *Tetrahedron Letters*, 3243 (1969); see also C. L. Perrin and D. J. Faulkner, *ibid.*, 2783 (1969)

[5] Prepared in 40–45% yield from 2-methyl-1-buten-3-yne, I. A. Favorskaya and N. N. Kopylov-Shakmatov, *Zh. Obshch. Khim.*, **27**, 2406 (1957) [*C.A.*, **52**, 7155 (1958)]

[6] W. S. Johnson, L. Werthemann, W. R. Bartlett, T. J. Brocksom, T. Li, D. J. Faulkner, and M. R. Petersen, *Am. Soc.*, **92**, 741 (1970)

Triethyloxonium fluoroborate, 1, 1210–1212; **2,** 430–431.

Nitrilium salts. Meerwein[1] showed that triethyloxonium fluoroborate reacts readily with nitriles to give N-ethylnitrilium salts (1) which, as expected, are readily hydrolyzed to amides. A more useful reaction is reduction with sodium borohydride to give secondary amines (2) in 70–75% yield.[2] The method cannot

$$RC\equiv N + (C_2H_5)_3\overset{+}{O}BF_4^- \xrightarrow[\text{reflux}]{CH_2Cl_2} RC\equiv \overset{+}{N}CH_2CH_3\,[BF_4]^- \xrightarrow[\text{In EtOH}]{NaBH_4} RCH_2NHCH_2CH_3$$

$$(1) \hspace{5cm} (2)$$

be extended to the synthesis of salts other than N-methyl or N-ethyl compounds because of the difficulty of preparation of the higher trialkyloxonium fluoroborates. Hence Borch[3] turned to dialkylcarbonium fluoroborates (3), which are

$$\left[HC\overset{OR}{\underset{OR}{\diagup}} \right]^+ BF_4^-$$

$$(3)$$

even stronger alkylating agents than trialkyloxonium salts. They are readily prepared by the reaction of orthoformate esters with boron trifluoride etherate at $-70°$. These react with nitriles in the same way to give N-alkylnitrilium salts in comparable yields.

Esterification. Acylamino acids or peptides can be esterified in an aqueous solution of sodium bicarbonate by triethyloxonium fluoroborate (excess).[4] The method is useful because ester groups of some peptides can be reduced by sodium borohydride in aqueous solution.[5]

Alkylation of sydnones. Mesoionic ring systems with exocyclic oxygen atoms ordinarily do not undergo O-alkylation with alkyl halides or sulfates. However, alkylation occurs readily with triethyloxonium fluoroborate. Thus 3-phenyl-sydnone (1) in methylene chloride is alkylated at room temperature to give (2).

$$(1) \hspace{5cm} (2)$$

The salts, when heated above the melting points, are reconverted into the corresponding sydnone.[6]

[1] H. Meerwein, P. Laasch, R. Mersch, and J. Spille, *Ber.,* **89,** 209 (1965)
[2] R. F. Borch, *Chem. Commun.,* 442 (1968)
[3] *Idem, J. Org.,* **34,** 627 (1969)
[4] O. Yonemitsu, T. Hamada, and Y. Kanaoka, *Tetrahedron Letters,* 1819 (1969)
[5] *Idem, ibid.,* 3575 (1968)
[6] K. T. Potts, E. Houghton, and S. Husain, *Chem. Commun.,* 1025 (1970)

Triethyl phosphite, 1, 1212–1216; **2,** 432–433.

Perkow reaction.[1] Perkow[2] noted that α-halo aldehydes and ketones react with trialkyl phosphites to give dialkyl vinyl phosphates:

$$(RO)_3P \ + \ O=\overset{|}{C}-\overset{|}{C}X \ \longrightarrow \ (RO)_2P\overset{|}{O}\overset{|}{C}=C< \ + \ RX$$

Fetizon *et al.*[3] used this reaction in a new synthesis of Δ²-steroids from 3-keto-steroids. For example, dihydrotestosterone acetate (1) is converted in high yield into the 2α-bromoketone (2) with phenyltrimethylammonium perbromide (**1,** 855). The α-bromoketone is then heated for 3 hours with freshly distilled triethyl phosphite to give the enol phosphate (3). On reduction of (3) with lithium in liquid

ammonia and *t*-butanol, Δ²-androstene-17β-ol (4) is obtained in 85% yield. This last step is an extension of G. W. Kenner and N. R. Williams' reduction of phenol diethyl phosphates to arenes (**1,** 1041).

[1]Reviews: F. W. Lichtenthaler, *Chem. Rev.,* **61,** 608 (1961); B. A. Arbusow, *Pure Appl. Chem.,* **9,** 307 (1964)
[2]N. Perkow, K. Ullerich, and F. Meyer, *Naturwiss.,* **39,** 353 (1952)
[3]M. Fetizon, M. Jurion, and N. T. Anh, *Chem. Commun.,* 112 (1969)

Triethylsilane, 1, 1218; **2,** 433.

Preparation. The reagent can also be prepared by reduction of triethylchloro-silane with LiAlH₄.[1]

RCOCl → RCHO. Under catalysis with 10% Pd-C, triethylsilane reacts with acyl chlorides at room temperatures to give aldehydes, isolated as the 2,4-dinitro-phenylhydrazones in yields of 50–70% generally. Yields are negligible in the case

$$RCOCl \ + \ (C_2H_5)_3SiH \ \xrightarrow{\ Pd\ } \ RCHO \ + \ (C_2H_5)_3SiCl$$

of branched acyl chlorides such as (CH₃)₂CHCOCl. Yields are in the range 40–70% in the case of aroyl chlorides. Triethylsilane is far more effective than other organosilicon hydrides.[2]

[1]J. D. Citron, J. E. Lyons, and L. H. Sommer, *J. Org.,* **34,** 638 (1969)
[2]J. D. Citron, *ibid.,* **34,** 1977 (1969)

Trifluoroacetic acid, 1, 1219–1221; **2,** 433–434.

Olefin cyclization. W. S. Johnson and co-workers[1] developed a new approach
to steroid total synthesis, based upon discovery of a remarkable biogenetic-like
cyclization applied in a first instance,[2] to the tetraenol (10), prepared in a sequence
starting with condensation of the dienic tosylate (1) with the lithium salt of 4-

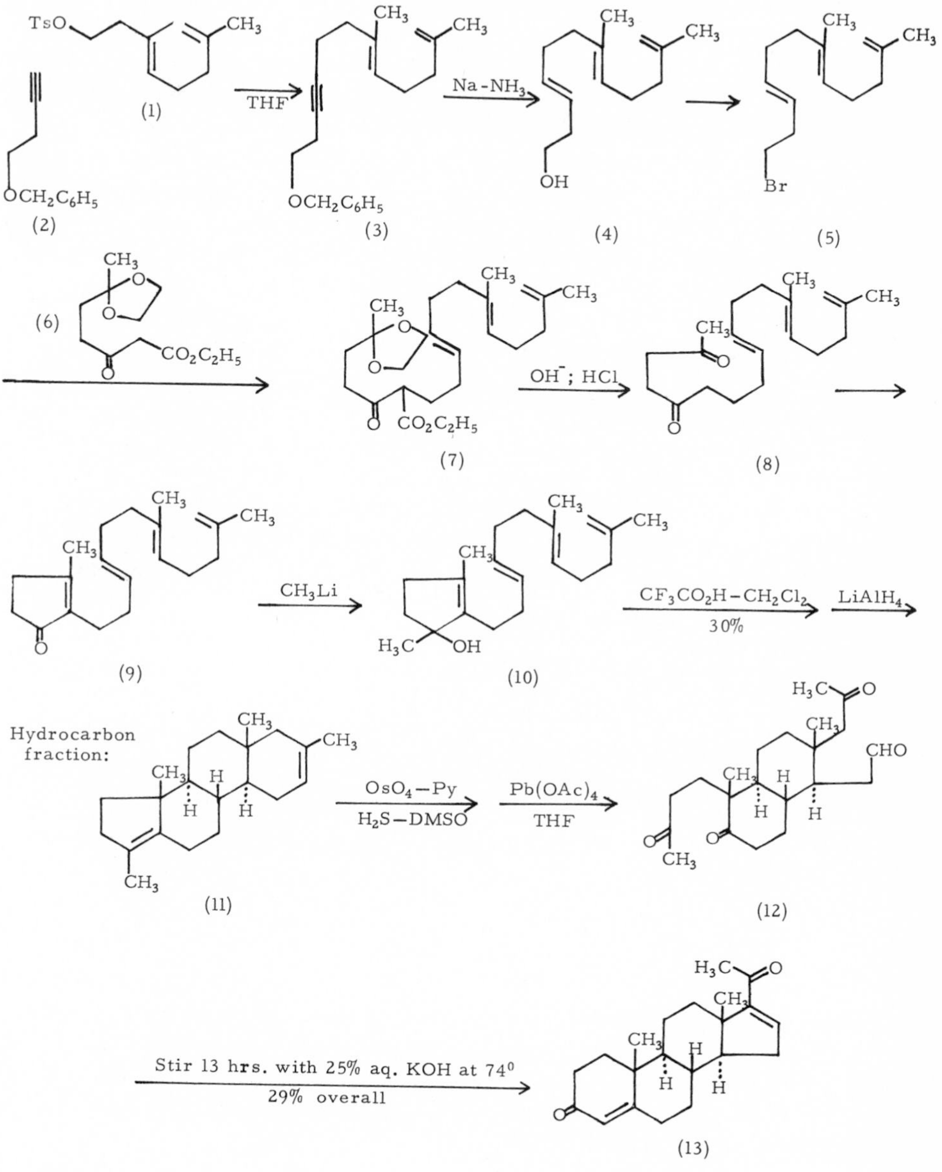

benzyloxy-1-butyne (2) to give the alkylation product (3), which on treatment with sodium in liquid ammonia was converted into the trienol (4). This last substance was transformed, by the action of lithium bromide on the tosylate, into the bromide (5), which was allowed to interact with the sodio derivative of the keto ester (6) in a warm 4:1 mixture of DMF and benzene. The resulting alkylation product (7), on treatment with aqueous-ethanolic barium hydroxide followed by acidification, underwent hydrolysis and decarboxylation to give the diketone (8), which on treatment with 2% sodium hydroxide in aqueous ethanol at 105–110° underwent cyclization to the tetraenone (9). Reaction of (9) with methyllithium then afforded the tetraenol (10), which was used for cyclization without purification. The reagent found most satisfactory for the cyclization was trifluoroacetic acid in methylene chloride at −78° for several hours. The product was treated with lithium aluminum hydride, and the resulting hydrocarbon fraction on crystallization afforded the pure product (11), m.p. 66–67.5°. Treatment of the crude tetracyclic diene (11) with excess osmium tetroxide in pyridine was followed by cleavage of the bis-osmate with hydrogen sulfide in dimethyl sulfoxide and oxidation with excess lead tetraacetate to give the triketoaldehyde (12), which when stirred with dilute aqueous potassium hydroxide was doubly cyclized to give dl-16,17-dehydroprogesterone (13) in 29% yield. The remarkable stereospecificity of the cyclization of (10) is evident from the fact that the reaction involves formation of no less than five centers of asymmetry.

A Sterling-Winthrop Research Institute group in cooperation with Johnson[3] applied the new scheme of biogenetic-like stereoselective olefin cyclization to the synthesis of 19-norsteroids and so gained access to a steroid type of importance in medicine.

The cyclization was also the key step in a total synthesis of dl-19-nor-10,17-dehydroprogesterone. Thus cyclization of the alcohol (14) gave the 19-nor-A-nor-

(14)   (15)

D-homosteroid (15) in 24% yield together with some of a double-bond isomer. Ozonolysis to a triketoaldehyde followed by acid cyclization gave the 19-norsteroid.

Johnson and Schaaf[4] applied the method of CF_3CO_2H-cyclization of polyolefinic allyl alcohols to the tetraenol (16) and obtained the two 13-epimeric alcohols (17) and (18) of the podocarpane series.

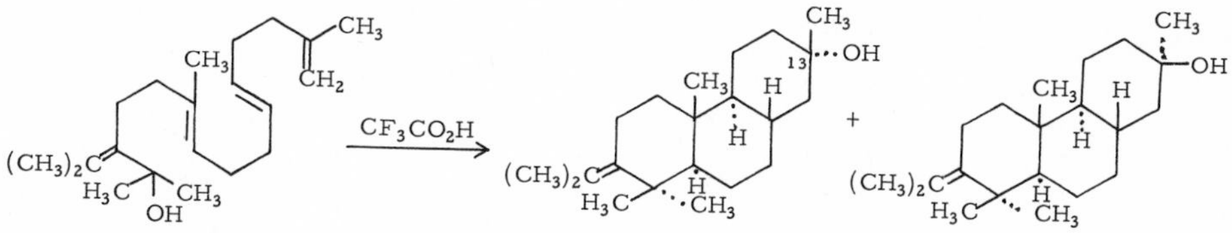

(16) (17) (18)

Synthesis of 2-methylcyclobutanone (5).[5,6] A solution of 5.29 g. of 3-pentyne-1-ol (1) in 70 ml. of 2,6-lutidine (2) was treated at 0° with 16.82 g. of 3,5-dinitro-benzenesulfonyl chloride (3). After addition of 120 ml. of toluene and stirring for 15 minutes, a mixture of 70 ml. of concentrated hydrochloric acid and 400 ml. of ice water was added. The toluene phase was separated and combined with two

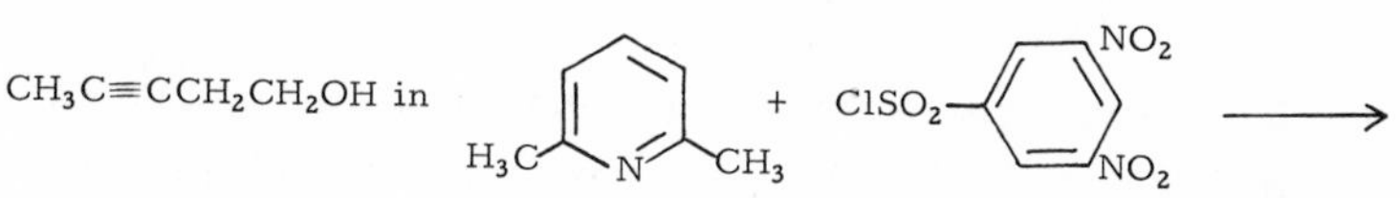

(1) (2) (3)

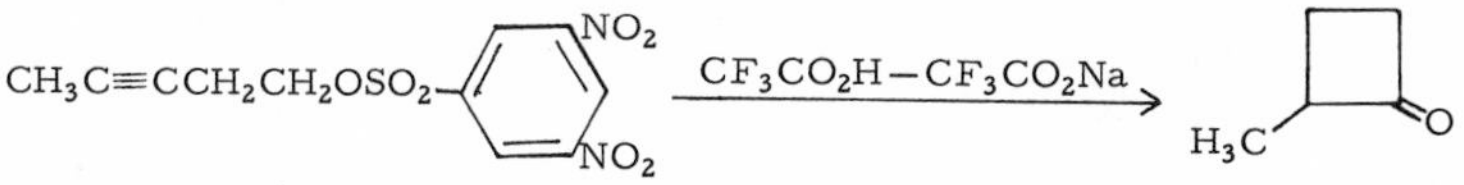

(4) (5)

toluene extracts of the aqueous phase. The combined extract was washed with dilute hydrochloric acid and water and dried over sodium sulfate. Removal of the solvent gave a red oil which crystallized from toluene–petroleum ether in yellow needles, m.p. 82°, yield 10 g. (51%).

Cyclization to the butanone (5) was effected by stirring 6.28 g. of the dinitro-benzenesulfonate (4) for 5 days at 45° in a mixture of 4.1 g. of sodium trifluoro-acetate and 50 g. of trifluoroacetic acid (if sodium trifluoroacetate is omitted, higher-boiling products are formed and the yield of 2-methylcyclobutanone is much lower).

Deuteration and tritiation.[7] Labeled trifluoroacetic acid has some advantages for deuteration and tritiation. It is obtained in quantitative yield by mixing tri-fluoroacetic acid and labeled water. It has excellent solvent properties and also functions as the acid catalyst. The uptake of label can be followed by nuclear magnetic resonance; the reagent is readily removed by evaporation *in vacuo*. In the case of olefins, addition to form the trifluoroacetate occurs simultaneously. It has been used for the labeling of Δ^8- and Δ^9-tetrahydrocannabinol.

[1]Review: W. S. Johnson, *Accts. Chem. Res.*, **1**, 1 (1968)
[2]W. S. Johnson, M. F. Semmelhack, M. U. S. Sultanbawa, and L. A. Dolak, *Am. Soc.*, **90**, 2994 (1968)

[3]S. J. Daum, R. L. Clarke, S. Archer, and W. S. Johnson, *Proc. Nat. Acad. Sci.*, **62**, 333 (1969)

[4]W. S. Johnson and T. K. Schaaf, *Chem. Commun.*, 611 (1969)

[5]M. Hanack, I. Hertrich, and V. Vött, *Tetrahedron Letters*, 3871 (1967)

[6]*Idem*, procedure submitted to *Org. Syn.* 1969

[7]M. L. Timmons, C. G. Pitt, and M. E. Wall, *Tetrahedron Letters*, 3129 (1969)

Trifluoroacetic anhydride, 1, 1221–1226.

Mannich reagent. French chemists reported that use of trifluoroacetic anhydride rather than acetic anhydride in the Polonovski reaction resulted in improved yields (2, 7, 8). They[1] now find that the reaction of trifluoroacetic anhydride with the N-oxide of trimethylamine leads to the trifluoroacetate of N,N-dimethylformaldimmonium (1). This ion is considered to be the actual reagent in

$$(CH_3)_3N^+-O^- \xrightarrow{(CF_3CO)_2O} [(CH_3)_2\overset{+}{N}=CH_2 \longleftrightarrow (CH_3)_2N-\overset{+}{C}H_2]CF_3CO_2^-$$

$$(1)$$

the Mannich reaction conducted with dimethylamine and formaldehyde, and indeed use of this salt gives higher yields than those obtained with the classical Mannich procedure.[2]

[1]A. Ahond, A. Cavé, C. Kan-Fan, and P. Potier, *Bull. soc.*, 2707 (1970)

[2]F. F. Blicke, *Org. Reactions*, **1**, 303 (1942)

α,α,α-Trifluoroacetophenone, 1, 1226–1227.

The procedure for the preparation of 2-phenylperfluoropropene has been published.[1]

[1]F. E. Herkes and D. J. Burton, *Org. Syn.*, **48**, 116 (1968)

2,4,6-Triisopropylbenzenesulfonyl chloride, TPS, 1, 1228–1229.

Glycerophospholipids. Phosphatidic acids, for example (1), can be esterified directly by use of TPS in the presence of pyridine to give glycerophospholipids (2).[1]

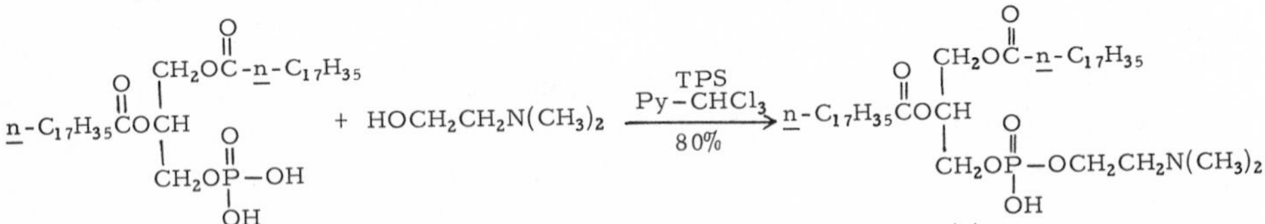

[1]R. Aneja, J. S. Chadha, and A. P. Davies, *Tetrahedron Letters*, 4183 (1969)

Trimesitylborane (TMB), $\left(H_3C\underset{CH_3}{\overset{CH_3}{\diamondsuit}}-\right)_3 B$ Mol. wt. 368.35.

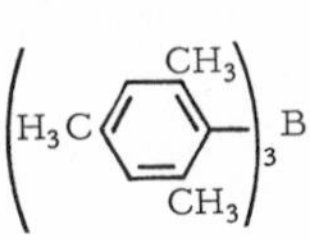

Reduction of α,β-unsaturated ketones.[1] This highly hindered borane accepts an electron from sodium metal to give a blue solution of the radical anion which can serve as a conductor of electrons for reduction of α,β-unsaturated ketones. Thus $\Delta^{1,9}$-octalone-2 (1) is reduced by TMB, sodium, and a proton source (*t*-butanol) in anhydrous 1,2-dimethoxyethane (nitrogen) in high yield (85–95%) to a mixture of *trans*-2-decalone (70%) and *cis*-2-decalone (30%), the equilibration point for this reduction. Overreduction to the alcohol is negligible. If the reduction

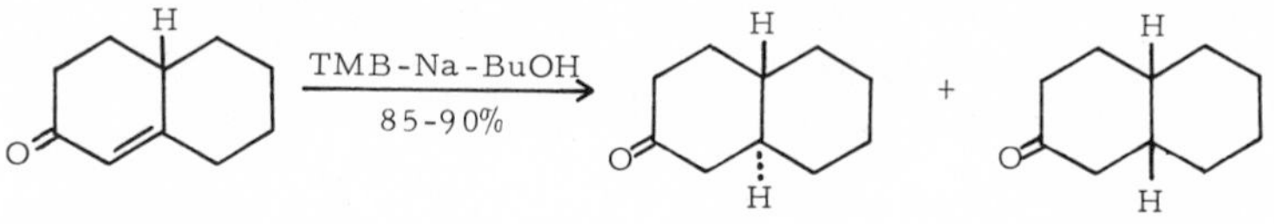

TMB-Na-BuOH
85-90%

(1)   (2, 70%)   (3, 30%)

is carried out in hexamethylphosphoric triamide with lithium and quenched with ammonium chloride, *cis*-2-decalone can be obtained as the predominant product (68% yield).

[1] S. D. Darling, O. N. Devgan, and R. E. Cosgrove, *Am. Soc.*, **92**, 696 (1970)

Trimethylamine oxide, $(CH_3)_3NO$, **1**, 1230–1231; **2**, 434.

Oxidation of organoboranes. The oxide quantitatively oxidizes a wide range of organoboranes to alkoxyboranes. Titration of the evolved trimethylamine can be

$$R-B< \; + \; (CH_3)_3NO \longrightarrow ROB< \; + \; (CH_3)_3N$$

used for estimation of boron–carbon bonds. This reagent may have some advantages over alkaline hydrogen peroxide in the case of base-sensitive boranes.[1]

For example, *trans*-2-methylcyclopropanol (4) was prepared from 1-methyl-cyclopropene (1).[2] The reaction of (1) with diborane leads to tris(*trans*-2-methyl-

B_2H_6
80%

$3\,(CH_3)_3NO$
70%

(1)   (2)   (3)

CH_3OH
$-B(OCH_3)_3$
75%

(4)

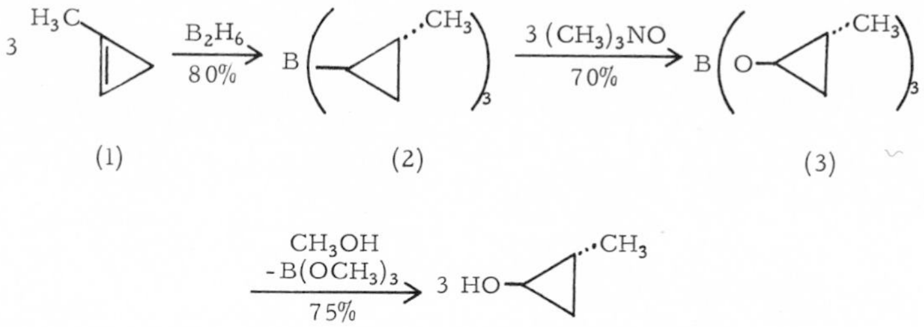

cyclopropyl)borane (2); oxidation of (2) with trimethylamine oxide gives tris-(*trans*-2-methylcyclopropyloxy)borane (3). Transesterification with methanol then gives *trans*-2-methylcyclopropanol. Oxidation with the usual reagent, hydrogen peroxide, cannot be used in the case of cyclopropanols. The starting

$$(5) \quad CH_2\!=\!\underset{\underset{CH_3}{|}}{C}\!-\!CH_2Cl \quad \xrightarrow{} \begin{array}{l} \xrightarrow[50\%]{NaNH_2} \quad H_3C\!\!-\!\!\triangleleft \quad + \; NH_3 \; + \; NaCl \\[2mm] \qquad\qquad\qquad (1) \\[2mm] \xrightarrow[36\%]{KNH_2} \quad CH_2\!\!=\!\!\triangleleft \quad + \; NH_3 \; + \; KCl \\[2mm] \qquad\qquad\qquad (6) \end{array}$$

material (1) is readily available by treatment of methallyl chloride (5) with commercial sodamide in gently refluxing THF.[3] Surprisingly, use of potassium amide gives methylenecyclopropane (6) as the only C_4-hydrocarbon.[2]

[1] R. Köster and Y. Morita, *Ann.*, **704**, 70 (1967)
[2] R. Köster, S. Arora, and P. Binger, *Angew. Chem., internat. Ed.*, **8**, 205 (1969) (Note: there are two papers on the same page.)
[3] F. Fisher and D. E. Applequist, *J. Org.*, **30**, 2089 (1965)

Trimethyl borate, 2, 435.

The procedure for the reaction of this ester with Grignard reagents has now been published.[1]

Reformatsky reaction. The Reformatsky reaction is normally conducted at reflux temperatures in benzene or benzene–ether as solvent. Rathke and Lindert[2] studied the Reformatsky reaction of acetaldehyde and ethyl bromoacetate under various conditions; the yield of ethyl 3-hydroxybutanoate obtained in benzene at

$$CH_3CHO \; + \; BrCH_2COOC_2H_5 \; \xrightarrow{Zn} \; CH_3\underset{\underset{OH}{|}}{C}HCH_2COOC_2H_5$$

75° (2 hours) was only 22%. They then found that the yield is increased to 65% if the reaction is run at 25° for 4 hours. The yield was further improved to 95% by using THF–trimethyl borate as solvent at a temperature of 25° (2 hours). The function of the mild acid trimethyl borate is presumably to neutralize zinc alkoxides formed in the reaction. The advantage of the trimethyl borate procedure is most apparent with reactive aldehydes and ketones. With less reactive carbonyl compounds comparable yields are obtained utilizing benzene as solvent at 25°.

[1] R. L. Kidwell, M. Murphy, and S. D. Darling, *Org. Syn.*, **49**, 90 (1969)
[2] M. W. Rathke and A. Lindert, *J. Org.*, **35**, 3966 (1970)

Trimethylchlorosilane, 1, 1232; 2, 435–438.

Trimethylsilyl enol ethers. House *et al.*[1] have described two procedures for the preparation of trimethylsilyl enol ethers from aldehydes and ketones. In one method triethylamine in THF is used as base. These conditions usually afford an equilibrium mixture, and in fact the method is very satisfactory for equilibration. The more highly substituted enol ether usually predominates, as in the example:

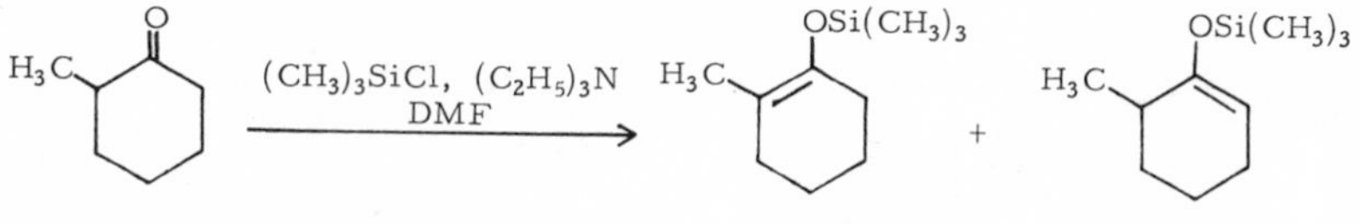

$$78\% \qquad\qquad 22\%$$

In the second method the enolate anion is generated under conditions of kinetic control. For this purpose the hindered lithium diisopropylamide in 1,2-dimethoxyethane was found to be the most satisfactory base. In these conditions the less highly substituted enol ether predominates.

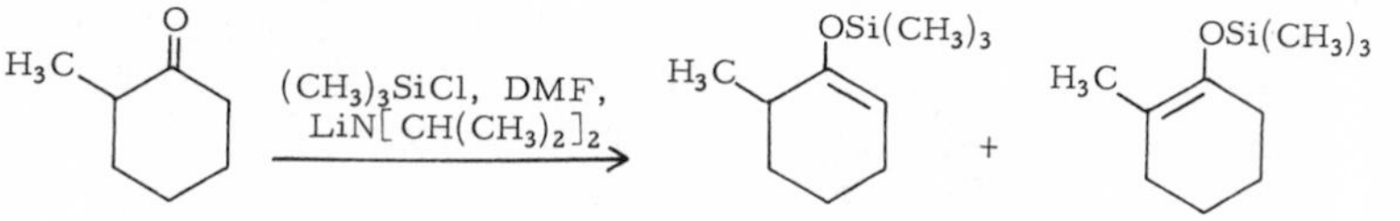

$$99\% \qquad\qquad 1\%$$

Hydroxylation. Maume and Horning[2] report that trimethylsilyl (TMSi) enol ethers of ketosteroids when irradiated with UV light (dibenzoyl peroxide can also be used) are converted into the trimethylsilyl derivative of an α-hydroxy ketone. Thus the TMSi enol ether of Δ⁵-androstene-3β-ol-17-one (1) gives (2) as the chief reaction product. Similarly, Δ⁵-androstene-3β,17β-diol-16-one is converted into

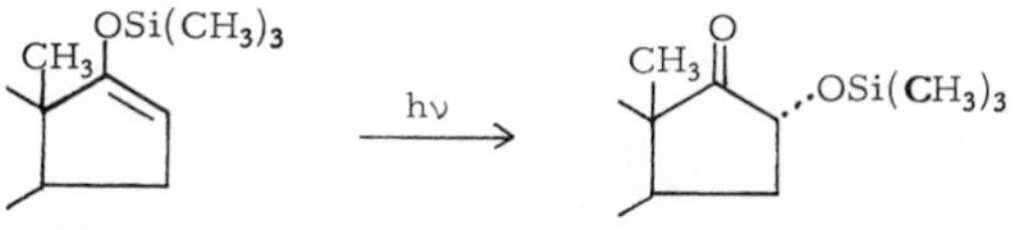

the tri-TMSi ether of Δ⁵-androstene-3β,15α,17β-triol-16-one. Presumably an intermediate trimethylsilyloxy free radical is generated and reacts with the TMSi enol ether to form the trimethylsilyl derivative of an α-hydroxy ketone. In the two cases reported the introduced hydroxyl group has the α-orientation.

Acyloin condensation (**2**, 435–436). Japanese chemists[3] have extended Bloomfield's procedure for ring enlargement of cycloalkane-1,2-dicarboxylic esters. Thus the esters (1) were treated with sodium in xylene at 110–120° in the presence of trimethylchlorosilane and the reaction mixture refluxed to ensure ring opening of the cyclobutenes (2) to 1,3-dienes (3). Acidic hydrolysis gave 1,2-diketones (4) in 71–74% yields. These can be reduced to acyloins (5) by alkaline hydrolysis with triethyl phosphite (74–79%).

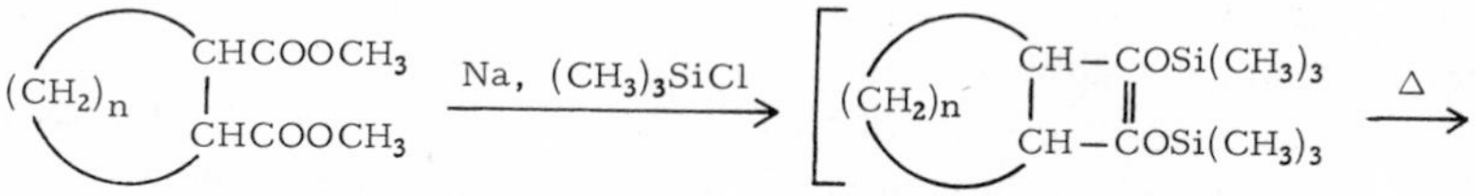

(1) n = 9, 10, 11   (2)

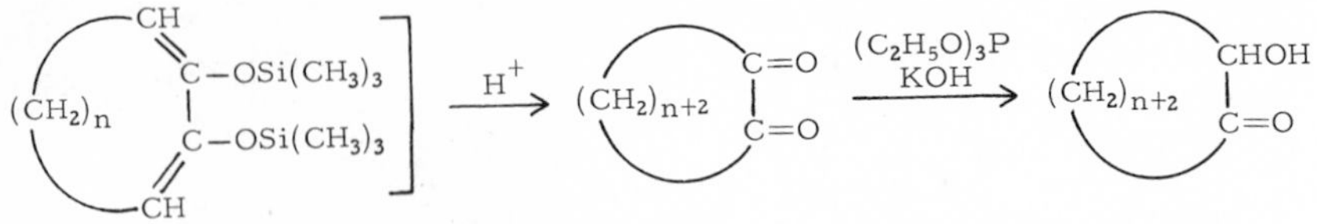

(3)   (4)   (5)

Wynberg *et al.*[4] have synthesized the interesting spiro[adamantane-2,1'-cyclo-butane-2',3'-dione] (8) from methyl 2(2-methoxycarbonyladamantyl)acetate (6) by acyloin condensation in the presence of trimethylchlorosilane to give the bis(trimethylsiloxy)cyclobutene derivative (7) in 75% yield. The key step, conversion of (7) into (8), was accomplished under non-hydrolytic conditions by addition of bromine to a solution of (7) in carbon tetrachloride.[5]

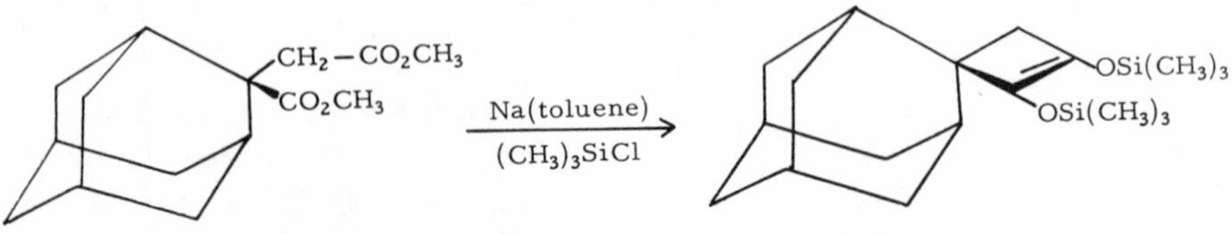

(6)   (7)

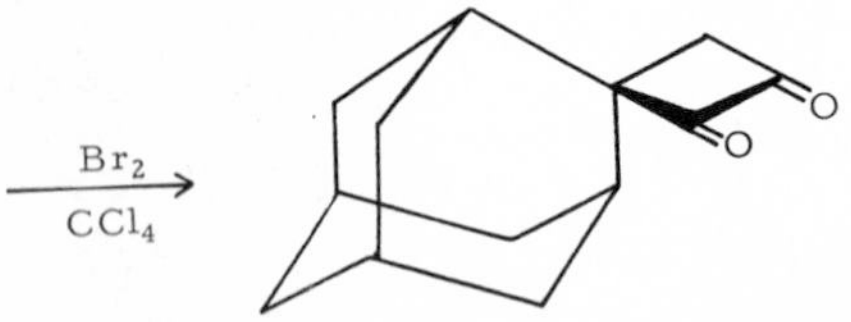

(8)

[1]H. O. House, L. J. Czuba, M. Gall, and H. D. Olmstead, *J. Org.*, **34**, 2324 (1969)
[2]G. M. Maume and E. C. Horning, *Tetrahedron Letters*, 343 (1969)
[3]T. Mori, T. Nakahara, and H. Nozaki, *Canad. J. Chem.*, **47**, 3266 (1969)
[4]H. Wynberg, S. Reiffers, and J. Strating, *Rec. trav.*, **89**, 982 (1970)
[5]This reaction had been used earlier by K. Rühlmann and S. Poredda [*J. prakt. Chem.*, [4], **12**, 18 (1960)]; for example, they obtained benzil in 98% yield from 1,2-diphenyl-1,2-bis(trimethylsiloxy)ethene by treatment with bromine.

Trimethylironlithium, $(CH_3)_3FeLi$. Mol. wt. 107.89, brown-black.

The reagent is prepared[1] by the reaction in ether of 3 equivalents of methyl-lithium with 1 equiv. of ferrous iodide ($-20°$, N_2 atmosphere). The reagent decomposes appreciably upon storage at $0°$ for several hours.

Cross coupling.[1] (Compare **Dimethylcopperlithium**, **2**, 151–153; This volume). The reagent reacts rapidly with vinylic and allylic bromides and iodides to give products of cross coupling in good yield. Only moderate yields of products of

cross coupling are obtained from primary alkyl iodides, the predominant products
arising from metal-halogen exchange.

Examples:

Iodobenzene → Toluene (50%)

3-Bromocyclohexene → 3-Methylcyclohexene (50%)

trans-1-Bromo-2-phenylethylene → *trans*-1-Phenylpropene (83%)

7,7-Dibromonorcarane → 7,7-Dimethylnorcarane (65%)

1-Iododecane → *n*-Undecane (2%) + 1-Decene (60%)

1-Iodocyclohexane → Methylcyclohexane (10%) + Cyclohexene

(Main product)

Trimethylironlithium (6 equivalents) was found to be superior to dimethylcopperlithium for the conversion of (1) into (2); in this case the copper reagent
gave (2) and about an equal amount of the isomeric product resulting from allylic
transposition.[2]

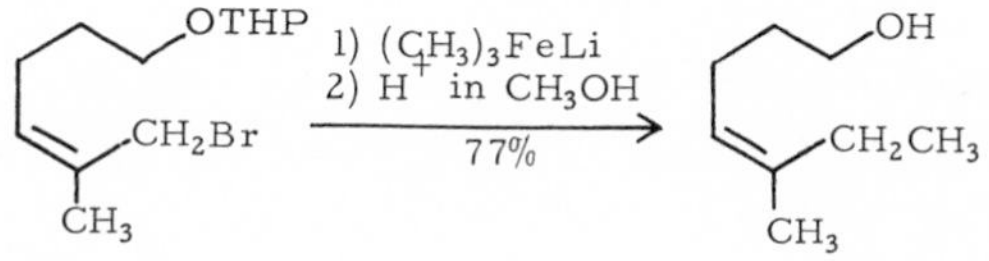

(1)   (2)

[1]E. J. Corey and G. H. Posner, *Tetrahedron Letters*, 315 (1970)
[2]E. J. Corey, H. Yamamoto, D. K. Herron, K. Achiwa, *Am. Soc.*, **92**, 6635 (1970)

Trimethyl orthoformate, $HC(OCH_3)_3$. Mol. wt. 106.12, b.p. 99–102°. Suppliers:
Aldrich, Eastman, Fisher, others.

β-Lactones. β-Hydroxy acids are not readily dehydrated directly to β-
lactones. The desired reaction, however, can be conducted by a two-step procedure. The hydroxy acid, for example hydroxypivalic acid (1), is heated with
trimethyl orthoformate in benzene with distillation of the benzene–methanol
azeotrope (58%). The 1,3-dioxane derivative (2) is obtained in 87% yield. This
product is converted into a β-lactone (3) when heated at 150–200°.[1]

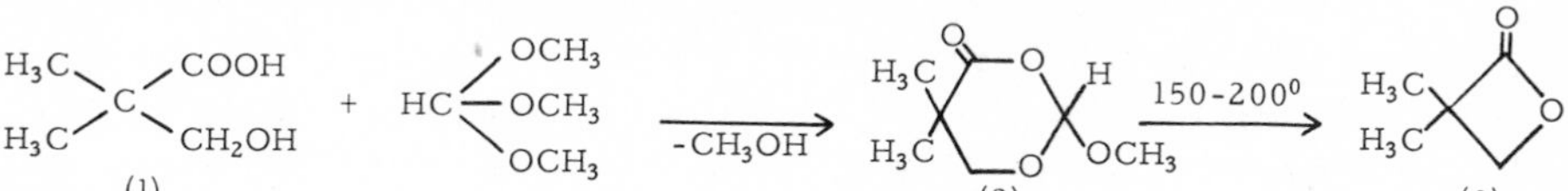

(1)   (2)   (3)

[1]R. C. Blume, *Tetrahedron Letters*, 1047 (1969)

2,4,4-Trimethyl-2-oxazoline, Mol. wt. 113.16; b.p. 112–113°,

$n^{26.2}$D 1.4186. Supplier: Columbia Organic Chemicals, Columbia, S. C.

Preparation.[1] The oxazoline (1) is prepared by refluxing glacial acetic acid and 2-amino-2-methyl-1-propanol:

(1)

Aliphatic carboxylic acids and esters.[2] The reagent (1) is converted into an anion which is alkylated at the C_2-methyl group (2). The 2-oxazoline ring is then hydrolyzed by heating in 5–7% ethanolic sulfuric acid to give the ethyl ester (3). Reaction of the lithio salt of (1) with an aldehyde, followed by hydrolysis, leads

(2)   (3)

to a β-hydroxy ester. Dialkylacetic acid esters can be obtained by treating (2) with *n*-butyllithium to give the anion and then with an alkyl halide to give (4). Hydrolysis of (4) as before gives the disubstituted acetic acid ester (5). Yields are for the most part about 70–80%. The overall effect, therefore, is to alkylate acetic

(4)   (5)

acid [from which (1) is prepared]. Of course, other carboxylic acids can be employed in place of acetic acid. (Compare **2,4,4,6-Tetramethyl-5,6-dihydro-1,3-(4H)-oxazine**, This volume).

[1]H. L. Wehrmeister, *J. Org.*, **27**, 4418 (1962); P. Allen and J. Ginos, *ibid.*, **28**, 2759 (1963)
[2]A. I. Meyers, D. L. Temple, Jr., *Am. Soc.*, **92**, 6644 (1970)

Trimethyloxonium fluoroborate, **1**, 1232; **2**, 438.

α-Acetoxy ketones. In a new method for α-acetoxylation of ketones,[1] the ketone, for example 4-heptanone (1), is converted into the oxime acetate (2). This is then alkylated to form an iminium salt (3). Alkylation with methyl iodide, dimethyl sulfate, or methyl tosylate is very slow even at elevated temperatures, but proceeds readily in nitromethane with trimethyloxonium fluoroborate (2–3 hours at 25°). Elimination of fluoroboric acid (triethylamine) then gives an enamine (4), which rapidly rearranges to an α-acetoxyimine (5). The final step is hydrolysis to the α-acetoxy ketone (6). The intermediate (3) can be isolated, with no particular advantages.

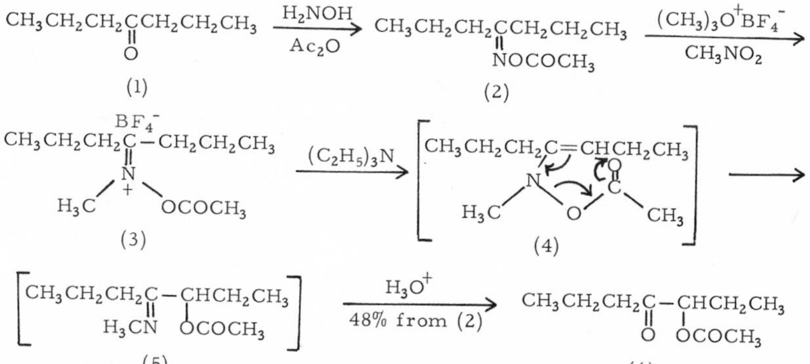

$$CH_3CH_2CH_2CCH_2CH_2CH_3 \xrightarrow[Ac_2O]{H_2NOH} CH_3CH_2CH_2CCH_2CH_2CH_3 \xrightarrow[CH_3NO_2]{(CH_3)_3O^+BF_4^-}$$

(1) (2)

[chemical structures (3), (4)]

(3) $\xrightarrow{(C_2H_5)_3N}$ (4) $\longrightarrow$

[chemical structures (5), (6)]

(5) $\xrightarrow[48\% \text{ from (2)}]{H_3O^+}$ (6)

The sequence can be applied to unsymmetrical ketones. The isomeric oxime acetates of 2-methylcyclohexanone (7) and (8), lead to mixtures of stereoisomeric acetoxy ketones (9) and (10) in 41–51% yield. Less than 1% of the tertiary acetate (11) is obtained. This result contrasts with acetoxylation of the ketone with lead tetraacetate which gives a substantial amount of the tertiary acetate as well as (9) and (10).

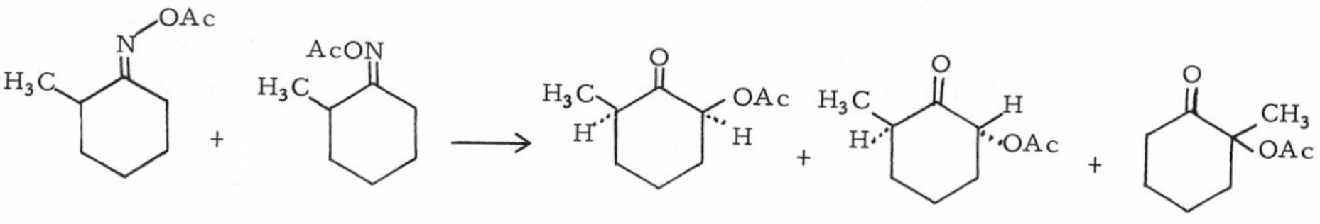

(7) (8) (9) (10) (11)

[1]H. O. House and F. A. Richey, Jr., *J. Org.*, **34**, 1430 (1969)

Trimethyl phosphite, 1, 1233–1235; 2, 439–441.

Olefin synthesis (**1**, 1233–1234). The Corey-Winter procedure was used for introduction of the double bond in the first synthesis of the highly strained twistene tricyclo[4.4.0.0³·⁸]dec-4-ene (6). A procedure involving dehydrohalogenation is ruled out by the sensitivity of the twistane skeleton to acid-catalyzed rearrangement. The complete synthesis[1] involved the Diels-Alder reaction of methyl 4,5-dihydrobenzoate with methyl acrylate to give two diesters, (2) and (3). After hydrogenation of the mixture, the diester (4) desired for the synthesis was obtained by fractional crystallization. Acyloin condensation followed directly by catalytic hydrogenation gave the diol (5), twistane-4,5-diol. The final steps

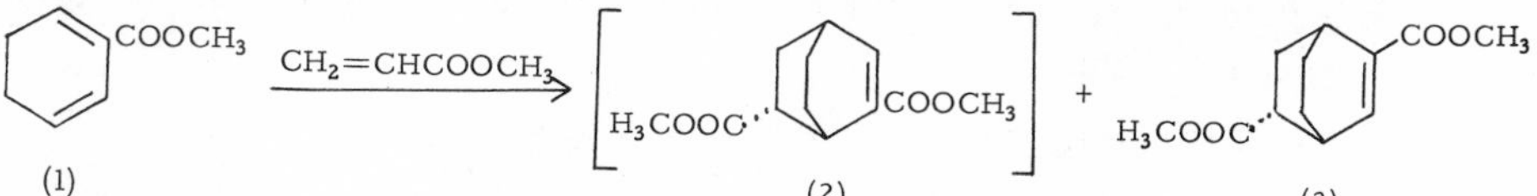

(1) (2) (3)

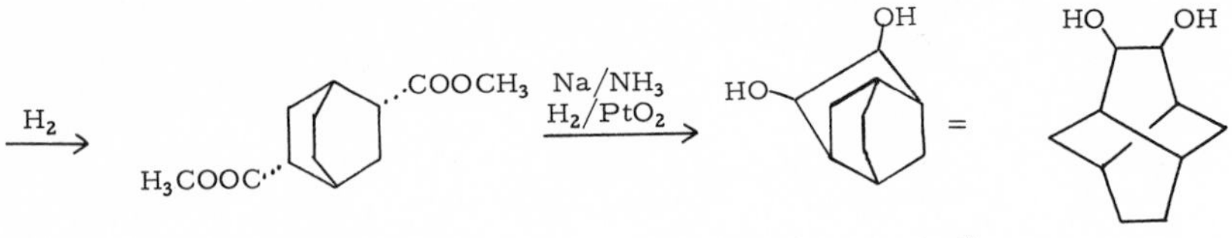

(4)   5)

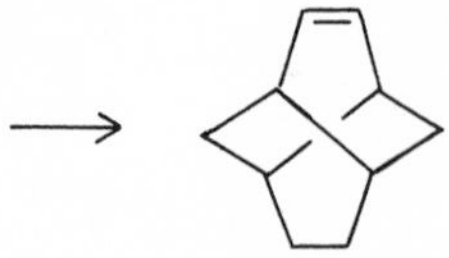

(6)

involved conversion into the cyclic thionocarbonate (thiocarbonyldiimidazole) and treatment with trimethyl phosphite. The synthetic scheme has the further advantage that the intermediate diacid corresponding to (4) can be resolved via the brucine salt to give (−)-(4), which, when carried through the synthesis, affords (+)-twistene (6).

The new olefin synthesis was also used for a simple synthesis of tetra-O-acetyl-conduritol-B, tetra-O-acetyl-(±)-cyclohexene-1,3/2,4-tetrol (9), from 1,4,5,6-tetra-O-acetylmyoinositol (7).[2]

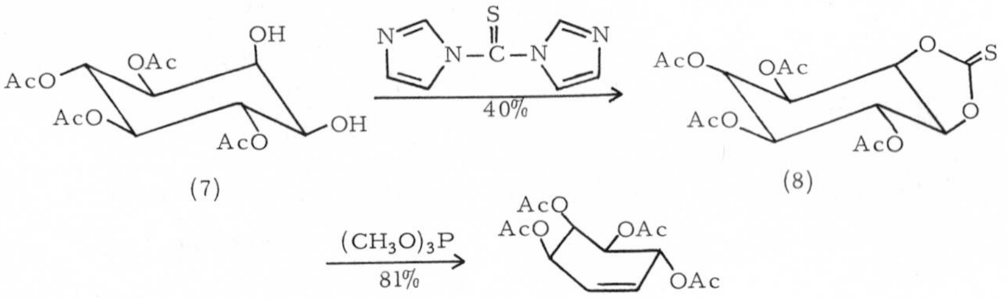

(7)   (8)

(9)

[1]M. Tichý and J. Sicher, *Tetrahedron Letters*, 4609 (1969)
[2]T. L. Nagabhushan, *Canad. J. Chem.*, **48**, 383 (1970)

Trimethylsilyl azide, 1, 1236.

Preparation. The preparation of this stable and safe substitute for hydrazoic acid from the reaction of trimethylchlorosilane and sodium azide has been described in detail.[1]

$$(CH_3)_3SiCl + NaN_3 \xrightarrow[85\%]{} (CH_3)_3SiN_3 + NaCl$$

[1]L. Birkofer and P. Wegner, *Org. Syn.*, **50**, 107 (1970)

Trimethylsilyldiethylamine (TSiD). 127°/738 mm. Supplier: Pierce.

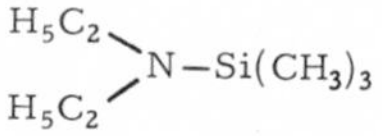

$$\begin{array}{c} H_5C_2 \\ \diagdown \\ H_5C_2 \diagup \end{array} N{-}Si(CH_3)_3 \qquad \text{Mol. wt. 145.33, b.p.}$$

Selective silylation of alcohols.[1] The reagent selectively silylates equatorial hydroxyl groups in quantitative yield within 4–10 hours at room temperature. Axial hydroxyl groups do not react under these conditions.

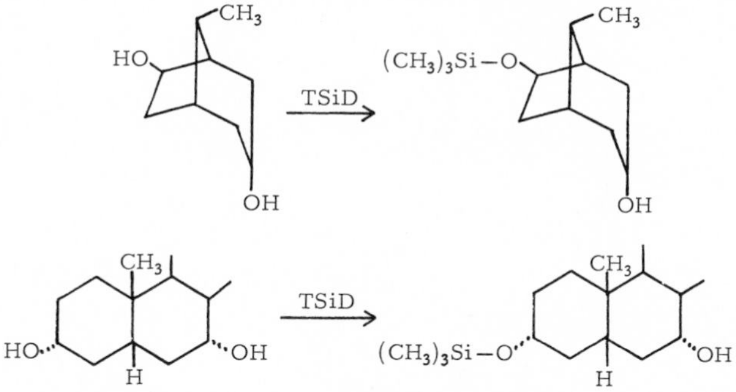

[1] I. Weisz, K. Felföldi, and K. Kovács, *Acta Chim. Acad. Sci. Hung.*, **58**, 189 (1968)

Triphenylphosphine, 1, 1238–1247; **2,** 443–445.

Desulfurization. Evans *et al.*[1] noted that dialkenyl disulfides, for example (I), are converted by reaction with triphenylphosphine into dialkenyl sulfides (II) with an accompanying allylic rearrangement. Ollis[2] used this reaction in one step

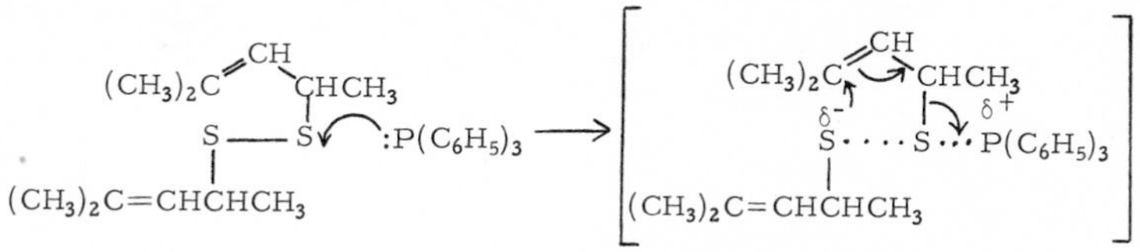

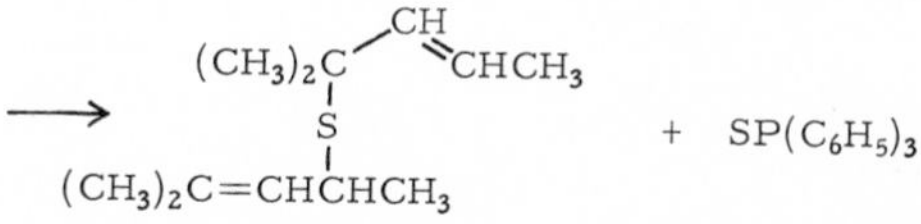

of a synthesis of squalene (6) from farnesyl chloride (1).

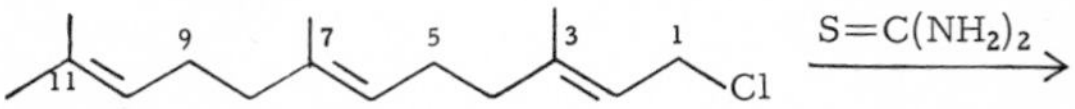

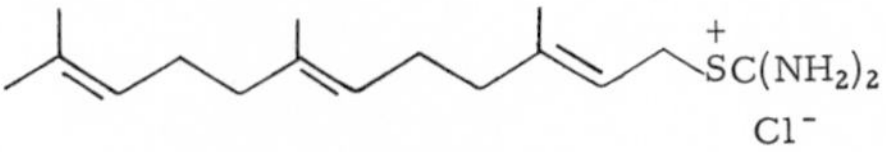

(2)

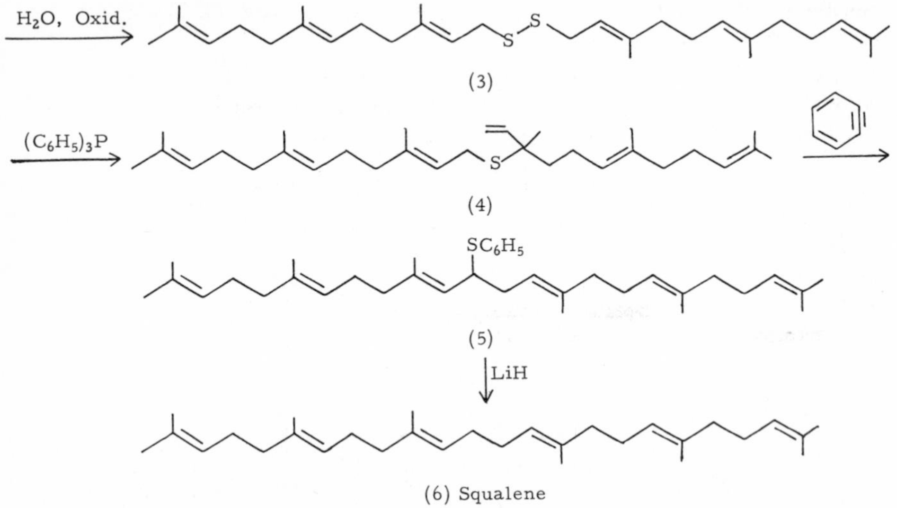

$$\xrightarrow{\text{H}_2\text{O, Oxid.}}$$

(3)

$$\xrightarrow{(C_6H_5)_3P}$$

(4)

SC$_6$H$_5$

(5)

$\downarrow$ LiH

(6) Squalene

Synthesis of polycyclic compounds. 2,2′-Bisbromomethyldiphenyl (1) reacts with 1 mole of triphenylphosphine to give the phosphonium salt (2), which is converted into the ylide (3) on treatment with base (sodium methoxide). The ylide (3) is not isolated but undergoes intramolecular C-alkylation to form the phosphonium salt (4). This salt on treatment with phenyllithium gives the ylide (5). This ylide is readily transformed into phenanthrene or 9-substituted phenanthrenes.[3]

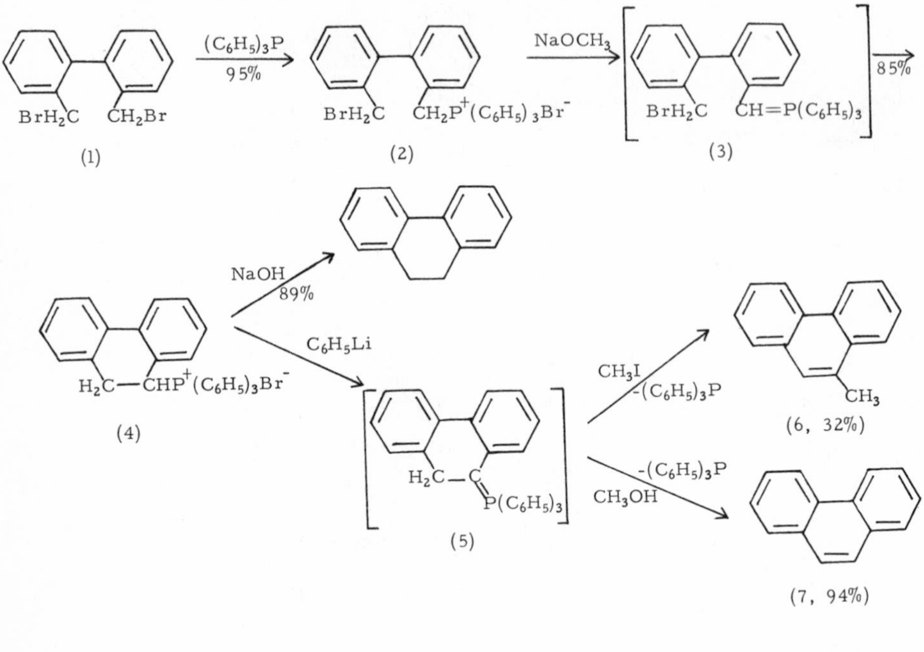

BrH$_2$C   CH$_2$Br

(1)

$$\xrightarrow[95\%]{(C_6H_5)_3P}$$

BrH$_2$C   CH$_2$P$^+$(C$_6$H$_5$)$_3$Br$^-$

(2)

$$\xrightarrow{\text{NaOCH}_3}$$

BrH$_2$C   CH=P(C$_6$H$_5$)$_3$

(3)   85%

H$_2$C—CHP$^+$(C$_6$H$_5$)$_3$Br$^-$

(4)

$$\xrightarrow[89\%]{\text{NaOH}}$$

$$\xrightarrow{C_6H_5Li}$$

H$_2$C—C

P(C$_6$H$_5$)$_3$

(5)

$$\xrightarrow[-(C_6H_5)_3P]{\text{CH}_3\text{I}}$$

CH$_3$

(6, 32%)

$$\xrightarrow[\text{CH}_3\text{OH}]{-(C_6H_5)_3P}$$

(7, 94%)

The method has been extended to the synthesis of acenaphthylene (8) and acenaphthene (9) as shown.[4] The scheme has also been used for six- and seven-

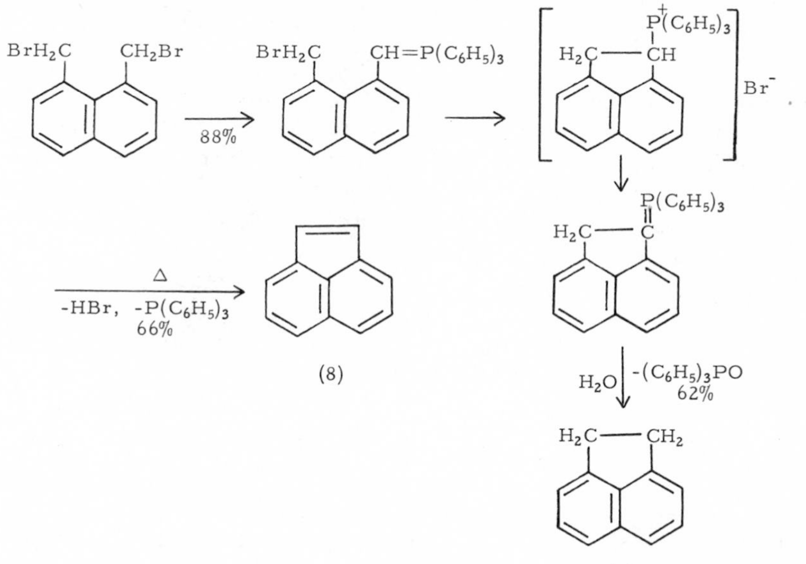

(8)

(9)

membered rings (tetralin and 1,2;3,4-dibenzocycloheptadiene-1,3).

Olefin synthesis by double extension. Barton *et al.*[5] have developed a useful olefin synthesis which is particularly applicable to highly hindered olefins. It depends on removal of two groups of atoms from between the two carbon atoms which form the new double bond:

In one example the process involves loss of nitrogen and a sulfur atom. Thus cyclohexanone is treated with hydrazine and hydrogen sulfide to give the tetra-hydrothiadiazole (1) in quantitative yield. This is then oxidized with lead tetra-acetate at 0° to give the azo sulfide (2) in 95% yield. When (2) is heated with triphenylphosphine (1.1 moles) at 100° for one hour, biscyclohexylidene (4) is obtained in 77% yield. The thiirane (3) is an intermediate; it has been obtained by pyrolysis of (2).[6]

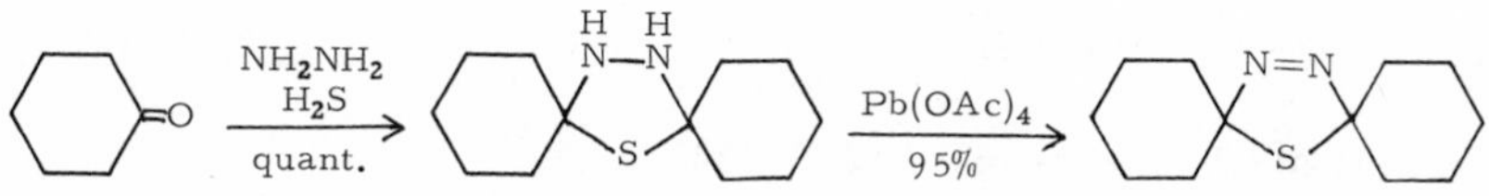

(1)

(2)

$$\xrightarrow[\text{-N}_2]{\substack{(C_6H_5)_3P \\ 100^\circ, \text{1 hr.}}} \left[\bigcirc\!\!\diagdown\!\!\underset{S}{\diagup}\!\!\bigcirc \right] \xrightarrow{77\%} \bigcirc\!\!=\!\!\bigcirc + (C_6H_5)_3PS$$

$$\qquad\qquad\qquad\qquad (3) \qquad\qquad\qquad\qquad\qquad (4)$$

In another example of the new process, which involves loss of CO_2 and S,[5] thiobenzilic acid,[7] is condensed with a ketone, for example cyclohexanone, to give an oxathiolan-5-one (5) in high yield. *p*-Toluenesulfonic acid or boron trifluoride is used as catalyst. When (5) is heated with hexaethylphosphorous triamide (**1**, 425), the corresponding olefin (6) is obtained in good yield.

$$(C_6H_5)_2C\!\!\underset{COOH}{\overset{SH}{\diagdown}} + O\!\!=\!\!\bigcirc \xrightarrow{\text{TsOH, benzene}} \underset{C_6H_5}{\overset{C_6H_5}{\diagdown}}\!\!\underset{O}{\overset{S}{\diagup}}\!\!\bigcirc$$

$$\qquad\qquad\qquad\qquad\qquad\qquad\qquad\qquad (5)$$

$$\xrightarrow[\substack{160\text{-}200^\circ, \text{5 hrs.} \\ 82\%}]{[(C_2H_5)_2N]_3P} \underset{C_6H_5}{\overset{C_6H_5}{\diagdown}}\!C\!\!=\!\!\bigcirc$$

$$\qquad\qquad\qquad\qquad\qquad (6)$$

[1]M. B. Evans, G. M. C. Higgins, C. G. Moore, M. Porter, B. Saville, J. F. Smith, B. R. Trego, and A. A. Watson, *Chem. Ind.*, 897 (1960)

[2]G. M. Blackburn, W. D. Ollis, C. Smith, and I. O. Sutherland, *Chem. Commun.*, 99 (1969)

[3]H. J. Bestmann, H. Häberlein and W. Eisele, *Ber.*, **99**, 28 (1966)

[4]H. J. Bestmann, R. Härtl, and H. Häberlein, *Ann*, **718**, 33 (1968)

[5]D. H. R. Barton and B. J. Willis, *Chem. Commun.*, 1225 (1970); D. H. R. Barton, E. H. Smith, and B. J. Willis, *ibid.*, 1226 (1970)

[6]R. M. Kellogg and S. Wassenaar, *Tetrahedron Letters*, 1987 (1970)

[7]H. Becker and A. Bistrzycki, *Ber.*, **47**, 3149 (1914)

Triphenylphosphine–Carbon tetrachloride, 1, 1247; **2**, 445.

Nitriles from primary amides. Benzamide is converted into benzonitrile when heated in THF with triphenylphosphine–carbon tetrachloride (yields up to 83% depending on conditions). The reaction probably proceeds as follows:

$$C_6H_5CONH_2 + P(C_6H_5)_3 + CCl_4 \longrightarrow C_6H_5C(Cl)\!\!=\!\!NH + O\!\!=\!\!P(C_6H_5)_3 + CHCl_3$$

$$C_6H_5C(Cl)\!\!=\!\!NH + P(C_6H_5)_3 \longrightarrow C_6H_5C\!\!\equiv\!\!N + P(C_6H_5)_3 \cdot HCl$$

However, addition of triethylamine to promote elimination of hydrogen chloride from the intermediate chloroimide leads to lower yields.[1]

[1]E. Yamato and S. Sugasawa, *Tetrahedron Letters*, 4383 (1970)

Triphenylphosphine dibromide, 1, 1247–1249; **2**, 446.

Reaction with substituted acid amides.[1] The reagent reacts with N,N'-disubstituted ureas in the presence of triethylamine to give carbodiimides in 65–75% yield.

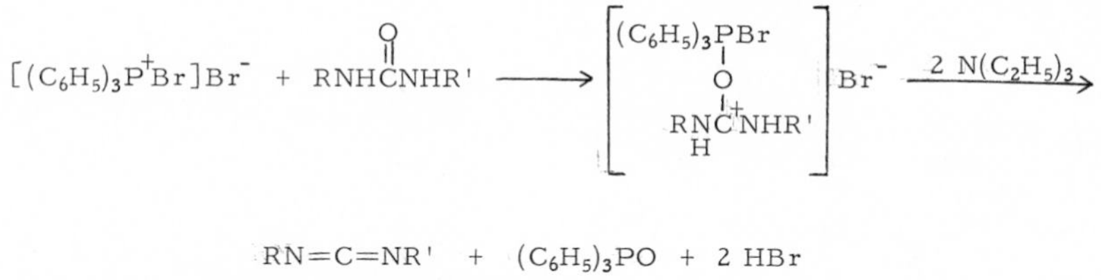

$$[(C_6H_5)_3\overset{+}{P}Br]Br^- \ + \ RNH\overset{\overset{O}{\|}}{C}NHR' \ \longrightarrow \ \left[\begin{array}{c}(C_6H_5)_3PBr \\ | \\ O \\ |+ \\ RN\overset{|+}{C}NHR' \\ | \\ H\end{array}\right]Br^- \ \xrightarrow{\ 2\ N(C_2H_5)_3\ }$$

$$RN{=}C{=}NR' \ + \ (C_6H_5)_3PO \ + \ 2\ HBr$$

Isonitriles are obtained in 50–70% yield by the reaction of triphenylphosphine dibromide with monosubstituted formamides:

$$[(C_6H_5)_3\overset{+}{P}Br]Br^- \ + \ OHCNHR \ \xrightarrow[-2\ HBr]{2\ N(C_2H_5)_3} \ :\!\ddot{C}{=}N{-}R \ + \ (C_6H_5)_3PO$$

Formylation.[1] A new variation of the Vilsmeier formylation reaction involves the reaction of triphenylphosphine dibromide with DMF in the presence of a suitably substituted substrate. The actual formylating reagent is considered to be (1).

$$\underset{\text{Br H}}{[(C_6H_5)_3P\ OC\overset{+}{N}(CH_3)_2]\ Br^-}$$

(1)

Cleavage of ethers. 7-Bromonorbornane (2) has been obtained in 74–83% yield by cleavage of the ether (1) by triphenylphosphine dibromide in refluxing acetonitrile.[2]

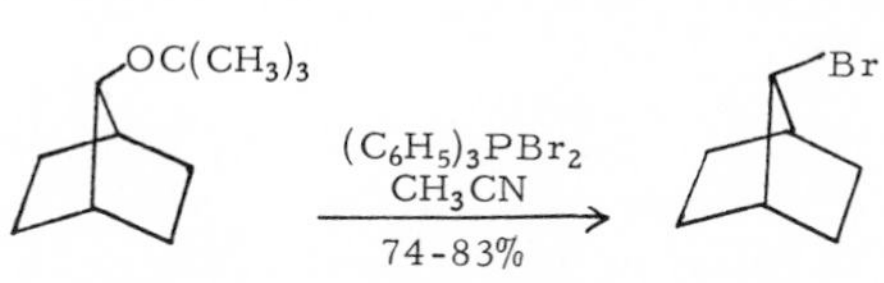

(1)   (2)

Reaction with a phenol.[3] For conversion of β-naphthol into β-bromonaphthalene, a mixture of 144 g. of triphenylphosphine and 125 ml. of acetonitrile is stirred with ice cooling and 88 g. of bromine is added dropwise in 20–30 minutes.

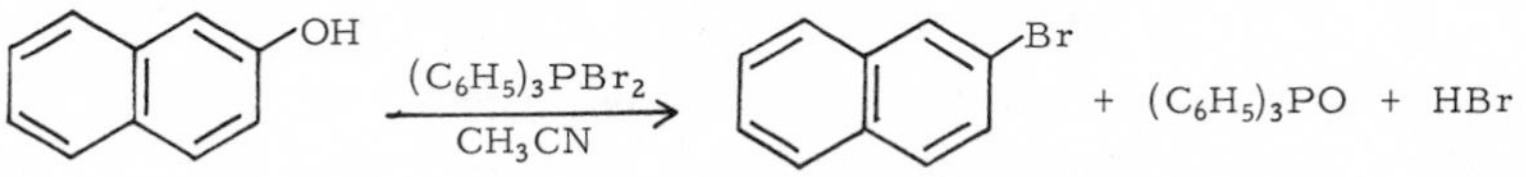

The ice bath is removed, 72 g. of β-naphthol in 100 ml. of acetonitrile is added, and the mixture is heated to 60–70° for at least 30 minutes. The submitters obtained β-bromonaphthalene as a white solid, m.p. 45–50°, yield 70–78% (for details of the workup, see the original).

Aziridines. The reaction of β-amino alcohols (1) with triphenylphosphine dibromide in acetonitrile (triethylamine) gives aziridines (2) in about 50% yield.[4] Piperazines (3) are formed as by-products.

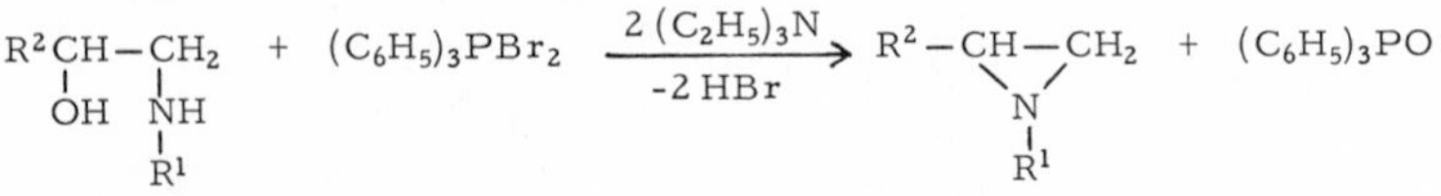

$$R^2CH-CH_2 \;+\; (C_6H_5)_3PBr_2 \;\xrightarrow[-2\,HBr]{2\,(C_2H_5)_3N}\; R^2-CH-CH_2 \;+\; (C_6H_5)_3PO$$
$$\underset{\underset{R^1}{|}}{\overset{|}{OH}}\;\; \overset{|}{NH}$$

(1)   (2)

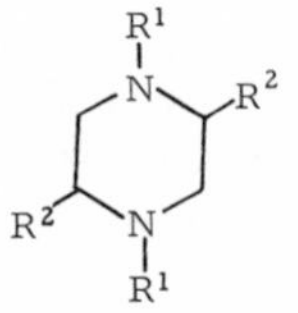

(3)

[1] H. J. Bestmann, J. Lienert, and L. Mott, *Ann.*, **718**, 24 (1968)
[2] A. P. Marchand and W. R. Weimar, Jr., *Chem. Ind.*, 200 (1969)
[3] J. P. Schaefer, J. Higgins, and P. K. Shenoy, *Org. Syn.*, **49**, 6 (1969)
[4] I. Okada, K. Ichimura, and R. Sudo, *Bull. Chem. Soc. Japan*, **43**, 1185 (1970)

Triphenylphosphine–2,2'-Dipyridyl disulfide.

Peptide coupling. Mukaiyama *et al.*[1] have used the combination of triphenylphosphine and 2,2'-dipyridyl disulfide (supplier: Pierce) to effect coupling of an N-protected amino acid with an amino acid ester. The condensation involves oxidation-reduction:

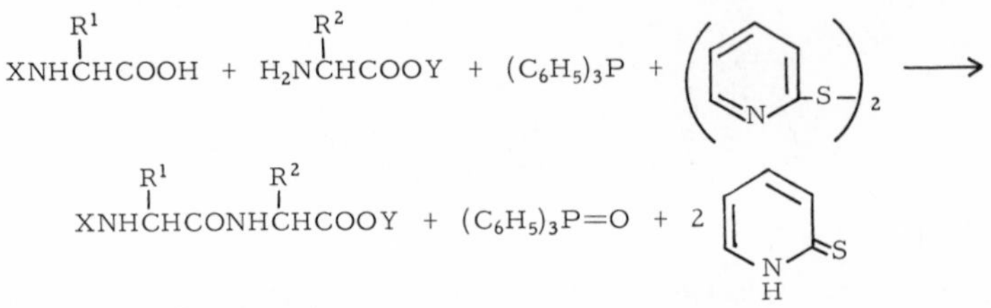

Bz-L-Leu-Gly-OC$_2$H$_5$ was produced by this procedure in methylene chloride solution (30 minutes, room temperature) in 91% yield and with high optical purity. The reaction is applicable to a wide variety of amino acids and requires only mixing of reactants. Side-chain protection is not required for hydroxyls of Tyr, Thr, and Ser, and no nitrile formation is noted with Glu and Asp.

[1] T. Mukaiyama, R. Matsueda, and M. Suzuki, *Tetrahedron Letters*, 1901 (1970)

Triphenyl phosphite, 1, 1249; 2, 446.

Peptide synthesis. Peptides can be synthesized from N-protected amino acids and amino acid esters by use of triphenyl phosphite.[1] The reaction is strongly catalyzed by imidazole (**1**, 492–494; **2**, 220); with this catalyst the reaction can be

$$\text{ZNHCHR'COOH} + \text{NH}_2\text{CHR''COOC}_2\text{H}_5 + \text{P(OC}_6\text{H}_5)_3 \xrightarrow{\text{H}}$$

$$\text{ZNHCHR'CONHCHR''COOC}_2\text{H}_5 + (\text{C}_6\text{H}_5\text{O})_2\text{POH} + \text{C}_6\text{H}_5\text{OH}$$

conducted in 85–99% yield at 40° in 18 hours. Racemization is negligible. The most suitable solvents are dioxane and DMF. Carbobenzoxy and t-butyloxy-carbonyl groups give the best results as N-protecting groups. The diphenyl phosphite and phenol which are formed are readily removed by washing with water and ether.

[1]Y. V. Mitin and O. V. Glinskaya, *Tetrahedron Letters*, 5267 (1969)

Triphenyl phosphite ozonide, $(\text{C}_6\text{H}_5\text{O})_3\text{PO}_3$.

Singlet oxygen. Thompson[1] reported that ozone and triaryl phosphites form 1 : 1 adducts which are stable at low temperatures; when the temperature is raised the adducts decompose to triaryl phosphates and molecular oxygen. Murray and Kaplan[2] reasoned that the oxygen formed should be in the singlet state and indeed found evidence for this assumption. Thus they were able to perform reactions

$$(\text{C}_6\text{H}_5\text{O})_3\text{P} + \text{O}_3 \xrightarrow[\text{CH}_2\text{Cl}_2]{-78^0} (\text{C}_6\text{H}_5\text{O})_3\text{P}\langle\text{O}_3\rangle \xrightarrow{> -35^0} (\text{C}_6\text{H}_5\text{O})_3\text{P}{=}\text{O} + {}^1\text{O}_2$$

characteristic of dye-photosensitized oxidations. One such reaction is Diels-Alder reaction with conjugated dienes to give 1,4-*endo*-peroxides. Thus, when the adduct is allowed to warm up in the presence of 1,3-cyclohexadiene (1), 5,6-dioxabicyclo[2.2.2]octene-2 (2) is obtained in 67% yield:

(1)   (2), 67%

A second characteristic reaction of singlet oxygen is the "ene" reaction with olefins with an allylic hydrogen to give allylic peroxides. Thus 2,3-dimethyl-butene-2 (3) is converted in 53% yield into the allylic peroxide 2,3-dimethyl-3-hydroperoxybutene-1 (4):

$$H_3C \diagdown C=C \diagup CH_3 \quad + \quad (C_6H_5O)_3P\diagup O \quad \xrightarrow{> -35°} \quad H_2C \diagdown C-\overset{\overset{CH_3}{|}}{\underset{|}{C}}-CH_3 \quad + \quad (C_6H_5O)_3P=O$$

(3)   (4)

Surprisingly, Bartlett and Mendenhall[3] have found that triphenyl phosphite ozonide reacts with (3) at −60° to give the same hydroperoxide (4). At this temperature triphenyl phosphite ozonide does not decompose to singlet oxygen.

The ozonide is also able to imitate another reaction of singlet oxygen: 1,2-cycloaddition to vinylene diethers to yield 1,2-dioxetanes.[4] However, there is one difference. Singlet oxygen reacts stereospecifically with cis- and trans-diethoxyethylene, (5) and (7), to give (6) and (8), respectively. Triphenyl phosphite

(5)   (6)

(7)   (8)

ozonide, on the other hand, reacts with (5) and (7) to give a mixture of (6) and (8) in which the less hindered (8) is the major component (81–83%).

[1] Q. E. Thompson, Am. Soc., 83, 845 (1961)
[2] R. W. Murray and M. L. Kaplan, ibid., 91, 5358 (1969)
[3] P. D. Bartlett and G. D. Mendenhall, ibid., 92, 210 (1970)
[4] A. P. Schaap and P. D. Bartlett, ibid., 92, 6055 (1970)

Triphenyltin hydride, 1, 1250–1251; 2, 448.

Reduction. $\Delta^{14,16}$-Pregnadiene-20-ones can be selectively reduced to Δ^{14}-pregnene-20-ones with triphenyltin hydride. The reaction was originally[1] carried out in a sealed tube with several additions of the reagent. The method has since been simplified by use of refluxing xylene under nitrogen.[2] Yields of 60–80% are obtained.

Hydrodechlorination. exo-2-Phenylthio-endo-3-chloronorbornane (1) was successfully hydrodechlorinated to exo-norbornyl phenyl sulfide (2) when heated for 1 hour at 80° with triphenyltin hydride in the presence of a catalytic amount of azobisisobutyronitrile.[3]

(1) $\xrightarrow[62\%]{(C_6H_5)_3SnH}$ (2)

[1]E. Yoshii and M. Yamasaki, *Chem. Pharm. Bull. Japan*, **16**, 1158 (1968)
[2]T. Nambara, K. Shimada, and S. Goya, *ibid.*, **18**, 453 (1970)
[3]H. C. Brown and K.-T. Liu, *Am. Soc.*, **92**, 3502 (1970)

Tris(2-hydroxypropyl)amine (1,1′,1″-Nitrilotri-2-propanol), [$CH_3CH(OH)CH_2$]$_3$N. Mol. wt. 191.27, m.p. 6–22°. Supplier: Eastman.

Esterification. Carboxylic acids can be converted into methyl esters in high yield with dimethyl sulfate in the presence of this base as the proton acceptor (10–20% excess amine).[1] Ethyl esters can be prepared in the same way using diethyl sulfate. Dicyclohexylethylamine is equally effective but is not available commercially. The method is simple and rapid and is useful when strongly acidic conditions must be avoided.

This procedure has been used recently by Rapoport[2] in a synthesis of the chemotactic hormone sirenin (3) for esterification of a mixture of the 6,10-dimethyl-5,9-undecadienoic acids (1) and (2) (94% yield).

(1)   (2)   (3)

[1]F. H. Stodola, *J. Org.*, **29**, 2490 (1964)
[2]U. T. Bhalerao, J. J. Plattner, and H. Rapoport, *Am. Soc.*, **92**, 3429 (1970)

Tris(triphenylphosphine)chlororhodium, 1, 1252; **2,** 448–453.

Henbest reduction (see Chloroiridic acid). Substitution of chloroiridic acid by tris(triphenylphosphine)chlororhodium results in even greater stereospecificity; thus 5β-androstane-3,17-dione is reduced to the 3β-alcohol and 3α-alcohol in the ratio 50:1, with no detectable reduction of the 17-keto group. The reaction is preferably run in a sealed tube at 82°. 2-Ketosteroids are reduced entirely to the axial 2β-alcohol (Henbest's conditions give 1% of the 2α-alcohol).[1]

Homogeneous hydrogenation catalyst. This catalyst is useful for reduction of 1,4-dihydroaromatic compounds. Aromatic compounds are not formed by disproportionation, as is the case when the usual heterogeneous catalyst are used. Moreover it is possible to reduce disubstituted double bonds in the presence of tetrasubstituted double bonds. Thus isotetralin (1) or 1,4-dihydrotetralin (2) is reduced with this catalyst to an identical 80:20 mixture of 9,10-octalin (3) and 1,9-octalin (4). The latter octalin evidently arises by isomerization of the double

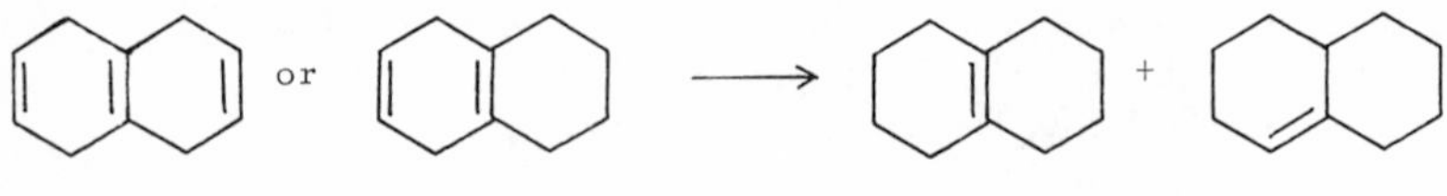

(1)   (2)   (3)   80:20   (4)

bond of (3). A further example is the reduction of santonin (5) to dihydrosantonin (6) in 90% yield.[2]

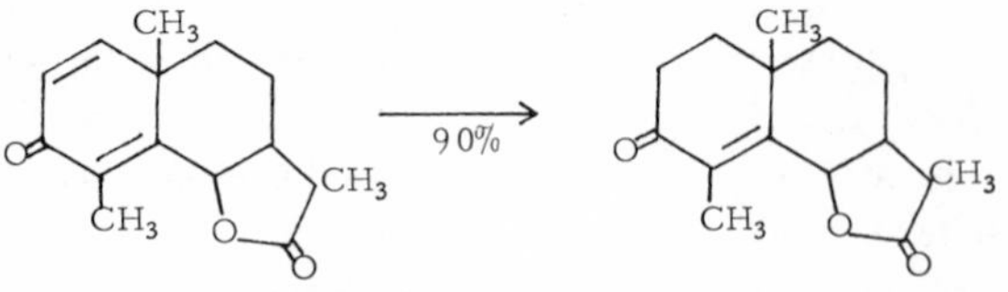

(5)   (6)

The catalyst can be activated for hydrogenation by small quantities of oxygen or by addition of hydrogen peroxide. Acceleration by factors varying from 1.5 to 4 can be realized. Presumably one triphenylphosphine ligand is displaced as triphenylphosphine oxide.[3]

The thiophene nucleus does not seriously poison this rhodium catalyst; Swedish chemists[4] consider this catalyst the best for hydrogenation of double bonds in thiophene derivatives.

This catalyst is useful for selective reduction of the carbon–carbon double bond of α,β-unsaturated acids, esters, ketones, nitriles, and nitro compounds. Sterically hindered double bonds, however, are not reduced. Selective hydrogenation of α,β-unsaturated aldehydes is hampered by concomitant decarbonylation; this reaction can be suppressed to some extent by carrying out the hydrogenation in absolute ethanol. The hydrogenations were conducted in benzene or ethanol at 40–60° and 60–100 psi pressure for 12–18 hours.[5]

Valence tautomerization. Tris(triphenylphosphine)chlororhodium (in the absence of hydrogen) catalyzes valence tautomerization.[6] Thus, when *exo*-tricyclo[3.2.1.0²,⁴]octene (1) is warmed at 90° with 1.3 moles % of the catalyst it is converted into a mixture of (2, 62%), (3, 32%), and (4, 6%). Rhodium on carbon

(1)   (2)   (3)   (4)

can also be used, but requires a higher temperature, and gives a dimer in addition. The rearrangement of (1) to (2) and (3) is intramolecular; thus a mixture of un-labeled and doubly labeled (1) gives only unlabeled and double labeled (2) and (3).

Moreover the conversion of (1) into (2) is stereospecific. Thus (1a) rearranges exclusively to (2a). Previously such valence tautomerizations have been assumed

(1a)   (2a)

to be concerted electrocyclic processes. However, in this case (2) is not related to (1) by an electrocyclic reaction. The original paper should be consulted for a suggested mechanism.

Cyclooligomerization of allene. Thermal oligomerization of allene gives a complex mixture of dimers, trimers, and higher oligomers. However, if allene is agitated at 70–80° in a pressure tube in the presence of a catalytic amount of tris(triphenylphosphine)chlororhodium, a single tetramer is obtained in 59% yield and a single pentamer is formed in traces (6%).[7] The structure of the pentamer is not known; structure (1) has been assigned to the tetramer on the basis of ozonization and reduction experiments. Note that the formation of a spiro ring

(1)

system is not usual. Tetramerization of allene to (1) has also been effected in 70–80% yield with a combination of halogenorhodium(I) complexes and triphenylphosphine.[8]

Synthesis of nitriles from secondary amides. Secondary amides are converted into nitriles when heated with the reagent for six hours at 250°:

$$ArCONHCH_2R \longrightarrow ArCN + RCH_2OH$$

Yields are in the range 25–90%. Primary amides are also converted into nitriles (65–80% yield). Tris(triphenylphosphine)chlororhodium is not the true catalyst in these reactions, since no nitrile is formed if the reaction mixture is heated in an atmosphere of argon unless the reagent is treated first with air or oxygen.[9]

Selective acetalation. The reagent catalyzes the acetalation of 3-ketosteroids; for example, 5α-pregnane-3,20-dione (1) is converted into 5α-pregnane-3,20-dione 3-dimethyl acetal in 78% yield. Keto groups at C_6, C_{12}, C_{17}, and C_{20} do not react; 2-ketosteroids react to some extent.[10]

Decarbonylation (2, 451). Decarbonylation reactions catalyzed by transition metal compounds have been reviewed by Tsuji and Ohno.[11]

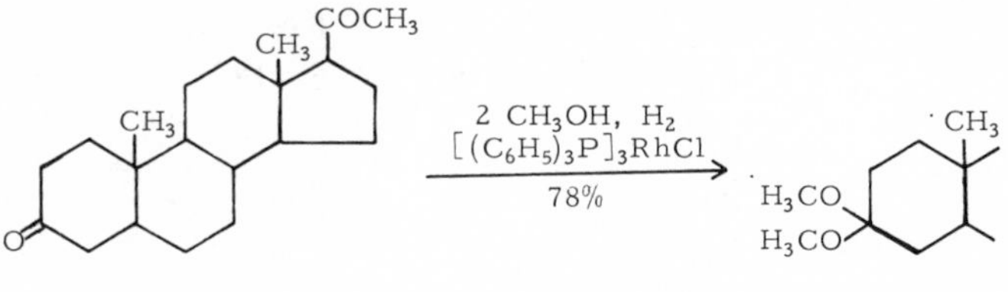

(1) (2)

Acetylenic ketones of type (1) are decarbonylated to give the coupled product (2) when heated for many hours with the rhodium complex in benzene or xylene solution. In contrast with decarbonylation of aldehydes, an equivalent of the rhodium complex is required.[12]

$$RC{\equiv}C{-}\overset{O}{\overset{\|}{C}}{-}C{\equiv}CR + RhCl[P(C_6H_5)_3]_3 \longrightarrow RC{\equiv}C{-}C{\equiv}CR + RhCl(CO)[P(C_6H_5)_3]_2 + P(C_6H_5)_3$$

(1)	(2)
R = C_6H_5	R = C_6H_5, 78%
R = $(CH_3)_3C$	R = $(CH_3)_3C$, 48%

The complex decarbonylates allylic alcohols at 110–115° in acetonitrile or benzonitrile. Thus cinnamyl alcohol is converted into ethylbenzene in 76% yield; styrene is obtained as a minor product (4% yield). The reaction may proceed

$$RCH{=}CHCH_2OH + [(C_6H_5)_3P]_3RhCl \longrightarrow \left\{\begin{array}{l} RCH_2CH_3 \\ \text{(major)} \\ RCH{=}CH_2 \\ \text{(minor)} \end{array}\right. + (C_6H_5)_3P + [(C_6H_5)_3P]_2RhCOCl$$

through allylic isomerization and subsequent decarbonylation of the aldehyde tautomer of the enol thus produced.[13]

Walborsky and Allen[14] have shown that the decarbonylation of aldehydes in which the carbonyl group is directly attached to a cyclopropyl as well as a trigonally and a tetrahedrally hybridized carbon atom proceeds with a high degree of stereoselectivity and with overall retention of configuration, as shown in the examples. The reaction is carried out in benzene, xylene, or acetonitrile at reflux

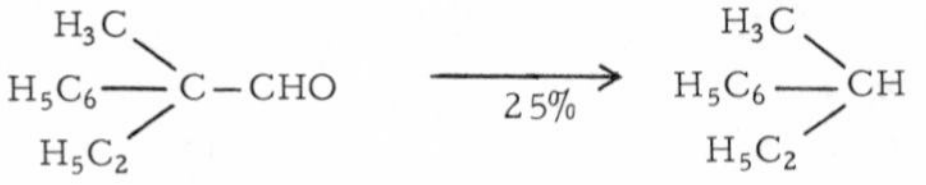

93% retention

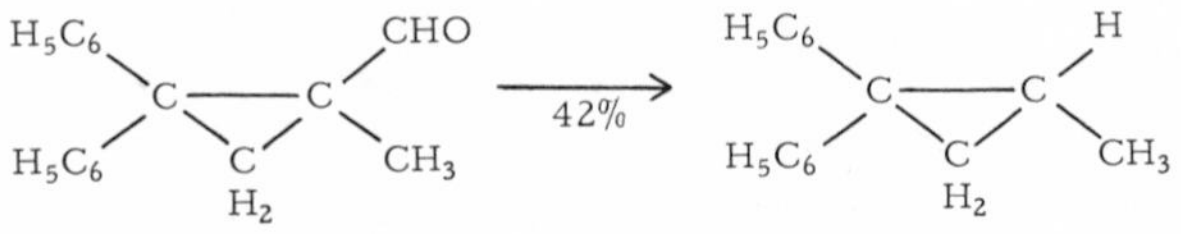

94% retention

$$\underset{32\%}{\longrightarrow}$$

100% retention

temperatures. The stereochemical results rule out radical or carbonium ion intermediates. When a C_1 deuterated aldehyde is decarbonylated, 100% of the deuterium is found in the hydrocarbon formed. The reaction is thus intramolecular.

Deuteration of olefins.[15] Deuterium adds specifically across the double bond of *n*-monoolefins when this homogeneous catalyst is used. In contrast, use of a heterogeneous catalyst such as platinum black leads to unspecific labeling.

[1]J. C. Orr, M. Mersereau, and A. Sanford, *Chem. Commun.*, 162 (1970)

[2]J. J. Sims, V. K. Honwad, and L. H. Selman, *Tetrahedron Letters*, 87 (1969)

[3]H. van Bekkum, F. van Rantwijk, and T. van de Putte, *ibid.*, 1 (1969)

[4]A.-B. Hörnfeldt, J. S. Gronowitz, and S. Gronowitz, *Chem. Scand.*, **22**, 2725 (1968)

[5]R. E. Harmon, J. L. Parsons, D. W. Cooke, S. K. Gupta, and J. Schoolenberg, *J. Org.*, **34**, 3684 (1969); R. E. Harmon, J. L. Parsons, and S. K. Gupta, *Org. Prep. Proc.*, **2**, 25 (1970)

[6]T. J. Katz and S. Cerefice, *Am. Soc.*, **91**, 2405 (1969)

[7]F. N. Jones and R. V. Lindsey, Jr., *J. Org.*, **33**, 3838 (1968)

[8]S. Otsuka, A. Nakamura, and H. Minameda, *Chem. Commun.*, 191 (1969)

[9]J. Blum and A. Fisher, *Tetrahedron Letters*, 1963 (1970)

[10]W. Voelter and C. Djerassi, *Ber.*, **101**, 1154 (1968)

[11]J. Tsuji and K. Ohno, *Synthesis*, **1**, 157 (1969)

[12]E. Müller, A. Segnitz, and E. Langer, *Tetrahedron Letters*, 1129 (1969)

[13]A. Emery, A. C. Oehlschlager, and A. M. Unrau, *ibid.*, 4401 (1970)

[14]H. M. Walborsky and L. E. Allen, *ibid.*, 823 (1970)

[15]J. R. Morandi and H. B. Jensen, *J. Org.*, **34**, 1889 (1969)

s-**Trithiane,** Mol. wt. 138.27, m.p. 216–218°. Suppliers: A, Columbia, E, F, Fluka, Pfaltz and Bauer.

Purification. Commercial *s*-trithiane is purified by extraction from a Soxhlet using 300 ml. of toluene for 39 g. of trithiane, followed by recrystallization from the same solvent.[1]

C—C Bond formation. *s*-Trithiane has been used very much like 1,3-dithiane (**2**, 182–187) in organic synthesis;[2,3] however, the carbanion derived from this reagent tends to undergo carbenoid decomposition. Metalated trithiane reacts readily with primary halides to give 2-alkyl-*s*-trithianes convertible into aldehydes, as shown for a preparation of *n*-pentadecanal:[1]

[1]D. Seebach and A. K. Beck, *Org. Syn.*, submitted 1970
[2]E. J. Corey and D. Seebach, unpublished.
[3]D. Seebach, *Angew. Chem., internat. Ed.*, **8**, 639 (1969); D. Seebach, *Synthesis*, **1**, 17 (1969)

Trityl hexachloroantimonate, $(C_6H_5)_3C^+SbCl_6^-$. Mol. wt. 577.81; yellow prisms, m.p. 218°.

The salt is prepared[1] in quantitative yield either by the reaction of triphenyl-methane with antimony pentachloride in carbon disulfide or by the reaction of trityl chloride and antimony pentachloride in carbon tetrachloride.

Like trityl perchlorate and trityl fluoroborate (**1**, 1256–1258; **2**, 454), this salt can be used for hydride abstraction; one advantage over the perchlorate salt is that it is not explosive. It has been used for the conversion of cycloheptatriene into tropylium hexachloroantimonate[1] and for conversion of 1,5-diketones into pyrylium salts.[2]

[1]J. Holmes and R. Pettit, *J. Org.*, **28**, 1695 (1963)
[2]D. Fǎrcaşiu, *Tetrahedron*, **25**, 1209 (1969)

L-Tyrosine hydrazide, **2**, 454. M.p. 196–198°.

Resolution. The reagent was "highly successful" for resolution of the synthetic DL-form of the natural imino acid, L-azetidine-2-carboxylic acid (1).[1]

(1)

[1]R. M. Rodebaugh and N. H. Cromwell, *J. Heterocyclic Chem.*, **6**, 993 (1969)

V

Vanadium oxyacetylacetonate (Vanadyl acetylacetonate), 2, 456.

Epoxidation.[1] Cyclohexene epoxide can be prepared in quantitative yield by the reaction of cyclohexene with *t*-butyl hydroperoxide catalyzed by vanadium oxyacetylacetonate.

Amine oxides. The procedure for the preparation of N,N-dimethyldodecyl-amine oxide (ref. 1, 2, 456) has now been published.[2]

Oxidation of aniline. Aniline is oxidized to nitrobenzene (39% conversion) by *t*-butyl hydroperoxide catalyzed by vanadium oxyacetylacetonate. Molybdenum compounds are somewhat less active; tungsten and cobalt compounds are inactive.[3]

[1]E. S. Gould, R. R. Hiatt, and K. C. Irwin, *Am. Soc.*, **90**, 4573 (1968)
[2]M. N. Sheng and J. G. Zajacek, *Org. Syn.*, **50**, 56 (1970)
[3]G. R. Howe and R. R. Hiatt, *J. Org.*, **35**, 4007 (1970)

Vanadium oxytrichloride, $VOCl_3$. Mol. wt. 173.32. Supplier: Alfa Inorganics.

Intramolecular oxidative phenol coupling. The 1,3-bis(hydroxyphenyl)propane (1) is converted into the phenolic dienone (2) in 76% yield by reaction with 2.5 molar equivalents of vanadium oxychloride in refluxing ether (10 hours).[1] Use

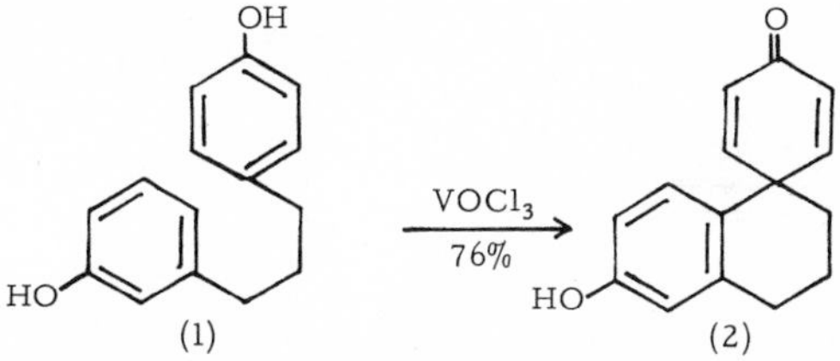

of alkaline potassium ferricyanide, ferric chloride, or manganic tris(acetyl-acetonate) gave yields of less than 10%. See also **Vanadium tetrachloride**, This volume.

The Amaryllidaceae alkaloids[2] probably arise by intramolecular oxidative phenol coupling. Indeed Barton[3] has effected a synthesis of (±)-narwedine (4) by oxidation of (3) with potassium ferricyanide; however, the yield of (4) was only

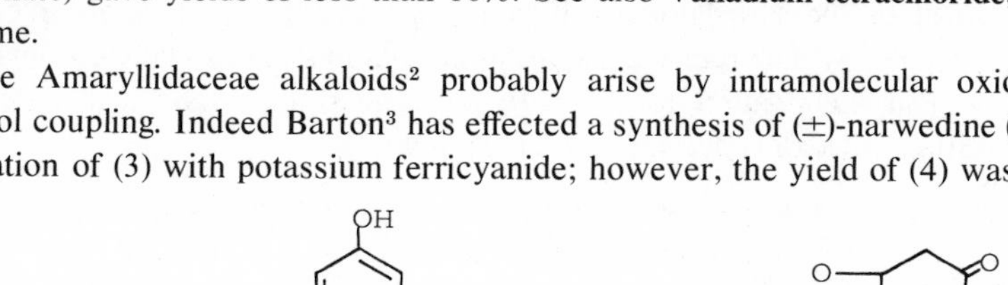

1.4%, the major product being a polymer. Yields obtained with manganese dioxide and lead dioxide were even lower.

Use of vanadium oxytrichloride for this type of oxidation results in much improved yields[4] and provides the key step in a biogenetic-type synthesis of the Amaryllidaceae alkaloid (±)-maritidine (10). The starting material is O-methyl-

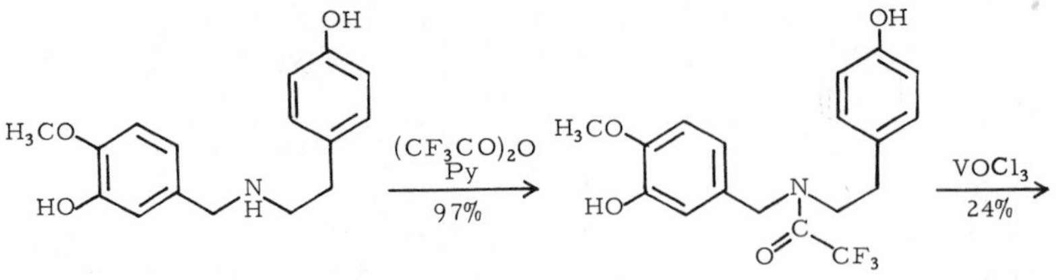

(5)   (6)

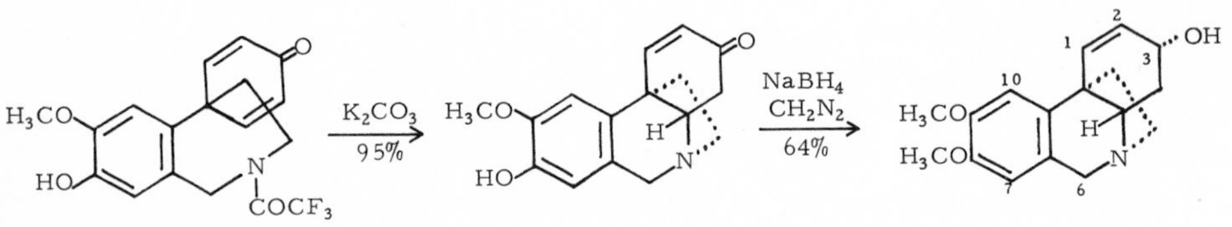

(7)   (8)   (9)

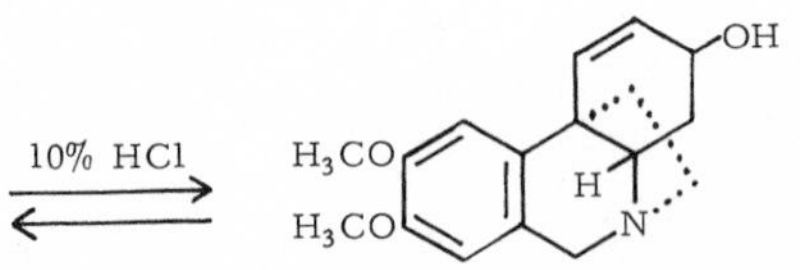

(10)

norbelladine (5); this is N-trifluoroacetylated (6) and this derivative is oxidized with excess vanadium oxytrichloride. The *para-para* coupled dienone (7) results. On removal of the N-trifluoroacetyl group, spontaneous cyclization gives the phenolic enone (8). The remaining steps in the synthesis are sodium borohydride reduction and methylation to give (9), (±)-epimaritidine, and acid-catalyzed epimerization at C_3 to give a mixture of (9) and (10).

[1]M. A. Schwartz, R. A. Holton, and S. W. Scott, *Am. Soc.*, **91**, 2800 (1969)
[2]Review: W. C. Wildman, *Alkaloids*, **11**, 387 (1968)
[3]D. H. R. Barton and G. W. Kirby, *J. Chem. Soc.*, 806 (1962)
[4]M. A. Schwartz and R. A. Holton, *Am. Soc.*, **92**, 1090 (1970)

Vanadium tetrachloride, VCl_4, mol. wt. 192.78; and **Vanadium oxytrichloride**, $VOCl_3$, mol. wt. 173.32, supplier: Alfa Inorganics.

Oxidative coupling of phenols.[1] Admixture of equimolar amounts of phenol and vanadium tetrachloride in an inert solvent, usually carbon tetrachloride, produces a black precipitate with concomitant evolution of hydrogen chloride. Acid hydrolysis gives three dimeric products, (2)–(4), in the ratio of 8 : 4 : 1 and an overall yield of 55–65%. The coupling occurs predominantly in the *para*-position.

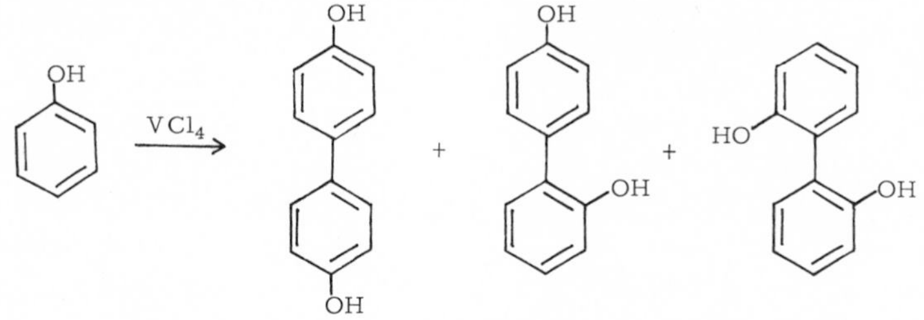

(1)   (2)   (3)   (4)

Both 1- and 2-naphthols react to afford 4,4′-dihydroxybinaphthyl (5) and 2,2′-

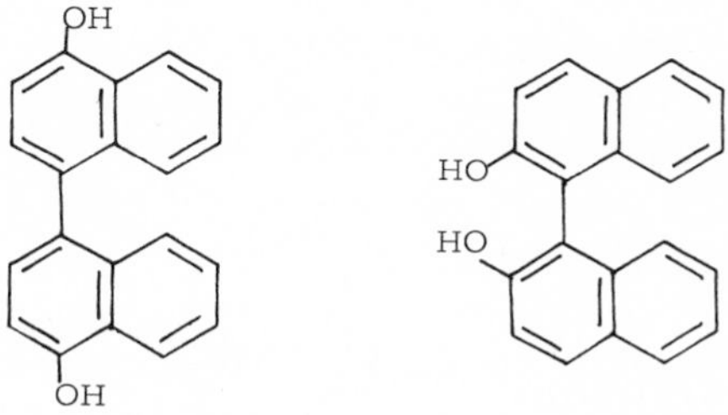

(5)   (6)

dihydroxybinaphthyl (6) in 40 and 38% yield, respectively. The reaction is widely applicable, but cresols give very low yields. Aniline and derivatives are oxidatively coupled.

Vanadium oxytrichloride is definitely less reactive, but it is useful in the case of reactive phenols such as 1- and 2-naphthol. In this case yields of (5) and (6) are somewhat higher than those obtained with vanadium tetrachloride.

Since aromatic hydrocarbons are relatively inert, the hydroxyl group is evidently necessary. Evidence has been obtained that complex vanadium phenoxides are formed during the reaction.

[1]W. L. Carrick, G. L. Karapinka, and G. T. Kwiatowski, *J. Org.*, **34**, 2388 (1969)

Vinyl triphenylphosphonium bromide, 1, 1274–1275; 2, 456–457.

Preparation. The procedure for the preparation of the reagent has been published.[1]

[1]E. E. Schweizer and R. D. Bach, *Org. Syn.*, **48**, 129 (1968)

Z

Zinc, 1, 1276–1284; **2**, 459–462.

Reformatsky reaction (**1**, 1285–1286). Frankenfeld and Werner[1] encountered certain difficulties in the Reformatsky reaction of aliphatic aldehydes with ethyl bromoacetate following the usual procedures as reviewed by Shriner.[2] Yields were low and the products difficult to purify. After a careful study of various conditions, they recommend the following modifications. The zinc[3] should be activated by washing with 20% hydrochloric acid and then washed until neutral with acetone and anhydrous ether. The zinc is then thoroughly air-dried and used immediately. Colored by-products result from traces of moisture. However, the temperature of reaction is even more critical. In the case of straight-chain aldehydes the optimum temperature is 80–85° and is regulated by first heating the suspension of zinc in benzene to vigorous reflux; small amounts of the aldehyde and ethyl bromoacetate are then added cautiously (with heat, if necessary) until a vigorous reaction is apparent from the reflux rate. Thereafter the temperature is maintained at 80–85° by addition of reagents. The reaction can be extremely exothermic. Yields obtained were generally 50–80%.

British chemists[4] have suggested a further improvement. Yields of β-hydroxy esters range from 10 to 90% (average *ca.* 60%) in the standard procedure. Likewise a wide range of yields have been reported for the hydrolysis of the β-hydroxy esters to the free acids. Low yields are largely due to concomitant β-elimination leading to the α,β-unsaturated acids. Ethyl and methyl esters have been employed usually. Cornforth *et al.*[4] report that use of *t*-butyl esters, which are readily dealkylated under mild acidic conditions, gives improved yields of β-hydroxy acids. They then found that β-hydroxy acids can be obtained directly in about 90% yield by the following modified procedure. The Reformatsky reaction is carried out in THF using a *t*-butyl halo ester; after adduct formation is complete the THF is exchanged for anhydrous benzene. The viscous adduct is only slightly soluble in benzene, but after 2 hours reflux is converted into a granular precipitate of the β-hydroxy acid. The zinc ion of the adduct is considered to act as an acid catalyst for hydrolysis of the ester.

Curé and Gaudemar[5] report that yields in the Reformatsky reaction are generally improved if the reaction is carried out in two steps. First, the α-bromo ester (1) is converted into an organozinc bromide (2) by reaction with zinc in pure and dry dimethoxymethane.[6] This derivative is apparently formed in almost quantitative yield. The aldehyde or ketone is then added at 0°. The β-hydroxy ester is obtained in about 60–90% yield.

$$BrCH_2COOC_2H_5 + Zn \xrightarrow{CH_2(OCH_3)_2} BrZnCH_2COOC_2H_5 \xrightarrow[60-90\%]{\underset{R^2}{\overset{R^1}{\diagdown}}C=O} \underset{R^2}{\overset{R^1}{\diagdown}}CCH_2COOC_2H_5$$

(1)   (2)   (3)

α-Methylene γ-lactones. α-(Bromomethyl)acrylic esters (1) readily form with zinc in THF organozinc compounds which react with aldehydes and ketones to give α-methylene γ-lactones (2) in yields varying from 42 to 100%.[7] However,

$$\underset{R^2}{\overset{R^1}{\diagdown}}C=O + BrCH_2\underset{\overset{\|}{CH_2}}{C}-COOR + Zn \rightarrow$$

(1)   (2)

formaldehyde affords a mixture of α-methylenebutyrolactone and γ-hydroxy-α-methylenebutyric ester in low yield.

Dreiding *et al.*[8] have also used essentially the same method for preparation of α-methylene-β-methylbutyrolactones. Thus β′-bromotiglic ester (3) reacts with cyclohexanone under Reformatsky conditions to give (4) in 78% yield. The ester

$$\text{cyclohexanone} + \underset{H_3C}{\overset{H}{\diagdown}}C=C\underset{CH_2Br}{\overset{COOCH_3}{\diagup}} \xrightarrow{Zn}$$

(3)   (4)

(3) is obtained together with γ-bromotiglic ester on treatment of either tiglic ester or angelic ester with NBS.

Dehalogenation (**1**, 1278–1279). Treatment of either dibromocyclobutene or diiodocyclobutene (1) with zinc dust in DMF results in a vigorous exothermic reaction and formation of the two dimers of cyclobutadiene (3) and (4) in 60% yield, the former predominating. The simplest explanation of the reaction is that dehalogenation occurs to yield cyclobutadiene (2), which then undergoes a Diels-Alder type of dimerization.[9]

(1) X = Br, I   (2)   (3)   (4)

Serini reaction (**1**, 1282; **2**, 459–460). Ghera[10] has published the full paper on the synthesis of ketones by zinc-catalyzed rearrangement of secondary esters of trisubstituted 1,2-glycols:

$$\underset{R_1R_2C-CHR_3}{\overset{OH\ (OCOR_4)}{|\qquad\quad|}} \longrightarrow R_1R_2CHCOR_3$$

Benzoates and p-nitrobenzoate esters can be used as well as acetates, higher yields being obtained with the former. Yields are in the range 50–95%. Replacement of metallic zinc by anhydrous zinc acetate gives similar results. Ghera discusses the mechanism and stereochemistry of the Serini reaction. He suggests that the catalyst functions as a complexing Lewis acid.

Clemmensen reduction (**2**, 461). The original procedure of reduction of un-hindered keto groups of steroids to methylene groups used active zinc dust and acetic anhydride saturated with hydrogen chloride as solvent. In more recent work[11] use of diethyl ether saturated with hydrogen chloride is recommended. The reduction occurs at 0° in one hour.

Phenylcarbene.[12] Zinc reduction of benzaldehyde in dry ether in the presence of boron trifluoride generates phenylcarbene (or a carbenoid species) which can be trapped by simple olefins. Thus phenylnorcarane ($endo:exo = 8:1$) can be prepared from cyclohexene in 35% yield (based on benzaldehyde). Enol acetates are converted into phenylcyclopropyl acetates.

$$C_6H_5CHO \xrightarrow{BF_3} C_6H_5CH{=}\overset{+}{O}\overset{-}{B}F_3 \xrightarrow{+e^-} C_6H_5\overset{\cdot}{C}H{-}\overset{-}{O}\overset{-}{B}F_3$$

$$\Big\downarrow{\substack{+e^- \\ -\overset{-}{O}\overset{-}{B}F_2,\ -F^-}}$$

$$C_6H_5CH{:}$$

Zinc dust and alkali. Reduction (**1**, 1282–1283). The addition of a catalytic amount of copper sulfate to the zinc-ammonia reduction of 2-(2'-thenoyl)benzoic acid (1) to 2-(2'-thenyl)benzoic acid (2) raises the yield to 90% from 30%.[13]

(1)   (2)

[1]J. W. Frankenfeld and J. J. Werner, *J. Org.*, **34**, 3689 (1969)

[2]R. L. Shriner, *Org. Reactions*, **1**, 1 (1942)

[3]Certified zinc metal dust (98.8%, Fisher) was used.

[4]D. A. Cornforth, A. E. Opara, and G. Read, *J. Chem. Soc.* (C), 2799 (1969)

[5]J. Curé and M. Gaudemar, *Bull. soc.*, 2471 (1969)

[6]Dimethoxymethane of good quality is treated with sodium wire over a period of about 24 hours, until there is no further reaction, and then distilled.

[7]E. Öhler, K. Reininger, and U. Schmidt, *Angew. Chem., internat. Ed.*, **9**, 457 (1970)

[8]A. Löffler, R. D. Pratt, J. Pucknat, G. Gelbard, and A. S. Dreiding, *Chimia*, **23**, 413 (1969)

[9]E. K. G. Schmidt, L. Brener, and R. Pettit, *Am. Soc.*, **92**, 3240 (1970)

[10]E. Ghera, *J. Org.*, **35**, 660 (1970)

[11]M. Toda, Y. Hirahata, and S. Yamamura, *Chem. Commun.*, 919 (1969)

[12]I. Elphimoff-Felkin and P. Sarda, *ibid.*, 1065 (1969)

[13]H. Wynberg, J. de Wit, and H. J. M. Sinnige, *J. Org.*, **35**, 711 (1970); see also H. E. Schroeder and V. Weinmayr, *Am. Soc.*, **74**, 4357 (1952)

Zinc borohydride, $Zn(BH_4)_2$. Mol. wt. 95.08, soluble in ether.

Preparation.[1] This complex borohydride is prepared conveniently by the reaction of sodium borohydride and zinc chloride in ether solution.

Reduction. Ethereal solutions of zinc borohydride are almost neutral; thus it is useful for reduction of substances that are sensitive in alkali. Thus Gensler *et al.*[1] wanted to reduce podophyllotoxone (1) to podophyllotoxin (2). Lithium

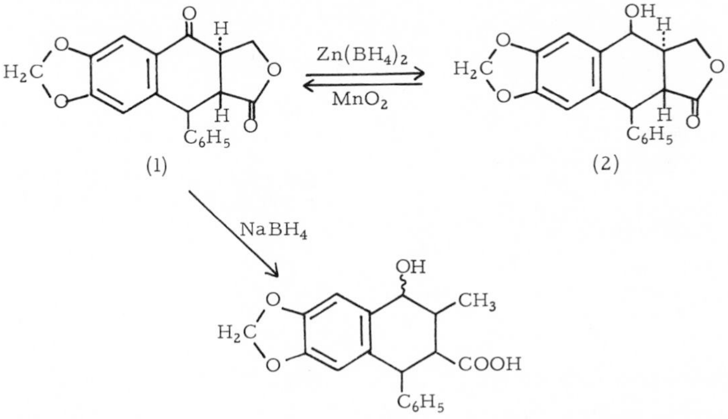

aluminum hydride could not be used, for it would also reduce the lactone function. Reduction with alcoholic sodium borohydride gave an acidic product, for which structure (3) is suggested. It was then found that the desired reduction could be effected in 82% yield by ethereal zinc borohydride. Note that the reverse reaction is effected with Attenburrow manganese dioxide in 78% yield. In this case the usual oxidation reagents (potassium dichromate, Oppenauer oxidation, silver carbonate, and potassium permanganate) failed.

The reagent was used by Corey *et al.*[2] in one step in a synthesis of prostaglandins, which are extremely sensitive to alkali. Thus the ketone (4) was successfully reduced to a mixture of epimeric alcohols.

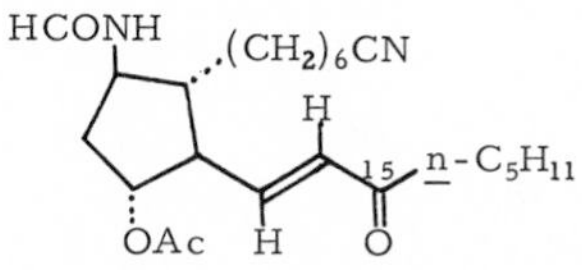

[1]W. J. Gensler, F. Johnson, and A. D. B. Sloan, *Am. Soc.*, **82**, 6074 (1960)
[2]E. J. Corey, N. H. Andersen, R. M. Carlson, J. Paust, E. Vedejs, I. Vlattas, and R. E. K. Winter, *ibid.*, **90**, 3245 (1968)

Zinc chloride, $ZnCl_2$, 1, 1289–1292; 2, 464.

Decarbonylation of cyclopropene acids. In a study of the synthesis of methyl sterculate (6) from methyl stearolate (1), Gensler *et al.*[1] were unable to repeat the apparently straightforward synthesis based on addition of the Simmons-Smith reagent described in **1**, 1021–1022. They were also unable to effect addition of methylene generated by cuprous bromide decomposition of diazomethane. However, the reaction of (1) with diazoacetic ester in the presence of copper bronze, followed by hydrolysis, gives the cyclopropene diacid (2) in 70–90% yield.

$$CH_3(CH_2)_7C \equiv C(CH_2)_7COOCH_3 \xrightarrow[\text{2) OH}^-]{\text{1) N}_2\text{CHCOOC}_2\text{H}_5} CH_3(CH_2)_7C \overset{\overset{\displaystyle COOH}{|}\diagup CH}{=\!=\!=} C(CH_2)_7COOH$$
$$(1) \qquad\qquad\qquad\qquad\qquad (2)$$

$$\xrightarrow{(COCl)_2} CH_3(CH_2)_7C \overset{\overset{\displaystyle COCl}{|}\diagup CH}{=\!=\!=} C(CH_2)_7COCl \xrightarrow{ZnCl_2} CH_3(CH_2)_7C \overset{\overset{+}{\diagup CH}\quad Cl^-}{=\!=\!=} C(CH_2)_7COCl$$
$$(3) \qquad\qquad\qquad\qquad\qquad (4)$$

$$\xrightarrow{CH_3OH} CH_3(CH_2)_7C \overset{\overset{+}{\diagup CH}\quad Cl^-}{=\!=\!=} C(CH_2)_7COOCH_3 \xrightarrow{NaBH_4} CH_3(CH_2)_7C \overset{\diagup CH_2}{=\!=\!=} C(CH_2)_7COOCH_3$$
$$(5) \qquad\qquad\qquad\qquad\qquad (6)$$

The next problem is to remove the carboxylic acid function of the cyclopropene ring. Direct decarbonylation of cyclopropene acids to give cyclopropenyl cations has been effected by reaction with acetic anhydride and perchloric acid or fluoroboric acid,[2] but these are potentially hazardous reagents. Instead Gensler converted the diacid into the diacid chloride (3, oxalyl chloride). Treatment of (3) with zinc chloride selectively decarbonylates the chlorocarbonyl group attached to the cyclopropene ring to give (4). In order to effect reduction of the cyclopropenium ion group without reduction of the acid chloride function, (4) was converted into the cyclopropenium ion ester (5). Sodium borohydride reduction then gave methyl sterculate (6) in 40% overall yield from (2).

[1]W. J. Gensler, M. B. Floyd, R. Yanase, and K. Pober, *Am. Soc.*, **91**, 2397 (1969)
[2]D. G. Farnum and M. Burr, *ibid.*, **82**, 2651 (1960); R. Breslow, H. Höver, and H. W. Chang, *ibid.*, **84**, 3168 (1962)

Zinc–Dimethylformamide.

1,4-Reduction of o-methylene bromides. Thiele[1] in 1899 reported that butadiene is formed from 1,4-dibromobutene-2 by 1,4-reduction with zinc dust in ethanol. Alder and Fremery[2] in a study of the preparation of 5,6-dimethylene-1,3-cyclo-hexadiene (2) (*o*-xylylene) applied this reaction to α,α'-dibromo-*o*-xylene (1), but

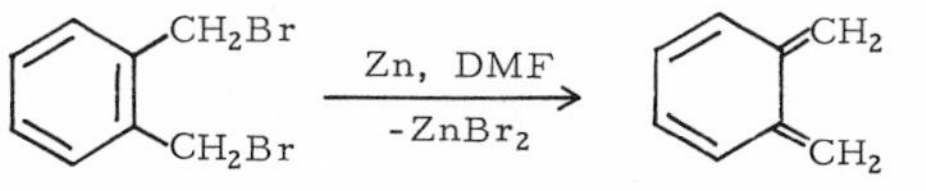

(1)   (2)

obtained very poor yields by the original procedure. They then turned to zinc powder in DMF and obtained excellent yields of (2). Actually (2) is too unstable for characterization, but its formation was demonstrated by preparation of Diels-Alder adducts from dienophiles in high yield.

This method has been applied recently for a similar preparation of 7,8-dimethylene-1,3,5-cyclooctatriene (4) from the dibromide (3) in nearly quantitative yield.[3] This dimethylenecyclopolyene is isolable, but it undergoes rapid reaction with atmospheric oxygen to form intractable products.

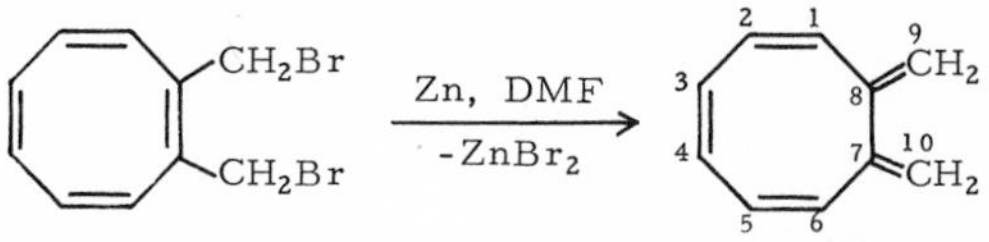

(3)   (4)

[1] J. Thiele, *Ann.*, **308**, 333 (1899)
[2] K. Alder and M. Fremery, *Tetrahedron*, **14**, 190 (1961)
[3] J. A. Elix, M. V. Sargent, and F. Sondheimer, *Am. Soc.*, **92**, 962 (1970)

Errata for Volume 2

Page 3, *Abbreviations for* HMPT, *read* Hexamethylphosphoric triamide.

Page 24, **2-Amino-2-methyl-1,3-propanediol**, following line, *read* [1, 37, before **2-Amino-2-methyl-1-propanol**]. Mol.

Page 25, **Aryldiazonium tetrahaloborates**, next to last line, *read* occurs near room temperature, the salt should never be allowed to dry completely and should be used

Page 64, **Ceric ammonium nitrate**, replace middle formulation with the <u>following</u>:

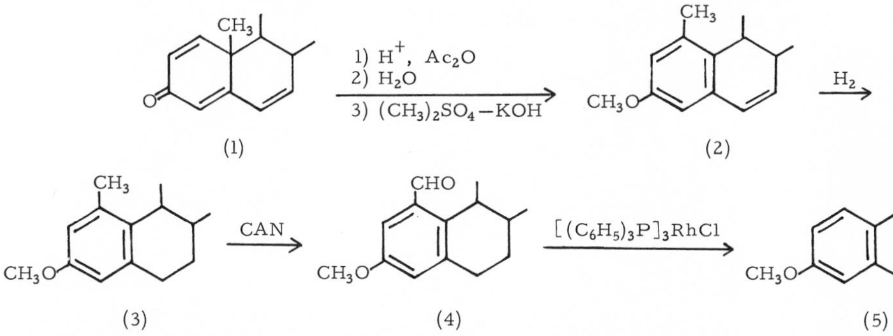

Page 67, **1-Chlorobenzotriazole**, ref. 1, *read* C. W. Rees and R. C. Storr

Page 105, **2-Diazopropane (Dimethyldiazomethane)**, last formulation, replace with the following:

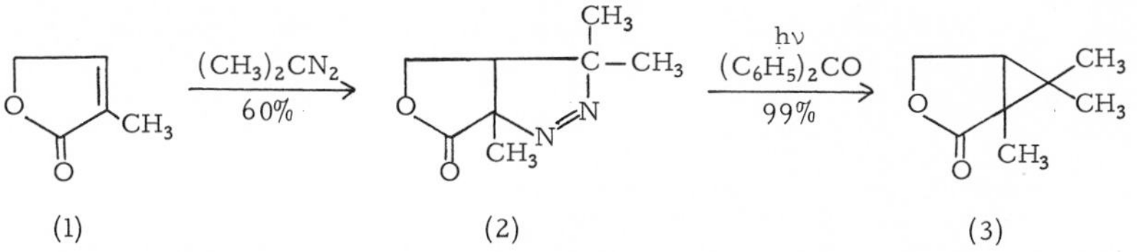

Page 106, **Diborane**, replace middle formulation with the following:

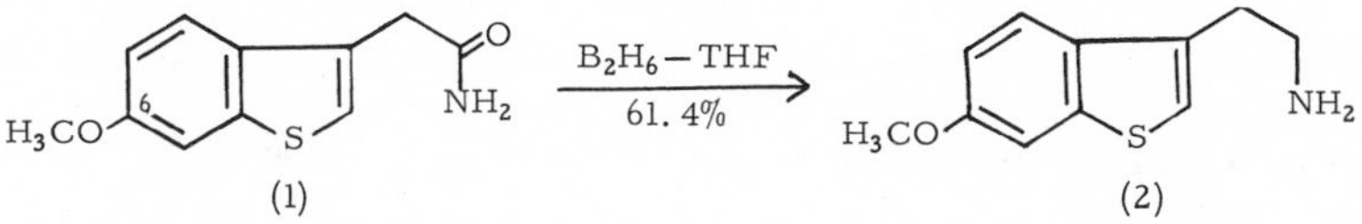

Page 109, **Di-μ-chloro-π-allyldipalladium**, *add* Supplier: Alfa Inorganics

Page 112, page heading, *read cis*-3,4-Dichlorocyclobutene

Page 126, **Dicyclohexylcarbodiimide**, line 12, *change* Corey et al,[30] *to* Corey et al.,[30]

Page 154, **Dimethylformamide diethyl acetal**, ref. 1a, *add* ; *ibid.*, **91**, 6470 (1969).

Page 169, **Dimethyl sulfoxide-derived reagent (a). Sodium methylsulfinylmethide**, line 9 after first formulation, *change* (7).[11j(b)] *to* (7).[11i(b)]

Page 171, **Dimethylsulfoxide-derived reagent (c). Dimethyloxosulfonium methylide**, line 6 from bottom, *change* α,β-unsaturated *to* α,β-unsaturated

Page 186, **1,3-Dithiane**, line 5 after first formulation, *change* to give methyl-6- *to* to give 2-methyl-6-

Page 196, **Ethyl (dimentylsulfuranylidine) acetate**, References, *read:*

[1] G. B. Payne, *J. Org.,* **32**, 3351 (1967); G. B. Payne and M. R. Johnson, *ibid.,* **33**, 1285 (1969)

[2] G. B. Payne, *ibid.,* **33**, 1284 (1968)

[3] *Idem, ibid.,* **33**, 3517 (1968)

Page 207, reagent title, *read,* **Hexaethylphosphorous triamide** [Tris(diethylamino)phosphine, [$(C_2H_5)_2N$] $_3$P] [**1**, 425, before references].

Page 227, **Ion-exchange resins**, bottom of page, Reference [35a], *read* [34a]

Page 261, **Manganese dioxide, (b) active,** middle formulation, replace with the following:

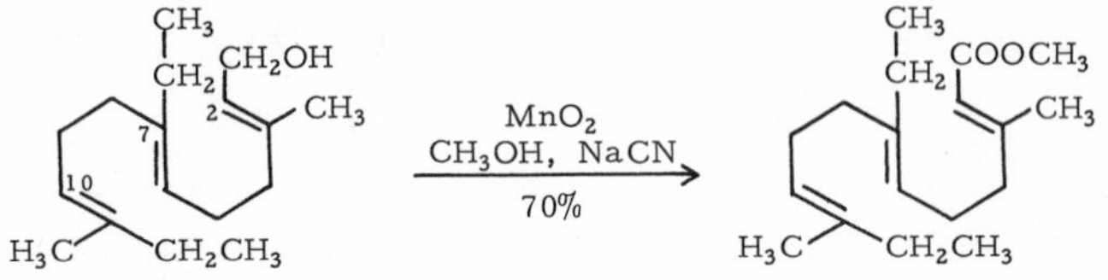

(10)   (11)

Page 274, **Methyl iodide,** *Desulfurization,* first line, *change* cis-2,3-dimethylthirane *to* cis-2,3-dimethylthiirane

Page 288, **Naphthalene—Lithium,** [$C_{10}H_8$] $^-$Na$^+$, *change* to **Naphthalene—Lithium,** ($C_{10}H_8$] $^-$Li$^+$. First formulation, *change* to:

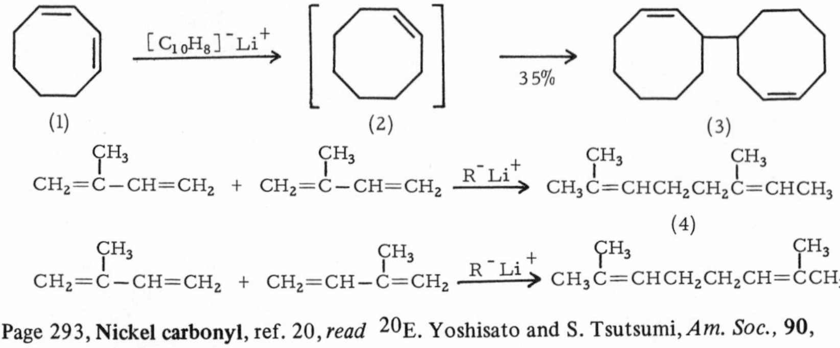

(1)   (2)   (3)

(4)

Page 293, **Nickel carbonyl,** ref. 20, *read* [20] E. Yoshisato and S. Tsutsumi, *Am. Soc.,* **90**, 4488 (1968)

Page 314, **Periodic acid**, bottom formulation, *read:*

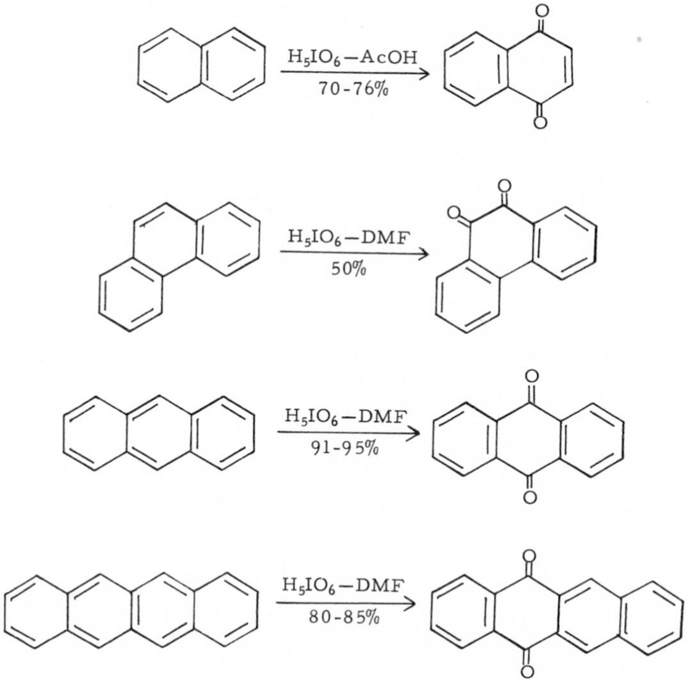

$$\text{naphthalene} \xrightarrow[70-76\%]{H_5IO_6-AcOH} \text{1,4-naphthoquinone}$$

$$\text{phenanthrene} \xrightarrow[50\%]{H_5IO_6-DMF} \text{phenanthrenequinone}$$

$$\text{anthracene} \xrightarrow[91-95\%]{H_5IO_6-DMF} \text{anthraquinone}$$

$$\text{naphthacene} \xrightarrow[80-85\%]{H_5IO_6-DMF} \text{naphthacenequinone}$$

Page 315, **Periodic acid–Chromic acid**, formulation, *read:*

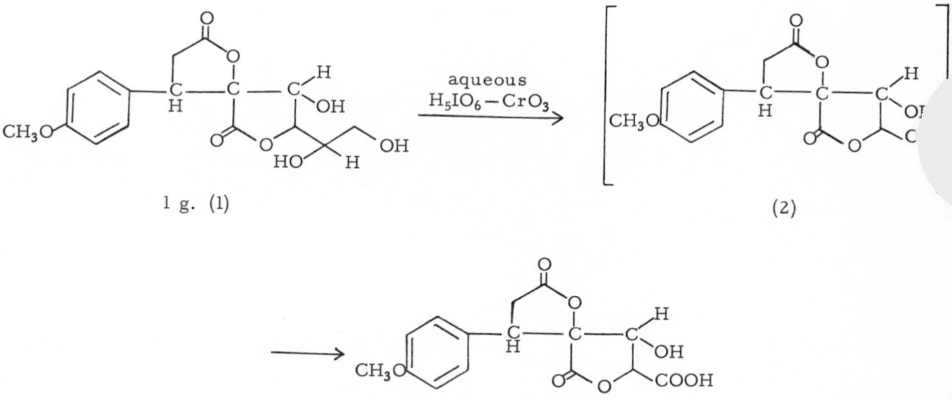

$$(1) \quad \xrightarrow[H_5IO_6-CrO_3]{\text{aqueous}} \quad (2)$$

1 g. (1)   (2)

$$\longrightarrow$$

843 mg. (3)

Page 368, **Silver nitrate**, ref. 17b, *read* A. C. Cope, R. A. Pike, and C. F. Spencer, *Am. Soc.,* **75,** 3213 (1953); A. C. Cope and R. D. Bach, *Org. Syn., 49,* 39 (1969)

Page 389, **Sodium trithiocarbonate,** *add* Supplier: Eastman

Page 397, **Tetramethyl pyrophosphite,** *read* **Tetraethyl pyrophosphite**

Page 406, **Thallium triacetate,** *add additional reference* Review: W. Kitching, *Organometallic Chem. Rev., 3,* 61 (1968)

Page 415, **Titanium trichloride,** *change* Mol. wt. 104.27 *to* Mol. wt. 154.27

Page 424, **Tri-*n*-butyltin hydride,** last formulation, *read:*

$$
\underset{(1)}{\underset{Cl}{\overset{Cl}{\diagdown}}\underset{Cl}{\overset{Cl}{\diagup}}}
\xrightarrow{\;2\ Bu_3SnH\;}
\underset{(2)}{\underset{Cl}{\overset{H}{\diagdown}}\underset{Cl}{\overset{H}{\diagup}}}
\xrightarrow{\;H_2O\;}
\underset{(3)}{\underset{O}{\overset{H\quad H}{\diagup}}}
$$

Page 430, **Triethyl orthoformate,** first formulation, *read:*

$$
\text{(azulene)} \;+\; HC(OC_2H_5)_3 \;+\; Pet.\ ether \;+\; BF_3\cdot(C_2H_5)_2O \xrightarrow{25^0} \text{(formyl-azulene, CHO)}
$$

10 g. (Blue)   10 g.   150 ml.   10 g.   (Red)

Page 451, **Tris(triphenylphosphine)chlororhodium,** *Angular methylation,* last line, *change* (271–271) *to* (1, 271–272)

Page 457, **Vinyl triphenylphosphonium bromide,** last formulation, *replace with:*

$$
\text{(salicylaldehyde, CHO, OH)} \xrightarrow[92-97\%]{\overset{CH_3OH}{NaOCH_3}} \text{(CHO, O}^-Na^+\text{)} \xrightarrow[54-58\%]{CH_2{=}CHP(C_6H_5)_3Br^-\ in\ DMF} \underset{(4)}{\text{(2H-chromene, O)}}
$$

Page 475, Index of reagents according to type, GRIGNARD REACTION, *change* Methyltri-*n*-butylphosphine)copper complex *to* Methyl(tri-*n*-butylphosphine)copper complex

Page 482, WITTIG REACTIONS, *change* Carbomethoxymethylenetriohenylphosphorane *to* Carbomethoxymethylenetriphenylphosphorane

Page 493, Author Index, *add* Johnson, M. R., 196

Page 494, Author Index, second column, *change to:*
Kierstead, R. C., 111, 112
Kierstead, R. W., 29

Page 504, Author Index, second column, *change* Storr, R. E., 67 *to* Storr, R. C., 67

Page 535, Subject Index, first column, *after* N,N,N′,N′-Tetraethyl methylphosphonous diamide, 105
add Tetraethyl pyrophosphite, 397

Page 535, Subject Index, second column, *remove* Tetramethyl pyrophosphite, 397

Page 538, Subject Index, first column, Vilsmeir reaction, 129
change to Vilsmeier reaction, 129

SUPPLIERS

The suppliers referred to most frequently, the abbreviations
which we have used for them, and their addresses are as follows:

(A)   Aldrich Chemical Co., 940 West St. Paul Ave., Milwaukee, Wis.
         53233

(B)   J. T. Baker Chemical Co., 222 Red School Lane, Phillipsburg,
         N. J. 08865

(E)   Eastman Organic Chemicals Dept., Rochester, N. Y. 14650

(F)   Fisher Scientific Co., 717 Forbes Ave., Pittsburgh, Pa. 15219

(Fl.)   Fluka AG, Buchs, Switzerland

(K and K)   K and K Laboratories, Inc., 121 Express St., Engineers Hill,
         Plainview, N. Y. 11803

(K.-L.)   Koch-Light Labs., Ltd., Colnbrook, Buckinghamshire, England

(MCB)   Matheson, Coleman and Bell, 2909 Highland Ave., Norwood
         (Cincinnati), Ohio 45212

(Sch.)   Schuchardt, Dr. Theodor, Postfach 801549, 8 Munich 80, Germany

Other suppliers of chemicals, solvents, or apparatus cited
are listed below:

Abbott Laboratories, Chemical Market-
ing Division, North Chicago, Ill. 60064

Ace Glass Inc., Vineland, N. J. 08360

Aceto Chemical Co., 126-02 Northern
Blvd., Flushing, N. Y. 11368

Aero Chemical Corp., 338 Wilson Ave.,
Newark, N. J.

Air Reduction Chemical and Carbide Co.,
150 East 42nd St., New York, N. Y.
10017

Alfa Inorganics, Inc., P.O. Box 159,
Beverly, Mass. 01915

Allied Chemical Co., Marcus Hook, Pa.
19061

Allied Chemical Corp., Baker and
Adamson Products, General Chemical
Division, 40 Rector St., New York,
N. Y. 10006

Allied Chemical Corp., Solvay Process
Division, 61 Broadway, New York,
N. Y. 10006

American Agricultural Chemical Co.,
Dept. E-10, 100 Church St., New
York 7, N. Y.

American Platinum Works, Newark, N. J.

American Potash and Chemical Corp.,
680 Wilshire Pl., Los Angeles, Calif. 90005

Ansul Chemical Co., 101 Stanton St., Marinette,
Wisc. 54143

Antara Chemicals, Division of General
Aniline and Film Corp., 435 Hudson St.,
New York 14, N. Y.

Arapahoe Chemicals, Division of Syntex
Corp., 2855 Walnut St., Boulder, Colo.
80302

Ayerst Laboratories, Inc., Rouses Point,
N. Y. 12979

Baker and Co., Inc., Newark, N. J.

Beacon Chemical Industries, Inc., 33 Rich-
dale Ave., Cambridge, Mass. 02140

Bishop, J., and Co. Platinum Works,
Malvern, Pa. 19355

Borden Chemical Co., Division of the
Borden Co., 350 Madison Ave., New York,
N. Y. 10017

Bower, Henry, Chemical Mfg. Co., Gray's
Ferry Rd. and 29th St., Philadelphia, 46,
Pa.

British Drug Houses Ltd., Laboratory Chemicals Division, Poole, Dorset, England. U. S. distributor: Gallard Schlessinger Chem. Co., 580/58 Mineola Ave., Carle Place, L.I., N. Y. 11514

Burdick and Jackson Laboratories, Muskegon, Mich. 49442

Calbiochem, Box 54283, Los Angeles, Calif. 90054

Callery Chemical Co., Callery, Pa. 16024

Carbide and Carbon Chemical Co., 270 Park Ave., N. Y., N. Y., 10017

Carborundum Co., Niagara Falls, N. Y.

Carlisle Chemical Works, Inc., Reading, Ohio 45215

Chattem Chemicals, Fine Chemicals Division of the Chattanooga Medical Co., Chattanooga, Tenn. 37409

Chemical Intermediates and Research Labs., Box 146, Cuyahoga Falls, Ohio

Chemical Procurement Laboratories, Inc., 18–17 130th St., College Point, N. Y. 11356

Chemical Samples Co., 4692 Kenney Rd., Columbus, Ohio 43220

Ciba Chemical and Dye Co., Division of Ciba Corporation, Fairlawn, N. J. 07410

Coaltar Chemicals Corp., 430 Lexington Ave., New York, N. Y.

Columbia Organic Chemicals, Inc., P.O. Box 5273, Columbia, S. C., 29205

Commercial Solvents Corp;, 245 Park Ave., New York, N. Y. 10017

Crown Zellerbach Corp., Chemical Products Division, Camas, Wash. 98607

Davidson Chemical Co., Division of W. R. Grace and Co., 101 N. Charles St., Baltimore, Md. 21203

Deere, John, Chemical Co., Tulsa and Pryor, Okla.

Delmar Scientific Laboratories, Inc., 317 Madison St., Maywood, Ill. 60153

Dow Chemical Company, Midland, Mich., 48640

du Pont de Nemours, E. I., and Co., Inc., Electrochemicals Department, Wilmington, Del. 19898

du Pont de Nemours, E. I., and Co., Inc., Industrial and Biochemicals Department, Wilmington, Del. 19898

Eastern Chemical Corp., Box P, Pequannock, N. J. 07440

Eastman Chemical Products, Inc., Subsidiary of Eastman Kodak Company, P.O. Box 431, Kingsport, Tenn. 37662

EGA-Chemie KG, Steinheim, Germany

Engelhard Industries, Inc., 113 Astor St., Newark, N. J. 07114

Enjay Chemical Company, A Division of Humble Oil and Refining Co., 60 West 49th St., New York, N. Y. 10020

Ethyl Corp., 100 Park Ave., New York, N. Y. 10017

Fairmont Chemical Co., Inc., 136 Liberty St., New York 6, N. Y.

Farchan Research Laboratories, 4703 E. 355 St., Willoughby, Ohio 44094

FMC Corporation, Inorganic Chemicals Division, 633 Third Ave., New York, N. Y. 10017

Foote Mineral Co., Route 100, Exton, Pa. 19341

Frinton Laboratories, P. O. Box 301, South Vineland, N. J. 08360

Fritzsche, Dodge, and Olcott, 76 Ninth Ave., New York, N. Y. 10011

GAF Corp., 435 Hudson St., New York 14, N. Y.

Gilman Paint and Varnish Co., 216 W. 8th St., Chattanooga, Tenn. 37402

Goodrich, B. F., Chemical Co., 3135-T Euclid Ave., Cleveland, Ohio 44115

Grace, W. R., and Co., Raney Catalyst Division, 4000 N. Hawthorn, Chattanooga, Tenn. 37406

Hammons, W. A. Drierite Co., Xenia, Ohio, 45385

Harshaw Chemical Co., 4540 Willow Pkway., Cleveland 6, Ohio 44125

Hengar Co., 1711 Spruce St., Philadelphia 3, Pa.

Henley and Co., Inc., 202 E. 44th St., New York, N. Y. 10017

Hercules, Inc., 910 Market St., Wilmington, Del. 19899

Hooker Chemical Corp., 1940 Ward St., Niagara Falls, N. Y. 14302

Houdry Chem. Co., 1339 Chestnut St., Philadelphia, Pa. 19107

Howe and French, Inc., (now Sci Chem. Co.), 99 Broad St., Boston, Mass. 02110

International Minerals and Chemicals Corp. 5401 Old Orchard Rd., Skokie, Ill. 60076

Jefferson Chemical Co., Inc., P. O. Box 53300, Houston, Texas, 77052

Johnson, Matthey, Ltd., 78 Hatton Garden, London, E.C.1, England (608 5th Ave., N. Y., N. Y. 10020)

Kaplop Laboratories, 6710 West Jefferson, Detroit 17, Mich.

Kay-Fries Chemicals, Inc., 360 Lexington Ave., New York, N. Y. 10017

Koppers Company, Inc., Chemicals and Dyestuffs Division, 1450 Koppers Bldg., Pittsburgh, Pa. 15219

Lithium Corporation of America, Inc., 2 Pennsylvania Plaza, New York, N. Y., 10001

Lucidol Division, Wallace and Tiernan Inc., 1741 Military Rd., Buffalo, N. Y., 14240

Mackay, A. D., Inc., 198 Broadway, New York, N. Y. 10038

McKesson Chemical Co., Inc., 708 3rd Ave., N. Y., N. Y. 10017

Mallinckrodt Chemical Works, P. O. Box 5439, 2nd and Mallinckrodt Sts., St. Louis, Mo. 63160

Mann Research Labs., Inc., (Schwarz-Mann), Mountain View Ave., Orangeburg, N. Y. 10962

Matheson Co., Inc., P. O. Box 85, East Rutherford, N. J. 07073

Merck Chemical Division, Rahway, N. J. 07065

Merck, E., AG, 61 Darmstadt, Germany

Metal and Thermit Corp., Rahway, N. J. 07065

Metalloy Corp., Rand Tower, Minneapolis, Minn.

Metalsalts Corp., (Division of Merck), Lincoln Ave., Rahway, N. J. 07065

Monsanto Chemical Co., 800 N. Lindbergh Blvd., St. Louis, Mo. 63166

MSA (Mine Safety Appliance) Research Corp., 2 Esmond St., Smithfield, R. I. 02917

Niacet Chemicals Division, Niagara Falls, N. Y.

Niagara Chlorine Products Co., Inc., Lockport, N. Y.

Oldbury Electrochemical Co., Niagara Falls, N. Y.

Olin Mathieson Chemical Corp., Chemicals Division, 745 Fifth Ave., New York 22, N. Y.

Orgmet, Inc., Ward Hill Industrial Park, Haverhill, Mass. 01830

Ott Chemical Co., 500 Agard Road, Muskegon, Mich. 49445

Ozark-Mahoning Co., 7870 South Boulder, Tulsa, Okla. 74119

Parr Instrument Co., 209 53rd St., Moline, Ill. 61265

Peninsular ChemResearch Inc., P. O. Box 1466, Gainesville, Fla. 32601

Penn Salt Chemicals Corp., Industrial Chemicals Division, 3 Penn Center Plaza, Philadelphia, Pa. 19102

Pfalts and Bauer, 126-02 Northern Blvd., Flushing, N. Y. 11368

Pfister Chemical Works, Inc., Ridgefield, N. J. 07657

Pfizer, Chas., and Co., Inc., Chemical Sales Division, 235 East 42nd St., New York, N. Y. 10017

Pharmacia Fine Chemicals, Inc., 800 Centennial Ave., Piscataway, N. J. 08854

Phillips Petroleum Co., Market Development Division, Bartlesville, Okla. 74004

Pierce Chemical Co., P. O. Box 117, Rockford, Ill. 61105

Pilot Chemicals, Inc. 36 Pleasant St., Watertown, Mass. 02172

Quaker Oats Co., Chemicals Division, 341 Merchandise Mart Plaza, Chicago, Ill. 60654

Rancy Catalyst Co., Inc., 924 Hamilton National Bank Bldg., Chattanooga, Tenn. 37402

Research Organic/Inorganic Chemical Corp., 11686 Sheldon St., Sun Valley, Calif. 91352

Riedel-de Haen AF, 3016 Seelze bei Hannover, Germany

Robeco Chemicals Inc., 25 E. 26th St., New York 10, N. Y.

Roberts Chemical Co., P. O. Box 546,
Nitro, W. Va. 25143

Rohm and Haas Co., Independence Mall,
West Philadelphia, Pa. 19105

Sharples Chemicals Co., 955 Mears Rd.,
Warminster, Pa. 18974

Shell Chemical Co., 59 West 50th St.,
New York, N. Y. 10020

Smith, G. Frederick, Chemical Co., P. O.
Box 5906 Station D, 867 McKinley Ave.,
Columbus, Ohio 43222

Standard Oil of Indiana, 910 So. Michigan
Ave., Chicago, Ill. 60605

Stauffer Chemical Co., 299 Park Ave.,
New York, N. Y. 10017

Stepan Chemical Co., Maywood Division,
100 West Hunter Ave., Maywood, N. J.
07607

Syntex S.A., Apartado Postal 2679,
Mexico, D. F., Mexico

Texaco Inc., 135 East 42nd St., New York,
N. Y. 10017

Texas Alkyls, Inc., 6910 Fannin St., Suite
300A, Houston, Texas 77025

Thallium Ltd., Carrow Road, Norwich,
England

Thomas, Arthur H., Co., Vine St. and
Third, Philadelphia, Pa. 19105

Titanium Alloy Manufacturing Division,
National Lead Co., 111 Broadway,
New York, N. Y. 10006

Troemer, Henry, Inc., 6824 Greenway Ave.,
Philadelphia, Pa. 19142

Trylon Chemicals, Inc., P. O. Box 600,
Mauldin, S. C. 29662

Union Carbide Chemicals Co., Div. of
Union Carbide Corp., 270 Park Ave.,
New York, N. Y. 10017

Universal Oil Products Co., 30 Algonquin
Rd., Des Plaines, Ill. 60018

U. S. Industrial Chemicals Co., Division of
National Distillers and Chemical Corp.,
99 Park Ave., New York, N. Y. 10016

United States Testing Co., 1941 Park Ave.,
Hoboken, N. J. 07030

Ventron Corp., 188 Congress St., Beverly,
Mass. 01915

Victor Chemical Works, Div. of Stauffer
Chemical Co., 155 North Wacker Drive,
Chicago 6, Ill.

Virginia-Carolina Chemical Corp., 401 E.
Main St., Richmond, Va. 23219

Virginia Chemicals Inc., Portsmouth,
Virginia 23703

Welsbach Corp., Ozone Process Division,
3340 Stokley St., Dept. F, Philadelpha,
Pa. 19129

Westvaco Mineral Products Corp., Chem.
Div., 299 Park Ave., N. Y., N. Y. 10017

Westville Chemical Corp., Route 110,
Monroe, Conn. 06468

Wilkens-Anderson Co., 4515 W. Division
St., Chicago, Ill. 60651

Wilson Laboratories, 4221 South Western
Blvd., Chicago, Ill. 60609

Wood Ridge Chemical Corp., Park Place
East, Wood Ridge, N. J. 07075

Yeda Research and Development Co. Ltd.,
P.O. Box 26, Rehovoth, Israel

INDEX OF REAGENTS ACCORDING TO TYPES

CARBONYLATION: 9-Borabicyclo[3.3.1]-nonane. Carbon monoxide. Dichlorobis-(triphenylphosphine)palladium.

CARBOXYLATION: Carbon dioxide. Magnesium methyl carbonate. Nickel carbonyl.

CHLORINATION: Dimethylformamide—Thionyl chloride. Iodobenzene dichloride. Nitrosyl chloride. Sulfuryl chloride. Trichloroisocyanuric acid.

CHLORINATION OF SULFIDES OR SULFOXIDES: Iodobenzene dichloride.

CHLOROCARBONYLATION: Oxalyl chloride.

CLAISEN REARRANGEMENT: Diethyl allylthiomethylphosphonate. Triethyl orthoacetate.

CLEAVAGE OF: ETHERS: Aluminum chloride. Boron tribromide. Bromine. Dimethylformamide. Methylmagnesium iodide. Triphenylphosphine dibromide.

α-GLYCOLS: Silver iododibenzoate
KETOXIMES: Chromous acetate
METHYL ESTERS: Lithium n-propyl mercaptide.
OXIMES AND SEMICARBAZONES: Ceric ammonium nitrate.

CLEMMENSEN REDUCTION: Zinc dust.

COMPLEX OF BENZYNE. Nickel carbonyl.

π-COMPLEXES: Diiron nonacarbonyl.

CONDENSING AGENT: Calcium carbide.

CONJUGATE ADDITION: Dimethylcopperlithium. Nickel carbonyl. Tetrakis[iodo-(tri-n-butylphosphine)copper(I)].

CONVERSION OF PHENOLS TO ARYL BROMIDES: Triphenylphosphine dibromide.

COUPLING OF: ALLYLIC COMPOUNDS: Dimethylcopperlithium. Iron. Nickel carbonyl. Trimethylironlithium.

GRIGNARD REAGENTS: Thallium(I)-bromide.
TERMINAL ACETYLENES: Cuprous chloride.

CROSS COUPLING: Dimethylcopperlithium. Trimethylironlithium.

CYANATION OF HALIDES: Sodium dicyanocuprate.

CYCLIZATION: Boron trifluoride. Polyphosphoric acid. Stannic chloride.

CYCLIZATION OF δ- OR ε-HALOKETONES: Di-n-butylcopperlithium.

CYCLIZATION OF OLEFINS: Trifluoroacetic acid.

CYCLOADDITIONS: Bis(acrylonitrile)nickel-(O). 2-Diazopropane. Dichloroketene. 5,5-Dimethoxy-1,2,3,4-tetrachlorocyclopentadiene. Ethoxyketene.

CYCLODEHYDROGENATION: 2,3-Dichloro-5,6-dicyano-1,4-benzoquinone. Polyphosphate ester.

CYCLOOLIGMERIZATION: Tris(triphenylphosphine)chlororhodium.

CYCLOPROPANATION: Cuprous chloride. 2-Diazopropane. (Dimethylamino)phenyloxosulfonium methylide. Dimethyl diazomethylphosphonate. Ethyl diazoacetate. Methylphenyl-N-p-toluenesulfonylsulfoximine. Simmons-Smith reagent.

DAKIN-WEST REACTION: N,N-Dimetnyl-4-pyridinamine.

DARZENS REACTION: Aluminum chloride.

DEALKYLATION: Boron tribromide. 2-Nitroprop-2-yl hydroperoxide. Pyridine hydrochloride. Sodium thiophenoxide.

DECARBONYLATION: Dicyclohexylcarbodiimide. Tris(triphenylphosphine)-chlorohodium. Zinc chloride.

DECARBOXYLATION: Copper powder.

DECHLOROHYDRINATION: Chromium II-amine complexes.

DEHALOGENATION: Cuprous chloride. Lithium amalgam. Zinc dust.

DEHYDRATION: Boron trioxide. p-Chlorophenyl chlorothionoformate. Dicyclohexylcarbodiimide. Diethyl(2-chloro-1,1,2-trifluoroethyl)amine. Phosphoryl chloride—N,N-Dimethylformamide. Polyphosphate ester. Pyridine hydrochloride.

DEHYDROCHLORINATION: Lithium amide.

DEHYDROGENATION: Chloranil. 2,3-Dichloro-5,6-dicyano-1,4-benzoquinone (DDQ). Potassium nitrosodisulfonate. Selenium dioxide. Sulfur.

DEHYDROHALOGENATION: Dimethyl sulfoxide. Lithium carbonate. Potassium t-butoxide.

DEHYDROTOSYLATION: Potassium t-butoxide.

DEMETHYLATION: Boron tribromide. Chromic anhydride. Dimethylformamide. Methylmagnesium iodide.

DEOXYGENATION: Hexachlorodisilane. Iron pentacarbonyl. Sulfur trioxide–Pyridine.

DESULFURIZATION: Hexaethylphosphosphorus triamide. Triphenylphosphine.

DEUTERATION: Trifluoroacetic acid. Tris(triphenylphosphine)chlororhodium.

DIAZO TRANSFER: p-Toluenesulfonyl azide.

DIELS-ALDER REACTION: Aluminum chloride. Cupric fluoroborate. 1,3-Diphenylnaphtho[2.3-c]furane. 4-Phenyl-1,2,4-triazoline-3,5-dione. 3-Sulfolene.

DIMERIZATION OF OLEFINS: Di(cobalt-tetracarbonyl)zinc.

DISPLACEMENTS: Diethyl(2-chloro-1,1,2-trifluorotriethyl amine)(OH→F). Tetra-n-butylammonium acetate.

ENAMINES: Pyrrolidine.

ENOL ACETYLATION: Hydrobromic acid.

EPOXIDATION: m-Chloroperbenzoic acid. Hydrogen peroxide, basic. Hydrogen peroxide–Phenylisocyanate. Hydrogen peroxide–Selenium dioxide. Iodine. Peracetic acid. Perlauric acid. Tetralin hydroperoxide. Vanadium oxyacetylacetonate.

ESTERIFICATION: Boron trifluoride etherate. Diazomethane. Lithium t-butoxide. Triethyloxonium fluoroborate. Tris(2-hydroxypropyl)amine.

ETHER CLEAVAGE: Aluminum chloride. Bromine. Triphenylphosphine dibromide.

FISCHER INDOLE SYNTHESIS: Polyphosphate ester.

FLUORINATION: Diethyl(2-chloro-1,1,2-trifluoroethyl)amine.

FORMYLOLEFINATION. Diethyl 2-(cyclohexylimino)ethylphosphonate.

FRAGMENTATION: Diborane. p-Toluenesulfonylhydrazine.

GABRIEL REACTION: Cuprous bromide.

GLYCOL CLEAVAGE: Silver iododibenzoate.

GRIGNARD REACTIONS: N-Benzenesulfonylformimidic ethyl ester.

HALOGENATION: Arsenic trichloride (tribromide).

HYDRATION OF OLEFINS: Mercuric acetate.

HYDRIDE ABSTRACTION: Trityl hexachloroantimonate.

HYDROBORATION: Bis-3-methyl-2-butylborane. Diborane. Dicyclohexylborane. Trimethylamine oxide.

HYDROCHLORINATION OF ALCOHOLS: Cyanuric chloride.

HYDRODECHLORINATION: Triphenyltin hydride.

HYDROGEN-DEUTERIUM EXCHANGE: Disodium platinum tetrachloride.

HYDROGENATION CATALYSTS: Chloroplatinic acid–Triethylsilane. Lindlar catalyst. Nickel boride. Osmium-on-carbon. Palladium hydroxide. Tris(triphenylphosphine)chlororhodium.

HYDROGENOLYSIS: Aluminum hydride.

HYDROIODINATION: Iodine.

HYDROXYLATION: Trimethylchlorosilane.

HYDROXYMETHYLATION: Hydroxylamine-O-sulfonic acid.

INCLUSION COMPLEXES: Adamantane. Thiourea.

INTERCONVERSION OF cis-trans OLEFINS: Iodine isocyanate.

IODINATION: Iodine. Iodine–Periodic acid.

ISOMERIZATION: Aluminum bromide. Aluminum chloride. Potassium t-butoxide. Silver fluoroborate.

KNOEVENAGEL CONDENSATION: Titanium tetrachloride.

MANNICH REACTION: Bis(demethylamino)methane. N,N-Dimethylformaldimmonium trifluoroacetate. Trifluoroacetic anhydride.

MEERWEIN SYNTHESIS: Boron trifluoride.

METHYLATION: Dimethyloxosulfonium methylide. Methyl fluoride–Antimony pentafluoride. Methyl iodide.

METHYLENATION: Magnesium.
METHYLENE TRANSFER REAGENTS:
(Dimethylamino)phenyloxosulfonium
methylide. Dimethyloxosulfonium methylide. Simmons-Smith reagent.

OLEFIN ADDITIONS: Benzenesulfonyl
azide. Chlorosulfonyl isocyanate. Chromous chloride. Cyanamide. Dichloroketene.
Dinitrogen tetroxide–Iodine. Diphenylketene. Ethoxyketene. Fluoroxytrifluoromethane. Iodine. Iodine azide. Iodine
isocyanate. Iodobenzene dichloride. Mercuric azide. Nitrosyl chloride. Nitrosyl
fluoride. Nitryl iodide. Potassium t-butoxide–Bromoform (dibromocarbene).
Sulfur monochloride. Trichloramine.
OLEFIN CYCLIZATION: Trifluoroacetic
acid.
OLEFIN ISOMERIZATION: Rhodium
trichloride hydrate.
OLEFIN SYNTHESIS: Diethyl phosphorochloridate. Trimethyl phosphite.
ORGANOLITHIUM COMPOUNDS: Dichloromethyllithium.
ORGANOLITHIUM COMPOUNDS, ACTIVA-
TION: N,N,N′,N′-Tetramethylenediamine.
OXIDATION, REAGENTS: Ammonium
persulfate–Silver nitrate. Argentic picolinate. Bromine. N-Bromosuccinimide. t-Butyl
chromate. Caro's acid. Ceric ammonium
nitrate. 1-Chlorobenzotriazole. m-Chloroperbenzoic acid. Chromic acid. Chromic
anhydride. Chromyl chloride. 3,5-Di-t-
butyl-1,2-benzoquinone. 2,3-Dichloro-5,6-
dicyano-1,4-benzoquinone, see N,N-Diethyl-
1-propynylamine. Dimethyl sulfoxide. Dimethyl sulfixode–Diphenylketene-p-
tolylimine. Hexafluoroacetone–Hydrogen
peroxide. Hexamethylphosphoric triamide.
Iodine–Dimethyl sulfoxide. Iodobenzene
dichloride. Iodosobenzene diacetate. Lead
dioxide. Lead tetraacetate. Manganese
dioxide. Manganic acetylacetonate. N-
Methylmorpholine oxide–Hydrogen
peroxide. Molybdenum hexacarbonyl.
Nitric acid. Notrosyl fluoride. Perbenzoic
acid. Periodic acid. Pertrifluoroacetic
acid. Phosphatolead(IV) acids. Potassium
nitrosodisulfonate. Pyridine dichromate.

Ruthenium tetroxide. Selenium dioxide.
Silver carbonate. Silver carbonate–Celite.
Silver iododibenzoate. Silver oxide. Sodi-
um dichromate. Sodium persulfate. Thal-
lium(III) trifluoroacetate. Trimethylamine
oxide. Vanadium oxyacetylacetonate.
OXIDATION, PFITZNER-MOFFATT TYPE:
Dimethyl sulfoxide.
OXIDATION OF SULFIDES TO SULFIXIDES:
1-Chlorobenzotriazole. Iodobenzene di-
chloride. Perbenzoic acid.
OXIDATION OF THIAXANTHONE TO THE
SULFOXIDE: Iodosobenzene diacetate.
OXIDATIVE AROMATIZATION: Man-
ganese dioxide.
OXIDATIVE CLEAVAGE: Selenium di-
oxide.
OXIDATIVE CONDENSATION: Man-
ganese dioxide.
OXIDATIVE COUPLING: Di-n-butylcopper-
lithium. Dimethylcopperlithium.
OXIDATIVE CYCLIZATION: Iodosobenzene
diacetate.
OXIDATIVE DECARBOXYLATION: Lead
tetraacetate. Potassium persulfate.
OXIDATIVE DEMETHYLATION: Chromic
anhydride.
OXIDATIVE PHENOL COUPLING: Ferric
chloride. Silver carbonate. Vanadium oxy-
chloride. Vanadium tetrachloride.
OXIDE CLEAVAGE: 1,3-Dithiane. Di-
methylcopperlithium.
N-OXIDES: m-Chloroperbenzoic acid.
OXYALKYLATION: Methoxymethyl-
sulfonate.
OXYGENATION: Cupric acetate–amine
complexes. 4-Ethyl-2,6,7-trioxa-1-phospha-
bicyclo[2.2.2]octane. Triphenyl phosphite–
Ozone.
OXYMERCURATION: Mercuric acetate.
OZONIZATION: Lindlar catalyst.

PEPTIDE SYNTHESIS: t-Amyl chloroformate. t-Butyl azidoformate. Diethyl
methylenemalonate. N-Ethoxycarbonyl-2-
ethoxy-1,3-dihydroquinoline. Hexamethylphosphoric triamide. 1-Hydroxybenzotriazole.
Triphenylphosphine–2,2′-Dipyridyl disulfide. Triphenyl phosphite.
PERKOW REACTION: Triethyl phosphite.

PEROXIDE OXIDATION: Manganic acetylacetonate.

PHENOLIC OXIDATIVE COUPLING: Ferric chloride. Silver carbonate. Vanadium oxytrichloride. Vanadium tetrachloride.

PHENYLATION: Naphthalene–Lithium.

PHOSPHORYLATION: 2-Chloromethyl-4-nitrophenyl phosphorochloridate. o-Phenylene phosphorochloridate. Phosphoryl chloride. Phosphoryl chloride–Trimethyl phosphate.

PHOTOCHEMICAL REACTION WITH OLEFINS: Iodoform.

PINACOL REDUCTION: Aluminum amalgam.

PREPARATION OF ENOLATES: Sodium bistrimethylsilylamide.

PROTECTION, ACIDS: p-Bromophenacyl bromide.

    AMIDE GROUP: 4,4'-Dimethoxybenzhydrol.

    AMINO GROUP: t-Amyl chloroformate. t-Butyl azidoformate. 9-Fluorenylmethyl chloroformate.

    CARBONYL GROUP: N,N-Dimethylhydrazine.

    CARBOXYL GROUP: 2-Amino-2-methyl-1-propanol. Trichloroethanol.

    DIHYDROXYACETONE SIDE CHAIN: Acetone.

    HYDROXYL GROUP: Bis(trimethylsilyl)acetamide. Boric acid. 4-Methoxy-5,6-dihydro-2H-pyrane. Phenylboronic acid. β,β,β-Trichloroethyl chloroformate.

    OF PHENOLS: p-Bromophenacyl bromide.

    OF STEROIDAL $\Delta^{5,7}$-DIENES: 4-Phenyl-1,2,4-triazoline-3,5-dione.

    OF SULFHYDRYL GROUP: Diethylmethylene malonate.

PROTECTIVE GROUP FOR NUCLEOSIDES: β-Benzoylpropionic acid. Tetramethyl orthocarbonate. Phenylboronic acid.

PSCHORR SYNTHESIS: Sodium iodide.

REARRANGEMENTS: Potassium t-butoxide. Silver fluoroborate.

REDUCTION OF α,β-unsaturated ketones: Birch reduction.

REDUCTION, REAGENTS: Aluminum hydride. Bis-3-methyl-2-butylborane. Chloroiridic acid. Chromium(II)-amine complexes. Chromous chloride. Chromous perchlorate. Diborane. Diimide. Diisobutylaluminum hydride. Diphenyltindihydride. Hexaethylphosphorous triamide. Hexamethylphosphorous triamide. Hydrido(tri-n-butylphosphine)copper(I). Hypophosphorous acid–Hydrogen iodide. Iridium tetrachloride. Lithium, see Hexamethylphosphorous triamide. Lindlar catalyst. Lithium–Alkylamine. Lithium aluminum hydride. Lithium aluminum hydride–Aluminum chloride. Lithium–Ammonia. Lithium cyanohydridoborate. Lithium–Diphenyl. Lithium–Ethylenediamine. Lithium perhydro-9b-phenalylhydride. Potassium hydroxide–Ethylene glycol. Sodium aluminum hydride. Sodium amalgam. Sodium bis-(2-methoxyethoxy)-aluminum hydride. Sodium borohydride. Sodium borohydride–Cobalt chloride. Sodium hydrosulfide. Sodium trimethoxyborohydride. Sulfides. Tri-n-butyltin hydride. Trichlorosilane. Triethylamine–Borane. Triethylammonium formate. Triethyloxonium fluoroborate. Trimesitylborane. Triphenyltin hydride. Zinc borohydride. Zinc–Dimethylformamide. Zinc dust.

REDUCTIVE ALKYLATION AND CARBOMETHOXYLATION: Lithium–Ammonia.

REDUCTIVE AMINATION: Lithium cyanohydridoborate.

REFORMATSKY REACTION: Naphthalene–Lithium. Trimethyl borate. Zinc.

RESOLUTION: Dehydroabietylamine. d- and l-α- Methylbenzylamine. (S)-1-Nitroso-2-methylindoline-2-carboxylic acid. L-Tyrosine hydrazide.

RING CONTRACTION: Chloramine.

RING EXPANSION: Cyanogen azide. Diazoethane. Silver sulfate.

SERINI REACTION: Zinc.

SILICONIDES: Dimethyldiacetoxysilane. Dimethyldichlorosilane.

SILYLATION. Dimethyldiacetoxysilane. Trichlorosilane. Trimethylsilyldiethylamine.

SIMMONS-SMITH REAGENT, which see.
SINGLET OXYGEN: 4-Ethyl-2,6,7-trioxa-1-phosphabicyclo[2.2.2]octane. Triphenyl-phosphate—Ozone.
SOLVOMERCURATION: Mercuric acetate.
SPIROANNELATION: Diphenylsulfonium cyclopropylide.
SYNTHESIS OF: ACETYLENES: Ethyli-denetriphenylphosphorane. 3-Lithio-1-trimethylsilylpropyne. p-Toluenesulfonyl-hydrazine.

ACID CHLORIDES, ALKYL CHLORIDES: Hexamethylphosphoric triamide—Thionyl chloride.
ALDEHYDES: 9-Borabicyclo[3.3.1]-nonane. Diazoacetaldehyde. Diethyl-allylthiomethylphosphonate, Diethyl phenyl orthoformate. Dithiane. Lith-ium—Alkylamine. Lithium diisopropyl-amide. Sodium tetracarbonylferrate-(-II). 1,1,3,3-Tetramethylbutylisonitrile. 2,4,4,6-Tetramethyl-5,6-dihydro-1,3-(4H)-oxazine. Triethylsilane. s-Trithiane.
ALKYLALLENES: Dimethylcopper-lithium.
ALKYLAMINES: Bisbenzenesulfeni-mide. Trichloramine.
ALKYL AZIDES: Mercuric Azide.
N-ALKYL BENZOPHENONE IMINES: Titanium tetrachloride.
ALKYL IODIDES AND BROMIDES: Iodine.
ALKYLMAGNESIUM FLUORIDES: Iodine.
ALKYLMERCURIC HALIDES: Dicyclo-hexylborane.
3-ALKYLPENTANE-2,4-DIONES: Di-methyl sulfoxide (solvent effects).
1-ALKYNES: Dimethyl sulfoxide—derived reagent (a). Sodium methyl-sulfinylmethide.
2-ALKYNES: Dimethyl sulfoxide—derived reagent (a). Sodium methylsulfinylmethide.
ALLENES: Bis-3-methyl-2-butylborane.
ALLENIC ALCOHOLS: Dihydropyrane.
ALLYLIC CHLORIDES: Tosyl chloride—Lithium chloride.
AMIDES: Formic acid. Mercuric nitrate. Tetrachlorosilane.

AMINE OXIDES: m-Chloroperbenzoic acid.
AMINES: N-Benzenesulfonylformamidic ester. Diborane. Lithium bisbenzene-sulfenimide. Sodium borohydride. Triethyloxonium fluoroborate.
AMINO ACIDS: 1-Amino-(S)-2-[(R)-hydroxyethyl]indoline. (S)-1-Amino-2-hydroxymethylindoline.
ANHYDRIDES: Phosgene. Sodium bi-carbonate.
AROMATIC BROMIDES: Thallium(III)-acetate.
AROMATIC IODIDES: Thallium(III)-trifluoroacetate.
ARYLSULFONIC ACIDS: Dimethyl-thiocarbamyl chloride.
1H-AZEPINES: Iodine isocyanate.
AZIDES: p-Toluenesulfonyl azide.
AZIRIDINES: Chloroacetyl chloride. (Dimethylamino)phenyloxosulfonium methylide. Iodine azide. Lead tetra-acetate. Triphenylphosphine di-bromide.
AZOXY COMPOUNDS: Thallium.
BENZENE DERIVATIVES: Dimethyl acetylenedicarboxylate.
BENZOXAZOLES: Polyphosphate ester.
BENZTHIAZOLES: Polyphosphate ester.
BIPHENYLS: Thallium(I) bromide. Thallium(III) trifluoroacetate.
α-BROMO ACIDS: N-Bromosuccinimide.
BROMOHYDRINS: N-Bromosuccinimide.
CARBAMATES: Iodine isocyanate.
CARBODIIMIDES: Iron pentacarbonyl. Triphenylphosphine dibromide.
CARBOXYLIC ACIDS: 2,4,4,6-Tetramethyl-5,6-dihydro-1,3-(4H)-oxazine. Tetraethyl dimethylamino-methylenediphosphonate. 2,4,4-Trimethyl-2-oxazoline.
β-CHLOROCARBAMATES: Chromous chloride.
CYCLIC α,β-UNSATURATED KETONES: Dimethylmethylphosphonate.
ω-CYANOALDEHYDES: Nitrosyl chloride.
N-CYANOAZIRIDINES: Cyanamide.
CYCLOBUTANOL: 9-Borabicyclo[3.3.1]-nonane.

CYCLOPROPANES: 9-Borabicyclo-
[3.3.1]nonane. Ethyl diazoacetate.
Simmons-Smith reagent.
CYCLOPROPANOLS: Lithium.
DECALIN-1,8-DIONES: Magnesium
methoxide.
DESOXY SUGARS: N,N-Dimethylthio-
carbamoyl chloride.
α-DIAZOALDEHYDES AND KETONES;
α-DIAZOCARBOXYLIC ACID ESTERS:
p-Toluenesulfonyl azide.
α-DIAZOKETONES: Chloramine. Dicyclohexylcarbodiimide.
vic-DICHLORIDES: Trichloramine.
1,3-DIENES: N,N'-Dimethyl-2-allyl-1,
3,2-diazaphospholidine 2-oxide.
1,5-DIENES: Hexamethylphosphoric
triamide.
DIKETO-1,3-DIAZETIDINES: Boron
trichloride.
β-DIKETONES: Boron trifluoride.
1,3-DIKETO-4-PENTENES: 2,2-Diethoxy-
vinylidenetriphenylphosphorane.
gem-DIMETHYLCYCLOPROPANES:
2-Diazopropane.
1,1'-DIMETHYLFERROCENE: Methylcyclopentadiene.
1,3-DIOXOLANES (ETHYLENEKETALS):
Ethylene oxide.
2,2-DIPHENYLCYCLOBUTANONES:
Diphenylketene.
ENOL ETHERS: Trimethylchlorosilane.
EPISULFIDES: Sulfur monochloride.
EPOXIDES: Dimethylformamide
dimethyl acetal. Dimethyloxosulfonium
methylide. Iodine. Methylene bromide–
Lithium. Methylphenyl-N-p-toluenesulfonylsulfoximine.
α,β-EPOXY SULFONES: Hydrogen
peroxide.
FORMAMIDES: N,N-Dimethylthio-
formamide.
GLYCEROPHOSPHOLIPIDS: 2,4,6-
Triisopropylbenzenesulfonyl chloride.
GLYCIDIC ESTERS: t-Butyl chloro-
acetate.
GLYCOSYL FLUORIDES: Silver tetrafluoroborate.
vic-HALOCYANOAMINES: Cyanamide.
2-HALOETHYL α-HALOCARBOXYL-
ATES: Ethylene oxide.

HETEROCYCLES: Dicyclohexyl-
carbodiimide. Formamidine acetate.
Thionyl chloride.
HYDRAZONES: Hydrazine.
β-HYDROXY ACIDS: Naphthalene–
Lithium.
β-HYDROXY ESTERS: Lithio ethyl
acetate.
INDOLES: Dimethylsulfonium methylide.
IODOHYDRINS: Iodine.
ISOCYANATES: Carbonium monoxide.
Dicyclohexylcarbodiimide.
ISOPROPYLIDINE DERIVATIVES:
Dimethyl sulfoxide. Perchloric acid.
KETIMINES: Molecular sieves.
α-KETO ACIDS: 1,1,3,3-Tetramethyl-
butylisonitrile.
β-KETO ESTERS: Ethyl diazoacetate.
KETONES: Dicobalt octacarbonyl.
Diethyl methylthiomethylphosphonate.
2,4,4,6-Tetramethyl-5,6-dihydro-1,3-
(4H)-oxazine.
β-LACTAMS: Chlorosulfonyl isocyanate.
β-LACTONES: Dichloroketene. Trimethyl
orthoformate.
LITHIUM ALDIMINES: 1,1,3,3-Tetra-
methylbutylisonitrile.
2-METHYLCYCLOBUTANONE: Trifluoroacetic acid.
α-METHYLENEBUTYROLACTONES:
Magnesium methyl carbonate.
METHYL KETONES: Dimethylcopperlithium. Dimethylzinc.
NITRILES: β-Chlorophenyl chlorothionoformate. Chlorosulfonyl
isocyanate. Diphenyl phosphorochloridate.
Thallium(III) trifluoroacetate. Triphenylphosphine–Carbon tetrachloride. Tris(triphenylphosphine)
chlororhodium.
NITROPARAFFINS: Lithium diisopropylamide.
NUCLEOSIDE DERIVATIVES: Tri(2-
chloroethyl) orthoformate.
NUCLEOSIDES: Mercuric cyanide.
OLEFINS: t-Butyl hydroperoxide.
Diethyl phosphorochloridate. Dimethylcopperlithium. N,N-Dimethyl-
thiocarbamoyl chloride. Trimethyl
phosphite. Triphenylphosphine.

ORTHOCARBONATES: Tetramethyl orthocarbonate.

OXIRANES: (Dimethylamino)phenyl-oxosulfonium methylide. Dimethyl-sulfonium methylide. (Dimethyl sulfoxide, derived reagent (b)). Methylene bromide–Lithium. Methylphenyl-N-p-toluenesulfonylsulfoximine.

PHENOLS: Thallium(III) trifluoro-acetate.

POLYCYCLIC COMPOUNDS: Triphenylphosphine.

SPIROCYCLOBUTANONES: Dimethyl sulfixide, derived reagent (c). Dimethyl-oxosulfonium methylide.

SPIROKETONES: Diethyl allylthio-methylphosphonate.

SPIROPENTANES: Methyllithium.

STYRENES: Simmons-Smith reagent.

TETRAHYDROFURANES: Silver oxide.

THIOKETONES: Phosphorus pentasulfide.

TOSYLATES: m-Chloroperbenzoic acid. p-Toluenesulfonyl chloride.

TRIALKYL ORTHOFORMATES: Formamide.

vic-TRIAZOLES: Ethyl azidoformate.

TRISUBSTITUTED OLEFINS: Dimethylcopperlithium. Triethyl orthoacetate.

α,β-UNSATURATED ACID CHLORIDES: Boron trichloride.

α,β-UNSATURATED ALDEHYDES: 1,1-Diethoxy-2-propyne.

α,β-UNSATURATED METHYL KETONES: 3-Butyne-2-one.

VINYL FLUORIDES: Fluoromethylene-triphenylphosphorane.

VINYL HALIDES: Lithium 2-methoxy-ethoxide.

VINYLSILANES: Lithium 2-methoxy-ethoxide.

THALLATION: Thallium(III) trifluoro-acetate.

TRANSANNULAR CYCLIZATION: Mercuric acetate.

TRANSMETALLATION: N,N,N',N'-Tetramethylethylenediamine

TRAPPING AGENT: 1,3-Diphenyl-naphtho[2,3-c]furane.

TRITIATION: Trifluoroacetic acid.

TSHUGAEFF REACTION: Dimethyl sulfoxide–derived reagent (à). Sodium methylsulfinylmethide.

ULLMANN ETHER SYNTHESIS: Cuprous chloride.

VALENCE TAUTOMERIZATION: Tris-triphenylphosphine)chlororhodium.

VILSMEIER REACTION: N,N-Dimethyl-thioformamide. Triphenylphosphine dibromide.

WILLGERODT-KINDLER REACTION: Sulfur.

WILLIAMSON ETHER SYNTHESIS: Dimethyl sulfoxide.

WITTIG REACTION: Boric acid.

WITTIG REAGENTS: 2,2-Diethoxyvinyli-denetriphenylphosphorane. Diethyl allylthiomethylphosphonate. Ethyl-idenetriphenylphosphorane. Fluoro-methylenetriphenylphosphorane.

WURTZ REACTION: Lithium amalgam.

WURTZ-TYPE CONDENSATION: Cuprous chloride.

AUTHOR INDEX

SUBJECT INDEX